T0331704

Laser Interaction
and Related Plasma Phenomena
Volume 9

Laser Interaction
and Related Plasma Phenomena

Volume 9

Edited by
Heinrich Hora
CERN
Geneva, Switzerland
and
George H. Miley

Fusion Studies Laboratory
University of Illinois
Urbana, Illinois

SPRINGER SCIENCE+BUSINESS MEDIA, LLC

Proceedings of the Ninth International Workshop on Laser Interaction
and Related Plasma Phenomena, held November 6–10, 1989, at the
Naval Postgraduate School, Monterey, California

Library of Congress Catalog Card Number 86-651543 (ISSN 0148-0987)

ISBN 978-0-306-43722-9   ISBN 978-1-4615-3804-2 (eBook)
DOI 10.1007/978-1-4615-3804-2

© 1991 Springer Science+Business Media New York
Originally published by Plenum Press, New York in 1991

PREFACE

The 9th International Workshop on "Laser Interaction and Related Plasma Phenomena" was held November 6-10, 1989, at the Naval Postgraduate School, Monterey, California. Starting in 1969, this represents a continuation of the longest series of meetings in this field in the United States. It is, in fact, the longest series anywhere with published Proceedings that document the advances and the growth of this dynamic field of physics and technology.

Following the discovery of the laser in 1960, the study of processes involved in laser beam interactions with materials opened a basically new dimension of physics. The energy densities and intensities generated are many orders of magnitude beyond those previously observed in laboratories. Simultaneously, the temporal dynamics of this interaction covers a broad range, only recently reaching ultra short times, of the order of a few femtoseconds.

Applications of this technology are of interest for many types of material treatments. Further, from the very beginning, a key ambitious goal has been to produce fusion energy by intense laser irradiation of a target containing appropriate fusion fuels. The various phenomena discovered during the ensuing research on laser-fusion are, indeed, much more complex than originally expected. However, in view of recent advances in physics understanding, a route to successful laser fusion can be seen. The development of fusion energy received a very strong stimulation since the last workshop due to the now partially publicized results of underground nuclear explosions. These experiments were designed to confirm certain basic physics required to achieve high gain targets for laser fusion operation. These developments have occurred at a most opportune time. Hopefully, this will help lead to a solution of the physics needed for an economically competitive scheme of energy production by lasers. If so, this represents an important route to overcome the growing problems associated with the greenhouse effect that results from the increasing use of fossil fuel power plants. The optimism growing out of these experiments was visible in the workshop by increased activity in laser fusion by the U. S. Department of Energy (US DOE). G. J. D'Alessio from the US DOE presented an overview of this comprehensive program. The growing optimism in the area was reflected by the statement attributed to the responsible director at DOE, S. Kahalas, that "laser fusion is no longer a question of 'what if,' but 'when.'"

In addition to laser-fusion, the present International Workshop also provided a strong coverage of various new applications of laser-produced plasmas for physics research. These include, for instance, space physics and new schemes for the acceleration of electrons to very high energies by lasers, and the experimental achievement of densities 1000 times higher than in the normal state.

The session on laser fusion included a comprehensive review by D. Kania from the Lawrence Livermore National Laboratory (LLNL) (not in these proceedings). He highlighted the news that the NOVA laser had just achieved laser pulse energies in the third harmonics of 50 KJ. This occurred just as NOVA was scheduled to approach full power operation. An interesting feature of these studies was the use of NOVA for certain direct drive experiments in addition to the usual indirect drive.

D. Cartwright, S. Figueira, et al. described the strong inertial confinement program at Los Alamos National Laboratory (LANL) which centers on the development of an efficient KrF laser driver. The results reveal the excellent beam quality and an operation at more than 100 MW/cm^2 laser operation. The presentation by S. Nakai from the Institute of Laser Engineering (ILE), Osaka University, reviewed a number of new achievements since the preceding workshop two years ago. One most significant result is the announcement that a polyethylene target has been compressed by lasers to 1000 times its solid state density. This corresponds to a specific weight of about 1000 grams/cm^3 and represents the highest density ever achieved in a laboratory experiment under reversible conditions. Other highlights of his presentation included a confirmation that values of ρR of 0.1 g/cm2 have been reached. It is estimated that a value 3 times higher at a temperature of 5 keV would be sufficient for break-even. It is remarkable that this was all achieved with laser pulse energies only up to 10 kJ. These developments indicate that the use of direct drive for laser fusion stands a good chance of succeeding.

Further, important results from laser-fusion related experiments were presented in the talk by J. Knauer, R. McCrory, et al. from Laboratory for Laser Energetics, (LLE), University of Rochester (not in these proceedings). One recent development that he described is a method to smooth the laser plasma interaction with an advanced interacting mechanism which extends the use of the random phase plate technique. A review by H. Peng, et al. from the Southwest Institute of Nuclear Physics, China, summarized research covering a number of years, culminating with recent results based on the installation of a 2-beam 20-cm diameter neodymium glass laser in Shanghai. Apart from fusion gain measurements based on neutron production, experiments with this laser for driving X-ray lasers were reported. The presentation also described calorimetric measurements outside of the pellet configuration employed for studies of indirect drive implosions.

The topic of pellet fusion reactors was covered by R. Peterson (U. of Wisconsin) who concentrated on ways to stop X-ray damage of the first wall in next step ICF experimental facilities. A number of conceptual aspects for the development of laser fusion power plants were presented in the paper by H. Hora, G. Miley, et al. (U. of New South Wales, U. of Illinois) based on the consequences of the rapid progress that has been made in solving various physics issues, including the spark ignition of targets and the use of indirect drive. Additional results were also presented concerning the volume ignition of targets in a fashion that requires less entropy production but gives higher gains. New results were presented to explain how to achieve smooth interaction by analyzing standing wave production of density ripples and their suppression by use of random phase plates or induced special incoherence. It is concluded that these results may ultimately enable improved designs of laser fusion power stations producing less expensive energy.

The use of lasers for muon catalyzed fusion was elaborated by S. Eliezer (SOREQ, Israel) based on the observation that today's experimental results of physics demonstrate near break-even for this type of fusion. One difficulty is still the sticking problem of muons by helium atoms. This might be solved using laser radiation at resonance conditions to release the trapped ("stuck") muons.

A panel discussion "Views on Future Directions in ICF" was moderated by D. Cartwright (LANL) with E. Storm (LLNL), S. Nakai (ILE, Osaka), J. Knauer (LLE-Rochester), and B. Ripin, Naval Research Laboratory (NRL) as panelists. A focus of the discussion revolved around our readiness to move directly to the high gain test facility (Laser Microfusion Facility). Some concern was expressed that an intermediate level facility would be less risky and still provide key scaling data. Other panel members felt the technology was ready for this step.

A special session at the Workshop contained a series of presentations on laser development beginning with a review of glass laser drivers for ICF by J. Murray (LLNL) and a paper by T. Kessler (LLE-Rochester) on glass laser technology (not in these proceedings). It is evident that improved glass lasers can build on the technology base developed over many years of experience. Large glass laser systems for MJ laser pulses (such as the Athena project at LLNL) provide a way to build on the developments. Indeed, the potential for such systems has been recognized since the presentation of K. Manes (LLNL) at the 7th Workshop four years ago. One new aspect of recent concepts, however, is the use of multipass operation.

The possibility of alternative laser drivers in the ultraviolet range is illustrated by recent developments in the development of a KrF laser driver. The most advanced system of this type was described by D. Cartwright, J. Figueira, et al. (LANL) (see the section before). They made evident that many of the conditions and capacities of this laser were underestimated in the past. A key advantage of a gas laser system is the avoidance of anisotropies in the laser medium. Other advances in the development of KrF lasers were highlighted in the report by H. Yoneda (ILE, Osaka).

As in the past, the workshop included discussions of various future laser concepts. The report by C. Collins, et al. from the U. of Texas - Dallas reviewed the current status and issues involved in the development of gamma ray lasers. The achievement of X-ray lasers from glass laser irradiated cyclindrical targets was covered by the contributions of D. Mayhall, Y. Lee, et al. (LLNL), and T. Boehly, et al. (LLE-Rochester). Lee reported about alternative schemes for modeling X-ray lasers with direct interactions and line radiation pumping of lithium-like ions in resonance with hydrogen-like ions. The presentation by T. Boehly reviewed recent developments in X-ray lasers where the measurement of 105 Å output from an aluminum target represents the shortest wavelength achieved at LLE.

Within the discussion of alternative concepts, the presentation by M. Prelas and F. P. Boody, U. of Missouri, on nuclear-driven solid-state lasers resulted in interesting estimations for this approach. Another intriguing result was presented in a paper by H. Fearn and M. Scully from the U. of New Mexico concerning the scheme of lasing without inversion. The laser mechanism is based on a quantum phase matching and resonances. This approach could be especially interesting for microwave and picosecond pulses. Potentially, it could also be used to drive X-ray lasers at relatively low power densities.

A series of presentations was devoted to laser-plasma instabilities, a topic which has been studied intensively for many years. K. Mizuno, et al. presented a paper titled "Ion Acoustic Parametric Decay Instabilities in Laser Produced Plasmas" which is the result of collaboration between U. of California-Davis, LLE-Rochester and LLNL. A key issue brought out in this presentation is that these decay instabilities are sensitive to the thermal conductivity and its abnormality in laser-produced plasmas. Based on this, it was suggested in the discussion of this paper that one use of this effect, relative to plasma diagnostics, would be the determination of the temperature and the effective ionization of the target.

Recent results of unique experiments by R. Dragila, B. Luther-Davies, and R. A. Maddever (Australian Nat. University) were presented. They concentrated on the experimental understanding of extensive studies of modulation in the spectra of the backscattered light. Further understanding was provided by measurement of a pulsating laser-interaction mechanism with about 10-15 picosecond duration. This was evident from the pulsation period of the variation of reflectivity between a few percent to nearly 100% during these cycles.

The presentation by R. Drake and R. Turner (LLNL) summarized their very detailed studies of Raman spectra from laser-produced plasmas. Two papers from A. Giulietti, et al., Istituto di Fisica Atomica e Molecolare, Italy (IFAM) were devoted to experimental studies of plasma instabilities in long scale-length laser-produced plasmas. This work included an analysis of the second harmonic and three and one-half harmonics spectra from laser-radiated thin films. Very intriguing results on the theory of instabilities were presented in papers by A. Bergmann and P. Mulser, and S. Hüller, P. Mulser, and H. Schnabl from the U. of Darmstadt. In the former, nonlinear Langmuir waves were studied, the waves being excited by resonance absorption using Vlasov simulation. In the second paper Brillouin backscattering was studied for cases involving inhomogeneous densities and nonstationary conditions. The problems of instabilities and turbulence were explored experimentally in considerable detail by J. Grun, et al. (NRL). In their work, mechanisms involved in laser-driven, interpenetrating foils were studied.

The next group of topics were devoted to different types of interaction properties. In view of the recent availability of extremely short laser pulses of very high intensities, such interaction provided a focus for discussions by P. Mulser (Darmstadt) and K. Niu of the Tokyo Institute of Technology (TIT) who discussed electron motion in plasma irradiated by strong laser light. The issue of early-time shine-through in laser irradiated targets was studied by D. Bradley, et al. (LLE-Rochester). In the same session, F. Schwirzke of the Naval Postgraduate School (NPS) reported on his experiments and models for understanding laser-induced breakdown and high-voltage-induced breakdown of metallic surfaces through the evolution of unipolar arcs.

Problems of laser-produced double layers were discussed by H. Hora, S. Eliezer, et al. (U. of New South Wales and SOREQ, Israel). They described a generalization of the plasma model of surface tension to include the degenerate electrons of metals and the positive charges in nuclei. Results indicate a remarkable agreement with the measured values.

The problems of the equation of state in laser-matter interaction were followed up in a paper by H. Szichman, S. Eliezer, and A. Zigler (SOREQ, Is) following many years of research work on this topic. For purposes of diagnostics and for materials processing using laser-produced plasmas, it is essential to understand the irradiation damage in single and polycrystalline CsI. This subject was discussed by G. Miley, et al. of Fusion Studies Laboratory of the U. of Illinois (FSL-Illinois). Cross-field interactions between laser and particle beam studies were explored by K. Kasuya, et al. (TIT). Active beam-control and active laser-diagnostics of intense, pulsed ion sources were studied, and numerous results were presented. The problem of saturated magnetics for laser and plasma interaction was presented by H. Kislev and G. Miley (FSL-Illinois). Also, the problem of nuclear-induced UV fluorescence was treated by W. Williams and G. Miley of the same laboratory. H. Figueroa (U. of S. California) reported on microwave plasma mirrors and their similarity to laser-produced plasma mirrors.

Another section was devoted to laser acceleration and microwaves. A paper by F. Caspers and E. Jensen (CERN at Geneva) discusses particle acceleration by laser beams employing the axial electrical fields in the optical beams. While this effect turns out to have a linear nature (via transverse fields production), it still becomes quite significant in selected ranges of parameters. The paper by L. Cicchitelli, et al. (U. of New South Wales) is again devoted to the acceleration of electrons in vacuum without plasma. They show how energies up to a TeV may be achieved with very intense laser pulses.

Based on their extensive experience with laser interaction with plasmas for beat wave acceleration (or "Wakefield" acceleration), C. Joshi, et al. (UCLA) demonstrated experimentally the predicted upshift of electromagnetic radiation in the microwave range where the rapid creation of the plasma is essential. This result published preliminarily at the conference for the first time (not in these proceedings) opens new areas for microwave techniques. These concepts are closely related to those reviewed by M. Gundersen (U. of Southern California) who concentrated on the backlighted thyratron. D. Mayhall, et al. (LLNL) reported on their extensive 2-dimensional calculations of electron lower formations where cross microwaves penetrate air at low pressure. A further application of laser-plasma interaction studies was presented by S. Eliezer (SOREQ, Israel) who reviewed a unique experimental study of laser-driven shock waves. These experiments showed how incident and reflected compression and rarefaction waves can produce strong amplitudes in special regions of plane parallel targets. As a result, a brittling of the material only occurs in these regions of high amplitudes. Knowing this, one can produce layers of material with enhanced resistances against shock wave breaking.

The application of laser produced plasmas to other fields of research was illustrated by the presentation of B. Ripin (NRL) (not in these proceedings) whose Laboratory has initiated a new effort to the study of space plasma physics problems in the laboratory by use of laser plasma interactions. This is a field for which experts have waited for a long time. It might be noted here that at a NASA conference on double layers in Huntsville, Alabama, March 1986, Nobel Laureate Hannes Alfven was intrigued by the possible use of studies of double layers in laser produced plasmas for understanding aspects of space plasmas. Such research is now being done at NRL. Ripin's report was devoted to studies of the scaling of supernova turbulence and the fractal geometry of flutelike boundaries. Even a simulation of the magnetosphere is possible using well-known structures from barium-induced jets.

The final section contained papers about inertial confinement fusion (ICF) calculations, both for laser and particle beam drivers with emphasis on issues related to particle beam fusion. A presentation by G. Velarde, et al. (Madrid Polytechnica) summarizes the extensive work of this research group on the theory of high-gain direct-drive ICF. It might be noted that the discussion touched on the question of how ignition temperatures of less than 4.5 keV are possible, even if the temperature is not controlled by re-absorption of bremsstrahlung. The paper by K. Niu and T. Aoki (TIT) concentrated on adiabatic target compression following recent developments with respect to spark and volume ignition for achieving the highest possible fusion gains. Two papers by X. T. He, M. Yu, and T. Q. Change, Centre for Nonlinear Studies (Beijing, China) reported on calculations of X-ray conversion and implosion characteristics for high-gain laser fusion. The paper by Y. E. Kim (Purdue University) considered some unique combinations of nuclear fission and fusion reactions in order to obtain an improvement in fusion gain by fission-induced effects. While these results were hypothetical, they are worthwhile when considering various routes for future studies.

One problem in laser fusion that has gained attention over the years is the stopping power of charged nuclear reaction products and other high energy ions in dense plasmas. This problem was quite controversial 10 years ago, and it still needs clarification in view of the importance of driving targets by particle beams as well as the self-heating of a target during a fusion burn by reaction productions.

C. Deutsch et al. (U. of Paris) presented advanced results from their extensive work on ion stopping in plasmas, especially with reference to heavy ion beam fusion. Other papers devoted to the mechanisms of particle beam driving of ICF included in the presentation by W. Jiang, et al. (Nagaoka University) who concentrated on the focusing of proton beams and their interaction with the target. The problem of charge neutralization of intense light ion beams was analyzed theoretically by T. Kaneda and K. Nui (TIT) while D. P. Kothe and C. Choi (LANL and Purdue University) presented an implicit fluid-particle model for the study of ion beam-plasma interactions.

In summary, the workshop demonstrated again the powerful progress that has occurred in the field of high-intensity laser-plasma interactions and related problems of plasma physics. Of the numerous highlights concerning laser fusion, the most significant may be the announcement of successful underground target tests as a signal that the physics needed for next step studies is in hand. Other achievements in laser-interactions include the realization of new applications for particle acceleration and astrophysics. The remarkable achievement of a density of 1000 g/cm^3 in a laboratory experiment was presented for the first time at this meeting.

The week of the workshop was coincident with the famous press conference of President George Bush where he indicated his strong concern about the increase of the average temperature of the atmosphere due to the "greenhouse" effect. This coincidence reinforced the feeling of conference attendees that fusion power should now be strongly considered a possible solution of this most threatening problem. If so, accelerated development is essential.

We are grateful for the support of the conference by the U. S. Department of Energy (Office of Inertial Confinement Fusion), the Fusion Power Associates (Stephen Dean, Director), the Fusion Studies Laboratory of the University of Illinois at Urbana, and the Department of

Theoretical Physics of the University of New South Wales. We also wish to thank the Superintendent of the Naval Postgraduate School, RADM Robert C. Austin USN, and Dr. Gordon Schacher, Dean of Science and Engineering, for their hospitality and encouragement. Special recognition is due to Professor Fred Schwirzke for his splendid arrangements and organization of the meeting site. The overwhelming work load was again managed splendidly by the Conference Secretary, Mrs. Chris Stalker, whose efficiency and enthusiasm are especially acknowledged with our thanks. Thanks are also due to Ms. Doris Bock at the University of New South Wales. Last, but not least, Mrs. Fred Schwirzke again had a most vital role in local arrangements. Thanks are also due to Dr. T. Kaneda, Tokyo Institute of Technology for video-taping the panel discussion.

We are also indebted to the advisors of the Workshop for their contributions to its planning and their participation, when possible, in the meeting. Without their continuous consultation, we would not have achieved the strong results presented in these proceedings. The advisors are: N. G. Basov (Lebedev Physical Institute, Moscow, USSR), R. Dautray (Limeil-Valenton, France), S. Eliezer (SOREQ, Israel), E. Storm (LLNL, USA), G. Velarde (Polytech. Univ. Madrid, Spain), A. H. Guenther (LANL, USA), R. Hofstadter (Stanford U., USA), M. H. Key (Rutherford Lab., England), P. Mulser (TH Darmstadt, Germany-W), M. C. Richardson (LLE-Rochester, USA), F. Schwirzke (Naval Postgr. School, USA), S. Singer (LANL, USA), and C. Yamanaka (Univ. Osaka, Japan).

Heinrich Hora
U. of New South Wales
Australia

George H. Miley
U. of Illinois
USA

CONTENTS

III.   LASER-PLASMA INSTABILITIES

IV.   INTERACTION PROPERTIES

U.S. INERTIAL CONFINEMENT FUSION PROGRAM EXPERIMENTS:

RESULTS & IMPLICATIONS

G. J. D'Alessio

Inertial Fusion Division
Defense Programs
U.S. Department of Energy
Washington, D.C. 20545

Abstract

During 1987 and 1988 several landmark experiments were performed by
a number of participants in the U.S. Inertial Confinement Fusion (ICF)
Program. As a result, for the first time in the history of the program,
there is a consensus that the basic concept of indirect-drive ICF has
been experimently validated. ICF experimental activities include the
classified Centurion/Halite program at the Nevada Test Site. In
addition, indirect-drive implosion experiments are performed on the NOVA
laser system at the Lawrence Livermore National Laboratory. High
convergence ratio implosions have been consistently obtained. At the
University of Rochester's Laboratory for Laser Energetics, direct drive
cryogenic target implosions on the Omega laser resulted in relatively
high density D-T fuel compressions. Neutron yields and other key
parameters measured in these experiments show good correlation with
results predicted by modeling, indicating that we have begun to
understand the underlying physical phenomena in a reasonably
comprehensive manner. To assist the program in evaluating this new state
of knowledge, DOE convened a group of independent reviewers who assessed
these results in late 1988. They identified additional areas which
require investigation before a decision can be made to construct the next
large ICF facility. They concluded that such physics data could be
acquired within the 5-year time frame with the appropriate upgrade of
present facilities. They also found that the ICF Program's present
efforts to plan for a high-gain Laboratory Microfusion Facility (LMF)
during the late 1990's are highly appropriate.

Introduction

The following overview of the ICF Program is taken directly from the
unclassified testimony of Brigadier General Paul F. Kavanaugh, USA,
former Deputy Assistant Secretary for Military Application on March 6,
1989, before the Department of Energy Nuclear Facilities Panel of the
House Committee on Armed Services.

"The inertial fusion program is a special part of the Weapons
Research, Development, and Testing technology base. In the near term,
the program contributes to the nuclear weapons program through increased

understanding of the phenomena that occur during a thermonuclear explosion; in the future, there is the potential of a civilian energy application to electric power generation."

"The program is conducted by the Department of Energy, Defense Programs through the three national weapons laboratories: Lawrence Livermore National Laboratory in Livermore, California; Los Alamos National Laboratory in Los Alamos, New Mexico; Sandia National Laboratories in Albuquerque, New Mexico; and the three support laboratories: KMS Fusion, Inc., in Ann Arbor, Michigan; Naval Research Laboratory in Washington, D.C.; and the University of Rochester in Rochester, New York. Each program participant has one or more distinct roles in the program, as we shall discuss."

"The ICF Program has made large strides in target physics, mainly in Centurion/Halite, a classified program that uses underground nuclear explosions in Nevada, and in NOVA laser experiments at the Lawrence Livermore National Laboratory in California, as well as in experiments on the Omega laser at the University of Rochester. Based on the remarkable progress of the last two years, the IF community has concluded that inertial fusion in the laboratory is technically feasible."

"This year, we have continued this progress with significant achievements at all of the major inertial fusion laboratories including progress in both target physics and driver development. In addition, we have completed three driver technology reviews (for krypton fluoride lasers, solid state lasers, and light ions) and a target physics review by a group of independent experts. A unified, long-range program plan has been developed focussing on the Laboratory Microfusion Facility (LMF), which would be the next major step in the program."

"Because of this progress, the Department has restructured the program to begin planning for an LMF, which would entail a 5 to 10 megajoule driver (laser or particle beam) and be capable of high gain (energy released by fusion reaction compared to driver energy input). However, it is too early to decide which laser or particle beam driver technology is required for the LMF, and it is necessary to intensify efforts in driver development to determine this. In addition, the experimental target physics program must continue in order to determine the lowest energy that will be capable of achieving high gain with reasonable confidence. A lower energy driver will allow lower overall projections of facility cost, a necessary prerequisite for a commitment to construction."

"This year, we are asking the National Academy of Sciences to carry out an independent review of the ICF Program, as requested by Congress. A draft report is due to the Congress by January 15, 1990 with a full report by September 15, 1990."

ICF PROGRAM

Program Objective: High-Gain Laboratory Microfusion and Its Applications

The concept of ICF was developed at the Lawrence Livermore National Laboratory in the early 1960's. The objective of the ICF Program, since its inception as a formal Federal program in the early 1970's, has been to obtain a high yield [from 200 up to 1000 megajoules (MJ), or up to nearly 500 pounds high-explosive equivalent] microfusion capability in the laboratory. This will require achieving high gain (output energy from 10 up to more than 100 times input energy) from an inertial fusion

target (a hollow spherical shell filled with deuterium
and tritium) driven by an approximately 10 MJ output driver with a power
pulse of ten or more nanoseconds (billionths of a second) in
duration. The goal of the program is to exploit the military and
civilian applications of high-gain laboratory microfusion.

ICF Feasibility Requirements: Three Demonstrations - Capsule Physics, Laboratory Physics, Driver Technology and Cost

High confidence in three technical areas are required before a
decision is made to proceed with the construction of a laboratory high-
gain fusion facility: (1) a demonstration that when the proper
conditions are created around a fusion capsule, high gain will result,
(2) a demonstration that a laboratory driver (laser or particle-beam
facility) can create the drive conditions necessary to obtain high gain,
and (3) a demonstration that the technology is available to build a
suitable driver at affordable cost. In 1986, the National Academy of
Sciences (NAS) ICF Review Committee (Happer Committee) Report recommended
emphasizing the first two areas (via underground nuclear tests and
exploitation of new facilities) and delaying the third, an aggressive
advanced driver development program, until the target physics results
justified a change in priorities.

Major Advances in ICF Physics (1987 and 1988): High Gain ICF is Feasible

Progress since 1986 has been quite rapid. Many of the prerequisites
of the NAS report have now been satisfied, and the program is moving much
closer to meeting the rest.

The consensus of the ICF community and DOE is that the program has
progressed much more rapidly than the Happer Committee or anyone else
anticipated in 1985. In fact, several significant scientific firsts have
been achieved in the last year. These results have presented the U.S.
with the opportunity to embark on a new phase in the ICF program, the
planning of a Laboratory Microfusion Facility (LMF).

Recent Centurion/Halite results were, in fact, an historic advance
in 40 years of fusion R&D. These physics results have allowed
demonstration of excellent performance, putting to rest certain
fundamental questions about the basic feasibility of achieving high gain.

During 1987 and 1988, NOVA demonstrated major progress toward
several Happer Committee physics goals. For the first time, NOVA
produced laser beam/fusion capsule interaction conditions that meet or
exceed many of those required for high gain in the laboratory. In
another important first, NOVA compressed the radius of an ICF capsule to
about 3/4 of the value required for ignition in high-gain capsules.
These results matched theoretical predictions and have markedly increased
confidence in our capability to produce physics conditions for high-gain
laboratory ignition.

Beam uniformity has been markedly improved on the OMEGA laser to
enable the program to meet another Happer Committee milestone. This is
the achievement of deuterium/tritium densities in compressed, directly
driven, cryogenic ICF capsules that exceed 100 times normal liquid DT
densities.

The laser technology and cryo-target operation on OMEGA will be
important contributions to high gain driver and target chamber
development.

Target Physics Review: 1988

A four day review to assess the status of these target physics results was held at U.S. Department of Energy (DOE) Headquarters on November 14-17, 1988. To provide DOE with unencumbered views, independent reviewers who are recognized experts in relevant fields, but who were not currently associated with the ICF program were utilized. They reached seven broad conclusions.

1. The Centurion/Halite experiments have laid to rest some fundamental questions about the basic fasibility of achieving high gain.

2. Present physics data base is inadequate for an LMF decision. The aim of the ICF program is to implode a capsule containing deuterium and tritium (D/T) with directed energy from a driver to achieve core ignition and propagation of the D/T burn to achieve high fusion yield. The original concept for ICF is by direct laser drive. There was concern, however, that the coupled requirements for driver illumination symmetry, suppression of plasma instabilities in the corona, and maintenance of the stability of the implosion required by the direct drive approach may be difficult to meet simultaneously. The indirect drive approach differs from the direct drive approach in the use of a hohlraum to convert the directed beam energy to X-rays that then drive the capsule. This approach was expected to relax and uncouple some of the critical requirements. A recognized disadvantage, however, is the somewhat higher energy requirement associated with indirect drive. For these reasons, on balance, indirect drive has been pursued in the last few years as the mainline approach with direct drive considered as a backup option. A variation of the indirect drive approach is to use a pulsed-power light-ion driver. However, a target design data base for light-ions is largely lacking. This approach must demonstrate adequate power density on target soon to remain a driver candidate.

3. Some basic plasma physics and energy scaling issues must still be addressed. For indirect drive, driver beam quality and beam-plasma interaction are expected to be issues of concern in the efficient conversion of driver beam energy to X-rays for symmetric illumination of the capsule. These basic physics issues have not been addressed adequately and still need to be resolved. NOVA can address many critical LMF issues if upgraded.

For the direct drive approach, significant advances have been realized by improving beam uniformity to achieve required target illumination symmetry, by understanding the suppression of beam-plasma instabilities to smooth beam profiles, and by pulse shaping to improve beam-to-target energy conversion efficiency. However, there is still the need to demonstrate scaling of these techniques to power levels required for LMF sized targets. An upgrade of the Omega laser direct drive system has been proposed.

Cryogenic targets are on the critical path for achieving high gain.

4. Continue direct drive program. The direct drive approach has been considered as a backup option because of concerns for meeting multiple plasma physics and driver technology requirements that are strongly coupled. In the last few years, the direct drive community has made significant progress in developing techniques and an understanding of the physics to alleviate some of these concerns which are now shared by the indirect drive approach. At the same

time, the driver energy required to achieve stable implosion by direct drive is edging toward levels comparable with the lower range of the indirect drive approach. For these reasons, efforts to develop the direct and indirect drive approaches are now more complementary than competitive. Therefore, the direct drive program should be continued, not only as a backup and potential alternative option, but also to complement the indirect drive program in addressing some of the common critical issues.

5. <u>A physics data base adequate for an LMF decision is 3-5 years away, with significant additional resources needed to increase probability of success</u>. At present, the progress that can be expected in resolving the remaining target physics issues is largely resource and facility limited, but the program's pace and plans in 1988 are on track to provide the necessary target physics data base for a decision on the LMF in a 3 to 5 year time frame. Additional resources would increase the probability of resolving key physics issues associated with the LMF decision in a timely way. Both the direct and indirect drive programs should address the physics issues that could reduce the minimum size of the LMF driver without additional technical risk.

6. <u>Program priorities should continue to buttress and not compromise the physics research</u>. Continued planning for the LMF at this time is well justified and the ICF community should develop a strategy of cooperation to resolve the remaining critical issues. In particular, significant additional resources are required to develop the data base on driver technology to determine energy, power and cost scaling that will also be needed for the LMF decision. However, <u>the additional resources for driver development cannot be generated at the expense of the target physics program without severely damaging the possibility of an LMF decision in the time frame desired</u>.

7. <u>LMF planning is highly appropriate</u>. A principal finding of the review was that the Centurion/Halite experiments have validated the basic concept of ICF. Even with these and NOVA experimental results, there is a unanimous consensus among the independent reviewers that the present target physics data base in not adequate for a decision to proceed with the design and construction of LMF at this time. However, in light of the promising results that have been achieved, continued planning for the LMF is highly appropriate. There is no significant advantage to an intermediate facility.

<u>Program Consensus: Continue Vigorous R&D Leading to a 10 MJ LMF</u>

The ICF Program participants now agree that comprehensive data which answer all remaining significant questions about achieving high-gain microfusion in the laboratory are not yet available. However, the recent physics accomplishments have given the ICF community greatly increased confidence that:

a) The ICF program should be continued vigorously now. (The Happer Committee's basic policy question was whether the ICF program could shed sufficient light on the ultimate feasibility of ICF by 1991 in order to justify continuing the program.)

b) A 10 MJ (or somewhat greater) output driver that couples the appropriate energy into a target ablator, with required pulse shaping, will achieve sufficient gain for a multitude of military applications.

Program Transition: From Basic Feasibility Research to LMF Data Base R&D

For these reasons the ICF Program has begun planning the orderly reorientation of its R&D activities toward establishing LMF capsule physics and driver technology data bases (including preliminary designs and cost estimates). The key technical issues raised by the NAS and earlier reviews have continued to be addressed during FY 1988 and FY 1989 and will continue to be addressed in the near term while LMF-specific activities begin to grow.

Target physics work in the FY 1989 and FY 1990 near-term program includes:

Capsule and Beam/Target Interaction Physics:

1. Preparation for another Centurion/Halite experiment.

2. Higher efficiency, pulse-shaped indirect-drive experiments on the NOVA glass laser, which achieved full design energy (120 kJ at 1 micron) in January, 1989. Scaled-down LMF hohlraum experiments and capsule convergence studies will eventually lead to validation of implosion targets at least partially scalable to LMF physics regimes.

3. Hohlraum experiments are to begin in FY 1990 on the PBFA-II pulsed-power, light-ion accelerator. PBFA-II plans in FY 1991 include initial light-ion ICF capsule experiments aimed at demonstrating pulsed-power's capability as a low-cost, high-power ICF driver.

4. On the Aurora gas laser, flat targets were tested in FY 1989 and will continue to be tested. In FY 1990, hohlraum experiments will be performed to begin demonstrating the basic feasibility of krypton fluoride lasers as ICF drivers.

5. The Omega glass laser will continue D-T gas and cryogenic capsule implosion experiments with its newly installed beam smoothing technology aimed at demonstrating the feasibility of the direct-drive approach to ICF. The Congressionally mandated Omega upgrade report was completed in October 1989. This facility is also used by National Laser User's Facility (NLUF) contractors, thereby providing an unclassified, high power laser/target experiments system to universities and others.

6. The Pharos flat target laser system completed direct-drive symmetry studies using its unique beam smoothing technique in March, 1989, after which it was phased out of the ICF program. Planning will continue for the high uniformity amplifier for the Nike KrF gas laser, using a more sophisticated application of the beam smoothing technique used on Pharos.

7. All of the above experimental activities necessarily include required in-house target design, development and fabrication, with some target component supply and support from outside contractors; implosion (X-ray, etc.) and ignition (neutron) as well as diagnostics development; theoretical and numerical modeling and experimental data analysis and comparisons are also included.

The future R&D program consists of two equally important technical elements that correspond to new out-year planning estimates:

(1) An enhanced target physics activity (a $30M-$100M/yr increase over 3 years) leading to an LMF capsule design and beam/target interaction physics data base, and

(2) A major, new driver technology development initiative (a $20M-$100M/yr increase over 3 years) to provide the engineering and cost data base for a high confidence LMF driver decision.

1.   Enhancing ICF Physics Activities

Implementation of the new strategy would enable the vigorous, physics program required in the laboratory and, as needed, at the Nevada Test Site (NTS) to acquire data necessary to design a Laboratory Microfusion Facility (LMF) target and the LMF beam/target interaction system.

o   This approach maintains the top priority accorded to physics studies by the NAS Committee Report (1986); specifically, that the two new major indirect drive facilities, NOVA and PBFA-II, should be fully exploited to obtain physics data. However, under the FY 1989 and FY 1990 Level-of-Funding, the real level of physics program activity has diminished notably at the national laboratories. These decreases in the real level of experimental program activity and data output can only be alleviated by the enhanced program resource options described in Figure 1.

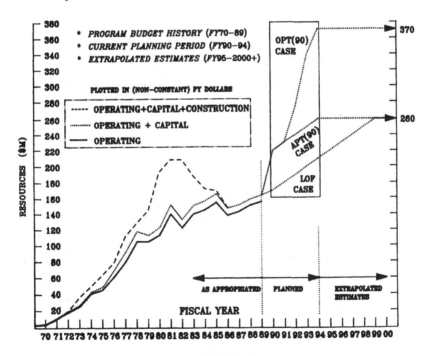

FIGURE 1

ICF PROGRAM RESOURCE REQUIREMENTS FOR THREE FY90-94 PROGRAM OPTIONS

o   In this way, the schedule for NOVA experimental objectives for
    cryogenic targets can be reduced to 4-5 years in the range at or
    above the accelerated physics and technology (APT) case, instead
    of the present 10-12 years associated with the present level of
    funding (LOF) case. Upgrades would include enhanced energy
    output at 0.35 micron, more precise beam output balance and
    diagnostics and a cryogenic capability.

o   Also, a reasonable PBFA-II light-ion target physics data base
    can be achieved by the mid-1990s in the range at or above the
    APT case instead of near the turn of the century in the LOF
    case, should the pulsed-power approach be successfully
    demonstrated a D-T target shooter as planned in FY 1991.

o   Aurora and Nike would also be able to accomplish their target
    physics goals several years sooner in the range at or above the
    APT case than would be possible with the present level of
    funding (LOF), and the next step in KrF laser target physics
    could be identified and implemented.

o   Similarly, the next Centurion/Halite experiment which has been
    delayed and scaled down in the LOF case, could be accelerated
    and restored with the enhanced resources in the range at or
    above the APT case.

o   In direct-drive experiments, recent successes on the Omega laser
    facility could be fully exploited with additional resources at
    or above the APT case level. This would enable the direct-
    drive approach to explore higher convergence ratio targets with
    greater compressions by upgrading Omega to 30 kJ total laser
    energy.

o   Target supply and supporting activities are well funded at
    present at the LOF level.

o   At level of funding (LOF), pursuit of indirectly driven
    cryotargets would be delayed as contrasted with cases at or
    above APT. This would entail significant delay in the useful
    high gain operation of the LMF, and would decrease the
    confidence in the physics data base available in advance of an
    LMF construction decision.

2.   Advanced Driver Initiative

     The other major near term program objective is determination of the
optimum ICF driver, based on performance and cost, to be used for the
LMF. The NAS Committee relegated this area to a secondary priority until
Centurion/Halite and NOVA results were sufficiently favorable and until
the potential of igniting D-T fuel on PBFA-II is evaluated. But, it did
recommend that an innovative engineering and design program on advanced
lasers be supported in the interim. These activities were to be
structured to provide affordable choices for a large laser driver by
about 1991 (The NAS Committee's estimate made in 1986). The new strategy
focuses on the following points:

o   The three leading ICF advanced driver candidates, the Nd glass
    laser, the KrF gas laser, and the pulsed-power light-ion
    accelerator, have various mixtures of technical and cost issues
    that need to be resolved before such a final decision can be
    made.

o   For each of three program resource options (OPT, APT, LOF), the
     FY 1990-1994 ICF Program Plan proposes a distinct sequence of
     events for arriving at a driver decision.

     - At the optimum physics and technology (OPT) level (ramp up to
        $370m/yr), full scale engineering modules for each of the
        three driver technologies could be built, tested and used to
        validate technical performance and certain cost estimates.

     - At the accelerated physics and technology (APT) level (ramp
        up to $260m/yr), only one full scale driver engineering
        module could be built, with the choice based on an early best
        estimate of the most suitable driver for the LMF. Should
        that driver approach fail to meet performance and cost
        estimates, one of the remaining two driver approaches would
        then be tried.

o   Increasing program resource levels to values between OPT and APT
     will enable a sound driver decision to be made during the mid to
     late 1990's, respectively. Higher confidence, earlier LMF
     decision milestones are associated with levels nearer the OPT
     case. The APT case is estimated to be the minimum level for
     positioning the U.S. to begin LMF construction during this
     century.

o   However, it is necessary to note that at the present FY 1989
     resource level (Level-of-Funding) extended into FY 1990, there
     will not be adequate resources to yield sufficient progress in
     the early to mid 1990's for a driver decision to be made based
     on even one type of driver engineering module.

o   On this basis, the plan recommends the OPT strategy to develop
     the three driver technologies to the point where a high
     confidence driver decision could be made within 3-4 years from
     the start of a major driver development resource initiative.

During the last three years, technical, management, and enhanced
funding strategies have been explored for acquiring, in parallel, the
target physics and driver technology data bases required to select a high
confidence design for the LMF. This plan positions the United States to
decide to build a very high confidence LMF as early as the mid-1990's in
the Optimum Physics and Technology (OPT) resource case, or by the end of
the century in the accelerated physics and technology (APT) resource
case.

This ICF Program strategy continues to adhere to the NAS Committee's
guiding principle, "the most urgent task is to study the physics involved
in pellet compression and ignition. This will be the case until the
physics issues identified by the NAS Committee are resolved." With
significant additional resources (in the ranges noted above) beyond those
needed to pursue the enhanced physics activities, the ICF Program could
construct prototype glass laser, gas laser and pulsed power driver
modules for an LMF. Based on the enhanced physics data base and the
three driver module performance data bases, a high confidence driver
selection decision could be made as the last step before a formal federal
decision to construct the LMF. The schedule for LMF construction and
utilization is given in Figure 2 as a function increased ICF R&D funding.

Summary

In summary, the inertial fusion program has achieved major target
physics demonstration goals over the past two years. Coupled with

9

A. EARLIEST POTENTIAL DATES FOR KEY LMF MILESTONES

CASES	DRIVER DECISION	PHYSICS ASSESSMENTS	MSA* DECISION	CONSTRUCTION		HIGH-GAIN USE
				START IN	COMPLETE BY	
OPT(90):	1993-4	1993-4	1994-5	1995	2001	2003-4
APT(90):	1995	1996-8	1997-9	1998	2004	2009-12
LOF:	2000	2001-2	2001-2	2004	2010	2020-30

*ALSO, EARLIEST POTENTIAL DATE FOR CONSTRUCTION FUNDS

B. KEY COMPARISONS OF LMF AVAILABILITY AND COST

CASE COMPARISONS:	FULL USE LMF IS AVAILABLE	FOR A TOTAL R&D COST:
OPT vs APT:	7-9 YEARS SOONER	$0.75 BILLION LESS
OPT vs LOF:	20-30 YEARS SOONER	$3-5 BILLION LESS
APT vs LOF:	10-20 YEARS SOONER	$2.25-$4.25 BILLION LESS

C. LMF CAPABILITY FOR MILITARY APPLICATIONS (MA) AND CIVILIAN APPLICATIONS (CA)

CASE:	CAPABILITY LMF:	R&D PROGRAM FOR:
OPT:	FULL	MA & CA
APT:	FULL	MA OR CA
LOF:	UNCERTAIN	MA

FIGURE 2 - KEY LMF MILESTONES AND COMPARISONS FOR THREE ICF PROGRAM OPTIONS

additional rapid progress in target physics and driver development, the program has begun planning for a new, large facility, the Laboratory Microfusion Facility, that could have significant military, civilian energy, and scientific applications and benefits. However, significant additional R&D resources are required to make the programmatic transition from scientific feasibility to LMF physics and engineering data base development. This is necessary because LMF cost, benefit, and technological risk will have to be clearly understood before a decision to proceed with construction can be made.

STATUS OF INERTIAL CONFINEMENT FUSION RESEARCH AT LOS ALAMOS NATIONAL LABORATORY

D. C. Cartwright, J. F. Figueira, T. E. McDonald, D. B. Harris, and A. A. Hauer

Los Alamos National Laboratory
Los Alamos, New Mexico 87545

Los Alamos National Laboratory is engaged in a long-range program to investigate the merits of Krypton-Flouride (KrF) lasers as inertial confinement fusion (ICF) drivers. Because of their intrinsic short wavelength (0.25 μm), broad bandwidth (~0.5%), smooth spatial beam profile, and precise and flexible temporal pulse-shaping capabilities, KrF lasers appear to be an attractive ICF driver candidate from the laser-target interaction physics standpoint. Additionally, the use of a gaseous lasing medium with the potential of high overall system efficiency preserves a direct path to energy production. Current cost projections for MJ-class KrF systems show it to be affordable.
The present Los Alamos KrF experimental facility, AURORA, has recently begun integrated system experiments. It has already achieved more than 100 TW/cm^2 on target with >1 kJ of energy. The facility is designed to deliver 5 kJ to the target with a 5-ns pulse length and a spot size ~200 μm. Shaped pulses from 2-20 ns can also be tested. The system employs angular multiplexing, where 96 beam pulses are overlapped into a single, 500-ns beam train. The 96 beam paths are offset and staggered thorough the amplifier chain to provide spatial and temporal separation of the individual beams. The energy from each amplifier is extracted continuously over the 500-ns electron-beam pumping duration. The appropriate time delay is then removed from each beam segment so that they are simultaneously recombined at the target plane. The first target physics experiments will be x-ray conversion experiments, followed by experiments on the effects of laser pulse shaping and bandwidth on target performance.

I. INTRODUCTION

Nuclear fusion, the process that produces the energy in all stars, has been the goal of U.S. laboratory experiments since the magnetic confinement fusion program was initiated in 1952. The national Inertial Confinement Fusion program began in earnest in the early 1970s and currently has seven participants: Los Alamos, Livermore and Sandia National Laboratories; KMS Fusion; the University of Rochester; the Naval Research Laboratory; and Lawrence Berkeley Laboratories.

The ICF mission statement for the Los Alamos National Laboratory is as follows:

The Los Alamos ICF program is one of the main efforts by the Department of Energy (DOE) to evaluate the scientific feasibility of inertially confined fusion, using intense lasers or particle beams to

compress and heat small masses of deuterium-tritium fuel to ther-
monuclear burn conditions. The goals of the national program are to
support research in high energy-density science and to conduct re-
search on the potential of inertial fusion for energy production.

The key technical elements within the ICF program are
- minimum input energy and
- the development of a laboratory driver suitable for driving such
  targets at an acceptable cost.

In their 1986 report on the US ICF Program, the National Academy
of Sciences recommended that the ICF program be continued at the 1985
level of effort for about five years and then be reevaluated with
respect to the progress it had made toward achieving its technical
goals. The purpose of this paper is to document the technical progress
made by Los Alamos since 1985 in four areas:
- materials technology for ICF,
- laser-matter interaction,
- KrF laser development for ICF applications, and
- plasma theory support of the Sandia light ion program.

A detailed summary of the technical progress in Los Alamos ICF
program since 1985 is contained in Cartwright et al. 1989.

A. General Requirements for ICF

Inertial confinement fusion attempts to rapidly heat a small amount
of deuterium and tritium (DT) fuel to temperatures high enough to
promote fusion reactions in the fuel and at the same time, the fuel is
compressed to densities high enough to facilitate reaction of a large
fraction of the fuel before it is cooled by hydrodynamic expansion. In
the sun, the primary fusion reaction occurs between two nuclei of the
lightest isotope of hydrogen, ^{1}H. The two nuclear reactions of interest
for fusion are:

$$D + D \rightarrow \begin{cases} T(1.01 \text{ MeV}) + p(3.02 \text{ Mev}) \\ \\ ^3He(0.82 \text{ MeV}) + n(2.45 \text{ MeV}) \end{cases} \quad (1)$$

$$D + T \rightarrow\ ^4He(3.5 \text{ MeV}) + n(14.1 \text{ MeV}) \quad . \quad (2)$$

Both of these fusion reactions are very energetic, but the DT
reaction is preferred because of the greater neutron kinetic energy. It
is also easier to initiate the DT reaction because the temperature
required to ignite the fusion fuel is lower by approximately a factor of
5 than for the DD reaction.

The temperature and density-confinement time product requirements for
ICF are the same as those required for magnetic confinement. In ICF,
one increases the density many orders of magnitude, and thus decreases
the required confinement time. Basically, the concept is that if the
density is high enough then the fuel can burn before the internal energy
from the burning causes it to disassemble. A spherical mass of fuel is
heated by energy impinging on the outside of the sphere. The hot
plasma, which expands outward, produces a compression wave that
propagates toward the center of the fuel. The compression of the DT is
designed so that the final fuel density is 1000 to 10,000 times the
original liquid liquid density of DT, 0.2 gm/cm^3. In DT at 1000 times
the density of liquid DT, the number density is 3×10^{25} nuclei/cm^3.
Therefore, a confinement time of only 30×10^{-12}s satisfies the Lawson

criteria. The diameter of the superdense compressed DT mixture is ~100 mm, and at the burn temperatures of 20-50 keV, the free expansion time of the material is on the order of tens of picoseconds (10^{-12}s).

The difficulty in accomplishing this in the laboratory arises from the requirement to contain the maximum energy released by the fusion reaction; this can be accomplished only if the mass of fuel is sufficiently low. For small fuel masses to support a thermonuclear burn in which the energy produced by fusion reaction sustains the burn, the fuel must be condensed to high density. A fundamental scaling parameter for ICF is the product of fuel density r and the radius R of the column containing the fuel. It can be shown that the density to which the fuel must be compressed increases as the reciprocal of the square root of the fuel mass, that is, $\rho \approx 1/\sqrt{M}$. Achieving the required compression for small fuel masses without expending excessive energy from the driver is the basic requirement of ICF.

B. Target Fabrication

A major component of a strong target fabrication capability is fast and effective reaction to requests for out-of-the-ordinary targets. The Los Alamos effort fosters this capability by maintaining a strong micromachining group in the same building that houses the rest of fusion target fabrication. This arrangement makes the machinist part of the interactive design team in an immediate way.

Our capabilities include: CNC diamond turning for optical surface finishes; lapping capability for spherical targets and mandrels; microscopic machining operations and hole drilling for target components of 0.1-mm size; electro-discharge machining (EDM), which allows machining of delicate pieces without deformation; and a CNC milling machine for limited run production pieces. These machining facilities are complemented by a well-equipped inspection group with large optical comparators, a Moore Precision Measuring Machine, and microgram balance. This shop is manned with some of the world's best machinists. It has the capability for safely carrying out hazardous operations such as Be machining and work on DT containing components.

ICF targets require many materials which are products of development efforts. Finding effective machining operations for these new materials is a development effort in itself. Laser micromachining is an integral part of our target fabrication capability. Our dye laser can bore micron-sized holes and hole arrays in a variety of substrates including glass, plastics, and metals. Our operation has been worked well in the Centurion program. We are anxious to put it to the test in the fabrication of targets for AURORA.

X-ray techniques are used on ICF targets to measure coating and wall thicknesses, variations in coating and wall thickness, gas fill pressures, and elemental composition. These techniques also measure the density and thickness uniformities of bulk materials from which smaller ICF targets are made. They are equally useful for both opaque and transparent materials. Microradiographic gauging, a development specifically for ICF target characterization, measures thicknesses and thickness variations in coatings. Standard x-ray fluorescence techniques detect gas fill pressures and elemental composition. Standard x-ray and gamma-gauging techniques provide information about both the absolute density and thickness and their uniformities for bulk samples.

High-gain target designs for an ICF fusion reactor rely on cryogenic fuel for their high calculated performance. Recent developments in the physics of solid DT layers has made possible the fabrication of cryogenic targets in mm-sizes suitable for reactor-type yields.

Solid DT self-heats because of the beta decay of tritium and the subsequent reabsorption of the beta particle (electron) in the frozen DT. Heat production at the rate of 0.3 W/gm of T is measured. The radioactively deposited energy drives selective sublimation in the frozen fuel. Regions where the DT-ice layer is thick have warmer interiors than regions where the layer is thin. Thus material sublimes from the thick regions and coats out on the regions where the ice is thin. This process proceeds exponentially with a characteristic time constant for DT of about 30 minutes.

Researchers at Los Alamos set out to verify that the process takes place as predicted. Experiments in a cylindrical geometry for pure T_2 (Hoffer and Foreman 1988) produced a symmetrization time constant for that material of 14.8 minutes. Later measurements on DT confirmed the model for fresh DT and also measured a slowing of the process that increased as the sample aged; each day of sample age added 12 minutes to the symmetrization time constant. Apparently ^{3}He, a non-condensable product of the beta decay of T, accumulates in the target interior and impedes the mass flow of the DT across the target. Further work on this phenomenon is in progress, but the beta-layering process appears to provide thick, symmetric layers of solid DT. Designs of beta-layered targets for the ICF reactor and means of fabricating them have been published (Foreman and Hoffer 1988, 1989).

C. Applications of ICF Technology to other Fields

The ICF program has spawned a rich diversity of technological advances and applications ranging from advances in laser/optical technology itself to applications in fields such as biology and materials science, as well as the more obvious plasma and atomic physics. Los Alamos has pursued a number of these applications and has assisted others in transferring technology. In some cases, the technology has been commercialized and has led to the formation of companies in the private sector. (See Cartwright et al. 1989 for more detail.)

One example is the program Los Alamos is currently building in biomedical applications of lasers. This will be a joint effort among Los Alamos, the University of New Mexico School of Medicine, and the Lovelace Research Foundation. A broad range of capabilities in lasers and optics, many deriving from developments in the ICF program, will be brought to bear upon a variety of medical problems.

Another example is the unique potential of laser plasma (LP) x-ray sources in non-fusion applications. Considerable improvement in parameters such as repetition rate and reliability for Nd:glass, CO_2, and KrF lasers has led to a substantial amount of work on non-fusion applications in recent years (Eason et al. 1986; Yaakobi et al. 1986; Hauer 1988). Most of the applications make use of the fast intense x-ray burst that can be produced with LPs. The energy available in a single burst is often adequate for detection and recording. LP x-ray sources are thus (in contrast to synchrotron radiation) uniquely suited to the study of nonrepetitive or irreversible events (such as some types of materials phase changes). Laser systems suitable as x-ray sources will soon be capable of >1-Hz repetition rate, and thus LPs will even be competitive on a quasi-cw basis. X-ray fluence available with existing laser systems ranges up to 10^{16} keV/keV/steradian with photon energies up to 10 keV or greater.

LP x-ray diffraction and radiography have also been applied to biological samples. The study of living cells is of obvious fundamental importance in medical and biological research. Although much can be learned using conventional microscopy in the optical region, a basic limitation is placed on the maximum spatial resolution by the wavelength of the light used. The resolution can be improved by probing with shorter wavelengths, that is, electron microscopy or soft x-ray contact microscopy. However, in electron microscopy, the sample must be dried and coated before observations can be made; the cell is thus killed and the sample may appear vastly different from its living counterpart. Similar problems occur with x-ray microscopy when a conventional x-ray source is used. A contact print of the sample is taken by irradiating it with x rays over a period of several hours. During this time, the cell suffers severe radiation damage and is killed during the course of the exposure. If a 1-ns burst of LP x-rays is used as the source, this problem is overcome; the radiation damage occurs on a microsecond time scale; thus a print of the living cell is obtained before cell death occurs. Laser plasma x-ray sources have thus made possible a valuable new technique for biological research.

II. THE KrF LASER FOR ICF APPLICATIONS

The Los Alamos National Laboratory has been actively engaged in the development of high-power gas lasers for inertial confinement fusion applications for twenty years. The Los Alamos invention of a scalable electron-beam pumping technology by Boyer, Fenstermacher, Leland, Nutter, and Rink in 1969 (Fenstermacher et al. 1971) revolutionized the scalability limits of gas lasers and allowed the construction of large-aperture, multikilojoule devices. The Los Alamos program began with highly efficient long wavelength molecular lasers based on CO_2. This very successful laser development program culminated in the operation in 1983 of the ANTARES laser at the 35-kJ level (40 TW of power). Based on results from the target experimental programs in the early 1980s, a national decision about the required laser wavelength occurred, and the Los Alamos ir laser program was terminated in favor of the short wavelength KrF gas laser.

The KrF laser is relatively new to the ICF community; it was demonstrated in 1975 (Brau and Ewing 1975), followed immediately by other confirmations of spectroscopy and lasing (Brau 1975; Ewing and Brau 1975; Ault et al. 1975; Mangano and Jacob 1975; Searles and Hart 1975; and Tisone et al. 1975). The development of high peak power KrF laser technology for Inertial Confinement Fusion (ICF) applications is actively in progress throughout the world with major facilities underway in Japan, Canada, England, the USSR, and the US. Because of the newness of the technology, most of these efforts are a strong mixture of facility engineering, advanced technology research and development, and advanced conceptual design studies. The Los Alamos National Laboratory KrF laser development program is probably the largest effort in the world. It addresses both near-term integrated laser demonstrations and the longer-term advanced design concepts and technology advancements required for larger fusion laser systems. In this section we will review the basic features of the KrF lasers, describe the status of the worldwide KrF technology program, provide an overview of the Los Alamos laser development program, describe current progress on the near-term technical activities, and discuss the future directions of the Los Alamos KrF laser development program.

A. Basic Features of the KrF Laser

KrF lasers operate by electrically pumping high-pressure gas mixtures of krypton (Kr), fluorine (F_2), and a ballast gas such as argon (Ar) with self-sustained electrical discharges or with high-energy electron-beams. The electrical excitation initiates a complex chain of reactions that result in the production of the krypton fluoride (KrF*) molecule and various absorbers. The KrF* molecule can then lase to the unbound lower level, emitting a photon at 248 nm. The upper state lifetime is very fast in the excited KrF molecule, and storage times are limited to approximately 5 ns by quenching and spontaneous emission. KrF is the second most efficient member of a class of excimer lasers that also include the well-studied XeCl (lasing at 308 nm), XeF (lasing at 351 nm), and ArF (lasing at 193 nm), which is the most efficient.

A unique combination of features appears to make the KrF laser well suited as a driver for inertial confinement fusion drivers. These features are summarized below. They will be discussed in more detail in the following sections of this chapter.

- The laser directly operates at 248 nm, optimizing the ICF target efficiency without added wavelength conversion complexity.
- Unlike other ICF lasers, the KrF laser is basically not a storage laser; it prefers to operate in the continuous energy extraction mode. Because of this feature, loaded KrF amplifiers tend to be very linear with little temporal pulse shape distortion. This allows for a very robust pulse-shaping capability that may prove absolutely essential for efficienttarget performance.
- Electron-beam pumped KrF lasers are scalable to large energies in a single module. This feature has been clearly demonstrated by amplifier architectures now under development. The AURORA large aperture module has already demonstrated 10 kJ extracted from a 2000:1 volume, and advanced Los Alamos designs will explore the 50-kJ to 250-kJ region.
- The laser operates with a broad lasing bandwidth in excess of 200 cm^{-1} that allows the use of spatial and temporal smoothing techniques for both direct and indirect target drive applications.
- Although all of the current ICF related KrF laser technology development programs are emphasizing single-shot facilities, the basic design features of the KrF gas laser will permit extensions to repetitively pulsed devices in the future, if commercial energy applications are pursued.
- The laser medium is a nondamaging gas, eliminating the need for extensive protection systems to insure the survivability of the laser medium. This feature also allows KrF to readily adapt to multiple-pulse operation for commercial applications.

B. Worldwide Progress

Since their invention in 1975 by Brau and Ewing, excimer lasers have been the subject of steady and increasing interest. Major progress has been made in the commercial development of high average power devices, with 1-KW devices being actively developed on several continents through private companies or industrial consortia. More recent interest has centered on the development of high peak power devices as potential replacements for harmonically converted glass lasers in inertial confinement fusion and atomic physics applications. Active research and technology programs are in progress in the United States and competitive programs are being pursued in Japan, Canada, Germany, the United Kingdom, and the Soviet Union. These programs have led to a series of first-generation, integrated laser-target systems that are in design, construction, or testing. Table 1 shows a current compilation of these laser systems. The Lawrence Livermore National Laboratory (LLNL) RAPIER system was operated briefly in the early 1980s and then

TABLE 1. KrF Laser Technology Is Being Pursued
Internationally for ICF Applications

Laser System	Status	Energy	Power
RAPIER (LLNL)	1982	800 J	1×10^{10}W
SPRITE (Rutherford)	1983	200 J	3×10^{9}W
AURORA (Los Alamos)	1985	10 kJ	2×10^{10}W
	1988	1 kJ	2×10^{11}W
	1989	4 kJ	10^{12}W
NIKE (NRL)	1989	10 J	-----
	1993	2 kJ	-----
RAPIER B (U of Alberta, Canada)	1988 (proposed)	100 J	-----
		1 kJ	-----
ASHURA (Electro-Technical Lab, Japan)	1988	500 J	5×10^{9}W
	(future)	1 kJ	2×10^{11}W
Euro-Laser (Rutherford)	1996	100 kJ	variable

decommissioned. The SPRITE laser, built at the Rutherford-Appleton
Laboratories in the United Kingdom, was the first truly high-power KrF
facility in operation. The UK government has approved an upgrade of
this facility to the 3-kJ level and designs are currently in progress
for the 100-kJ Euro-laser. The Naval Research Laboratory (NRL) is the
other DOE US participant in the KrF laser development program. NRL has
started construction of the NIKE laser that will produce 2-4 kJ. The
University of Alberta at Edmonton is proposing to build a 1-kJ
facility, utilizing some of the components from the LLNL RAPIER laser
that have been provided by the DOE through a joint US/Canada protocol.
Several Japanese universities are actively involved in the pursuit of
KrF LASERS; the major operating facility in operation is the ASHURA
laser at the Electro-Technical Laboratory of the University of Tokyo.
In addition, the Kurchatov Institute of the USSR is constructing a kJ-
class KrF facility in collaboration with Evremov Electro-Technical
Institute and is pursuing conceptual designs for a 10-kJ class device.
All of these facilities are based on electron-beam pumped KrF laser
technology; they employ several different optical architectures and
different design philosophies. This broad array of activities will
continue to enrich the technology available to KrF lasers and to
improve the performance and reduce the cost of many system components.

C. Overview of the Los Alamos KrF Laser Development Program

In the preceding sections we have discussed both the potential
advantages of KrF lasers and the international effort now underway in
KrF laser technology. The potential advantages of the laser are well
recognized, but they must be demonstrated at current scale size in

integrated laser-target systems. More importantly, these advantages must be shown to scale economically to the 100-kJ to 10-MJ sizes required for future progress in the ICF program. The Los Alamos laser development program is composed of three major elements that are intended both to aggressively address near-term feasibility of the KrF laser concept and to show the way to the cost and performance improvements required for future laser facilities.

- The AURORA Laser Facility is a 1-TW KrF laser designed as an integrated performance demonstration of a target-qualified excimer laser system.
- An advanced design effort evaluates the concepts that offer the improved performance and lower cost that will be essential for the construction of future lasers in the 0.1- to 10 MJ class.
- A laser technology program addresses both performance and cost issues that will be important in advanced laser system designs.

Each of these programs is briefly described in this section and will be described in much more detail later in this chapter.

1. AURORA. The near-term goal for Los Alamos is the successful integration and operation of the of the AURORA Laser Facility at the multikilojoule level with powers approaching 1 TW. AURORA is a short-pulse, high-power, krypton-fluoride laser system. It serves as an end-to-end technology demonstration prototype for large-scale excimer laser systems of interest to short wavelength ICF investigations. The system employs optical angular multiplexing and serial amplification by electron-beam driven KrF laser amplifiers to deliver multikilojoule laser pulses of 248 nm and 5-ns duration to ICF relevant targets. The design goal for the complete system is 5 KJ in 48 laser beams. Figure 1 shows a schematic diagram of the laser system.

Substantial progress has been made on the facility in the last several years including the following highlights:
- demonstration of 96-beam multiplexing and amplified energy extraction, as evidenced by the integrated operation of the front end, the multiplexer (12-fold and 8-fold encoders), the optical relay train, and three electron-beam driven amplifiers;
- assembly and installation of the demultiplexer optical hardware, which consists of over 300 optical components ranging in size from several centimeters square to over a meter square;
- completion of pulsed-power and electron-beam pumping upgrades on the LAM (Large Aperture Module), PA (Pre-Amplifier), and Small Aperture Module (SAM). The SAM shows a 40% increase in deposited electron beam energy, and the PA deposited energy has been increased by a factor of two; and
- integration of the entire laser system; the extraction of 4 kJ from the laser in 96 beams; and the delivery of 1.2-kJ, 5-ns pulses to the target chamber in 36 beams with intensity of 100 TW/cm^2 on target.

A major milestone was achieved recently at Los Alamos National Laboratory with the beginning of fully integrated operations of the AURORA krypton-fluoride (KrF) laser at the 100 TW/cm^2. Ultraviolet light at 1/4-micron wavelength is now being focused routinely onto targets at greater than 100 TW per square centimeter to produce strong plasma heating and intense soft x-ray emission. The light is focused onto targets under irradiation conditions that apply directly to ICF research and the study of other high-energy-density processes. This experiment demonstrates the viability of excimer lasers for fusion research and marks the beginning of a concerted research program in laser fusion using the KrF laser.

The AURORA laser system is a prototypical ICF driver on which KrF technology issues are being addressed in a system environment. The laser is designed to produce 5-10 kJ in 48 beams having a variable pulse length from 2-10 ns. Laser beam alignment and focusing onto the target is maintained in real time by a self- adaptive, automatic control system. The laser beam alignment process achieves total system alignment within 5 minutes and has a pointing accuracy of 5 microradians in focusing the 48 beams into one side of the target chamber shown in Fig. 2.

In the initial series of experiments with AURORA, the heating of gold targets by the laser was diagnosed with x-ray imaging (illustrated in Fig. 3) and x-ray spectroscopy. The image in Fig. 3 was obtained with an instrument whose spectral bandpass was centered at 1.5 kV. The image indicates strong plasma heating over the 500- to 600-micron extent of the heated region.

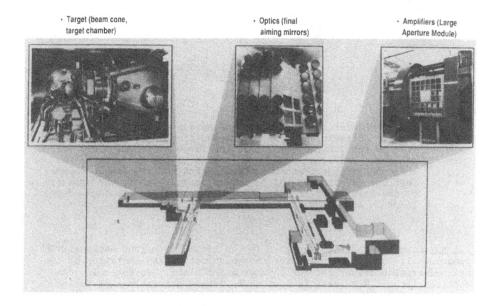

Fig. 1.   Schematic of the three major AURORA subsystems: target, optics, and amplifier (l to r).

Subsequent experiments will operate at much higher intensity levels (greater than 200 TW/cm^2) and will concentrate on the critical ICF issues of energy deposition and transport, x-ray conversion, and hydrodynamic instabilities. Key measurements will include: time-resolved measurements of absolute soft x-ray emission (20 eV-2 keV), soft x-ray pinhole camera imaging of target emission, soft x-ray spectroscopy, and scattered light amplitude and angular distribution.

*Fig. 2. Target chamber and final focusing system on the AURORA KrF
laser. Ultraviolet light is focused from the right onto one
side of the ICF targets inside the large spherical metal vacuum
chamber. Focusing lenses are located on the far side of the
cone-shaped object attached to the right of the target chamber.
Instruments attached to the target chamber detect the effects
of the heating of the target and thus the effectiveness of the
laser in achieving the predicted performance.*

2. Advanced Laser System Design. In the longer term, the national ICF
program will continue to plan for the construction and operation of the
next generation driver for ICF physics experiments. To determine the
applicability of KrF laser technology to future generations of fusion
drivers, Los Alamos has begun a design effort to explore systems in the
100kJ to 10MJ range. This Advanced KrF Laser Design effort provides
information to the KrF program on the design and cost of future KrF
laser-fusion systems and provides directions and goals to the KrF
technology development effort. Because no current ICF driver has
demonstrated both the required cost and the performance scaling, and
because uncertainties exist in laser-matter interactions and target
physics, Los Alamos is currently pursuing a range of advanced KrF laser
design activities: work is currently in progress to scope a 10-MJ
Laboratory Microfusion Facility (LMF), a 720-kJ LMF Prototype Beam Line,
a 250-kJ Amplifier Module (AM), and a 100-KJ Laser Target Test Facility.

The purpose of the scoping study by the Department of Energy for a
LMF is to examine a facility with a capability of producing a target
yield of 1000 MJ in a single-pulse mode. An example of a KrF design
that requires only minor extrapolations in pulsed-power technology is
shown in Fig. 4. This system uses angularly multiplexed amplifier
modules $1.3 \times 3.9 \times 3.8$ m^3, each of which produces 250 kJ of 248-nm
radiation. These units are then arranged in a trifold cluster, to form

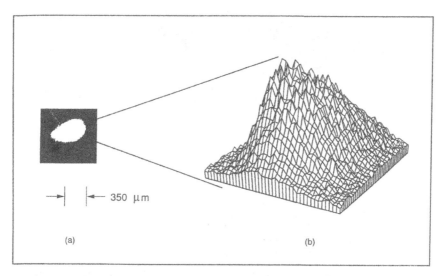

350 μm

(a)   (b)

Fig. 3. (a) X-ray image of the heating of a gold target by ultraviolet
light from the AURORA laser. The image was produced by x-rays
with energy of approximately 1.5 kilovolts, which indicates
strong heating of the target. (b) Surface plot of the image
showing a smooth distribution of irradiances from the laser
beams.

an LMF beam line that produces 720 kJ. These beam lines can then be
arranged to produce the required energy ranging from 720 kJ to 10 MJ.

3. KrF Technology. The current costs of all laser drivers are
unacceptable for an LMF scale system. To reduce these costs to an
acceptable level, we are currently structuring our advanced technology
programs to address the major cost drivers identified by the design
studies. Figure 5 shows the laser driver cost by system for a design
based on current AURORA technology, compared to the design using the

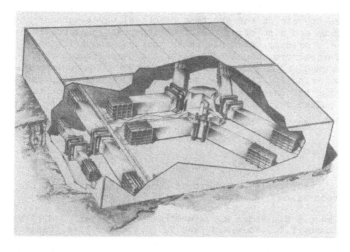

Fig. 4. LMF beam line showing a trifold array of 240-kJ modules.

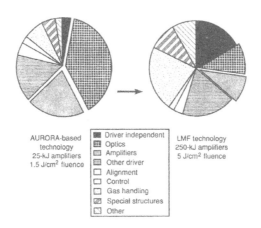

AURORA-based
technology
25-kJ amplifiers
1.5 J/cm² fluence

■ Driver independent
⊞ Optics
▦ Amplifiers
▤ Other driver
☐ Alignment
☐ Control
☐ Gas handling
▨ Special structures
☐ Other

LMF technology
250-kJ amplifiers
5 J/cm² fluence

Fig. 5. Distribution of costs for current
(AURORA) and future (LMF) technologies.

250-kJ laser module described above. The optics and pulse power account
for 60% of the total cost for the AURORA-based designs; this produces an
unacceptanble high laser system cost. Advanced designs concepts have
identified technology areas that can be improved to reduce the overall
system costs. The optics and pulse power costs have been reduced to 19%
of the total system cost, andd the total system cost is reduced by a
factor of 5. The KrF laser technology development program addresses
performance improvements and cost reductions for the LMF designs in the
areas of optics, pulse power, and laser performance, as well as those
technical issues effecting system reliability and modeling accuracy.

4. Future Directions. The National ICF program has achieved notable
progress over the last four years and is currently evaluating its status
and future directions. The long-term goal of the national program is
the Laboratory Microfusion Facility.

The Los Alamos National Laboratory and the Naval Research Laboratory
(NRL) are evaluating the potential performance and cost advantages of
the KrF gas lasers for ICF for both defense and civilian applications.
It is the strong consensus in the field of laser fusion that short
wavelength lasers are the most effective in transferring energy from the
laser beam into compression of the DT (deuterium-tritium) fuel capsule.
Excimer lasers operate at the shortest practical wavelength of any ICF
driver candidate and also offer the other important characteristics of
broad bandwidth; precise, flexible pulse shaping; and efficient
operation. Furthermore, the AURORA laser has demonstrated that
electron-beam pumped KrF lasers can be scaled to large energies in a
single module. Finally, the basic design features and efficient
operation of the KrF laser permit straight-forward extension to
repetitively pulsed devises for future commercial energy applications.

REFERENCES

Cartwright,D.C. et al., 1989, "Inertial Confinement Fusion at Los
Alamos, Progress Since 1985," LA-UR-89-2675.

Eason,R.W., Cheng,P.C., Feder R., Michette,A.J., Rosser, R.J.,O'Neill,J,
Owadano,Y., Rumsby,P.T., Shaw,M.J.,and Turcu,I.C.E., 1986,"Laser X-ray
Microscopy," Opt. Acta 33:501.

Foreman, L. R., and Hoffer, J.K., 1988,"Solid Fuel Targets for the ICF Reactor," _Nucl. Fusion_ 28:1609.

Foreman, L. R., and Hoffer, J.K., 1989, "Fabrication of ICF Reactor Targets Based on Symmetrization of Solid Fuel," LA-UR-88-3273, Los Alamos National Laboratory, Los Alamos, NM. (To be published in _Laser and Particle Beams Journal_.)

Fenstermacher, C. A., Nutter, M., Rink, J.P.,and Boyer, H, 1971, 1971, "Electron Beam Initiation of Large Volume Electric Discharges in CO_2 Laser Media," _Bull. Am. Phys. Soc_., 16:42.

Hauer, A. A., Forslund, D.W., McKinstrie, C.J., Wark, J.S, Hargis,Jr., P.J., Hamil, R.A., Kindel, J.M., 1988,"Current New Applications of Laser Plasmas," LA-11296-MS, Los Alamos National Laboratory, Los Alamos, NM.

Yaakobi, B., Bourke, P, Conturie, Y, Deletrez, J, Forsyth, J.M., Frankel, R.D., Goldman, L.M., McCrory, R.L., Seka, W,, Soures,J.M., Burek, A.J.,and Deslattes, R.E.,1981,"High X-ray Conversion Efficiency with Target Irradiation by a Frequency Tripled Nd:Glass Laser," _Opt. Comm_ 38:196.

HIGH DENSITY COMPRESSION OF HOLLOW-SHELL TARGET BY GEKKO XII AND LASER

FUSION RESEARCH AT ILE, OSAKA UNIVERSITY

S. Nakai, K. Mima, M. Yamanaka, H. Azechi, N. Miyanaga,
A. Nishiguchi, H. Nakaishi, Y.-W. Chen, Y. Setsuhara,
P. A. Norreys, T. Yamanaka, K. Nishihara, K. A. Tanaka,
M. Nakai, R. Kodama, M. Katayama, Y. Kato, H. Takabe,
H. Nishimura, H. Shiraga, T. Endo, K. Kondo, M. Nakatsuka,
T. Sasaki, T. Jitsuno, K. Yoshida, T. Kanabe[*], A. Yokotani[*],
T. Norimatsu, M. Takagi, H. Katayama, Y. Izawa, and C. Yamanaka[*]

Institute of Laser Engineering, Osaka University, Osaka 565,
Japan

[*]Institute for Laser Technology, Osaka 565, Japan

I. INTRODUCTION

High density compression of over 600 times solid density has been achieved by the laser implosion of a hollow shell pellet of deuterated and tritiated plastics (CDT). This is the highest density of material which has ever been realized in a laboratory and corresponds to the density four times higher than the center of the sun.

The ignition condition of laser fusion is estimated to be $\rho/\rho_s = 500 \sim 1000$, $T = 5 \sim 10$ keV and $\rho R \sim 0.3$ g/cm^2, where ρ/ρ_s is normalized density of compressed core by a solid fuel density ρ_s, T is the temperature and ρR is the areal mass density of the compressed core with radius R[1].

The high temperature compression was demonstrated by using LHART (Large High Aspect Ratio Target) of gas filled Glass Micro-Balloon (GMB) achieving the fuel temperature more than 10 keV. The physics of shock multiplexing and stagnation free compression have been investigated in detail[2,3].

The high density compression approaching to 1,000 times solid density is another key issue for achieving the ignition condition with relatively smaller driver energy of order 100 kJ. Note that volume ignition concept[4] has been also proposed which does not necessarily require such a high density. The physics of high density hollow shell implosion has been investigated by using CDT[5] or cryogenic D$_2$ fuel hollow shell pellet[6].

The advanced technologies which enable the high density compression are the uniformity of the driving pressure on the surface of pellet, the sphericity and uniformity of pellets, and diagnostics of high density ρ and ρR.

GEKKO XII glass laser has been improved in performance on the output energy balance of the twelve beams, on the power balance especially at the rising part of the tailored pulse. The nonuniformity due to the near field irregularity of each beam has been improved by using the random phasing technique[7].

New fabrication techniques of plastic hollow shell pellets of CD, CDTH and CDTSi have been developed, where D, T and Si are Deuteron, Triton and Silicon in plastic chemical components, respectively.

A comprehensive set of diagnostics has been applied to characterize the implosion dynamics and the compressed fuel conditions. A high speed x-ray multi-frame camera with exposure time of 80 ps, together with a x-ray streak camera, is a powerful tool for observing the implosion process.

For characterizing the compressed core, a special material design of a plastic shell target has been adopted. A CD pellet is used for the ρR measurement by means of secondary nuclear fusion reaction[8]. A CDTH pellet is for the use of elastically scattered protons by 14 MeV neutron[9]. A CDTSi pellet is used to diagnose higher ρR value by means of Si activation method.

The $\rho_D R$ = 25 mg/cm^2 and ρ_{total} = 44 g/cm3 was obtained in July 1987 using the secondary reaction measurement, where ρ_D g/cm^3 and ρ_{total} g/cm^3 is the density of Deuteron in CD and total mass density including carbon of CD plastics[5].

The $\rho_D R$ = 30 mg/cm^2 and ρ_{total} = 100 g/cm^3 was demonstrated in June 1988 by the knocked-on proton measurement[10]. The Si activation method was introduced in October 1988, which showed high density compression more than 600 times initial solid density[11]. Since then the calibration of each methods has been carefully proceeded and cross checking with different methods has been performed. All these procedures have given us the confidence of the accuracy of the experimental results.

Following the investigation of a hollow shell implosion physics by plastic fuel, a implosion experiment of cryogenic D_2 fuel has been performed. A low density foam of hollow shell configuration are immersed in the liquid D_2 to form a hollow shell fuel pellet. Detailed characteristics of cryotargets have been investigated with respect to the interaction, the energy transport, and the preliminary implosion performance.

Cannonball targets have been investigated to obtain uniform driving pressure on the pellet. The fundamental processes such as the x-ray conversion characteristics from laser, energy confinement in a cavity and x-ray driven burn through of a thin foil have been examined.

High power laser development toward the construction of 100 kJ blue laser are proceeded. Ignition and breakeven experiments are planned in the near future.

II. HIGH DENSITY IMPLOSION OF PLASTIC HOLLOW SHELL TARGET

1. Experimental Condition and Simulation

1-1 Laser

The GEKKO XII glass laser system with twelve beams at 0.53 μm wavelength has been used for the high density implosion experiment. The frequency conversion efficiency of GEKKO XII is measured to be higher than 80 %. The energy imbalance between each beam at 0.53 μm is less than ± 3 % for peak to valley or ± 1 % for rms. This extremely good performance in high energy balance has been achieved by the improvement of the energy monitoring system (Second Harmonic Monitor) which can reduce the errors due to the electronic digitizing circuit and the optical detector system. The power imbalance of 12 beams outputs is estimated to be less than 10 % through the laser pulse according to the measurement of energy balance at different energy levels of wide range. This good power balance is due to almost same values of the frequency conversion efficiency in 12 beams.

In order to improve the target performance, several modifications of operation condition have been made. The first one is the use of a long laser pulse duration for reducing the power density of the irradiating laser beam on the target to decrease the hot electron generation in the target plasma. The laser pulse duration is increased from 1 ns to 1.7 ns at FWHM using a pulse stacking method. The output pulse of the oscillator (1 ns) is divided into two beams and stacked again to one pulse by mirror arrays to provide a 1.7 ns rectangular pulse shape after saturated amplification. The second attempt is the use of random phase plates (RPP) to improve the energy distribution in the irradiation beam pattern. Randomly distributed 2 mm square segments of $\lambda/2$ optical thickness for 0.53 μm laser light coated on a 400 mm diameter glass provide many beamlets. The segmented laser beam is focused on to the target at an in-focus condition (d/R = -5; d is the distance between the focal point and the center of target, and R is the target radius).

The target irradiation uniformity is evaluated with an aid of the computer simulation using measured beam patterns. The focused beam pattern at the on-target plane is measured with the actual focusing optics and then the energy distribution on the spherical target can be calculated. The RMS deviation (σ) of the spherically expanded harmonics are calculated for the beam without (normal beam) and with RPP (random phased plate) as shown in Fig. II-1 for the focusing condition of d/R = -5 on the target of 500 μm diameter. The RMS values with the thermal smoothing of 0.05 in green beam irradiation (σ_{th}) are given.

1-2 Pellet fabrication

We have developed a technique for fabricating CD, CDT, CDTH and CDTSi hollow shell pellets using a hydrogen exchange reaction. The starting materials for CDT and CDTSi are the deuterated polystyrene ($-C_8D_8-)_n$, and deuterated poly-p-trimethylsilylstyrene ($-C_{11}D_{16}Si-)_n$, respectively.

The hollow shell pellets used in the high density compression experiments were fabricated by a modified W/O/W emulsion method which was developed by U. Kubo[1]. In the improved W/O/W emulsion method, the specific density of O phase is adjusted to be equal to that of W phase to make a high quality polymer shell whose sphericity is $(R_{max}-R_{av})/R_{av} < 2$ %, and uniformity of wall thickness is $(t_{max}-t_{av})/t_{av} < 2$ % (Fig. II-2).

The tritiated-deuterated ploystyrene shells with high specific radio activity are very difficult to fabricate by a synthetic method since the initial inventory of tritium becomes very high (1000 ci/g for $C_8D_4T_4$). Moreover the degree of polymerization is difficult to control during polymerization process due to the β-rays generated from tritium disintegrations. Therefore tritium was incorporated into shell material after it is shaped by means of isotope exchange reaction method enhanced by UV irradiation[2], which is a modification of the Wiltzbach method[3].

A pure tritium gas of ~20 atm. in pressure were employed to expose

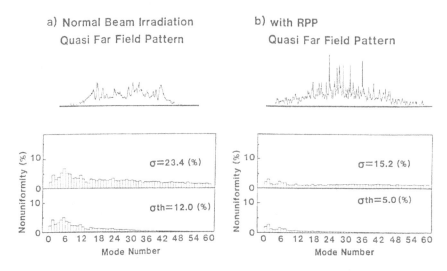

a) Normal Beam Irradiation
   Quasi Far Field Pattern

b) with RPP
   Quasi Far Field Pattern

$\sigma = 23.4$ (%)

σth$= 12.0$ (%)

$\sigma = 15.2$ (%)

σth$= 5.0$ (%)

Nonuniformity (%)

Mode Number

Fig. II-1   Nonuniformity of target irradiation on GEKKO XII.

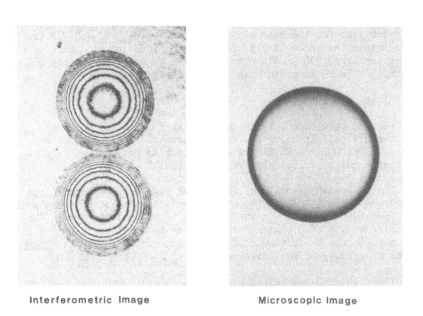

Interferometric Image

Microscopic Image

Fig. II-2   Uniformity of polymer shell.

the CD, CHD or CDSi hollow shell pellets. In order to enhance the exchange rate, the pellets were irradiated by UV light from a high pressure mercury lamp of 450 Watts or low pressure mercury lamp of 70 Watts. The latter was a spiral type one which can reduce the spatial nonuniformity of tritium exchange rate due to nonuniform UV irradiation. The uniformity of tritium distribution in depth was examined by separate experiments using laminated layers of same material.

After the exchange, the contents of tritium were measured by two methods. The tritiated CD_xT_y and $CH_xD_yT_z$ shells were burnt by a platinum heater in a closed vessel filled with water and oxygen. Tritium concentration in water was measured to determine the specific activity of tritium in the polymer shells. The specific activity of each shooting target was determined by measuring the characteristic X-rays of Si exited by the β-rays from tritium disintegrations. Tabel 1 shows the results for CD_xT_y, $CD_xD_yT_z$ and CD_xT_ySi shells exposed to the tritium gas of 2-30 atm. with and without UV irradiation.

To reduce the influence of a stalk to the implosion symmetry, we used a 2.5 μm diameter polyester fiber in the high density experiments instead of a usual 7 μm fiber.

1-3 Diagnostics

X-ray, plasma and unclear particle diagnostic instruments were used to characterize the implosion performance of plastic hollow shell targets.

The implosion dynamics is one of good measures to clarify the reliability of hydrodynamic code calculations. The shell trajectory and the time of peak compression were observed using an x-ray streak camera and the mulit-frame x-ray framing cameras[4].

The implosion core was characterized by the fusion particle measurements because the energetic fusion particles have long mean free paths in comparison with x-rays and the fusion particles can emerge from the compression plasma carrying the information at the time of peak fusion reaction. To measure the ρR of the compressed fuel, we have developed the secondary reaction method[5], the knocked-on proton method [6] and the silicon activation method[7][8].

A. x-ray streak camera

The implosion of the hollow shell requires the extremely wide intensity dynamic range of x-ray measurements, since the weak x-ray emission from the ablating plasma is followed by the intense emission from the compresed core. A pair of a 15μm pinhole and a 15-μm slit created a two dimensionally and a one dimensionally resolved x-ray images on a photo cathode (25 mm in diamter) of the streak camera. The pinhole image with a thin x-ray filter (combination of 25μm Be and 0.8μm Al) was streaked giving the informations about the shell trajectory until the peak compression occurs. The acceleration and terminal velocity of the shell can be measured from this streaked picture. Then we can obtain the hydrodynamic efficiency by estimating the residual mass which collapses at the center of the target. On the other hand the slit image was greatly filtered to measure the time of peak compression. The typical streaked x-ray image is shown in Fig. II-3. The temporal and spatial resolution was approximately 20ps and 20μm, respectively.

Table 1 Specific activity and exchange rate of tritium in CDT, CHDT and CDTSi shells.

Shell Material	Exposure Time (hr)	Specific Activity (Ci/g)	Exchange Rate (Ci/g atm hr)
CDT *1	168	0.95	0.00057
	432	2.40	0.00056
CDT *2	42	138	0.165
CHDT *2	47	91.5	0.0975
CDTSi *2	30	83.8	0.0931
CDT *3	50	150	0.15
CDTSi *3	16~35	40~100	0.125~0.143

T_2 gas pressure : 2~20 atm

Distance from UV lamp:10~15mm

#1 : Without UV irradiation

#2 : With a high pressure Hg lamp of 450W

#3 : With a spiral low pressure Hg lamp of 70W

B. multi-frame x-ray framing camera

The two dimensional observation of x-ray emmision is important for the qualitative estimation of uniformity of the accelerating shell. For this purpose the six and nine-frame x-ray framing cameras have been developed. As shown in Fig. II-4, the framing camera consists of a gated microchannel plate (MCP), electron multiplier and a phosphor screen. An array of three pinholes was coupled to the framing camera creating an array of three x-ray images on the front surface (Au photo cathode) of the gated MCP (8 mm x 55 mm). This MCP acts as a strip line along which an electric gate pulse travels[9], thus one can obtain the sequential gated images with the time interval of 170ps as the result of the correlations of photo electron signals and the gate pulse. The six (nine)-frame camera has two (three) sets of the gated MCP and MCP electron multiplier. The temporal resolution is 80ps, full width at half maximum (FWHM), with the spatial resolution of 15 1p/mm on the cathode. The optical image of the phosphor was recorded using an electrically cooled CCD camera, then stored in a frame memory system. The overall spatial resolution including the pinhole size, the image relay optics and the pixel size of CCD camera was about 20µm on the target plane. Using these framing cameras we observed the time history of the long-wavelength nonuniformity of the shell during the implosion process. Furthermore we noted the uniform shell acceleration followed by the symmetric expansion after the peak compression, which was the indispensable phenomena in the high ρR target shot.

C. secondary nuclear reaction method

The secondary reaction method is one of the most useful diagnostic technique for the measurement of the fuel ρR because the detection threshold, which can be determined as a product of the areal mass density and the neutron yield, is fairly low. And the remarkable advantage of this method is that the energetic secondary particles (usually neutrons of around 14 MeV are measured) can be easily distinguished from the primary

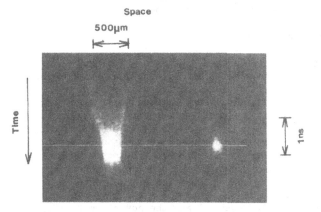

Fig. II-3 Typical x-ray streak picture of the implosion of plastic
hollow shell target. Shown in the left and right sides are
a pinhole image and a slit image, respectively.

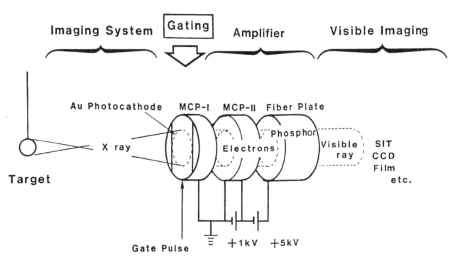

Fig. II-4 Each of the pinhole images on a gold photo cathode is
sequentially gated by an electric pulse traveling along a
strip line of micro channel plate (MCP-I). The electrons
are multiplied by the second MCP creating bright frame
images on a phosphor screen.

particles. This method has been applied to the implosion of CD shell target, and the maximum total ρR value which can be measured is about 30mg/cm^2 (range of 1MeV primary triton) including both carbon and deuterium at 1keV electron temperature[5]. The detection sensitivity is $Y_{N,DD} \cdot \rho_t R > 10^5$, where $Y_{N,DD}$ is the DD neutron yield and $\rho_t R$ is the total areal density in g/cm^2 including carbon.

D. knocked-on proton method

As the energetic charged particles which have lower nuclear charge are suitable for the dense plasma diagnostics, the measurement of protons elastically scattered by 14MeV neutrons is a candidate. For this purpose, we used the hydrogen-contained deutrated polystylene targets. The small part of the hydrogen and deutrium was replaced by the tritium as described in the section 1-2. To optimize the signal to noise ratio, here the most of the background is due to secondary reaction proton (12.5-17.4MeV), protons from D (n,2n) p reactions and knocked-on deutrons, the number density ratio between deutrium and hydrogen in the target was controlled to be 1:3. The resultant composition of the target was $CD_x T_y H_z$, where x=0.24, y=0.04 and z=0.72. Thus the high cross sections both of the D-T reaction and the proton recoil are the advantage to increase the detection sensitivity. The knocked-on particles were measured using a CR-39 track detector. By carefully adjusting the thicknesses both of a tantalum filter (130-μm-thick) and the CR-39 (150-μm-thick), the proton signal tracks of energies between 9.0MeV and 14.1Mev can be easily distinguished from the background tracks using both the track diameter analysis and the spatial coincidence between the front and rear surfaces[10]. The head-on collision deuteron (12.5MeV) creates almost the same track as 9.0MeV proton. As the number of protons in the above energy range saturates when the areal density become large enough to reduce energy of proton down to 9.0MeV, the ρR limit is estimated to be $\rho_t R \sim 200$mg/cm^2 as shown in Fig.II-5. The detection sensitivity is $Y_{N,DT} \cdot \rho_t R > 10^4$, when we use the target composition described above and require five counts on the CR-39 track detector subtending the solid angle of 10%.

E. silicon activation method

To prepare the diagnostic technique which has a higher ρR limit up to the ignition condition, we have adopted the neutron activation. The most appropriate tracer is sillion because of its low nuclear charge, the fairly large neutron cross section (0.26b)[11] for ^{28}Si(n,P)^{28}Al, the capability of efficient radioactivity measurement with a low background and its good dopability into plastic shell targets (see section 1-2). The activation yield in the target assuming central burn can be expressed as,

$$N^* = (f\beta N_A \sigma/A)\rho_t R Y_{N,DT} \qquad (1.1)$$

where f is the weight fraction of tracer, β is the natural abundance (0.92), and others have usual meanings. The number of signal counts is given by,

$$N_C = \eta_d\ \eta_c\ \exp(-\lambda t_d)\ \{1-\exp(-\lambda\Delta t)\}N^* \qquad (1.2)$$

Where η_d is the detection efficiency, η_c is the collection efficiency of activated isotopes, t_d is the delay time from the time of implosion to the start of counting, Δt is the counting time, and λ (5.26x10^{-3}sec^{-1}) is the decay constant. In the radioactivity measurement, the use of coincidence of β-ray (maximum energy is 2.86MeV) disintegration with the following γ-ray emission (1.78MeV) is a powerful technique to reduce the background noise.

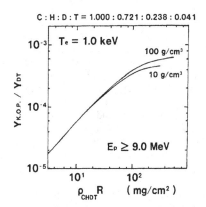

C : H : D : T = 1.000 : 0.721 : 0.238 : 0.041

$T_e = 1.0$ keV

100 g/cm³

10 g/cm³

$E_p \geq 9.0$ MeV

$Y_{K.O.P.} / Y_{DT}$

10^{-3}

10^{-4}

10^{-5}

10^1

10^2

$\rho_{CHDT} R$ (mg/cm²)

Fig. II-5   The knocked-on proton yield in the energy range of interest
(9.0–14.1 MeV) as a function of the total ρR of hydrogen,
deuterium, tritium-contained plastic (CHDT).

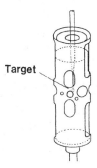

Target

Fig. II-6   The geometrical configuration of debris collection. This
collector covers an 85% solid angle over the target.

The target debris was collected using a niobium cap in an aluminum housing which covered an 85% solid angle over the target. Niobium is one of the metals which have a high melting point and whose neutron-induced reactions do not influence the measurement of ^{28}Al decay by a β-γ coincidence technique. The niobium collector was 45-mm-diam., 100-mm-long, 1-mm-thick cup which had twelve holes for the laser beams, four small holes for target monitors, and several small windows for other diagnostics as shown in Fig. II-6. After the laser irradiation, the activity of the collected debris was measured by using β-γ coincidence counter[12] which consists of a gas flow type proportional counter and a 6-in.-diam, 5-in.-long NaI(Tl) scintillator. During the counting the gas flow counter was covered with the debris collector cup, and then they were inserted into a 2-in.-diam, 4-in.-long well of the NaI(Tl) scintillator. The electric signals of β-ray and γ-ray counters were amplified, and were acquired into a counting system which supports not only the individual decays of β-rays, γ-rays and β-γ coincidence counts but also a time integrated spectrum of γ-rays coincided with β-ray signals. The detection efficiency of β-γ coincidence of ^{28}Al was calibrated to be 3.97% according to the principle of coincidence method. The natural background noise was extremely as low as 0.27 counts with a Poison distribution in the counting time of 600sec, and the back ground from the neutron- and charged particle-induced activation of niobium collector was experimentally clarified to be negligible.

The collection efficiency of the residual part of the shell target was calibrated precisely using ^{24}Na as a tracer. The adequacy to use ^{24}Na considering the atomic weigh dependece on the collection efficiency will be reported elsewhere[13]. The other uncertainties in the collection efficiency are due to the target mass (i,e, the residual mass) and to the surface condition of the collector. The collection efficiency does not so greatly depend on the target mass[7]. The surface roughness influences the collection efficiency, and the efficiency increases to be saturated (initially 15 %, finally 38 % in this experimental condition) as the collector is trained by the erosion and sputtering in the target implosion experiments[13].

The detection sensitivity estimated is $Y_{N,DT} \cdot \rho_t R > 10^6$, if we require the five signal counts of β-γ coincidence for 600sec assuming f=0.146 ($C_{11}D_{16}Si$ composition), $\eta_c = 0.38$, $\eta_d = 0.04$, and $t_d = 60$ sec. The feasibility of this activation method has been demonstrated[7]. A target used was CDTSi hollow shell whose diameter and wall thickness were 883 μm and 4.09 μm, respectively. The specific activity of tritium was 10.1 TBq/g. This target was irradiated by 0.53 μm laser light with an energy of 9.7 kJ in pulse width of 1ns yielding the neutrons of 9.7×10^{10}. Fig. II-7 shows the decay characteristics of β-rays, γ-rays and β-γ coincidence counts. The delay time t_d was 75 sec. The decay curve of β-γ coincidence shows a half life (2.24 min.) of 28 Al, thus the $\rho_t R$ was concluded to be 16.9 ± 2.5 mg/cm^2. This $\rho_t R$ value agreed well with that (17 ± 3 mg/cm^2) simultaneously measured by the knocked-on deuteron method[14], suggesting the reliability of our silicon activation measurement.

1-4 Simulation modeling and results

The implosion of a solid fuel target is much more sensitive to the preheating than that of a gas fuel target[15], since the initial specific entropy of a solid density plasma is much lower than that of a gas plasma. Both of the high energy electron and the radiation cause the preheating. In a low Z target like a plastic shell target, the electron preheating is dominant in comparison with the radiation preheating.

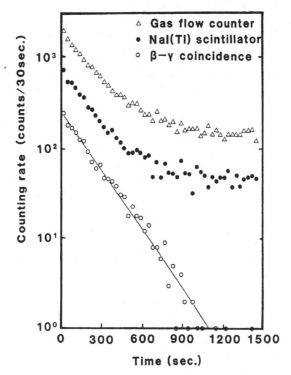

Fig. II-7   The decay curves of β-rays(Δ), γ-rays(•) and β-γ coincidence
counts (o) clearly show the disintegrations of ^{28}Al produced
by ^{28}Si $(n,p)^{28}$Al reaction in the target.

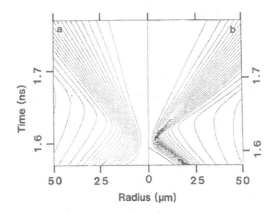

Fig. II-8   Flow diagrams obtained by 1-D hydrodynamic simulation with
a) the Fokker-Planck and b) the flux-limited Spitzer-Harm (F
= 0.1). The CD shell diameter is 486 μm, the shell
thickness is 3.5 μm, and the input laser energy is 8.56 kJ
in 1.57 nsec.

There are two mechanisms of high energy electron generation. The one of them is hot electron generation by the anomalous laser absorption, say, the resonance absorption, stimulated Raman scattering and so on. The other one is maxwellian tail formation by the electron-electron collisional relaxation. In the short wavelength laser irradiation with a moderate intensity, the hot electron generation by the anomalous absorption can be reduced to the level which is insignificant for the preheating.

However, we found that the latter process is not negligible even in a short wavelength laser irradiation as far as the corona temperature is higher than 1 keV. Namely, a significant amount of energy is included in the group of electrons whose energy is higher than 5 keV. Those electrons have a long mean free path and may heat the inside of the ablation front. We call this the nonlocal effects or the kinetic effects of the electron heat transport[16-23].

In order to simulate the shell target implosion, we introduced the Fokker-Planck equation in the 1D fluid code 'HIMICO' for describing electron dynamics. The Fokker-Planck equation which is used is as follows,

$$\frac{\partial f}{\partial t} + v \cdot \nabla f - eE \cdot \frac{\partial f}{\partial v} = \left(\frac{\delta f}{\delta t}\right)_c + \left(\frac{\delta f}{\delta t}\right)_h \qquad , \qquad (1.3)$$

where f is the velocity distribution function, v is the particle velocity, and E the electric field. The collision operator (the first term of the right hand side of Eq. (1.3) is

$$\left(\frac{\delta f}{\delta t}\right)_c = \frac{C}{v^2}\frac{\partial f}{\partial v} + \frac{D_{11}}{v^2}\frac{\partial}{\partial v}\left(\frac{1}{v}\frac{\partial f}{\partial v}\right) + \frac{D_\perp}{v^3}\frac{\partial}{\partial \mu}(1-\mu^2)\frac{\partial f}{\partial \mu} \qquad , \qquad (1.4)$$

where $C = V_e^4 / (z+1) \lambda_0$, $D_{11} = C \cdot (T_e/m_e + ZT_i/M_i)$, $D_\perp = (Z+1) C / 2$ and $\lambda_0 = T_e^2 / [4\pi n_e (z+1) e^4 \ln \Lambda]$ is the electron mean free path for 90° scattering. The second term of the r.h.s of Eq.(1.4) denotes the laser heating by the inverse-bremsstrahlung. This operator[24] is given by

$$\left(\frac{\delta f}{\delta t}\right)_h = \frac{Av_o^2}{3}\frac{1}{v^2}\frac{\partial}{\partial v}\left(\frac{g}{v}\frac{\partial f}{\partial v}\right). \qquad (1.5)$$

The notations of Eq.(1.5) is the same as the reference of B. Langdon[24]. We solve the above Fokker-Planck equation by approximating the electron distribution function by

$$f = f_0 + \mu f_1, \qquad (1.6)$$

where $\mu = \cos\theta$ and θ is the angle between electron velocity and the radial direction of the spherical coordinate.

In Figs. II-8 (a) and (b), we compare the implosion flow diagrams around the maximum compression between the Fokker-Planck case and the flux-limited Spitzer-Härm case, where the flux limitter f is decided so as to adjust the implosion time of the simulation to that of the experiment. In Figs. II-8 (a) and (b), the CD shell diameter is 486 μm, the shell thickness is 3.5 μm, and the input laser energy is 8.56 kJ in 1.57 nsec. In the Fokker-Planck case, the non-ablated part of the target shell is preheated and expands during implosion in comparison with the Spitzer-Härm case. The Fig. II-9 shows the compressed plasma area mass density v.s. target shell thickness. Because of the nonlocal heat transport by the

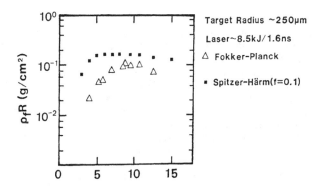

Fig. II-9 Compressed core ρR is plotted as a function of target shell
thickness. Triangles and filled squares show the simulation
results by the Fokker-Planck and the flux-limited Spitzer-
Härm (F=0.1), respectively.

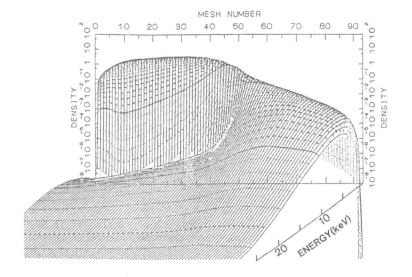

Fig. II-10 The space and energy distribution of electron at t = 1.4
nsec in Fig. II-8 (a) case. The number of multigroup is 75
from 0 eV to 88 keV (Shown only for <25 keV in the figure).
The energy width of multigroup $\Delta E_g = E_{g+1} - E_g$ with $E_1 = 1$ eV
and $E_{g+1} = 1.08^2 E_g$. The density is normalized by the cut-off
density of the laser wavelength $\lambda_L = 1.06$ μm.

Maxwellian tail electron, the area mass density ρR decreases by one order of magnitude when the shell thickness decreases from 8 μm to 4 μm in the Fokker-Planck case. On the other hand, the ρR does not decrease in the Spitzer-Härm case even if the target shell thickness decreases down to 5 μm.

The detail of the electron transport in the imploding shell is found in Fig. II-10. Here, the space and energy distribution of electron is shown. The space coordinate is expressed by the Lagrangian mesh number of the simulation. The critical surface is in between mesh number 72 and 73. The ablation front is around the mesh number 55. It is found that the high energy tail electrons are mainly generated between the critical surface and the ablation front. As shown is Fig. II-10, the energetic electrons whose energy is higher than 5 keV penetrate deeply into the inside of the ablation front. Actually, a double hump energy distribution is formed in the rear side of the ablation front.

We conclude that the energetic electron preheating is important for the shell target implosin even if there is no anomalous production of hot electrons. Therefore, the ablator layer has to be carefully designed to reduce the preheating by the nonlocal heat transport. The shiedling layer will be required thicker than 1 mg/cm^2 when the plastic ablator layer is used.

We compare the radiation heat transport in the CDT shell target implosion with that in the cryogenic fuel target. It may affect the ablation process and the compression.

In the ablation process, x-rays are emitted from the overdense plasma between the critical surface and the ablation front. Because of the energy loss by the X-ray radiation, the mass ablation rate is reduced. Therefore, the ablation processes for the CDT shell target and the DT foam cryogenic target are different. The mass ablation rate M relates to the radiation energy loss rate as follows,

$$M/m_i = U = \frac{q_e - q_r}{\frac{5}{2} n_i k_B T (1 + \bar{z})}$$

which is nothing but the energy balance equation near the ablation front. Here q_e is the inward electron heat flux, q_r is the outward radiation energy flux, z is the average ion charge, and m_i and n_i are the average ion mass and number density, respectively. According to a simulation for the CDT shell target, M increases by 1.5 times when the radiation energy loss is artifitially killed in the simulation. Namely, q_r is not negligible near the ablation front in comparison with q_e.

However, the radiation cooling due to carbon impurity ions is not effective in high density and high ρR plasmas. Since the average temperature of the compressed plasma is less than 0.5 keV, the Rossland mean free path of the radiation l_r which is given by

$$\rho l_r = 4 \times 10^2 \frac{1}{\rho} T^{7/2} \frac{\Lambda}{\bar{z}\,\bar{z^2}} \qquad g/cm^2$$

is less than 0.01 g/cm^2 for a plastic plasma of a several handred times solid density. Here ρ is the mass density in the unit of g/cm^3, T in keV. Namely, the x-ray radiations are trapped in the plasma when ρR is greater than 0.1 g/cm^2. Then, the radiation cooling is effective only in the thin

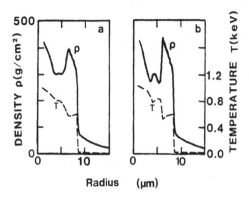

Fig. II-11   The density (solid line) and temperature (broken line)
             profiles a) with non-LTE radiation transport and b) without
             radiation loss.

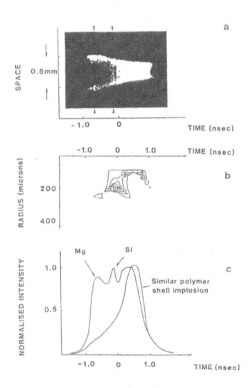

Fig. II-12   (a) Space and time-resolved image of imploding multilayer
             target with peak emission times marked by arrows, (b)
             simulation of the expected x-ray emission history as a
             function of space and time, and (c) spatially integrated
             x-ray emission over the bottom half of the image.

surface layer of the compressed plasma. The simulation results of Figs.
II-11(a) and (b) show that the compressed plasma density and temperature
profiles with radiation loss and without radiation loss, respectively. In
this case, the total ρR and the averaged core density is reduced by 1%
with artificial cut of radiation loss. The profiles of Figs. II-11 (a)
and 11 (b) are obtained by the hydrodynamic simulations. The initial
density profile is given by the stepwise approximation of density profile
on the way of implosion (at 500 psec before the collapse) of CD shell,
which diameter is 494 μm and thickness is 8.71 μm, irradiated by the laser
of 8.14 kJ in 1.91 nsec. The initial temperature and velocity are 6eV and
3×10^7 cm/sec, respectively.

In conclusion, the ablation and acceleration processes of the target
shell depend upon the target material because of the radiation cooling.
However, the deceleration and compression of the shell are not sensitive
to the radiation transport as far as the plasma temperature is lower than
1 keV, the total ρR is greater than 0.1 g/cm^2, and the impurity ion z is
not high.

2. Experimental Results

2-1 Implosion dynamics, energy transport and mass ablation

As for the implosion dynamics, energy transport and mass ablation, an
extensive array of diagnostics was applied. Plasma calorimeters measured
the coupling of the laser energy to the target. It was found to be 40 ±
5 % for the shot of a target with an outer diameter of 505 μm which was
irradiated with 8.33 kJ in a flat top pulse of 1.69 nsec FWHM duration.
The maximum and minimum energy imbalance between the beams on this shot
were +2.4 % and −3.9 %, respectively. The initial target was a CD shell
of 2.43 μm thickness overcoated with (from this inner lyer outwards): CH
1.37 μm; Al 0.299 μm; CH 0.97 μm; Si 0.217 μm; CH 1.17 μm; Mg 0.269 μm and
CH 1.86 μm.

The energy transport from the absorbed region to ablation surface was
evaluated from the x-ray emission history of buried marker layers, which
was observed by using an x-ray pinhole camera coupled to an x-ray streak
camera. Fig. II-12 (a) shows a streaked x-ray image of the above
described shot. Fig. II-12 (b) shows a simulation from the Fokker-Planck
hydrodynamic code of the expected x-ray emission as a function of both
space and time. Time is with respect to the center of the laser peak.
Figure II-12 (c) shows a plot of the intensity which is spatially
integrated over the bottom half of the image (where the exposure level is
better), together with a plot of intensity of a simple polymer shell of
the same mass that was also integrated over the bottom half of its image.
The mass ablation rate for the first two emission peak is estimated to be
3.2 (±0.2) × 10^5 gcm^{-2}s^{-1}. The areal density ablated is obtained by
integration of $\dot{m}$ as a function of the time of the peak x-ray emission.
Implosion times and also the terminal velocity just before the maximum
compression are derived from the x-ray streak picture.

Characteristic parameters of implosion such as absorption, ablated
areal mass density, implosion time and implosion velocity derived from the
experiment are compared with the simulation results. The Fokker-Planck
simulation shows close agreement with the experiment for all the implosion
parameters. The simulation results with the Spitzer-Harm description of
heat flow could not be adjusted to hit all the implosion parameters
simultaneously even with the change of adjustable parameter of flux
limiter f.

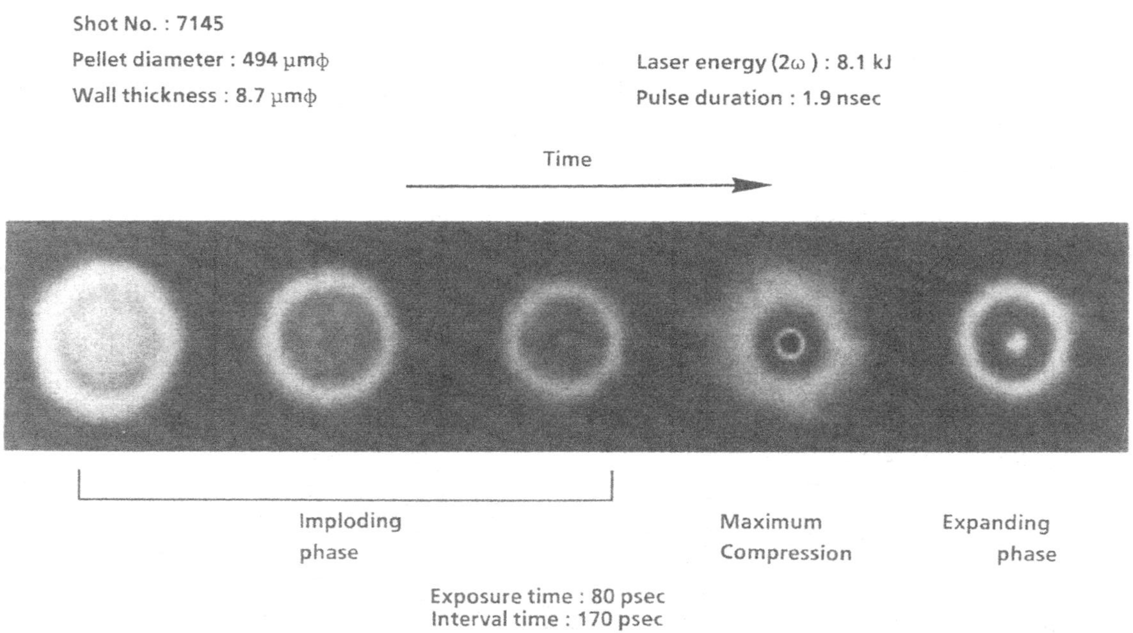

Shot No. : 7145

Pellet diameter : 494 μmφ

Wall thickness : 8.7 μmφ

Laser energy (2ω) : 8.1 kJ

Pulse duration : 1.9 nsec

Time

Imploding
phase

Exposure time : 80 psec

Interval time : 170 psec

Maximum
Compression

Expanding
phase

Fig. II-13   A series of X-ray pictures of a plastic shell target implosion taken by an x-ray multi-framing camera. Annular images are observed before and after the maximum compression without significant deformation. Fairly uniform implosion is achieved.

The implosion dynamics, stability and symmetry were observed by high speed multi-frame x-ray cameras. A successive frame picture is shown in fig. II-13. Good energy balance of twelve beams, and the use of the RPP enables us to obtain a symmetric implosion of uniform pellet. The growth of irregularity in the acceleration phase were analyzed as a function of laser irradiation nonuniformity due to the beam energy imbalance and/or due to the change of D/R.

2-2 Fuel ρR and convergence ratio

The density radius product of compressed core $(\rho R)_{fuel}$ is shown as a function of shell thickness in Fig. II-14, where the mass density of only the hydrogen isotope components in a plastic pellet are counted in. The values of $(\rho R)_{fuel}$ were measured by the secondary reaction method (solid circles in Fig. II-14) using CD shell pellet, by the proton knocked-on method (solid squares) using CDTH shell pellet. The highest value of ρR being measured by proton knocked-on method is almost the upper limit of the applicability. The data points plotted by stars were obtained by the Si activation method. In Fig. II-14 it is clearly seen that the ρR value of implosion with RPP is noticeably larger than that without RPP (hatched area).

One of the most important results of the shell implosion experiments is the agreement between the experimental results and 1D simulation, which is achieved by introducing a kinetic description of the nonlocal heat transport treated by Fokker-Planck equation. The hollow shell pellet implosion is very sensitive to the preheat. The proper treatment of the high energy tail of a single Maxwellian distribution is important in prediction, especially of the core parameters. If the classical energy transport with flux limiter f is used in the simulation code, the results show much larger value of ρR and do not reproduce the dependence of ρR on the thickness of shell pellet as described in section 1-4.

The characteristic feature of the discrepancy between the ρR values with RPP and without RPP (normal beam) has been examined. Most of the data points with normal beam lies on the shaded region in Fig. II-14. They are more than one order lower than that with RPP at around 7 to 10 μm of shell thickness. The difference becomes smaller with decreasing and increasing thickness from this region. To see the feature more clearly, the experimental values of ρR normalized by the simulation results, $\rho R(exp)/\rho R(sim)$, ∎ are plotted in Fig. II-15 with solid squares as a function of shell thickness. It is clearly seen that the one dimensional behavior of implosion is kept in the results with RPP through the differnt thickness, whereas the normal beam results show large degradation in obtaining the predicted ρR values. This might be the effect of the improved uniformity of laser intensity on the pellet surface. In Fig. II-15, the convergence ratio R_0/R_f is plotted as a function of shell thickness where R_0 is the initial pellet radius and R_f is the final compressed core radius in simulation. With the implosion of large convergence, ρR and ρ are large and the coupling efficiency is also high (70~80% of the initial mass is ablated at the shell thickness of around 6-10 μm). At this region of implosion, the uniformity of irradiation becomes more important because of higher convergence. At the region of shell thickness of 4 to 5 μm, where the difference between ρRs with and without RPP becomes smaller, the implosion is nearly the exploding mode and stable. At thicker region more than 12 μm, the implosion seems to be stable due to the low aspect ratio.

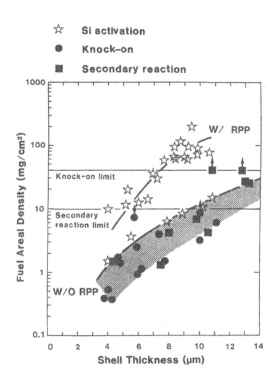

Fig. II-14 Fuel areal density $(\rho R)_{fuel}$ as a function of shell thickness measured by secondary reaction, proton knocked-on and Si activation method. The pellet diameters were 500-900 μm.

2-3 Derivation of compressed fuel density

The fuel density of the compressed core is derived by two different ways from the measured ρR. The one is to use the residual mass conservation with following equation

$$\frac{4\pi}{3} \frac{<\rho R>^3}{\rho^2} = 4\pi R_o^2 (\Delta R_o - \Delta R_{ablated}) = M_{core}$$

where R_0 and ΔR_0 are the initial pellet radius and thickness, $\Delta R_{ablated}$ is the ablated thickness of shell before the timing of maximum compression, which is derived by $\int_{-\infty}^{\tau_i} \dot{m} dt$, where τ_i is the implosion time.

The other is to use the compressed core size R to get ρ with $<\rho R>/R$. In the evaluation of R from an X-ray image for the high density compression, the effect of x-ray opacity must be carefully examined to give the correct core size. The core densities which were derived by two methods above described gave almost same values. As a cross check of the results, completely different derivation of the density was applied, which used the residual mass conservation and compressed core size with the relation of $(4\pi/3)R^3\rho = M_{core}$ to get ρ value. They support high value of the density derived by $<\rho R>$ measurements.

The achieved normalized densities are plotted in Fig. II-16 as a function of shell thickness. The total mass density of the initial pellet is 1.0 g/cm^3. Therefore, 600 times solid density means 600 g/cm^3. The mass density of D fuel component in pellet material is 0.16 g/cm^3, which is almost same density with solid D$_2$. It can be said that the required fuel density for ignition has been achieved.

2-4 Neutron yield, fuel temperature and coupling efficiency

Neutron yields were measured with several standard techniques: Al and Ag activation detectors and scintillation detectors. All detectors were absolutely calibrated with Al activation technique for DT neutrons and with DD proton counting technique for DD neutrons. The neutron yields plotted in Fig. II-17 are equivalent yields which are expected yields for the case of the equimolar mixture of deuterons and tritons. This normalization is necessary as the deuteron-to-triton ratio varies in each target. The equivalent yield for DD neutrons was also determined from the observed yield by multiplying a factor of ~30. The neutron yields for the RPP case was significantly lower than that for the normal beam case by an order of magnitude. This decrease of the neutron yield is due to that only 60-70 % of the incident laser energy can irradiate the target for the RPP case.

The fuel ion temperature in several shots was measured with the time-of-flight technique, but in most shots the neutron yields were not high enough to use this technique. The ion temperature shown in Fig. II-18 was therefore estimated from the observed yield, areal mass density and calculated residual mass. The directly measured temperature, which is also plotted for comparison, was higher than the estimated temperature. This is not surprising because the measured temperarue is neutron-weighted and hence indicate higher temperature than the average. The estimated temperature of ~0.3 keV for the high density compression targets (~500 μm dia. 8-10 ьm thick.) is comparable to the Fermi temperature (~1 keV) calculated from the measured density of ~600 g/cc. The compressed core is therefore expexcted to be partially degenerated. The degeneracy effect

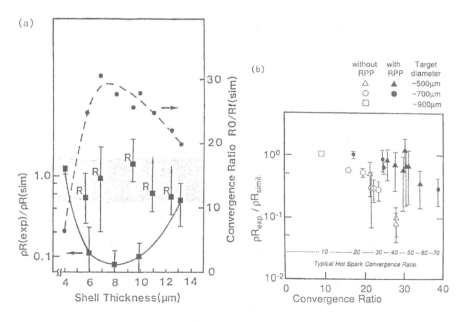

Fig. II-15 (a) Shell thickness dependence of ρR and convergence. The data points with suffix R represent the ρR value with RPP.

(b) The ratio of measured ρR to that simulated as a function of the convergence ratio. The hot spark means the region which contributes to 90% of neutron yield.

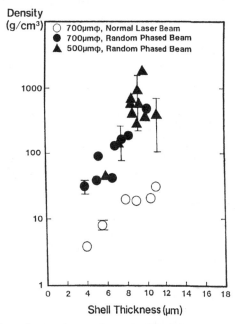

Fig. II-16 Random phase plates drastically increase the density of compressed plasma to achieve 600 times initial density (600 g/cm^3). The fuel part is about 1/7-1/6 of the total mass which depends on the target material composition.

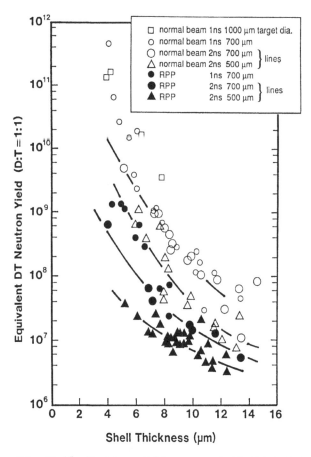

Fig. II-17 Neutron yield versus shell thickness.

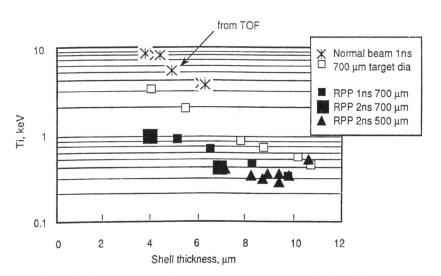

Fig. II-18 Estimated ion temperature versus shell thickness.

is necessary to be taken into account to estimate the energy of the compressed core.

It would be worthwhile to see whether the coupling efficiency from the absorbed energy to the core energy is reasonably high for the high density compression. The compressed core energy E_{core} may be estimated by

$$E_{core} = N_i [Z\varepsilon_e + \frac{3}{2} T_i + X_i - E_{corr}]$$

where N_i the ion number, Z the ion charge, ε_e (n_e, T_e) the electron kinetic energy averaged by the Fermi distribution as a function of the electron density n_e and the electron temperature T_e, T_i the ion temperature, X_i the mean ionization energy, E_{corr} the correlation energy given by $0.9 (Ze)^2/a$ for the mean ion distance a. For $T_e = T_i = 0.3$ keV, $\varepsilon_e \sim 0.9$ keV ($T_e = 0.3$ keV and $n_e = 2 \times 10^{26}$ cm^{-3}) $Z = 3.5$, $X_i = 0.52$ keV (fully ionization), $E_{corr} \sim 0.5$ keV from the measured density and $N_i \sim 2 \times 10^{17}$ from the residual mass, the core energy estimated fromm this expression was 4-6 % of the laser energy absorbed before the maximum compression. The coupling efficiency was also estimated independently from the residual mass and the implosion velocity to be the comparable value as above.

2-5 Degradation of implosion performance

The implosion performance is greatly improved by installing the random phase plates as is shown in the previous section. In this section, the imperfection of the target and the nonuniformity of the laser irradiation which affect the implosion uniformity are described.

Although the precision of the hollow shell fabrication itself is satisfactory, the support structure may perturb the implosion process. In the high density compression experiments, the obvious degradation of implosion performance due to the target support was observed. Figure II-19 shows the intensity contours of x-ray pinhole images of the targets sustained by a 7-μm-diam. plastic fiber (a) and a 2.5-μm fiber (b). The deformation of isointensity contours marked by an arrow in Fig. II-19 is due to the large diameter supporting fiber. The measured ρR values for the targets supported by 7 μm fiber were several times lower than those for 2.5 μm fiber. This phenomena indicates that the implosion uniformity is strongly affected by the imperfection of the target surface.

Concerning to the laser irradiation uniformity, we compared the x-ray intensity distributions with and without RPPs. Fig. II-20 shows a result of the x-ray computed tomography[25] which was obtained without RPP for the focusing condition of $d/R = -5$. Fig. II-20 (a) is a intensity contour of x-ray emitted from the surface of the target, and (b) is a line out in the center plane. The nonuniformity of the heated plasma with relatively low mode numbers is clearly seen, which is predicted by the laser illumination calculation described in the section 1-1. A mode expansion of this intensity distribution into the pherical harmonics is shown in fig. II-21 (a) which agrees well with that of the calculated laser intensity distribution (Fig. II-21 (b)). The most dangerous mode number would be around 6 in the twelve beam illumination system if the Rayleigh-Taylor instability could be supressed at higher mode numbers by the effects of the lateral thermal smoothing and ablative stabilization. And the absorbed laser intensity distribution may changes due to the expansion and

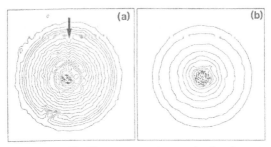

7µm plastic fiber   2.5µm plastic fiber

Fig. II-19   The intensity contours of typical x-ray pinhole images for
             random phased irradiation. A 7 µm plastic fiber (a) and a
             2.5 µm fiber (b) were used to support the target.   The
             position of the support is the top of each image.

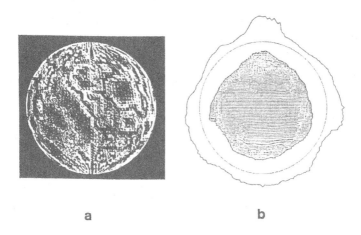

a   b

Fig. II-20   The nonuniformity of a plasma heated by laser beams without
             random phase plates. (a) is a result of computer tomography
             of the surface emission of x-rays, and (b) is its line out in
             the center plane.

the shrink of critical density surface. Thus implosion asymmetry was clearly characterized by a mode of ℓ = 6 as shown in Fig. II-22. The similar hexagonal deformation of the compressed core was observed in an α-particle imaging[26] and in a penumbral imaging of D-D reaction protons[27].

To check the influence of the laser illumination nonuniformity onto the implosion performance with RPP, the laser focusing condition was changed as d/R = -4, -5 and -6. The calculated nonuniformities (rms of spherical harmonics) arc 6.8 % (for d/R = -4), 5.4 % (-5) and 3.8 % (-6), respectively. An x-ray framing camera was used to observe the two dimensional symmetry of the accelerating shell (probably the ablation surface). Figure II-23 shows the estimated amplitudes of shell deformation as a function of the mode number. Figure (a), (b) and (c) correspond to the focusing conditions of d/R = -4, -5 and -6, respectively, for the targets of about 650 μm diameter and about 8 μm wall thickness. These deformations were obtained from the framing pictures in which target radii shrinked to about 40 % of their initial values. As shown in this figure, the symmetry of the imploding shell was improved with increasing |d/R|. This is consistent well with the calculated laser irradiation nonuniformity. ρR (including C, D, T and Si) of the imploded shell was simultaneously measured by Si activation method. Fig. II-24 shows the summary of the achieved ρR, the laser irradiation nonuniformity and the asymmetry of the accelerating shell as a function of d/R. This figure indicates that the focusing condition of d/R = -5 ~ -6 is favorable to minimize the implosion nonuniformity, when we use the present random phase plates. In addition the reason why there was no improvement of ρR at d/R = -6 comparing with d/R = -5 would be the fact that the overall nonuniformity including the imperfection of target and the laser power imbalance is comparable to the calculated irradiation nonuniformity (σ_{th} = 3.8 %).

3. Discussion

3-1 Physical meanings of 600 times solid density compression

We will discuss here the imploded plasma conditions which are required for achieving 600 times compression. Since 65% of the target total mass is ablated as for 500 μmφ and 8~10 μm^t shell target (see §II-2-1), the compressed core plasma radius R_f (μm) is required to be less than $R_0^{2/3}/4$ μm for the intial target radius R_0 (μm). Thus, the almost all the residual mass of the target shell may converge to the radius of less than $R_0/25$ for $R_0 \simeq 250$ μm in the experiment. Therefore, even if we do not take into account the hydrodynamic instabilities, non-uniformities of the ablation pressure P_a and the target shell thickness ΔR and the radius R are required to be less than 4%. This irregularity stands for the square root of the mean square average of the fluctuations.

As for the Rayleigh-Taylor modes triggered by the ablation pressure non-uniformity, the radial displacement of a shell surface ξ may satisfy the following equation,

$$\frac{d^2 \xi_l}{dt^2} = r_l^2 \xi_l + \delta g_l \qquad (3.1)$$

where r_1 is the Rayleigh-Taylor growth rate of the l mode and δg/g = $\delta P_a/P_a$ is the acceleration irregularity. The temporal evolution of ξ_l is given by

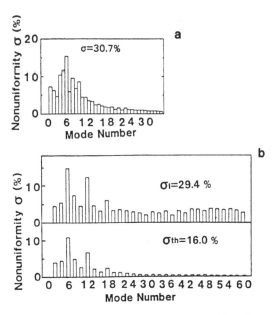

Fig. II-21 The comparison of the measured nonuniformity of x-ray
emission (a) and the calculated nonuniformity of laser
irradiation (b). $\ell = 6$ is a dominant mode number for
the dodecahedral irradiation system, if the laser beam
has a flat top intensity distribution.

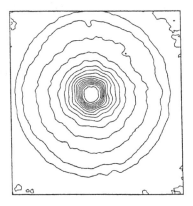

Fig. II-22 The low mode nonuniformity clearly appears in the x-ray
pinhole picture when a target is driven by normal laser
beams without random phase plates.

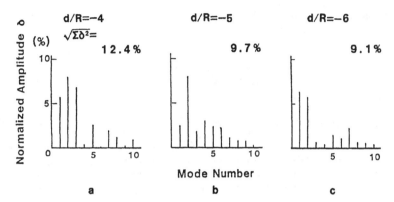

Fig. II-23 The shell deformation amplitudes measured by a multi-frame
x-ray framing camera as a function of mode number. The
laser focusing condition was changed as d/R = -4 (a), -5
(b) and -6 (c).

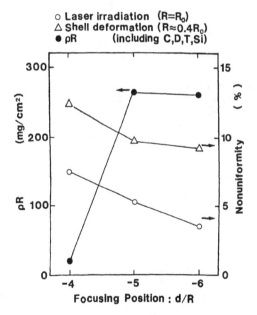

Fig. II-24 The laser irradiation nonuniformity, asymmetry of accelerating
shell and achieved ρR as a function of d/R.

$$\xi_l(t) = \frac{\delta g_l}{2\gamma_l^2}\left(e^{\frac{1}{2}\gamma_l t} - e^{-\frac{1}{2}\gamma_l t}\right)2 \qquad (3.2)$$

When we use the classical value of the Rayleigh-Taylor growth rate, γ_l $=\sqrt{lg/R_0}$, Eq.3-2 yields

$$\xi_l(t)/R_o = \frac{1}{2l}\cdot\frac{\delta g_l}{g}\left(e^{\frac{1}{2}\gamma_l t} - e^{\frac{1}{2}\gamma_l t}\right)2 \qquad (3.3)$$

In the case of the dominant low l modes for GEKKO XII laser (because of the irradiation symmetry δg_1 is maximum for l $\simeq$ 6), we can use the classical value for γ_1 and $\gamma_1 \tau_{imp} \simeq \sqrt{1} \simeq$ 2~3 (1=4~9). According to the irradiation uniformity which were measured recently, δg_1 /g < 0.02. The equation (3.3) yield the root mean square average (r. m. s. a.) of the low ℓ mode shell surface distortion is less than 0.02 ~ 0.03.

This result almost meets with the expected distortion 4% for the high density compression.

The effects of higher l modes are much more serious if the classical Rayleigh-Taylor growth rate is preserved for those modes. However, many theoretical and experiment works indicate that the growth rate is significantly reduced from the classical value. According to the result of (ref.28), the maximum growing mode is

$$l_{max} \approx 30$$

and the maximum growth rate is about 3 × 10^9 1/sec. The distortion ξ/R_0 is approximately 1%, according to eq. (3.2) and the irradiation uniformity measurement.

Let us conclude here about the implosion nonuniformity due to irregularity of the irradiation laser intensity. The most dangerous mode of the present experiment is the mode number l = 6. The r. m. s. a. of the shell surface distortion will be less than 4% after the shell acceleration. This value is consistent with the 600 times compression.

As for the target shell radius and thickness irregularity, the distortion grows as

$$\xi_l/R_o \simeq \xi_{lo}/R_o e^{\gamma_l \tau_{imp}}. \qquad (3.4)$$

Since $(\gamma_1 \tau_{imp})_{max} \simeq 3$, the initial pellet irregularity has to be less than 0.2% for the maximum growing mode l $\simeq$ 30. Since the measured target surface roughness is less than 1% for the peak to valley fluctuation, namely, $(\Delta R_{max} - \Delta R_{min}) / \Delta R_{av}$. Therefore, the required upper limit of the initial amplitude , 0.2% may be satisfied.

Note that the initial laser non-uniformity printing depends upon instantaneous intensity irregularities at the leading edge of the laser pulse. This laser printing effect can be attributed to the initial shell thickness fluctuations. We have also investigated those effects. The symmetry requirement is also found to be met for this case.

In conclusion, the green laser irradiation symmetry of the GEKKO XII laser and the plastic pellets used for the high density compression were found to be good enough for achieving 600 times compression. Those results suggest that the irradiation nonuniformity of GEKKO XII is small enough to achieve ignition.

3-2 Hot spark formation and neutron yield

In the 1D simulation, the hot spark is formed at the center of the target. The hot spark radius R_h is about a half of the total core plasma radius. Therefore, the convergence ratio R_0/R_h is required to be greater than 40~50. As it is discussed in the previous section, our laser and pellet symmetries are not good enough for achieving this high convergence. Therefore, in the experiments, the one dimensional hot spark may not be formed and the plasma density and temperature may be averaged out over the total core radius of $R_0/20 \sim R_0/30$.

As it is shown in the section II-2-4,(See Fig. II-17) the experimental neutron yield is reduced by two to three orders of magnitude in comparison with the simulation value. This reduction is partly explained by the collapse of the hot spark.

Since the implosion symmetry is marginal for the convergence ratio of 25~30, the plasma flow velocity may be very irregular at the final stage. Therefore, we can not expect that the all of the imploding shell kinetic energy is converted into the compressed plasma internal energy. Probably, a half of the shell kinetic energy will remain as the turbulent flow energy. Both of the collapse of hot spark and the turbulent energy loss will explain the reduction of the neutron yield in the high density compression case.

For achieving ignition and high gain, the hot spark formation is essential. The present experiment suggests to us that the factor 2 improvements of the laser and pellet symmetry is necessary for the hot spark ignition. In the up-grade GEKKO XII 24 beam laser system, it may be achieved by introducing the temporal smoothing of the laser beam, and the pellet levitation. Furthermore, we have to evaluate the symmetry of the blue laser case. After the blue laser direct irradiation experiment, we will decide whether the laser wavelength is 0.53μm or 0.35μm for the ignition experiment by the GEKKO XII up-grade system.

III. CRYOGENIC HOLLOW SHELL IMPLOSION

Cryogenic deuterium-tritium hollow shell target is essentially important for laser fusion. Cryogenic targets with glass microballoons (GMB) which act as sustainers of DT fuel and pusher-ablator have been studied at several laboratories[1-3]. The University of Rochester has achieved recently 100-200 times liquid densities using this target. The cryogenic DT shell with GMB has problems related to the hydrodynamic stability of the implosion. Most serious one may be the fuel-pusher mixing at the stagnation phase which cools down the fuel temperature.

A cryogenic foam shell target which consists of a liquid or solid DT shell supported by the low-density plastic-porous hollow shell (refer as foam shell here after) could be superior to the cryogenic target with GMB since the hydrodynamic problems could be suppressed. We have developed the cryogenic foam shell targets using a porous plastic hollow shell and studied the laser plasma interaction, energy transport and implosion physics.

1. Fabrication of the Target

1-1 Fabrication and characterization of foam shell

Foam shells were fabricated by the phase separation method in conjunction with the double-nozzle droplet producing method.[4] A polystyrene solution dissolved in dioxiane was mainly used as a starting material. The mass density of foam shell was controlled by adjusting the concentration of solute. The droplets produced by the double nozzle were fallen into the freezing tower for phase separation where the temperature of environmental gas was controlled to be less than -100 °C. The dioxiane in the porous polystyrene was evacuated in the low temperature vacuum vessel.

The sphericity and uniformity of shell walls were measured using the x-ray photogram obtained by contact radiography. The resolution is estimated to be about 5 μm. In order to use them for the experiment we judged the uniformity of wall thickness inspecting the enlarged photogram. The porous structure and surface smoothness were measured with the electron scanning microscope picture. The cell size changed with freezing speed in the freezing tower. Typical cell sizes of the foam shell used in the experiments were 3~5 μm in diameter and ~10 μm long. The mass density was controlled from 30 mg/cm^3 to 150 mg/cm^3.

1-2 Cryogenic techniques

A deuterium cryogenic foam shell target was made in the following way.[5] A liquid helium cryostat and a target suspender with a liquid nitrogen container are set at the bottom and top of the target chamber respectively. The diameter of target chamber is 1.7 m. The cryostat contains a small quartz pot where the liquid deuterium is filled in as shown in Fig III-1. The foam shell target suspended with a 5 μm stalk was immersed in the liquid deuterium. The liquid deuterium was filled both in the cells of wall and the cavity in this way. However the liquid deuterium was pushed out from the cavity and was filled only in the cells of wall when we made use of boiling bubbles came out from the target. A heater surrounding the pot controlled the boiling.

The filled liquid deuterium mass was measured using a resonance frequency of vitration of the stalk and shell target system.[6] After those processes the surrounding deuterium gas was evacuated to a pressure of the triple point keeping the temperature of outside of pot at 5~7 K in order to hold deuterium ice. The laser shot was performed after the fuel layer thickness was measured again and then the quick removement of the shroud following the evacuation of the residual gas. Note that in the above processes only a little amount (5 % of the total deutrium) of deuterium in target sublimates and the specific volume of the solid deuterium is smaller than that of the liquid.

2. Implosion Experiments

The physical properties of cryogenic foam targets are different from conventional targets used for implosion experiments such as GMB and

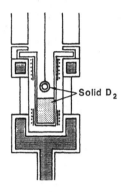

Freeze D$_2$

Fig. III-1 Schematic picture of the cryogenic system for the implosion
experiment.

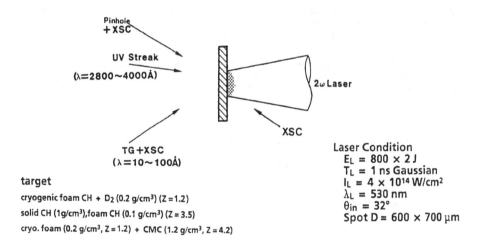

Pinhole
+ XSC

UV Streak
(λ=2800~4000Å)

2ω Laser

XSC

TG + XSC
(λ=10~100Å)

target
cryogenic foam CH + D$_2$ (0.2 g/cm^3) (Z = 1.2)
solid CH (1g/cm^3),foam CH (0.1 g/cm^3) (Z = 3.5)
cryo. foam (0.2 g/cm^3, Z = 1.2) + CMC (1.2 g/cm^3, Z = 4.2)

Laser Condition
E_L = 800 × 2 J
T_L = 1 ns Gaussian
I_L = 4 × 10^{14} W/cm^2
λ_L = 530 nm
θ_{in} = 32°
Spot D = 600 × 700 µm

Fig. III-2 Schematic arrangement of energy transport experiment.

plastic CH targets. They have nonuniform distributions of mass density and charge number (Z) of ions in a few micrometer scale. In such targets the filamentation of the laser beam and hydrodynamic instabilities could take place in the corona region and at near the ablation front respectively. The low average atomic mass causes a high speed shock and produces long scale length corona plasmas. The later may enhance nonlinear laser-plasma interactions such as stimulated Brillouin scattering (SBS), stimulated Raman scattering (SRS) and two plasmon decay instability. The low Z number of the target helps deep penetrations of energetic electrons so that the preheating of fuel might be serious.

We have studied the laser-plasma interaction, energy transport as well as the implosion physics using the planar and spherical cryogenic foam targets whose average mass density and Z are ~0.2 g/cm^3 and 1.2, respectively. The absorption of laser is about 40 % at a laser intensity of 4×10^{14} W/cm^2 and no enhancement of SBS, SRS and filamentation were observed[7].

Here we discuss the energy transport and implosion experiments.

2-1 Energy transport

Energy transport was studied by illuminating the 0.53 μm, 1 ns Gaussian pulse onto planar targets. The neighboring two beams of the GEKKO XII were used to achieve the average laser intensity of $\sim 4 \times 10^{14}$ W/cm^2 in the area of 0.6 X 0.7 mm^2. The incident angle of the both beam to the target normal was 32 degrees. The polarization directions of two beams were different with 45°. The schematic diagram of experimental arrangement is shown in Fig. III-2. The visible emission from the rear side was measured using a visible streak camera with a spectrometer. The spectrum response of the system was calibrated using a standard tungsten lamp. The x-ray emissions from both side were measured using calibrated x-ray streak cameras. SBS and SRS and the second and three halves harmonics of the incident light were also measured.

Fig. III-3 shows the temporal change of visible and x-ray emissions from rear side of the cryogenic foam target and the normal density polystyrene target having same areal density. The detection sensitivity for both target was same. The both emissions in visible and x-ray regions from the cryogenic foam target were stronger than those from the polystyrene target. The first shoulder in visible emission is due to the preheating by the electrons and/or soft x-ray. The second fast rise signal is due to the shock heating. The cryogenic foam target is more sensitive to the preheating and shock heating.

Fig. III-4 is the temperature of the rear side obtained by changing the arelal mass density of the cryogenic foam target. The data with CMC is obtained using a cryogenic foam target coated with sodium carboxyl-methyl-cellulose (CMC) at the front side. The thickness was 5 μm. The temperature was estimated assuming the black body radiation[8]. The temperature estimated from the first rise signal height increases with decreasing the target thickness. While the temperature from the second signal height that corresponds to the maximum temperature does not change so much even if the target thickness increases. The temperature for CMC coated target was lower than that of the without CMC coating. This result is very importnat to understand the mechanism of preheating corresponding to the first rise signal and to suppress the preheating.

We can estimate the relative values of preheating by x rays using the measured x-ray emission from the front and rear sides. The theoretically

● Cryogenic Target showed stronger emission than normal density CH target.
Preheating level of cryogenic target is higher than the CH.

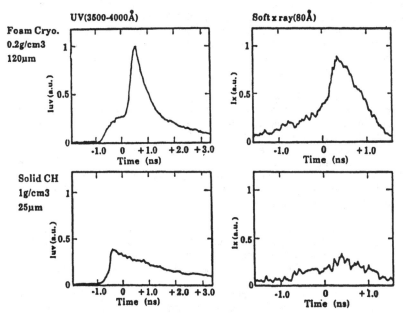

Fig. III-3   Temporal change of UV and x-ray emissions from the rear side
of planar cryogenic and plastic target for the same areal
mass.   (a) Cryogenic foam target (0.2 gm/cm^3) and (b) normal
density plastic (1 g/cm^3).

● Preheating and shock heating are suppressed with over-coating of CMC at front side.

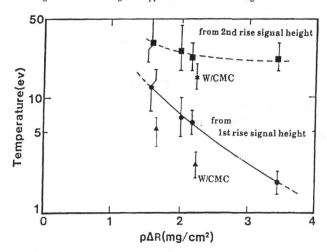

Fig. III-4   Temperatures of the rear side.   ■ indicates the total amount
of preheat by both shock and precursor preheat.   ● indicates
the precursor preheat.   W/CMC means the measured values of
the total and precursor preheats with cryogenic planar
targets overcoated with plastic, showing reduced preheat
levels.

estimated temperatures corresponding to the first rise signal did not agree with the experimental results. When relative transmitted x rays agree well with experimental results for both targets, the calcualted electron temperature of the target with CMC is higher than the experimental values of without CMC. The calculation was a relative fit to the measured preheated value without CMC. The actual preheat level with CMC coat was even lower than the value without CMC. Therefore x ray is not the major preheating source.

We can also estimate theoretically the preheating due to energetic electrons such as the ones of Maxwellian high energy tail and those produced by the resonance absorption and two plasmon decay. From x-ray spectrum measurement hot electrons of 6 keV were confirmed so that we used those electrons to estimate the preheating. The observed temporal change of preheating temperatures for the different thickness agreed reasonably well with the simulation results of 1 D hydro code HISHO where the temperature and energy fraction of hot electrons are 10 keV and 20 % of absorbed energy respectively

The origin of the second rise signal was confirmed to be due to the shock wave from the measurements of appearance time of the signal peak. The rear side temperature when shock wave arrived was ~40 eV for the target thickness of 70 μm or areal mass density $\rho\triangle R$ of 14 mg/cm^2.

From these experimental results the cryogenic foam target is found to be sinsitive to the preheating due to high energy electrons and also shock heating. The preheating, however, can be suppressed reasonably by adding the normal density CH layer as an ablator.

2-2 Implosion experiments

Targets used for the implosion experiment are 0.6~1.0 mm in diameter and 35~60 μm in wall thickness. The areal mass density $\rho\triangle R$ corresponds to the ones of normal density CH targets having the wall thickness of 7~12 μm. The implosion velocity was measured by a x-ray streak camera. The implosion symmetry was measured by x-ray framing cameras with temporal resolution of 80 ps which can take successive 6~9 images with time interval of 100~150 ps. The compressed core ρR was determined from the ratio of 14 MeV and 2.4 MeV neutron yields. The ion temperature was measured by the neutron time of flight method.

The compressed core size obtained using a time integrated x-ray pinehole camera was considerably bigger than that of the normal density CH shell target having the same initial diameter and areal mass density $\rho\triangle R$. The x-ray framing images in Fig. III-5 shows the center emission appears more than 300 ps before the maximum compression. This appearance time was much earlier than the time of normal density CH target. This early center emission can be explained as the effects of preheating and delayed arrival time of shock wave to the inner surface of the shell wall. The shock velocity in the cryogenic foam target should be $\sqrt{5}$ times faster than that of the normal density CH target since the mass density is 5 times lower. The traverse time of the shock wave in a shell wall with the same areal mass density becomes longer. By those effects precursor plasmas arrive at the shell center considerably earlier than the main body. This phenomena could be confirmed by the simulation. If we avoid the preheating the implosion dynamics becomes almost same as the normal density CH target. We could not find the separation of precursor plasma front and main body

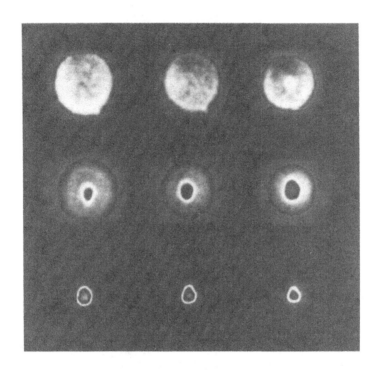

Fig. III-5 X-ray pictures of a cryogenic foam shell target implosion.
This was taken with a 9 frame x-ray framing camera. The
initial diameter and thickness of the shell were 740 μm and
45 μm, respectively. The separation time of frames was 170
psec and the exposure time of each frame was 80 psec. The
preceding x-ray emissions from the target center appears far
before the maximum compression.

WO/ Preheat   CHFDL

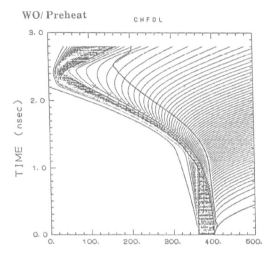

W/ Preheat

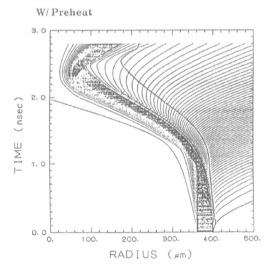

RADIUS (μm)

Fig. III-6   1D Simulation of cryogenic foam target. With and without
preheat, the implosions show distinctive difference. From
our planar target experiment, the preheat level was
confirmed to be the same order of the simulation with the
preheat. 20 % of the absorbed energy was divided to the hot
electrons due to resonance absorption with preheat.

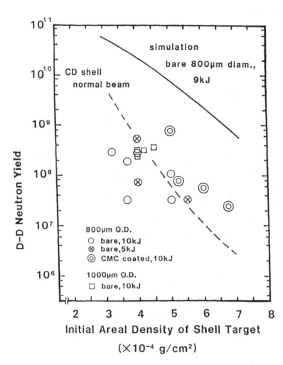

Fig. III-7   Neutron yield of cryogenic hollow shell v.s. the initial areal density. The solid line shows the result of the one-dimensional hydrodynamic simulation (HISHO) assuming that the hot electrons have 20 % of the absorbed energy. The dashed line is the results of the CD shell implosion without using the random phase plates.

as shown in Fig. III-6. The implosion velocity measured by x-ray streak camera agreed also with the simulation results with preheating.

The neutron yield and the ρR of the compressed core are shown in Fig. III-7, III-8. The experimental results were lower than the simulation results but they were almost same as the results of normal density targets irradiated by laser beams without random phase plate. The ion temperature was almost same as the average of the inner 4 meshes of the 1 D simulation. The inner 4 meshes correspond to a part of the precursor plasma.

From the results through laser-plasma interaction, energy transport and implosion experiments we have the counterplans for increase of absorption rate, suppression of preheating such as the overcoating of normal density CH layer and use of shorter wavelength laser with well designed pulse shape.

IV. CANNONBALL TARGET IMPLOSION

1. Radiation Driven Cannonball Target

In a radiation-driven cannonball target, laser energy is converted to x-ray radiation which then irradiates a fuel capsule. Compared with direct illumination of the fuel capsule with multiple laser beams, irradiation nonuniformities of high orders are smaller due to irradiation with incoherent x-rays. Therefore irradiation uniformity can be controlled by minimizing the nonuniformities of low orders which are determined by geometrical arrangement of the x-ray emitters. As an example we show in Fig. IV-1 a calculation on x-ray irradiation on a fuel capsule placed inside a cylinder which is irradiated by 10 laser beams. When the location L of the laser irradiation inside the cylinder is

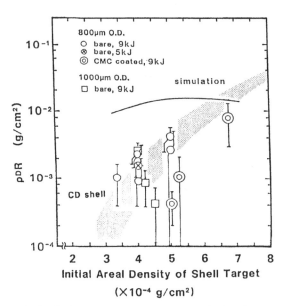

Fig. III-8 Final fuel density-radius product (ρR) of cryogenic hollow shell vs. the initial areal density. The solid line is the simulation result of the one-dimensional hydrodynamic code (HISHO). The shaded region shows the results obtained by the CD shell implosion without using random phase plate.

changed, irradiation distribution of the fuel capsule changes strongly. At an optimum configuration, irradiation uniformity close to 1 % is attained. Since the nonuniformity is mainly of mode 4, we can concentrate our attention to the consequences of this low order irradiation nonuniformity to the growth of the initial perturbation due to fluid instability and the pusher fuel mixing at the compression stage.

The efficiency that the laser energy is coupled to the fuel capsule is less with the indirectly-driven target. This coupling efficiency will be improved when x-ray radiation is confined inside a cavity in which a fuel capsule is placed.

2. Cannonball Target Research at ILE

A cannonball target was proposed originally as an indirectly-driven target where a plasma confined in a cavity applies a pressure on a fuel capsule[1]. Experiments were performed at either 1 μm or 0.5 μm laser radiation using the 2-beam GEKKO M-II laser[2] and the 12-beam GEKKO XII laser[3,4]. In these experiments both the plasma pressure and the x-ray radiation contributed to the implosion of the fuel target. High energy x-ray and high-energy particles were generated due to interaction of a cavity plasma with high-intensity, long wavelength laser light.

In July 1988 third harmonic conversion of 12 laser beams of GEKKO XII was completed for the second target chamber, and target irradiation with high energy 351 nm laser radiation became available since August 1988. With the short wavelength irradiation on high-Z materials, higher x-ray conversion efficiency is obtained and generation of high energy plasma component is reduced. Therefore a fuel capsule placed in a cavity is driven mostly by x-ray radiation.

Using the blue GEKKO XII, we have studied on basic components of radiation-driven cannonball target, such as x-ray emission, energy transport in a high-Z material, and radiation-driven acceleration of low-Z materials. In February 1989, a joint experiment between the Max-Planck-Institut für Quantenoptik, at Garching, FRG and the Institute of Laser Engineering, Osaka University was performed, and radiation confinement in a cavity[5] composed of a high-Z material (Au) was studied. In this experiment radiation confinement was demonstrated and x-ray flux of 2 X 10^{14} W/cm^2 corresponding to a radiation temperature of 230 eV was attained. Also radiation transport through a high-Z (Au) and a low-Z (Al) foil was investigated.

3. Cannonball Implosion with Blue-GEKKO XII

A radiation-driven cannonball target implosion was tested using the blue GEKKO XII. A fuel capsule was placed in an elliptical Au cavity witch was irradiated by 10 laser beams. The total laser energy was 5 kJ at 351 nm and the pulse shape was Gaussian with the pulse width of 0.7 ns. The fuel capsule was either a DT-filled glass microballoon or an argon-filled CH shell.

Implosion dynamics was studied with a space-resolving x-ray streak camera. A multiple-frame x-ray framing camera was used to observe the spatial distribution of x-ray emission from the fuel capsule when it was imploded. Figure IV-2 shows x-ray framing images of an argon-filled plastic shell taken at 100 ps interval with 80 ps exposure time. The irradiation location of the laser beams on the inside surface of the elliptical cavity was shifted outward in shot #7297 compared with shot

#7283. Consequently the x-ray irradiation distribution changed and the elongation direction of the Ar emission changed. In these two shots, the major mode numbers of the nonuniformity in the compression are 2 and 4. Deformation of the shell during acceleration phase was also observed by the x-ray back lighting technique and using the x-ray framing camera.

So far we have observed neutron yield of 9×10^8 from a DT-filled glass microballoon target. The fuel density estimated from the x-ray image was 4 g/cm^3 corresponding to 20 X liquid density. The fuel areal density evaluated from x-ray image and the secondary neutron yield was 7-9 mg/cm^2. Further experiment is in progress in which implosion uniformities during acceleration and compression phases will be studied in controlled conditions.

V. SUMMARY

Three series of experiments have been conducted by using GEKKO XII green (2ω) and blue (3ω) laser of 12 beams. They are the high density compression of hollow shell plastic pellet, the development of cryogenic target and the implosion of a hollow shell of cryogenic pellet, and the fundamental investigation of a Cannonball implosion.

The achievement of super-high density compression of 600 times solid density and the physics investigation on hollow shell target give us the confidence of reaching the ignition condition.

The key issues for this are the further improvement of the driving pressure uniformity, the pulse tailoring, the development of cryogenic hollow shell pellet with good uniformity and sphericity. 100 kJ blue laser of 24 beams is the reasonable test bed for the ignition experiment. GEKKO XII could be upgraded by adding booster amplifiers after splitting the present 12 beams into 24 beams. We are proceeding to new phase of

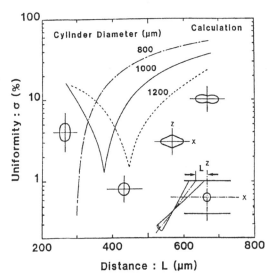

Fig. IV-1   Irradiation nonuniformity versus position of X-ray emitter.

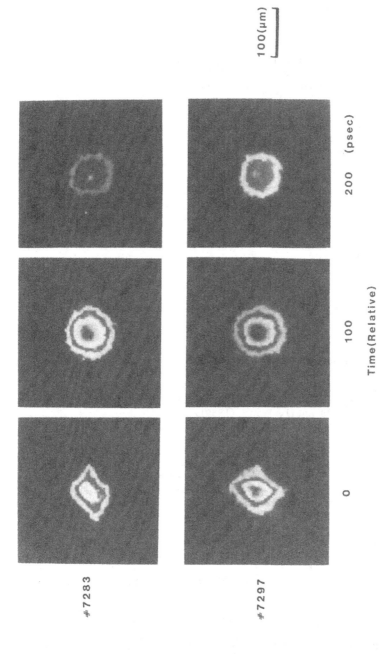

Fig. IV-2 X-ray framing images of Ar-filled plastic shell driven by soft x-ray radiation.

laser fusion with the developments of a high performance experimental laser, a high efficiency and high repetition operation laser, and the conceptual design of a reactor toward the energy development by laser fusion.

References

I. Introduction

1) Laser Fusion Ignition Experiment "KONGOH" Project: Report of Research Planning Committee at Institute of Laser Engineering, Osaka University, June, (1987).
2) C. Yamanaka and S. Nakai, Nature 319, 757 (1986).
3) H. Takabe et al., Phys. Fluids 31, 2884 (1988).
4) H. Hora, Z. Naturforsch 42a, 1239 (1987).
5) S. Nakai, ILE Quarterly Progress Report, July 1987-Sep. 1987 23, 7 (ILE-QPR-88-25) (1989).
6) K. A. Tanaka, ILE Quarterly Progress Report Jan. 1988-March 1988 25, 11 (ILE-QPR-88-25) (1989).
7) Y. Kato et al., Phys. Rev. Lett. 53, 1057 (1984).
8) H. Azechi et al., Appl. Phys. Lett. 49, 555 (1986).
9) H. Nakaishi et al., Appl. Phys. Lett. 54, 1308 (1989).
10) S. Nakai et al., 12th Int. Conf. on Plasma Phys. and Controlled Nuclear Fusion Res. IAEA-CN-50/B-1-1, Nice, France 12-19 Oct. 1988.
11) K. Mima et al., Invited Talk at APS Plasma Physics Division Meeting, 2I4, Hollywood, Florida, Oct. 1988.

II. High density implosion of plastic hollow shell target

1) U. Kubo and H. Tsubakihara, J. Vac. Soc. Technol. A4(3), 1134 (1986).
2) N. A. Ghanem and T. Westermark, J. Am. Chem. Soc. 82, 4432 (1960).
3) K. E. Willzbach, J. Am. Chem. Soc. 79, 1013 (1957).
4) M. Katayama et al., Proceeding of European Conference on Optics, 24-28 April, Paris (1989).
5) H. Azechi et at., Appl. Phys. Lett., 49, 555 (1986).
6) H. Nakaishi et al., Appl. Phys. Lett., 54, 1308 (1989).
7) H. Nakaishi et al., Appl. Phys. Lett., 55, 2072 (1989).
8) E. M. Campbell et al., J. Appl. Phys., 51, 6062 (1980).
9) D. J. Christie et al., Rev. Sci. Instrum., 56, 818 (1985).
10) Y. Setsuhara et al., to be submitted.
11) D. I. Garber and R. R. Kinsey, Neutron Cross Sections (Brookhaven National Laboratory, New York, 1976) Vol. II.
12) N. Miyanaga et al., Rev. Sci. Instrum., 57, 1731 (1986).
13) H. Nakaishi et al., (submitted to Rev. Sci. Instrum.).
14) F. J. Marshall et al., Phys. Rev. A40, 2547 (1989).
15) K. Mima, 'Inertial Confinement Fusion' edited by A. Caruso and E. Sindoni, Varenna, 6-16 Sept., 1988, p.125-137.
16) A. R. Bell et al., Phys. Rev. Lett., 46, 243 (1981).
17) J. P. Matte and J. Virmont, Phys. Rev. Lett., 49, 1936 (1982).
18) J. R. Albritton, Phys. Rev. Lett., 50, 2078 (1983).
19) J. P. Matte et al., Phys. Rev. Lett., 53, 1461 (1984).
20) A. R. Bell, Phys. Fluids, 28, 2007 (1985).
21) T. H. Kho, and M. G. Haines, Phys. Fluids, 29, 2665.
22) J. Delettrez, Con. J. Phys., 64, 932 (1986).
23) Y. Kishimoto et al., J. Phys. Soc. Japan, 57, 1972 (1988).

24) A. B. Langdon. Phys. Rev. Lett., 44, 575 (1980).
25) Y.-W. Chen et al., Opt. Comm. 71, 249 (1989).
26) Y.-W. Chen et al., (submitted to J. Appl. Phys).
27) Y.-W. Chen et al., Opt. Comm. (in press); N. Miyanaga et al., Proceeding of the Third World Conference on Neutron Radiography, May, Osaka (1989) (in press).
28) H. Takabe, K. Mima, L. Montier and R. L. Morse, Phys. Fluids, (1983).

III. Cryogenic hollow shell implosion

1) F. J. Marshall et al., Phys. Rev. A. 40, 2547 (1989)
2) E. K. Storm: Proceeding of Japan-US Seminar on "Thory and Application of Multiply-Ionized Plasma Produced by Laser and Particle Beams" May, 1982 Nara, ed. C. Yamanaka (1982) p.97-114.
3) C. Yamanaka, ILE Quartary Progress Rep. ILE Osaka ILE-QPR-85-13 (1985) p.2-9.
4) H. Katayama et al., J. Vac. Sci. Technol. to be published
5) T. Norimatsu et al., J. Vac. Sci. Technol., A6, 3144 (1987)
6) H. Katayama et al., Appl. Phys. Lett., 55, to be published (Dec 18, 1989 issue)
7) K. A. Tanaka, QPR, Osaka Univ. ILE, Quarterly Progress Report, ILE-QPR-88-25, p.11; T. Yamanaka et al., Laser Interaction with Matter edited by G. Velarde, E. Minguez, J. M. Perlado (World Sci. 1989, Teaneck, NJ, USA)
8) K. A. Tanaka et al., J. Appl. Phys. 65, 5068 (1989)

IV. Cannonball target implosion

1) H. Azechi et al., Jpn. J. Appl. Phys. 20, L477 (1981).
2) N. Miyanaga et al., Jpn. J. Appl. Phys. 22, L551 (1983).
3) Y. Kato et al., ILE Quarterly Progress Rept., No. 14, ILE-QPR-85-14, 1985, p.2 (unpublished).
4) Y. Kato et al., ILE Quarterly Progress Rept., No. 18, ILE-QPR-86-18, 1986, p.5 (unpublished).
5) R. Sigel et al., Phys. Rev. 38, 5779 (1988).

STATUS OF EXPERIMENTAL INVESTIGATIONS

OF ICF AND X-RAY LASERS IN SINPC

H.S.Peng, H.Z.Shen, Z.J.Zheng, Y.Cun, D.Y.Tang, and J.G.Yang

Southwest Institute of
Nuclear Physics and Chemistry
China

INTRODUCTION

SINPC is the only place where experimental investigations of laser fusion are conducted in China. To date two Nd_iglass laser facilities, LF-11(1.06μ m, 0.53μ m, 70J, 0.3-0.8ns) and LF-12(1.053μ m, 2x800J, 0.1-1.0 ns), completed and installed in 1985 and 1986, respectively, have been routinely operated for physical experiments to study basic processes of laser-plasma interactions. A great variety of diagnostics, such as XRD arrays, filter-fluorecser, x-ray pinhole cameras, plasma calorimeters, photodiodes, optical and x-ray streak cameras, OMA and neutron detectors, have been developed and activated. Simultaneuosly, we have established a microfabrication laboratory to develop target materials and target technology.

In 1986, we carried out direct-drive DT-filled glass capsules experiments on the LF-12 facility, as a simple demonstration of time synchronization and energy balance of the two beams. However, our research emphasizes indirect-drive regime. Therefore, most of the experiments performed in the past have involved hohlraum physics, including laser energy absorption, x-ray conversion, x-ray spectra, radiation temperature, suprathermal electrons, stimulated Raman scattering, plasma closure effects and so on. Indirect-drive implosions have been planned to be performed for hydrodynamics studies in the near future.

In recent years, we have also been involved in x-ray laser research using laser-produced plasma columns and succeeded in demonstrating soft x-ray pumped 108.9nm XeIII Auger laser, 10.57 and 15.47nm Li-like AlXI recombination pumped lasers and five 20-30nm Ne-like GeXXIII collisionally excited lasers with significant gain coefficients.

IMPLOSIONS OF LASER-DRIVEN EXPLODING PUSHER TARGETS

Direct drive, as one of the two fundamental approaches to ICF, have been actively investigated. In the past few years, great efforts have been focuced onto developing techniques, such as induced spatial incoherence(ISI) at NRL [1, 2] and random phasing at Rochester[3] and Osaka [4] to meet the requirement on target surface pressure distribution uniformity (<1-2%), so that high radial convergence, for example, of 30 can be reached. During the same period of time, exploding pusher targets have been imploded to produce neutrons with a neutron yield record of 2×10^{13} on Nova[5].

In 1986, when the LF-12 laser facility first went into operation, we once imploded also DT-filled glass capsules, in an exploding pusher manner, irradiated symmetrically by two beams from opposite directions, each delivering 50-100J in an 80-105ps pulse, with an energy symmetry to 14%. The glass balloons, 80-106 μm in diameter and 0.7 μm in wall thickness, were filled with 10-atm DT mixture ($D,T=2,1$), while some of them with 6-atm nitrogen gas to determine the electron density using Stark broadening of Ly-β line. The diagnostics consisted of scitillator-photomultiplier detectors for neutron time-of-flight measurements, x-ray pinhole cameras and an x-ray streak camera for obtaining compression and symmetry informations both time -integrated and -dependent, RAP (0.49-1.2nm) and KAP (0.75-1.3nm) crystal spectrometers, plasma calorimeters, photodiodes, and so on.

The fusion neutron yield recorded was 5×10^{8}. If DT content had been 1,1, the yield would have been higher than 10^{9}. As is shown in Fig.1, the implosion was fairly symmetric and the volume compression ratio was about 80, leading to a core fuel density of 1×liquid deuterium. The imploding time was roughly 150ps and an averaged imploding velocity was 3.3×10^{7} cm/sec. The absorption efficiency was only about 28% due to the high power density (7.5×10^{15} W/cm) and the low Z material.

ENERGY BALANCE IN HOHLRAUM TARGETS

Indirect drive ICF[6], which has significant advantages of relexing the requirements of direct drive on the optical uniformity and laser geometry, lessening the sensitivity of target implosions to the effects of hydrodynamic instability, and applicability of the obtained capsule implosion data base, has been successfully studied at Livermore. The progress is so promising that a 10MJ Laboratory Microfusion Facility has been planned to be built for high gain implsions.

The indirect-drive program, being conducted in SINPC, has only concentrated on basic phenomena of hohlraum physics due to the limited specifications of the laser facilities. Therefore, we first carried out experiments to see laser energy deposition and x-ray conversion. The hohlraum targets used in these experiments were configurated both spherical and cylindrical with no capsule inside. The dimension of the

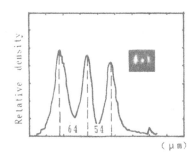

Fig. 1. Time -integrated x-ray image and density
distribution of an imploded glass capsule.

cavities made of gold ranged   from 400 to 600μm in diameter   with   a
wall   thickness of 15μm. Two inlet holes 190-250μm in diameter   were
opened   oppositely,while a small hole or a narrow slit for   diagnosing
at the 90° direction. The two beams with a divergency of 0.1mrad   each
delivered a   300-450J, 1.053μm, 800ps pulse to the target through   a
f/1.7 lens.The signal to noise ratio was larger than 10' and ASE<0.8mJ.
The diagnostics   included 30 photodiodes for scattered 1ω„,3/2ω„,2ω„
laser light distributions,two calorimeters for reflected light by   SBS
and   SRS   subtended   by   the lens, respectively, 12   calorimeters   for
plasma, energetic   electrons and x rays(<10keV) distributions   and   15
flat response XRDs and two XRD arrays(Dante spectrometers) for x   rays
measurements.

   The hohlraum targets showed a high absorption efficience. For most
of the shots, η_{α} was about 80% and   defined   as (E-E')/E,where E-laser
energy delivered to the target,and E'-escaped portion of laser   energy
from   the   inlet and diagnostic holes. We also took a quite number of
shots for Au disk targets at the same power densities,but η, was only
about   40% .   The main portion which dominated   the   energy balance
consisted   of   the plasma and scattered light(SBS,SRS and ω„). They
were 70-80% and 20-30% of the laser energy,respectively. The angular
distributions   of them are shown in Fig.2,where θ is the   angle with
respsct   to   the normal of the inlet hole. The part subtended   by   the
target   lens   is not included and,therefore,the   curves   terminate   at
small   angle. The scattered light shows a sharp distribution implying
that   the light comes mostly from the hole. However,we can still   see
the   flat   part   of   the curves at large   angle, which could   be   the
background   light or the light emitted by the disassembled plasma.   On
the   contrary,plasma curves are almost flat,except that the inlet   and
diagnostic holes make some extra contributions. The fairly symmetrical
distribution of plasma energy seems to be reasonable,if we assume that
the target disassembly should follow the momentum conservation law.

   X-ray   conversion in hohlraum targets should be much   higher   than
those of plane targets due to the confinement of plasms by the cavity.
However, one can not directly measure the x-ray amount as in the   case

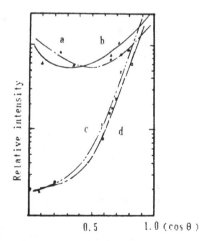

Fig. 2. Angular distributions of plasma and scattered light.
a) and b) plasma distributions of the two beams;
c) and d) scattered light of the two beams.

of plane targets. Furthermore, the x rays heat the inner
wall, converting back into plasmas again. We have been trying to
develop models for numerical simulations to fit the data obtained
experimentally. As the physics of conversion involves complecated
processes, it is not easy to quantitatively derive the values. At
present , we are inferring the conversion either from the measured
x-ray flux emitted from the holes, radiation temperature and its time
profile, or from the energy necessary for heating a thin layer of Au
cavity inner wall(for example, 0.5-1.0 μ m) to the measured radiation
temperature by radiation transport of subkeV x rays. These two
approximations bring us an x-ray conversion of about 40-50% of the
absorbed energy.

SUBKEV X RAYS

In the indirect drive regime, subkeV x rays play the important role
in driving capsules as laser beams do in the direct drive regime.
Therefore, we have been pursuing a detailed understanding of subkeV
x-ray physics itself and the effects on diagnostic accuracy.

The experimental setup is the same as described in Section 2. We
fielded sophisticated diagnostics to get more informations about
radiation temperature and its time profile, x-ray energy and time
spectra, x-ray flux, and plasma closure effects. In addition to those
named previously, we employed an x-ray streak camera, x-ray pinhole
cameras, and XRDs with an absorption foil[7].

The subkeV energy spectra measured both from the inlet and
diagnostic holes are very similar to a black body spectrum, as depicted
in Fig.3. There exists a hump on curve a) at about 0.7 keV, which, we
believe, could be contributed by Au 5-4 transitions. We first simply
derive the radiation temperature as the peak energy divided by 2.82, if
we assume a Planck's spectrum. The radiation temperature taken from
the inlet hole gives the information of a region, where x rays are
converted from the incident laser energy, and have an averaged value of
155eV, while the radiation temperature measured from the diagnostic
hole corresponds to an environment, where a capsule will be placed and
imploded, and is about 130eV. We also calculate the temperature from
the measured x-ray flux. To do this, we have absolutely calibrated the
sensitivities of the XRDs using proton-induced fluorescence sources. We
once drilled a small hole(50 μ m diameter) for measuring x rays, but we
could hardly get any available signal. This reminded us of plasma
closure effects and, then, bigger holes(>150 μ m) were opened. Even so, a
correction for the effective diameter still has to be made using a
suitable plasma closure velocity. The third approach is the
application of a pair of XRDs with different absorption foils. The
signal ratio of the two XRDs depends only on the temperature. Actually,
a quite few pairs were used to make a better measurement. To our
surprise, the results obtained from the three methods are in very good
agreement. The temperature is a fuction of the incident laser energy, as
shown in Fig.4. The curve gives a scaling relation, $T_\gamma = 50.6E^{0.183}$ for our
experiments, where T_γ is the radiation temperature in eV, while E the
laser energy in J. We have also derived a time dependent temperature, as
given in Fig.5, from the temporal spectra of the x rays.

The temporal spectra detected from holes(>150 μ m) were broadened
to about 1.5ns, the laser pulselength being 800ps. While the spectra
observed from a small diagnostic hole (50 μ m) were shortened, as shown
in Fig.6. We attribute this to plasma closure effects. X-ray images
also give a hint. Fig.7 shows time-integrated x-ray images of a slit

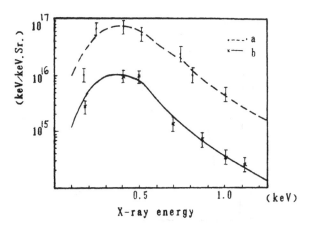

Fig. 3. SubkeV x-ray spectra of a hohlraum target.
a) x-ray spectrum from laser inlet hole;
b) x-ray spectrum from diagnostic hole.

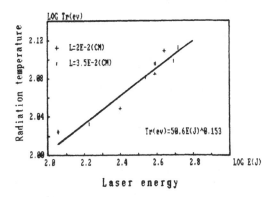

Fig. 4. Radiation temperature vs incident laser energy.

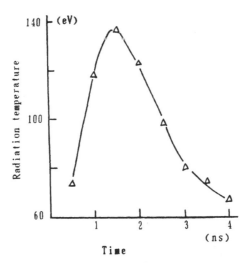

Fig. 5. Time-dependent radiation temperature.

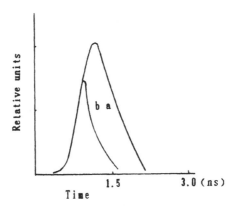

Fig. 6. Temporal x-ray spectrum.
a) x-ray profile of planar targets;
b) x-ray profile from a 50-μm hole.

(40μm wide and 700μm long) cut in a cylindrical wall. The left part corresponds to the region where the laser pulse directly strikes the wall, but has a darker and narrower(<40μm) image., The further from the laser beam, the brighter and wider. The only explanation we can suggest at present is the closure of the slit by plasmas created by the laser light. Thus, we conclude that in order to improve x-ray diagnosis for hohlraum targets a bigger hole must be opened and the results must be corrected by numerical simulations.

NONLINEAR PHENOMENA AND SUPRATHERMAL ELECTRON PRODUCTION

Many experiments have proved that the third harmonic of Nd,glass laser is effective for suppressing suprathermal electron production due to collisional stabilization. As we can only operate our laser facilities at $1.053\mu m$, hot electron preheat is still the major concern. In hohlraum targets, longer plasma scale lengths exist and severe instabilities can take place, resulting in a high hot electron yield. Therefore, a good understanding of nonlinear processes excited in underdense plasmas is a critical subject. Many laboratories have actively carried out investigations in this area both theoretically and experimentally[8-11], but very few results have been reported for hohlraum targets[12].

In our experiments, the scattered light from the target inlet hole was collected by the focus lens and then transported and splitted for Raman light energy measurements using a calorimeter and for spectroscopy with a 0.25m grating spectrograph, which was coupled with either a 128-element pyroelectric laserprobe for backward Raman light or a 1024-silicon photodiode array for $3/2\omega$,, 2ω, and forward Raman light. For some shots, an optical streak camera was installed to see temporal profiles. In addition, a filter-fluorescer spectrometer was used for hard x rays(5-88keV) converted from hot electrons.

The whole spectrum of the light produced by nonlinear instabilities in the plasmas has been recorded and is shown in Fig.8. To our knowledge, this is the first time to report such a spectrum for hohlraum targets. By analyzing the spectrum, we can have numerous informations about the physical processes. Most of the backward SRS light signal are spectrally located in a wavelength range of $1.5-1.9\mu m$. The corresponding plasma density is $0.09-0.20n_c$. As

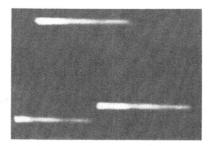

Fig. 7. Time-integrated x-ray images of a slit
(40μm×700μm) in a cylindrical wall
taken with a three-pinhole camera.

has been reported by many authors, we do not see much light scattered at $0.25n_c$. The sharp signal at 2.16μm is the second order diffraction of 1.053μm light for wavelength calibration. Resonance absorption produced $2\omega_o$ and two plasmon decay produced $3/2\omega_o$ light are clearly recorded and identified at their right wavelengths. Besides, there is another signal comparable to the $3/2\omega_o$ light with its peak at 0.75μm. We consider it to be the upshifted anti-Stokes light by forward Raman scattering[13]. Observation of this light in the forward direction have been made in a few laboratories. In our experiments, we can only put the spectrometer in the backward direction and fortunately have this light signal beyond our expectation. By looking at the Stokes and anti-Stokes light wavelengths, we find that the frequencies exactly meet the equation, $\omega_s = \omega_o \pm \omega_{epw}$. This indicates that the forward Raman light originates in the same density region as that for backward Raman light and is seeded by none other than the electrostatic wave excited by the backward Raman scattering. The upshifted light first propagates forward into dense plasmas and is refleced backward at its critical density. In Fig.9, streak images, taken from the camera screen, of the $2\omega_o$, $3/2\omega_o$ and forward Raman light are given. These light last about the same duration of the laser pulse, but with different onset times. $2\omega_o$ starts about 100ps after the laser onset time, while $3/2\omega_o$ and forward SRS about 300ps, which may be the time necessary for creating long scale length underdense plasmas. The streak images of $3/2\omega_o$ and RFS light are time-modulated in intensity, implying the transient behaviors of the bulk plasmas. The modulation frequencies change from shot to shot, ranging from 300 to 400ps mostly.

We assume that SRS light collected by the f/1.7 lens is only one third of the total amount. The downshifted Raman energy inferred in this way is a strong function of the incident laser energy and is about 5-8% of the latter, while the $3/2\omega$ and $2\omega_o$ are 0.2-1.5% and 0.05-0.3%, respectively. The anti-Stokes light is at the same energy level as $3/2\omega_o$.

The energy of suprathermal electrons is closely related to SRS(see Fig.10) and is about 4-5% of the laser energy. Therefore, the same conclusion that SRS is the major mechanism for producing suprathermal electrons, as has been proved in many earlier experiments, can be drawn again.

The temperature of suprathermal electrons derived from the x-ray spectra, as plotted in Fig.11, is as high as 40-50keV. Because the spectrograph was limited to 5-88keV energy region, we did not measure harder x rays. Forward Raman scattering can accelerate electrons to a much higher temperature because of its high phase velocity. Thus, we will develop detectors for high energy(>100keV) x-ray measurements.

X-RAY LASERS

In 1988, we conducted gain measurements of a XeIII 108.9nm laser[14, 15]. A laesr beam at 1.06μm and 20J in a 600ps pulse was focused onto a Ta foil target, through a cylindrical-spherical lens system, to produce soft x rays, which then was used to pump Xe gas(about 300Pa) in a U-shaped channel. The soft x rays cause photoejection of an inner-shell 4d electron from neutral xenon. The produced ion, XeII, undergoes Auger decay, yielding various excited states of XeIII. The 108.9nm transition then takes place between two of the XeIII levels.

A 1-m normal incidence vaccum spectrometer, coupled to an x-ray

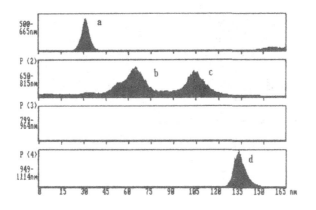

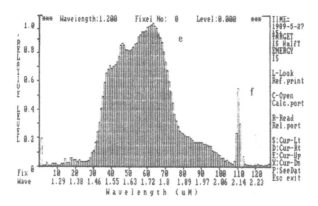

Fig. 8. The whole spectrum of the light produced
by nonlinear plasma instabilities.
a) $2\omega_0$;
b) $3/2 \omega_0$;
c) forward Raman light;
d) SBS light;
e) backward Raman light;
f) second order light of $1\omega_0$ for
wavelength calibration.

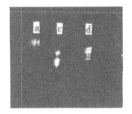

Fig. 9. Streaked images of $2\omega_0, 3/2\omega_0,$
and forward Raman light.
a) ω_0;
b) $2\omega_0$;
c) $3/2\omega_0$;
d) forward Raman light.

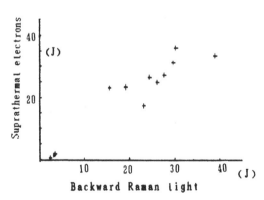

Fig. 10. Correlation of backward Raman light
with suprathermal electrons.

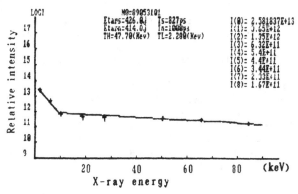

Fig. 11. The hard x-ray spectrum of a hohlraum target.

image intensifier (a microchannel plate with a screen) and then, to a microchannel photomultiplier, was used to measure a time-dependent 108.9nm laser signal. A gain coefficient of 2.38/cm and a 500ps UV laser pulse(FWHM) were obtained.

Recombination pumped Li-like AlXI lasers have been reported at 10.54 and 15.47nm(5f-3d and 4f-3d) by a few laboratories[16,17]. We have performed similar experiments recently using the same laser facility as mentioned above,but with a pulselength of 200ps and Al-coated(80nm) Formvar foils(50nm),the length being 6-10mm. Two 1-m grazing-incidence spectrographs with a instrumental spectral resolution of 0.006nm for a 7-μm slit were positioned to view the plasma both axially and transversely. The data were recorded on Kodak soft x-ray film. The gain coefficients obtained for 10.57 and 15.47nm were 3.18/cm and 2.26/cm,respectively.

Following the successful experiments in demonstrating soft x-ray 3p-3s lasing in the Ne-like GeXXIII ions performed at Naval Research Laboratory[18], we have made investigations too. Germanium targets(8-18mm long) used in the experiments were irradiated by a 1.5ns laser pulse at 1.053μm wavelength over an energy range of 550-650J with a 185μmx20mm line focus. The signals were recorded with the same two grazing-incidence spectrometers and Kodak SWR film. In addition, pinhole cameras and crystal spectrometers were placed to observe the basic behaviors of plasma columns.

We also succeeded in obtaining five transition lines with significant gain coefficients as listed in Table 1 in comparision with the results of NRL. The relative line intensities (see Fig. 12) were derived from the line densities with a density-exposure curve calibrated on a Henke x-ray source at the institute.

Table 1.Gain coefficients of GeXXIII lasers

Transitions	Measured wavelength (nm)		Predicted gain (1/cm)	Measured gain (1/cm)	
	SINPC	NRL		SINPC	NRL
J=0-1	19.6	19.61	5	3.0	3.1
J=2-1	23.2	23.22	6	4.2	4.1
J=2-1	23.6	23.63	6	4.0	4.1
J=1-1	24.7	24.73	3	1.6	2.7
J=2-1	28.6	28.65	5	4.6	4.1

The gain coefficients were experimentally found to be the function of laser power densities, which increased as the power densities changed from 0.5 to 1.2×10^{13} W/cm^2. However, we were not be able to reach the optimum condition due to the limited laser output.

The lasing lines on the film were much shorter than those spontaneous emission lines and the densities were spatially distributed as shown in Fig.13. From the profiles we inferred the divergence of the laser beams to be 10mrad, which could be caused by either plasma refraction or the geometrical dimensions. If we took only the latter into account, the maximum width in the pumping laser direction was less than 200μm.

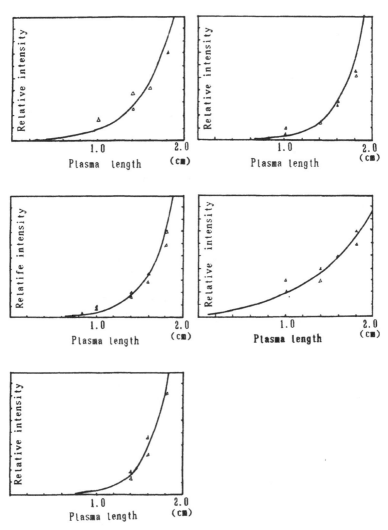

Fig. 12. Relative intensities of Ne-like GeXXIII lasers vs plasma lengths.
a) 19.84nm laser; b) 23.22nm laser;
c) 23.62nm laser; d) 24.74nm laser;
e) 28.64nm laser.

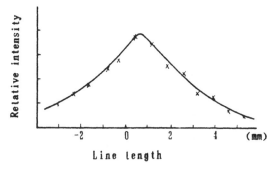

Fig. 13. Density profile of a laser line.
(0 corresponds to target surface).

ACKNOWLEDGEMENT

The authors thank M. Yu, R. Y. Hu, Y. B. Fu, and Z. C. Tao for support, T. X. Chang and his co-workers for helpful discussions, the groups at LF-11 and LF-12 laser facilities for excellent operations, and our colleques for assistance.

REFERENCES

1. R.Lemburg,A.Schmitt,and S.Bodner,J.Appl.Phys.62,2680(1987).
2. R.Lemburg,and J.Goldhar,Fusion Technology 11,532(1987).
3. T. Kessler,et al. ,Technical Digest of Conference on Laser and Electro-Optics,April 1988,Anahaim,Calif.,(1988) pp.25-29.
4. Y.Kato et al.,Phys.Rev.Lett.53,1057(1984).
5. E.M.Campbell et al.,Rev. of Sci.Inst.57,(8),2(1986).
6. E.Storm,et al,"Progress in Laboratory high gain ICF,Prospects for The Future, " Eighth Session of The International Seminars on Nuclear War,Erice,Sicily,Italy,August 20-23,1988.
7. H.Pepin et al.,Phys. Fluids 28,3393(1985).
8. C.S.Lui,M.N.Rosenbluth,and R.B.White,Phys. Fluids 17,121(1974).
9. W.Seka et al.,Phys. Fluids 28,2570(1985).
10. H.Fiqueroa,C.Joshi,and C.E.Clayton,Phys. Fluids 30,586(1987).
11. R.E.Turner et al.,UCRL 50021-86,3-21.
12. Y.Sakawa et al., Phys. Fluids 30,3276(1987).
13. K.G.Estabrook and W.L.Kruer,Phys.Fluids 26,1892(1983).
14. H.C.Kapteyn,R.W.Lee,and R.W.Folcon,Phys.Rev.Lett.57,2939(1986).
15. G.Y.Yin et al.,Opt.Lett.12,311(1987).
16. P.Jeagle et al,J.De Physique 47,c6-31(1986).
17. C.L.S.Lewis et al.,Plasma Physics and Controlled Fusion 30,35(1988).
18. T.N.Lee,E.A.Mclean,and R.C.Elton,Phys.Rev.Lett.59,1185(1987).

INVESTIGATIONS INTO X-RAY DAMAGE TO THE FIRST WALL OF THE

INERTIAL CONFINEMENT FUSION LABORATORY MICROFUSION FACILITY

Robert R. Peterson

Fusion Technology Institute
University of Wisconsin-Madison
Madison, WI 53706

INTRODUCTION

The target chamber of an Inertial Confinement Fusion (ICF) power plant or of an ICF Laboratory Microfusion Facility (LMF)[1] must survive repetitive blasts from microexplosions of targets. The LMF would explode perhaps as many as 1500 targets, each with a yield of 1000 MJ, over its 30 year lifetime, and several thousand more at lower yields. A typical ICF power plant design might explode 10^8 targets per year. One challenge of ICF target chamber design is the mitigation of the effects of the target-generated x-rays on the first surface. The design criteria for the LMF and for an ICF power plant differ significantly. Because of the large number of explosions, the first surface for a power plant must have essentially no vaporization of the solid wall or erosion of the wall will limit the lifetime. Wall erosion is a minor issue for the LMF, so significant vaporization of the first wall material could occur. One consequence of significant vaporization is the launching of shock waves into the solid wall. These vaporization driven shocks are the subject of this paper.

In an LMF target chamber, tens or hundreds of MJ of x-rays will be released by the burning target over a pulse width of a few ns. If x-ray absorbing structures or gases are placed between the target and the first wall, then the energy of the x-rays can be re-radiated to the wall over a time that is long compared to the thermal response time of the wall and vaporization of the surface of the wall may be avoided. A gas of high enough density and atomic number may prevent the propagation of the driver beam, though there may be solutions to this problem as well. In the absence of something to absorb the target generated x-rays, the x-ray power intensity on the first wall will be high enough to vaporize the first wall surface.

I will begin this paper with a study of the response of LMF first walls to target x-rays. I have used computer simulations to study the effect of the x-ray pulse width on the strength of shock waves in the wall material. I will also show the results of computer simulations of possible experiments to mimic the x-ray damage to potential first wall materials. I will then discuss such experiments done using x-rays from gas pinches generated on the SATURN accelerator at Sandia National Laboratories (SNL) in Albuquerque, New Mexico.

FIRST WALL

The first wall responds to target x-rays through very rapid energy deposition in a thin layer of material. This leads to volumetric vaporization of from a few to a few tens of microns of material and the generation of shock waves moving into the material. The volumetric vaporization has been a topic of study for several years[2] and will not be discussed in this report. I will concentrate on the generation of shocks.

An important aspect of this investigation of x-ray vaporization is computer simulation. I have used two different sets of computer codes in this, which have compensating strengths and weaknesses. I have used these codes to consider x-ray vaporization in ICF target chambers. Part of this has been to study the dependence of the wall response on x-ray pulse width. Finally, I have used computer simulations to help design and understand x-ray vaporization experiments.

Computer Codes

I have used two different sets of computer codes to study the launching of shocks by intense x-ray deposition and the subsequent propagation of these shocks into the material. The first set is the IONMIX code[3] coupled to the CONRAD code.[4] These were developed and are being maintained at the University of Wisconsin. CONRAD is a one-dimensional Lagrangian hydro- dynamics code with multigroup radiation diffusion. Equations- of-state and multigroup opacities are provided by the IONMIX code in tables. CONRAD includes time-dependent x-ray and ion sources and models energy deposition, thermal conduction, and phase transition in a solid or liquid wall. CONRAD has the advantage that one can directly calculate the mass of material vaporized, which is important both to target chamber design and to validation of the physical models assumed in the vaporiza- tion process. The heats of melting and vaporization can be a significant part of the energy budget and care has been taken to include them in CONRAD. In codes designed for use at higher energy densities, the heats of melting and vaporization are only included through the equation-of-state tables, and one is often not sure of the details. The other set is a radiation hydrodynamics code coupled to CSQ. I have used the radiation hydrodynamics code to simulate the deposition of x-rays in the material. This calculation is then coupled to the CSQ computer code. CSQ is a code written and maintained at SNL, that uses two-dimensional Eularian hydrodynamics and has sophisticated modeling of phase transitions and crush physics that are prob- ably important to shock attenuation in materials.[5] CSQ has rather limited radiation transport modelling, which makes such coupling to another computer code advisable when doing x-ray vaporization simulations.

First Wall Simulations

I have used these computer codes to simulate the responses of LMF first walls to the direct deposition of target x-rays. I have used LMF concepts devised both at SNL, applicable to light ion driven fusion, and Lawrence Livermore National Laboratory (LLNL), more tied to laser driven inertial fusion. The parameters used for the calculations and the results are summarized in Table 1. The SNL concepts often require a short distance between the last elements in the beam generation hardware and the target. The present baseline design invokes ballistic focussing of the ions with lens magnets.[6] The beam divergence places an upper limit on the distance between the lens and the target, which is currently believed to be 150 cm. The first wall of the target chamber is placed at the lens position. LLNL concepts using lasers have the final driver components many meters from the target, so there is greater freedom in positioning the first wall of the target chamber. I have considered wall radii of 4 and 5 m, respectively for calculations #4 and 5. For all calculations I have assumed that the target is releasing 220 MJ of x-rays from a total yield of 1000 MJ in 1 ns. I assume that the x-ray spectrum is as shown in Fig. 1. These are all consistent with the HIBALL target[7] and there will be some variation from this in the LMF due to different target designs.

Table 1. X-Ray Vaporization in LMF First Walls

Calculation #	1	2	3	4	5
Code	CONRAD	CONRAD	CSQ	CSQ	CONRAD
Concept	SNL	SNL	SNL	LLNL	LLNL
X-Ray Fluence (J/cm^2)	780	780	780	70	110
Wall Material	Al	C	Al	Al	Frost
Vaporized Mass (kg)	2.8	1.8	*	*	12.6
Peak Pressure in Vapor (GPa)	150	84	*	50	1.2
Impulse on Wall (Pa-s)	310	257	300[a]	100[b]	90.2

*Not Calculated
[a] 5×10^3 cm in back of surface
[b] 5×10^2 cm in back of surface

Response versus Pulse Width

I have tested the scaling of pressure with x-ray power with computer simulations. In all the calculations, the x-ray fluence is 780 J/cm^2 and the spectrum is as in Fig. 1. Only the pulse width of the x-rays on the wall is varied. I found that the vaporized mass and the total impulse are not much affected by the pulse width. However, as is shown in Fig. 2, the peak pressure on the wall is very much affected. I have proposed a scaling law

$$P = P(\Delta t = 1 \text{ ns})/\Delta t^n , \tag{1}$$

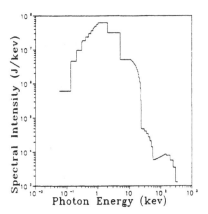

Fig. 1. HIBALL target x-ray spectrum.

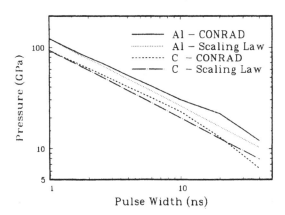

Fig. 2. Pressure on aluminum and graphite walls versus x-ray
pulse width. The energy fluence is 780 J/cm^2 and the
calculations were done with CONRAD. Scaling laws are
also shown.

where P is the peak pressure, Δt is the pulse width, and n is some real number. I have also plotted this scaling law for n = 2/3, and one can see that there is a reasonable fit.

I have also looked at the dependence on pulse width of the peak pressure inside the material with CSQ simulations. The results of these simulations are shown in Fig. 3. One can see that as one considers the pressure at greater distances from the surface, the dependence on the pulse width becomes weaker. Therefore, whether the x-ray pulse width is important becomes a question of whether or not one is interested in the material response near to the surface. The issue of pulse widths can be important when considering experiments to simulate the response of LMF first wall materials, which is the topic of the next section.

Simulation of Experiments

I have used CSQ to study how sample first wall materials might behave in experiments that mimic target chamber x-ray conditions. I have done such computer calculations for samples of aluminum, a thin layer of alumina on aluminum, and graphite. Parameters for three experimental environments are shown in Table 2 along with LMF conditions for SNL and LLNL concepts, where in all cases the wall or sample material is aluminum. PROTO-II and SATURN are electron accelerators at SNL that have been used for a number of years to create pulses of x-rays with gas pinches.[8] Specifically, gas puff pinches of neon produce the spectrum shown in Fig. 4.[9] When one compares this spectrum with the HIBALL target spectrum in Fig. 1 it is seen both have peaks at about 1 keV in photon energy. One must also compare the time dependence of the pinch generated x-rays to what the target emits. Here one finds that the HIBALL target emits x-rays over a period of 1 to 2 ns, while a neon gas pinch radiates 1 keV x-rays over 15 to 20 ns and lower energy photons over 100 ns. The pinch is created in the center of a circle of current return posts and the closest that a sample can be placed to the x-ray source is just outside these posts. GAMBLE-II is a machine at NRL that can accelerate protons in a beam to simulate x-ray deposition. One should note that the pulse width of the ion beam on GAMBLE-II is more than 40 ns while the gas pinch x-ray sources have less than half the pulse width. If one is only interested in stresses in the center of the material so that the energy density is important, then experiments on all three machines can be relevant to the LMF. If, however, stresses near the surface are important, the power density (power deposited per unit mass) is the important parameter and only SATURN can do LMF relevant experiments. Even SATURN can only provide a power at one half the LLNL LMF value. The most direct measure is the achievable stress in the material, which I have calculated with CSQ for PROTO-II, SATURN, and SNL and LLNL versions of the LMF. In aluminum, PROTO-II can provide stresses of 1 GPa 0.05 cm in back of the first surface and SATURN can provide 7.5 GPa. I have not yet calculated the stresses that GAMBLE-II could generate in aluminum, though based on the power density one would expect about 1 GPa. I calculated the stresses in a LLNL and SNL LMF aluminum wall to be 7.5 GPa and 14.0 GPa respectively. The calculation of the PROTO-II stresses was rather interesting

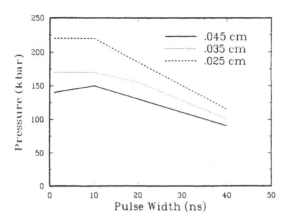

Fig. 3. Stresses at various positions in an aluminum wall
versus x-ray pulse width. The energy fluence is 780
J/cm² and the calculations were done with CSQ.

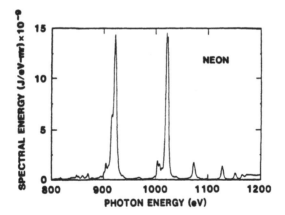

Fig. 4. X-ray spectrum from a neon gas pinch on PROTO-II. Only
the component above 900 eV in photon energy is shown.
There is another component to the spectrum below a few
hundred eV that has about 4 times the energy but more
than 5 times the pulse width.

because here the stresses are only a factor of a few larger than the yield stress and the stresses are non-isotropic. The longitudinal stresses at 0.05 cm peaked at 1.0 GPa while the transverse stresses peaked at 0.7 GPa. These simulations show that experiments on SATURN have the potential to much more closely mimic the conditions in the LMF target chamber than do experiments on GAMBLE-II or PROTO-II.

Table 2. X-Ray Driven Stresses in Aluminum

	PROTO-II (gas pinch)	SATURN (gas pinch)	GAMBLE-II (ions)	LMF/LLNL	LMF/SNL
Range in Al (mg/cm^2)	0.83^a	0.83^a	3.9^b	0.83^a	0.83^a
X-Ray Energy (MJ)	0.008	0.100	0.017	220	220
Distance (cm)	3.8	3.8	N.A	500	150
Energy Fluence (J/cm^2)	42	550	400	68	780
Energy Density (kJ/g)	51	660	108	82	940
Pulse Width (ns)	20	15	43	1	1
Power Intensity (GW/cm^2)	2.6	37	9.3	68	780
Power Density (GW/g)	2.5	44	2.5	82	940
Calculated Stress (@ 0.05 cm) (GPa)	1	7.5	not calculated	7.5	14

[a]Assuming 1 keV photons
[b]Assuming 1 MeV protons and no range shortening

I have simulated the response of four different materials to x-rays from SATURN with CSQ. The results are summarized in Table 3. In all cases, the samples are assumed to be 3.8 cm from the pinch, which is assumed to generate 100 kJ of x-rays in the lines shown in Fig. 4. There is assumed to be another 400 kJ in x-rays below about 200 eV in photon energy, making a total of 500 kJ in x-rays. I have assumed that the x-rays above 900 eV are emitted in 20 ns in these simulations and that the low energy component is radiated over 100 ns. I have done simulations for aluminum, graphite and aluminum coated with a 100 micron thick layer of alumina.

I have considered the effects of these low energy photons. I have done simulations where these photons are filtered out, perhaps with an aluminum foil, and where they are allowed to irradiated the sample. The ranges of 200 eV x-rays in aluminum and alumina are more than an order of magnitude less than the ranges of 1 keV x-rays and should be mostly absorbed in the blow-off plasma and not contribute to the launching of a shock in the material. Therefore, I only show results for these materials were the low energy photons have

not been filtered out; the results with filtering are
essentially the same. This is not the case for graphite
because the range of 200 eV x-rays is only a little shorter
than that for 1 keV x-rays. Both unfiltered and filtered
simulations are shown for graphite.

Table 3. Stresses in Various Materials
Generated with SATURN X-Rays

	Aluminum unfiltered	Graphite unfiltered	Graphite filtered	Alumina/ Aluminum unfiltered
Range of 1 keV x-rays (mg/cm^2)	0.83	0.50[a]	0.50	0.38
Mass Density (g/cm^3)	2.7	1.7	1.7	3.5
Energy Fluence (J/cm^2)	550[b]	2750	550	550[b]
Energy Density (kJ/g)	660	5500	1100	1440
Power Density (GW/g)	44	367	73	96
Calculated Stress @ 0.05 cm (GPa)	7.5	36.0	10.2	3.5
Calculated Stress @ 0.15 cm (GPa)	4.8	12.5	5.0	1.8
Calculated Stress @ 0.25 cm (GPa)	2.7	8.0	N.A	1.0

[a]The ranges of 1 keV and 300 eV x-rays in carbon are the same
The range of 100 eV x-rays is 0.05 mg/cm^2.
[b]Because of the much shorter range of low energy photons, the
part of the spectrum below 900 eV is ignored and is not
included in the energy fluence.

Except in the case of alumina on aluminum, one can see
that the stress increases with energy density and power
density. The stress is recorded at three positions in the
material, and one sees that the shock is attenuated in all
cases. For the case of alumina on aluminum, the calculated
stresses are much lower than the energy and power would
predict. There is a mismatch in the speed of sound, mass
density, and material strength at the alumina/aluminum interface. This leads to poor transmission of the shock across the
interface and a great reduction of the shock strength in the
aluminum, where the calculated stress is measured.

EXPERIMENTS ON SATURN

During May, 1989 I fielded some x-ray vaporization experiments on SATURN like those described in the previous section.
All of the samples were donated by LLNL or SNL. The space on

the machine was just what remained on experiments that were already planned. The exception to this was shot 669 which only had my samples on it and was donated by SNL. I did not have any active diagnostics to measure the stress levels. The sample holders were loaned to me by other experimenters at SNL. The samples were held in stainless steel 316 Swaglok fittings that held the samples in place with an annular lip. The backs of the samples were supported with carbon foam that was, in turn, supported with a thin aluminum disk. I am still in the process of analyzing these experiments. The samples after they were irradiated with x-rays are shown in Fig. 5. One can see in Fig. 5 that all samples except 2 and 5 were utterly destroyed. Sample 5, a two-directionally woven graphite in a carbon matrix called K-Karb, was not damaged in the plane of the graphite fibers but these planes became delaminated. Sample 2, aluminum 6061 with a layer of alumina blasted on its surface, survived well except that the alumina was removed. All of the other samples were fine grained graphites or graphites with short fibers. Sample 3, Graphnol, was a fine grained graphite that survived the best of these as it was broken into about 6 pieces. The others were turned into powder. I could not even find any pieces of sample 4.

In August of 1989, I fielded another set of experiments on SATURN. These used argon gas pinches as an x-ray source. The spectra from these pinches differ from those for neon in the photon energy of the dominant lines; neon has lines at about 0.9 and 1.0 keV, while argon emits lines in the 3 to 4 keV

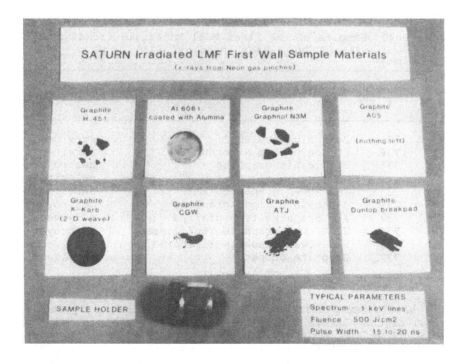

Fig. 5. Photograph of eight samples shot on SATURN during May 1989.

range. Also, argon has about 40 kJ in these lines, while neon can have as much as 100 kJ in its lines. The pulse widths of the x-rays can be as low as 10 ns for argon gas pinches. For these experiments, I used 3 and 4 directionally woven graphites in a solid carbon matrix, bare aluminum 6061 and aluminum 6061 coated with a layer of carbon, a loose carpet material made of graphite, and two samples of 2-directionally woven graphite, where the x-rays were unfiltered and then filtered with a thin aluminum foil. The 3 and 4 directional graphites were an attempt to stop the delamination seen in K-Karb. The aluminum experiments are an extension of the previous experiments with alumina on aluminum in that they use a sacrificial layer to protect the aluminum, while carbon would be much easier to spray onto the wall of an LMF before each shot. The carbon carpet is a relatively new idea for LMF target chamber wall protection,[10] which uses the looseness of a long fibered carpet to prevent the generation of a shock. The filtering of x-rays is an experimental test of the low energy photon effects examined computationally.

The results of these experiments are given in Table 4. One can see that the aluminum survived both with and without the carbon protection. The 4-directional weave was successful in combating delamination, though the 3-directional random weave was not. The graphite carpet was almost totally undamaged by the x-rays. The unfiltered 2-directional weave was destroyed, while the filtered sample survived. I have no quantitative results yet as to the performance of the gas pinches, but preliminary indications are that there were in excess of 350 kJ of x-rays on all shots.

Table 4. Samples of LMF First Wall Materials Irradiated with SATURN X-Rays in August 1989

Sample #	Shot #	Material	Result
1	736	Bare aluminum 6061	survived
2	736	Carbon coated aluminum 6061	survived
3	736	Stapleknit graphite	destroyed delaminated
4	737	4-D woven graphite (FMI)	survived
5	739	3-D random fiber graphite	destroyed
6	737	A05 graphite fine grained	survived
7A	739	2-d woven graphite (unfiltered)	destroyed
7B	739	2-d woven graphite (filtered)	survived
8	737	Graphite carpet	survived

I should reemphasize here that this work is in progress. Several of the numbers quoted in Table 4 are still preliminary; I am still working on the fluences and spectra for these shots. I have done no post-shot analysis of the samples yet. I plan to study those that survived with a scanning electron microscope to see if the shocks caused any changes to the materials. I need to run computer simulations for the exact fluence and

I have used a simple scaling law and computer simulations to show that target x-rays will generate shocks in the first surfaces of unprotected LMF target chamber walls whose strength depends on the x-ray fluence, power intensity, and fluence. I have shown that gas pinches on the SATURN electron accelerator can provide x-rays that are of relevance to some LMF concepts. I have performed preliminary experiments on SATURN that have shown bare aluminum, aluminum coated with a thin layer of alumina or graphite, four-directionally woven graphite, and graphite carpet all survive a single pulse of x-rays. I have begun to study the effects of low energy photons in the experiments done on graphite.

Several issues need to be studied before a first wall material is chosen for the LMF. The experimental results reported in this paper are still preliminary in nature and much more data is needed. Samples need to be analyzed with electron or optical microscopy. The changes in material properties, such as elastic modulus and yield strength, brought about by the shocks need to be measured. Samples need to be repetitively irradiated with x-rays to study how the changes in properties will affect the response to shocks. Additional computer simulations will be needed as more information on properties of the material is obtained. The effect of debris in the SATURN experimental chamber needs to be addressed as does damage to the samples not related to the passage of shocks. Finally the techniques developed in this project should be applied to other materials that may be more relevant to ICF power plant designs.

ACKNOWLEDGEMENT

This work was supported by Sandia National Laboratories and Lawrence Livermore National Laboratory. Computer simulations were performed on computers at SNL and both the experimental and computational work were performed while the author was on an extended visit to SNL.

REFERENCES

1. U.S. Department of Energy Inertial Fusion Division, "LMF-Laboratory Microfusion Capability Study, Phase I Summary," DOE/DP-0069.
2. R.R. Peterson, "Gas Condensation Phenomena in Inertial Confinement Fusion Reaction Chambers," Laser Interaction and Related Plasma Phenomena, vol. 7 (Plenum Press, New York, 1986), H. Hora and G. Miley, editors.
3. J.J. MacFarlane, "IONMIX - A Code for Computing the Equation of State and Radiative Properties of LTE and Non-LTE Plasmas," University of Wisconsin Fusion Technology Institute Report UWFDM-750 (December, 1987).

4. R.R. Peterson, J.J. MacFarlane, and G.A. Moses, "CONRAD – A Combined Hydrodynamics – Condensation/Vaporization Computer Code," University of Wisconsin Fusion Technology Institute Report UWFDM-670 (January, 1986; revised July, 1988).

5. S.L. Thompson and J.M. McGlaun, "CSQIII An Eularian Finite Difference Program for Two-Dimensional Material Response: Users Manual," Sandia National Laboratories Report SAND87-2763 (January, 1988).

6. C.L. Olson, "Achromatic Magnetic Lens Systems for High Current Ion Beams," Proceedings of the 1988 Linear Accelerator Conference, Williamsburg, VA, October 3-7, 1988, to be published.

7. G.A. Moses, R.R. Peterson, M.E. Sawan, and W.F. Vogelsang, "High Gain Target Spectra and Energy Partitioning for Ion Beam Reactor Design Studies," University of Wisconsin Fusion Technology Institute Report UWFDM-396 (1980).

8. R.B. Spielman, et al., "Efficient X-Ray Production From Ultrafast Gas-Puff Z Pinches," J. Appl. Phys. 57, 830 (1985).

9. "Progress Report: Narya Pulsed-Power-Driven X-Ray Laser Program -- January 1984 through June 1985," M.K. Matzen, ed., Sandia National Laboratories Report SAND85-1151, 1986.

10. M.J. Monsler and W.R. Meier, "A Carbon-Carpet First Wall for the Laboratory Microfusion Facility," Fusion Technology 15, 595 (1989).

LEADING ROLE OF LASER FUSION AND ITS ADVANCES BY

VOLUME IGNITION AND BY PULSATIONFREE DIRECT DRIVE

Heinrich Hora, Lorenzo Cicchitelli, Gu Min,
George H. Miley*, Gregory Kasotakis and Robert J. Stening

Department of Theoretical Physics
University of New South Wales
Kensington-Sydney 2033, Australia**

INTRODUCTION

In the search for large scale energy sources without fossil fuels, the option of laser fusion energy provides a complete physical solution at competitive economic cost. This projection is based on the advancing state of laser technology, on the results of underground nuclear tests (Centurion-Halite project) and on the concepts of indirect drive and spark ignition. While there is an urgent need to proceed with experiments of this kind without any modification as a first generation development, improvements for the next generation can be envisioned which could <u>result in energy at a lower cost than light water reactors.</u> Two of these improvements will be presented here: a) <u>volume ignition</u> of laser driven capsules and b) a <u>basically</u> <u>new analysis</u> of the difficulties of the pulsating laser-plasma interaction, based on the experiments of Maddever and Luther-Davies. The latter could provide an understanding of the necessity for <u>smoothing of the interaction for a reliable and efficient direct driving</u> of the capsules.

THE ENERGY PROBLEM

The splendid achievements of modern civilization are strongly based on a large scale energy consumption. Thanks to the theoretical revelation of Newton's mechanics, of his gravitational force and the result that energy or work is given by the product of force times length of directed action, James Watt ingeniously understood that which people saw since thousands of years before: when a lid of a kettle is lifted up by the steam of the boiling water, work is done. The steam produces energy by driving the gravitational weight of the lid against the height along which it is lifted up. (Preceding machines based on pumps did not need Newton's gravitation law; Watt also used the kinetic energy of rotation.) The invention of the steam engine and later of the internal combustion engine and the electrical engines released muscles to do work and provided a nearly paradisical standard of living for all people - even with growing population - in the industrialized countries now and in the developing countries next.

---

* Fusion Studies Laboratory, University of Illinois, Urbana Ill. 61801, USA.

** One of the authors (Heinrich Hora) was supported by the Wilhelm Heinrich and Else Heraeus Foundation, Germany, during his visiting appointment at the University of Giessen, Germany West.

Energy sources utilized to date have mostly been coal and petrol stored as fossil fuels in the earth. It turns out that further burning of the fossil fuel and loading of the atmosphere with the product of burning, carbon dioxide, may lead to a catastrophic change of the earth's climate. First, one should not condemn fossil energy but rather be grateful that it was the key to industry and to our modern life: but one has to be aware that its further use has to be stopped very quickly. The only solution is to look for an alternative energy source. For large scale use the presently available choice is only the nuclear fission reactor - despite concern about safety and waste disposal - at least for a transition period. The cheapest energy available we ever have is light water nuclear reactors. These have to be replaced in future, however, not only because of the desire to have better safety, to avoid possible nuclear weapons production, and to lessen problems of radioactive waste disposal, but also because the supply of uranium on earth will be absorbed within 100 years.

How critical it is to stop using fossil energy is immediately evident from the fact that energy production of over the next 100 years from coal and petrol would convert a considerable amount of these stored materials into atmospheric gases. This will lead to climatic changes of an unknown but very probably undesirable nature. To avoid such catastrophes a solution of the energy problem has to be found by intensive work involving a very large and long term effort. Fortunately, the leaders of the world are becoming aware of such problems known as the heating of the atmosphere [1] and the future of energy generation [2]. There are numerous hopes to find an alternative to fossil fuels. Increased public attention is being given to renewable energies, especially to solar energy. This will be a most profitable industry for providing up to 5% of all energy generation, measured in Billions of dollars enterprises in future, but its costs are too high. Even most optimistic estimations show renewable energy ten times more expensive than energy produced from light water reactors. The reason is simply the entropy involved (as has been shown by Niu [3]) and this is a principle that cannot be overcome.

For nuclear fission reactors, the technology for low cost energy production - even less expensive than fossil fuels [4] - is well established and the use of the safest new systems [5], even with absolutely safe passive protection systems [6] may be a solution only for a limited period. However there is a clear need for a long term solution.

Again and again hopes and promises have arisen that some very unexpected phenomenon will be discovered for solving the energy problem. There is for example the published expectation of an electrical discharge mechanism in hydrogen isotopes to produce exothermal energy by a rather low cost procedure [7] or the indication that some as yet undiscovered nuclear reaction mechanisms (e.g. cold fusion, so named because of the highly energetic exchange of neutrons between nuclei of very low thermal energy [8][9]) may be developed as an energy source. But the details of these are as yet so uncertain and unexplored that one has to wait for further results before more substantial decisions are possible.

There remains the option to reproduce on earth, in an economic way, the mechanism by which the huge amount of energy is generated in the interior of the sun. This process has been known since the beginning of the 1930s and consists of the <u>fusion of light atomic nuclei</u> to produce larger ones. An overview of the possible options is given in the next section.

PHYSICS SOLUTION OF ECONOMIC FUSION ENERGY BY LASERS

Large scale economic fusion energy production can be based on the nuclear fusion reaction of the heavy hydrogen isotope deuterium, D, with the superheavy radioactive hydrogen isotope tritium, T[10].

$$D + T \rightarrow {}^4He + n + 17.6 \text{ MeV} \tag{1}$$

producing a neutron of 14.1 MeV energy. In order to avoid the use of radioactive tritium and undesired radioactivity generated by the neutrons (most of which will be used for breeding tritium in lithium surrounding the reaction which will absorb most of the neutrons) an ideal reaction is [11]

$$H + {}^{11}B \rightarrow 3 {}^{4}He \tag{2}$$

where less radioactivity is produced per unit energy gained than by burning coal [12] (arising from the 2 parts per million uranium in all coal) and heat pollution is strongly reduced since the kinetic energy of the helium nuclei can be converted directly into electricity. It is very difficult, however, to make reaction (2) occur and the following considerations are mainly based on the reaction (1).

Of the other options for the large scale energy production, solar and wind energy are more than ten times more expensive than that from light water reactors at present even under the most favourable assumptions [4], confirming the entropy theorem of Niu [3]. The fusion reactions (1) or (2) require the plasma state with temperatures T above 10 or 100 million degrees at which the plasma of a sufficient density n (in atoms per cm^3) has to be confined for a sufficient time τ given by the Lawson criterion [13],

$$n\tau > 10^{14} \sec cm^{-3} \tag{3}$$

One method to confine plasma is to use magnetic fields and some very big projects study toroidal plasma configurations with high currents, called tokamaks. Apart from the fact that adequate confinement in a reasonable sized device seems doubtful, a very thorough but well optimistic estimation shows that the cost of the energy produced may be more than ten times of that from light water reactors [14]. A further problem to be noted is that the walls of the tokamak will be eroded about 1 cm per 24 hours operation according to the estimations for the ITOR (International Torus for Magnetic Confinement Fusion) projects. One "out" for magnetic confinement may be to burn aneutronic fuels in alternate confinement systems such as reversed field configurations. However, a sound experimental base for this approach is lacking.

The aforementioned physical solution for an economic fusion reactor by laser irradiation of fuel capsules has been derived on the basis of indirect driving [15]. This means that irradiated laser energy must be converted into x-rays in the capsule surface. These x-ray photons go onto the interior, producing a smooth compression and heating of the nuclear fuel such that a higher yield of fusion energy per total energy input results. This has been found possible by using the indirect driving of the intense x-rays experimentally where the irradiation energies are of about 100 MJ from nuclear explosions.The result of this work including the underground nuclear explosions is outlined in Ref. [16]. With more sophisticated x-ray generation by lasers, pulses of 10 MJ are designed to produce 1000 MJ energy from a capsule [18][19]. The result of this work including the underground explosions is [16]:

FUSION ENERGY BY LASERS IS NO LONGER A QUESTION OF "IF" BUT OF "WHEN".

Adding up the cost of a present day neodymium glass laser driver for producing 10 MJ pulses as less than $1 Billion [18,19] the technological development of a 1 pulse per second glass laser system even with higher efficiency than presently available is less than 2 or 3 Billion Dollars using present day experimental techniques and suitable engineering developments. The entire reactor for driving the DT fuel capsules for the laser fusion reactor is very simple [20], consisting of a rotating vessel of bi-concial shape with a half meter thick layer of ceramic pebbles of a lithium compound held to the interior wall by centrifugal forces. The capsules, ignited by the laser in the center, emit helium and neutrons into the pebbles. As they pass through this layer (of 0.5 meter thickness) the neutrons are slowed down or react, heating the pebbles and breeding lithium into helium and tritium.

The hot pebbles are removed giving their energy by heat exchangers to electrical power generators, while the tritium is recovered as fuel for further use in fusion capsules. If any radiation damage appears in the pebbles this can be repaired easily before feeding the cold pebbles again into the rotating vessel. Therefore the devastating wall problem, as encountered in the tokamak reactor with its 1 cm erosion per day, does not affect the CASCADE reactor [20]. The costs of this basically very simple system would be less than a tokamak with its superconducting coils and first wall-blanket, systems which are difficult to maintain and which will operate over an extreme range of temperatures complex. Adding up the low costs of a CASCADE reactor and of a laser driver using advanced glass technology, one can project that:

A SUITABLY ENGINEERED ATHENA [19] AND CASCADE [20] TYPE LASER FUSION REACTOR CAN BE EXPECTED TO PRODUCE ENERGY AT THE COST COMPETITIVE WITH LIGHT WATER REACTORS

where installation costs of about $5 Billion are estimated for a Gigawatt reactor unit. Maintenance costs remain uncertain until a demonstration plant is achieved. The technological development of such a reactor indeed will require at least 10 to 15 years even if a crash program is started now comparable to that for the flight to the moon. One can say that the physics problems are essentially solved; only the technology has to be developed. It is the same situation as in 1960 when the physics was to send a person to the moon and fly them back was known, but the technology had to be developed. Nobody knew in 1960 how the Saturn-5 Rocket would look. It should be mentioned that at that time Nobel Laureate Max Born stated that he did not expect manned flights to the moon within this century.

While no time should be lost prior to starting a crash program for the development of an Athena-Cascade type laser fusion reactor, one should simultaneously develop the physical basis for the next generation of laser fusion reactor systems. Apart from attention to the concepts of the heavy ion fusion drivers [21,22] or light ion fusion drivers [23], the capabilities of the KrF excimer laser were widely underestimated in the past and MJ pulse systems seem to be possible [24]. The level of development of this laser does not yet seem to be at the stage of the glass laser of the Athena project [19] so that a decision for "a second best solution in time may be preferred instead of the best solution too late". Though still in a much earlier stage of development, the scheme of a 100 MJ pulse hydrogen fluoride laser may be mentioned [25].

As already mentioned the Athena project [19] uses indirect drive and central spark ignition. We have demonstrated that improved efficiency may be obtained using volume ignition and direct drive accompanied by appropriate smoothing of the interaction. Further developments in theory and technology in the future suggest the possibility of the statement:

SECOND AND FURTHER GENERATION LASER FUSION REACTORS HAVE THE POTENTIAL WITH ANTICIPATED ADVANCES OF PHYSICS TO PRODUCE ELECTRICITY AT EVEN LOWER COSTS WITH INEXHAUSTIBLE, NON-RADIOACTIVE FUEL

in 50 years from now. This would really be the basis of the golden age which was expected from the use of nuclear energy.

The following sections will present recent developments on volume ignition and on smooth direct drive as examples of improvements in laser fusion physics for the next reactor generations. It will also be shown how - if laser fusion were pursued vigorously the most attractive hydrogen boron reaction, (Eq. (2)) may be used if the laser techniques will permit 100 times higher compression and 100 times larger laser pulses than in the present days Athena [19] scheme. Such improvements of techniques may well be possible within about 80 years so that an ideal, non-radioactive, inexhaustible, low cost energy source would be available.

HISTORICAL REMARKS ON VOLUME IGNITION

The fusion gain calculations for ideal adiabatic compression and expansion of the plasma according to the self similarity model [25] assume the initial conditions that an energy E_0 is deposited into a fully ionized deuterium tritium 50:50 plasma mixture of uniform density n_0 and spherical volume V_0 (having a radius R_0) resulting in a temperature T_0 and a self-similarity velocity v_0 of the pellet radius with a linear velocity profile into the interior of the pellet [27]. Then one calculates the gain G as the ratio of fusion energy generated to the input energy as the adiabatic self-similarity dynamics proceeds. Using a zero initial velocity it was found that the gain G very sensitively depended on the initial parameters. There was agreement with the initially reported values for the gain [24], investigation of a much wider range of parameters permitted the identification of an optimum initial temperature with higher gains resulting.

Using the conditions obtained above the fusion gain could be derived from the numerical results of the following formula for an optimum temperature T_{opt} at zero velocity:

$$G_0 = \left(\frac{E_0}{E_{BE}}\right)^{1/3} \left(\frac{n_0}{n_s}\right)^{2/3} \tag{4}$$

where E_{BE} is the break even energy and n_s is the solid state density. Equation 4 was derived from the computations [25] and first published in 1970 [28]. It is algebraically identical with the following equation derived by Kidder [29].

$$G_0 = Cn_0R_0 = D\rho_0R_0 \tag{5}$$

where ρ_0 is the initial density of the DT mixture. There has been a long discussion of the constants involved since the first results appeared in 1964 [25]. After an updating of the experimental cross sections in the computations and a clarification of the definition of the temperatures, with the relations (4) and (5) are given for a simple fusion burn of a plasma sphere with adiabatic compression to the optimum temperature and subsequent adiabatic expansion as:

$$T_{opt} = 17.2 \text{ keV}; E_{BE} = 6.9 \text{ MJ}; n_s = 5.9 \times 10^{22} \text{ cm}^{-3};$$

$$C = 8.78 \times 10^{-23} \text{ cm}^2; D = 21.07 \text{g}^{-1} \text{ cm}^2 \tag{6}$$

where n_s is the solid state density of the DT mixture and E_{BE} is the break even energy.

These calculations included a number of simplifications. The losses of bremsstrahlung were ignored, any additional gains from the self heat due to the slowing down of alphas and neutrons in the plasma were not included and the depletion of the fuel by the reaction was not considered. The basic result was discouraging from the beginning. It showed that laser pulse energies in the range of megajoules are necessary. Furthermore, if a 25% burn of the fuel is assumed, and the 17.2 keV optimum temperature is obtained, there results a gain of 41 only, much too low to compensate for the losses of inefficient laser systems.

The way out was proposed by Nuckolls [30]: the dynamics of the laser interaction with the capsule could be controlled in such a way that only a small part of plasma should be compressed in the center to densities of 1000 to 10,000 times the solid state at temperatures above 10 keV and that this central plasma would ignite a self sustained fusion combustion wave to travel through the less dense low temperature plasma surrounding the spark in order to achieve gains of several hundreds. Apart from the fact that the fusion wave is a very complex mechanism and the control of the parameters is extremely sensitive [31], a reliable model has been constructed to generate spark ignition, producing 1000 MJ fusion energy from a 10 MJ laser pulse [19].

When the calculations included the alpha particle reheat and fuel depletion, a volume ignition was found [32] for gains above 8, corresponding to ρ_0R_0 of 0.4 resulting in a gain formula including the fusion burn process below gains of 8 (equation 1) and volume ignition [33]

$$G = G_0 \frac{7.2 \times 10^{-4} G_0{}^3}{[1+1.44 \times 10^{-3} G_0{}^3]^{1/2}-1} \qquad (7)$$

This formula is correct only up to gains of 200. Above that value, fuel depletion lowers the increase in G with the input energy, approaching saturation, see Fig. 1. The optimum temperature is given by

$$T_{opt} = 17.2 \frac{3.52\{[1+0.142G_0]^{1/2}-1\}}{G_0} \qquad (8)$$

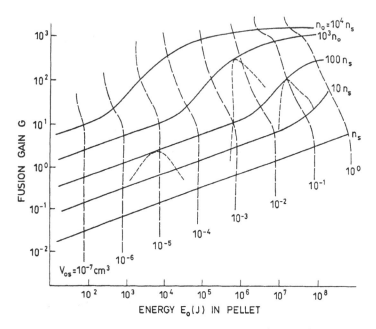

Fig. 1. Optimum fusion gains G calculated for an adiabatic compression and expansion of a DT plasma of an initial volume at solid state density with a maximum compression to a density n_0 in multiples of the solid state density. When reaching the density n_0, the compression velocity is zero and the input laser energy E_0 is fully converted into thermal energy representing the optimum temperature as given by Eq. (7). The dashed lines are the gains for a fixed volume for various input energy touching the optimum gains at the straight lines. The dashed lines are parabels for low gains and show a jump-like increase in the case of volume ignition [32],[33].

The computations improved as follows:

1) Latest values for the fusion cross sections were used, though no final decision was made as to which values were most reliable [34].

2) The depletion of fuel, initially included in the number of the reacting atoms only, has been taken into account now for the adiabatic compression and expansion hydrodynamics. The corrections are not very large but the model can now be considered to be complete with respect to fuel depletion.

3) The inclusion of the losses by bremsstrahlung has confirmed that this does not affect the results of equation (1) because of the high optimum burn temperature but becomes important at the lower temperatures applying to volume ignition. The re-absorption of bremsstrahlung was included in the following calculations where the classical collision frequency was used. In place of the approximate formula used by other authors [35], we used the detailed spectral distribution of the bremsstrahlung and its density and temperature dependence to calculate the mean free path of the photons for each time step of the plasma dynamics.

The main advantage of volume ignition, with accompanying gains much higher than the mentioned pure burn gain of 41, is that the stopping power and self-heating mechanism is understood more clearly than the fusion wave of spark ignition.

VOLUME IGNITION

In order to test the results of the computations of volume ignition [31] [34] with regard to simplified earlier known facts, figure 2 shows the time dependence of the plasma

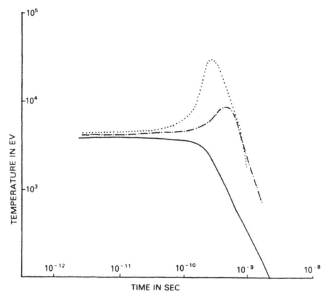

Fig. 2. Dependence of the temperature T of a pellet on the time t at initial compression to 100 times the solid state and initial volume of 10^{-3} cm^3. For the cases of the (fully drawn; dashed; dotted) lines respectively, the input energies were 11, 12, 13 MJ, the fusion gains G were 2.01, 22.38, 103.3, the initial temperatures were 3.95, 4.31, 4.67 keV and the deuterium tritium fuel depletion was 0.28, 3.33, 16.65%.

temperature T for the case of an initial density of 100 times the solid state with an initial volume of one cubic millimeter. Three cases are given in figure 1 where the input laser energy was 11, 12, 13 MJ respectively. The fusion gains G were 2.01, 22.38 and 103.3. The ignition process can be seen clearly, not only from the strong increase of the gain G accompanying little increase of the input energy, but from the increase of the temperature with time. In the first case, the temperature is nearly constant for a long time, since the temperature loss by bremsstrahlung and by the adiabatic expansion (sometimes called adiabatic loss) is nearly compensated by the alpha self-heat until a strong drop of the temperature occurs with the fast expansion. We call such a case of monotonically decreasing temperature a 'simple burn', similar to the pure burn cases of the early computations [25] [28], Eq. (4), where the contribution of self-heat is not strong.

A ten percent higher energy input results in the appearance of ignition instead of the simple burn: the alphas inject more heat into the plasma than the losses, and the temperature of the pellet increases after the input of the driver energy of 12 MJ. A maximum temperature of nearly 10 keV is reached before the fast expansion, and later adiabatic cooling drops the temperature. Increasing the input again by about 10 percent results in strong ignition with a maximum temperature of 31 keV and a high gain above 100.

The initial temperature in the three cases of figure 2 are 3.95, 4.31 and 4.67 keV. This is just a little bit less than the simple case [36] where the generated fusion energy is equal to the generated bremsstrahlung.

Obviously a small part of the bremsstrahlung is re-absorbed in the pellet. This fact can be expected since the maximum wavelength of bremsstrahlung is at 4.5 keV so the absorption length in the pellet of 100 times solid-state density is small compared to the typical diameter.

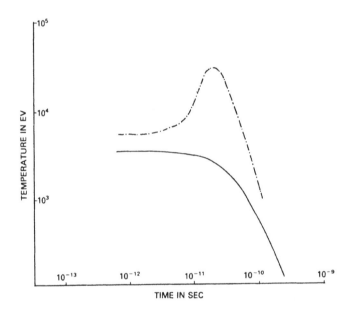

Fig. 3. Same as in figure 2 for a compression to 1000 times solid state density and initial volume of 10^{-6} cm^3. The (fully drawn; dashed) lines respectively correspond to initial laser energies of 100; 150 KJ, the gains G of 1.69, 99.55, the initial temperatures were 3.59, 5.39 keV and the fuel depletion was 0.21, 18.51%.

Figure 3 reports the results of a case as in figure 2 but for a compression to 1000 times the solid state density using an initial volume of 10^{-6} cm^3. There are only two cases shown, one with an input laser energy of 100 KJ and another one with 150 KJ. The respective gains G are 1.69 for simple burn and 99.5 for ignition. The initial temperatures were 3.59 and 5.39 keV. Again as in the preceding case no strong re-absorption of the bremsstrahlung can be observed with an initial temperature of about 5 keV even for ignition.

This can be confirmed by examining the initial radius and density and taking the maximum x-ray wave length for the 5 keV temperature: the absorption length is not shorter than the initial radius. The fuel depletion in the two cases of this figure were 0.21% and 18.51% respectively [37].

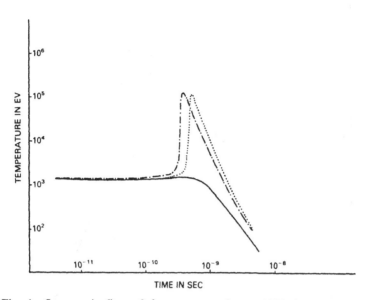

Fig. 4. Same as in figure 2 for a compression to 1000 times solid state density and initial volume of 10^{-3} cm^3. The (fully drawn; dashed; dotted) lines respectively correspond to initial laser energies of 40, 41, 42 MJ; the gains G of 2.41, 1170, 1199, the initial temperatures were 1.44, 1.47, 1.51 keV and the fuel depletion was 0.12, 59.52 and 62.39%.

A case with strong re-absorption of bremsstrahlung is reported in figure 4 with similar time dependence of the pellet temperature T(t) as in figure 2, after driver energy of 40, 41 and 42 MJ has been deposited, with an initial volume of one cubic millimeter and initial density of thousand times the solid state. The fusion gains G in these cases are 2.41, 1170 and 1199 respectively. The first case is practically a case of simple burn while the second and third cases are volume ignition as seen from the strong rise of the gain G by a factor of nearly 500 and with the strong increase of the pellet temperature to a maximum above 100 keV.

It is important to note that the initial temperatures are only 1.44, 1.47 and 1.51 keV respectively. The re-absorption of the bremsstrahlung is then so strong that the former threshold of 4.5 keV - when the bremsstrahlung is just compensated by the fusion energy generation - is strongly undercut. The further result is that the evaluation of the fuel depletion results in the respective percentages of depletion by the reactions of 0.12, 59.52 and 62.39%.

Another case with strong re-absorption is seen in figure 5. This is indeed an extreme case of 10 000 times the solid state density with an initial volume of 10^{-5} cm^3. The input laser energies are 3, 3.5 and 4 MJ with gains G of 0.5 (simple burn), 1731 and 1585 (volume ignition). The initial temperatures had the remarkably low values of 1.08, 1.26 and 1.44 keV. The fuel depletions were 0.02, 75.08 and 78.59%.

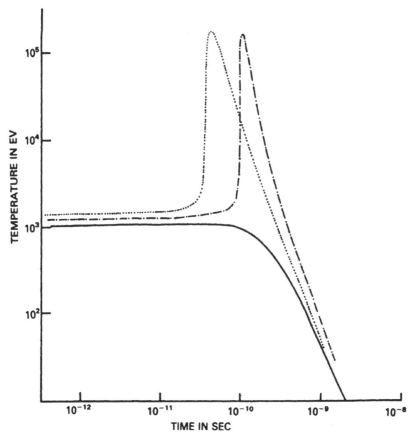

Fig. 5. Same as in figure 2 for a compression to 10 000 times the solid state density and initial volume of 10^{-5} cm^3. The (fully drawn; dashed; dotted) lines respectively correspond to initial laser energies of 3.0, 3.5, 4.0 MJ, the gains G of 0.5, 1731, 1585, the initial temperatures were 1.08, 1.26, 1.44 keV and the fuel depletion was 0.02, 75.08, 78.59%.

To show the cases with a fuel depletion above 80%, we plotted the calculations for a set of volumes for the cases of 10 000 times solid state density in figure 6.

COMPARISON AND CONCLUSIONS ON VOLUME IGNITION

The experiments on laser compression of pellets at ILE Osaka [38] have clearly proved how the ideal conditions of volume compression can be achieved by avoiding shocks and stagnation. The conditions required for the best fusion gains have been calculated by other authors [39] and so a comparison of our results with specific computations by Mima et al [40] is of interest. Their computations are much more detailed than ours since the laser interaction with the pellet plasma and the ablation is included at least with mechanisms known at present. Recent knowledge about the pulsation and smoothing of the interaction may cause modifications in the basic importance of mechanisms which, however, may not change the usual results of the energy transfer.

The calculations of Mima et al [40] show the amount of plasma involved in the ablation and how much energy enters the compressed pellet (determined by the hydrodynamic efficiency). From studies of the shock mechanisms in the compressed

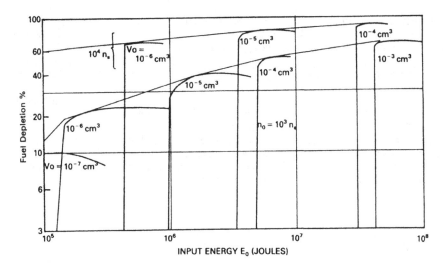

Fig. 6. Dependence of the ionic depletion of the fuel on the input laser energy for 1000 and 10 000 times the solid state density and initial pellet volumes V_0 from 10^{-7} cm^3.

plasma, a very detailed treatment of the dynamics of each partial shell is included, taking into account thermal conduction and radiation mechanisms.

Our calculations start from the energy input into the pellet. The laser energy is first reduced by the hydrodynamic efficiency; further, adiabatic compression and expansion dynamics are assumed so that no details about the shock effect in each shell are necessary. The experimental conditions [38] may be close to these cases so that the shorter computation time for each case permits a more detailed investigation of the dependence on the various parameters as derived from the large number of treated cases.

One important point of the general behaviour is that for fusion gains larger than 8 (corresponding to a $\rho_0 R_0$ of 0.4, above which the volume ignition occurs), our result for the exponent of the fusion gain dependence on the input energy

$$G = E_0^a$$

results in a value a = 0.9 - derived from our numerical plots - while Mima et al (1987) derived a value a = 1. The subsequent number N of neutrons generated per shot is then

$$N = E_0^{a+1}$$

where the exponent in our case is 1.9 and the value obtained by Mima et al [40] is 2.

For the case evaluated by Mima et al [40] for a total gain of 3 (related to the total input laser energy of 100 kJ) and 100 times compression of the pellet after an ablation given by 7% hydrodynamic efficiency, a $\rho_0 R_0$ value of equation (5) of 0.9 was found by Mima et al. (1987) while our result is 0.8, based on achieving the same total gain of 3 with a compression of the core of 100 n_s. This example indicates what gains may be expected from the next generation of experiments with 100 kJ pulses of laser energy if volume compression is applied.

If the above mentioned conditions can be verified experimentally, the present parametric framework for pellet fusion, based on volume ignition provides (Cicchitelli et al.

1988) fusion reactor pellet operation with 1 MJ driver pulses with a moderate compression to only 100 times the solid state density, and with the much less concern to avoid Rayleigh-Taylor instabilities with the larger geometry of volume ignition than for spark ignition.

There is another example which shows how our calculations with the rather simplified adiabatic capsule dynamics, give results which are very close to the most sophisticated calculations with detailed ablation and energy transport and the full shock dynamics [19]. 10 MJ of laser energy produce 1000 MJ fusion energy in a capsule with compression to 1000 times the solid state. In the compressed stage of the center of the capsule, Fig. 7, the density and temperature profile basically show a spark ignition case, where the center is of high density and temperature, igniting a fusion wave into the outer part of the plasma. However, the spark ignition case is less typical since the diameter of the center is larger nearly half way to the volume ignition case.

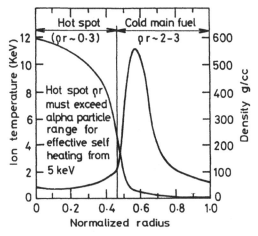

Fig. 7. Radial profile of density and ion temperature in compressed pellet at "central spark" ignition (Storm et al [19].

It is then not surprising when the very transparent results of our calculations with a rather simple and widely reproducible code are very close to the reported computation of Fig. 7. Starting from the results of Fig. 1, for a density of 1000 times the solid state and an energy E_0 of 1.5 MJ (corresponding to a hydroefficiency of 15% relating the 10 MJ laser energy converted into the reacting plasma after subtraction of 85% of the laser energy which went into the ablated plasma) the resulting gain G in the definition of Fig. 1 is 600, corresponding to an output of total fusion energy of 600 times 1.5 MJ = 900 MJ which is quite close to the result of Fig. 7 [19] with 1000 MJ.

The shock free and stagnation free ideal adiabatic volume compression and volume ignition provides a very transparent and reliable model since it is free from the uncertainties of the spark generation and from the complex - and rather incompletely understood - mechanisms of the self sustained fusion combustion wave. It corresponds to the thermodynamically ideal adiabatic compression dynamics and the ignition is similar to that of a Diesel engine which has the desirable feature of working within the whole volume simultaneously in contrast the combustion front mechanisms of the Otto engine. The initially expected very complicated mechanism of laser interaction, required to achieve a shockfree compression, has been solved by C. Yamanaka et al experimentally [38], [39] using a simple burn only without ignition and low applied laser pulse energies in the range of 12 KJ.

For the next generation design of laser fusion after the Athena scheme, volume ignition may therefore offer a very transparently understandable, very efficient, and experimentally more easily realized scheme (reduced accuracy of pulse parameters and less problems with the Rayleigh-Taylor instabilities [31]), especially if the later described more efficient direct drive is used where smaller capsules and low laser pulse energies down to 1 MJ may be applied simplifying the basic reactor conditions.

ESTIMATIONS OF FUTURE HYDROGEN-BORON LASER FUSION

The hydrogen boron reaction, Eq. (2), has such low cross sections that the results of a treatment of simple burn according to Eq. (4) are most disappointing. The optimum temperatures are beyond 150 keV and the energies E_0 are in the range of terajoules. When the first estimations of this kind were discussed [Ref. 30, p.402], or [41], R.E. Kidder commented that "hydrogen boron is not for the year 2000 but for the year 4000". Nevertheless there were indications that the self heat does reduce the optimum initial temperature, reducing the input energy and increasing the gain due to volume ignition in hydrogen boron [41] [42] as first seen from the result of volume ignition of DT [32].

We report here on recent results of volume ignition calculations for hydrogen boron where laser pulses of 1 GJ produce gains of 24 for compression to 100,000 times the solid state. These values are indeed far beyond the present technology. However, if laser fusion would be the main focus of large scale energy production with a huge amount of research and development next century, the difference in values between today's physical-technical achievements in laser fusion [19] are only a factor of 100 in density above the experimental value just reached [43] and a further factor of 100 above the magnitude of laser pulses obtained in the Athena project [19].

When our volume ignition code was extended to the case of p-^{11}B, we found that the initial adiabatic burning during compression and expansion, required an optimum temperature of about 150 keV. This can be reduced by self heating of an exceedingly high density plasma (10^5 times solid) such that the optimum initial temperature for volume ignition due to reheat and self-absorption of bremsstrahlung is only about 20 keV. The results of Fig. 8 immediately show the conditions required for volume ignition. The time dependence of the plasma temperature during the reaction is shown. As before for DT, we have also illustrated the conditions at a slightly lower initial temperature which did not permit ignition, so that the temperature dropped monotonously with time. With a little higher input energy, the temperature rose to a sharp maximum due to volume ignition. The gain achieved was 21.6 with respect to the input energy. This is indeed a low value and could only be considered in the far future if much improved drivers with high efficiency are developed. At the same time, the methods used to obtain uniform irradiation must be drastically improved if there is to be any hope of achieving the enormous compressed density implied here, approaching 100,000 times solid state density. The third case of Fig. 8 shows that a gain of 24.2 is achieved. The conditions for the future would then be that a laser of at least 30% efficiency is needed and that ways must be found to achieve the ultra high compression of 10^5 times solid state. The conversion of the output energy into electricity might be performed with low thermal pollution using high efficiency Mosely-type direct conversion.

The principle is known for laser amplifiers that an efficiency of 80% or more can be expected, while providing high gains in the visible or far UV range from cluster injection FEL [44]. The kinetic energy of condensed speckles with Mach 300 velocity is converted into optical energy.

The p-^{11}B parameters discussed here are indeed for quite futuristic designs. Still, these results should be of interest to identify directions and goals for future research.

THE MISERY OF LASER DIRECT DRIVE AND THE MADDEVER-LUTHER-DAVIES EXPERIMENT

In the first stages of laser fusion [24][25], direct driving of the pellet by the laser radiation was considered [30]. When it became evident that the interaction mechanism of laser radiation is complicated due to the unexpected nonlinear dynamics [45][46] and the discovery of a complex pulsation of reflected light with about 20 psec duration [47], the conclusion was drawn [15] that one should avoid all these difficulties by converting the laser radiation into x-rays and let these drive the capsule (called indirect drive).

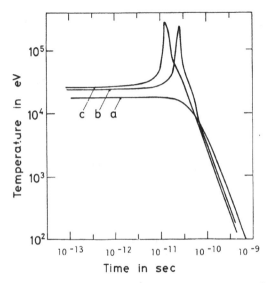

Fig. 8. Time dependence of the temperature of H-B pellets of 10^5 times solid state density with intake of the following energies E_0, a: 1 GJ, b: 1.34 GJ; c: 1.5 GJ, resulting in gains 0.07, 24.2; 21.9 respectively with a fuel burn of 0.18%; 81.5; 82.6% and initial temperatures of 17.9; 24; and 26.9 keV respectively.

Nevertheless an understanding the difficulties of the direct laser-plasma interactions was sought with the aim to overcome them. First it was assumed that the generation of parametric instabilities was the basic problem. Based on Landau's work on exponentially growing parametric oscillations [48], the parametric instabilities at interaction of electromagnetic waves with plasmas, were first studied by Orajevski and Sagdeev [49] and the subsequent papers were summarized by F.F. Chen [50] using an analysis based on the nonlinear force [48]. The essential processes are that the transverse laser waves are either converted into longitudinal Langmuir oscillations of the electrons (SRS: stimulated Raman scattering) or into acoustic ion oscillation and waves (SBS: stimulated Brillouin scattering).

After a few hundred papers had been published on SRS and SBS Drake [51] succeeded in proving experimentally that while SRS certainly existed and was possibly one of the reasons for observing (3/2) harmonics backscattered from the plasma, it is not a dominant reason for the transfer of the laser energy into the plasma; only a few percent of the energy may be exchanged in this way. The same was observed for SBS [52]: only a few percent of the laser energy is exchanged in this way. It has even been observed that these instabilities sometimes contribute much less than one percent [53].

The question then is what is the reason for all the observed complicated interaction properties if these are not the instabilities? One first has to exclude trivial cases as non-uniformities of laser irradiation or "hot spots" in the laser beams and the generation of self-focusing and filamentation in the plasma by nonlinear forces [54], or by thermal self-focusing [55] or relativistic self-focusing [56]. A nontrivial process is the rather neglected density rippling within the whole corona, which occurs a few picoseconds after the laser is turned on and is due to standing wave field components as first observed numerically, (see Fig. 7.7a of Ref. [57]), measured vaguely [47] and confirmed experimentally from a most complex analysis [58]. This proved that within a few picoseconds after switching on, the reflection of light moves from the critical density (as expected) to a very low density in the outer corona with nearly total reflection, changing to and fro. The motivation for these experiments came from the 4-Angstrom periodic spectra for backscattered light (fundamental and 2ω) seen as a pulsating acceleration and as a simple Doppler effect (Maddever-Luther-Davies experiment). The pulsation of reflectivity between the critical density and the corona periphery can be explained by numerical calculations as will be shown below in more detail [59]. Further it can be shown how and why the earlier used Random Phase Plate, developed by Kato and Mima [60], and other methods (broad band irradiation [61] or induced spatial incoherence (ISI) [62]) produce a very smooth laser-plasma interaction. This smoothing [53] further reduces the SRS by orders of magnitude.

We report here on improved numerical computations performed with a one-dimensional macroscopic hydrodynamic two-fluid code. It is shown that the observed difficulties of direct drive are caused by plasma density ripples and are produced by the nonlinear force of the laser standing waves within few pico seconds after switching on, cutting off laser interaction at the outermost periphery of the corona. The density ripple in the corona then relaxes hydrodynamically, and a few picoseconds later again permits optical propagation to the critical density until partially standing waves again produce density ripples and optical cut-off in the outer periphery of the corona. This results in the periodicity of the laser light reflectivity with a duration of 6 to 10 ps, which is in fair agreement with the observations of many experiments [47] [58] [63]. This pulsation can then be suppressed experimentally by using Kato's random phase plate [60], broad band lasers [61] or ISI [62]. This can also be explained by our numerical results where the local phase shifts of the standing waves in each laser beamlet are laterally washing out the density ripple.

We report here how the pulsation process can be examined numerically in detail by a very general hydrodynamic code, initially developed for studying the high electric fields in a plasma with laser interaction [64,65] and the generation of longitudinal oscillations with more than 10^8 V/cm amplitude in the cavitons with a sequence of double layers [64 to 69]. These computations have now been extended by special numerical techniques to 20 picoseconds despite the necessarily very small time steps of less than 10^{-16} s required to simulate the plasma oscillations.

The numerical results agree with the observed 10 ps pulsations [4] [58] [63] in such a way that the initial optical penetration of the laser light to the critical density ablates the plasma to velocities in the corona above 10^7cm s^{-1} within 2 ps. The partly standing wave field of the laser moves the plasma within these 2 ps to the nodes of the standing waves and produces a density ripple. The generated Laue-Bragg grating then causes very high reflection at the outermost periphery of the corona until the ripple relaxes hydrodynamically. A detailed computation gives a periodicity of about 7 to 10 ps as measured.

It should be noted that the pulsating reflection between the critical density and the outermost corona was numerically observed in 1974 [Fig. 7.7a of Ref. 57]. After the light first penetrates to the critical density (including swelling and nonlinear-force produced cavilon generation) the ripple produced by the standing wave field terminates the interaction at 2 ps and the light is nearly totally reflected at the outermost corona. Experiments by M. Lubin et al have confirmed this by showing that reflectivity changes from very low to high values during this time and with a pulsation period of about 10 ps [47]. The computation however could not simulate what happened at later times.

In order to study the generation of double layers and their associated dynamic electric field in laser-produced plasma without the formerly used space charge neutrality, a general physical model describing two macroscopic electron and ion fluids coupled by Poisson's equation and interacting with an external electromagnetic field was developed [64-69]. The problem is treated in one-dimension only and uses the x direction. This two-fluid model gives rise to seven basic equations (9) - (15) as follows:

Continuity equations for electrons (index e) and ions (index i)

$$\frac{\partial n_e m_e}{\partial t} + \frac{\partial m_e n_e v_e}{\partial x} = 0 \tag{9}$$

$$\frac{\partial n_i m_i}{\partial t} + \frac{\partial m_i n_i v_i}{\partial x} = 0 \tag{10}$$

Conservation of momentum

$$\frac{\partial(n_e m_e v_e)}{\partial t} = -\frac{\partial(n_e m_e v_e^2)}{\partial x} - \frac{\partial P_e}{\partial x} - n_e e E - n_e m_e \nu (v_e \text{-} v_i) + f_{NL} \tag{11}$$

$$\frac{\partial(n_i m_i v_i)}{\partial t} = -\frac{\partial(n_i m_i v_i^2)}{\partial x} - \frac{\partial P_i}{\partial x} - n_i Z e E + n_e m_e \nu (v_e \text{-} v_i) + \frac{m_e}{m_i} f_{NL} \tag{12}$$

Conservation of energy

$$\frac{\partial(m_e n_e \epsilon_e)}{\partial t} = -\frac{\partial(m_e n_e \epsilon_e v_e)}{\partial x} - P_e \frac{\partial v_e}{\partial x} - \frac{3}{2} \tau^{-1} k n_e (T_e \text{-} T_i) + \frac{\partial}{\partial x}(\kappa_e \frac{\partial T_e}{\partial x}) + W_L \tag{13}$$

$$\frac{\partial(m_i n_i \epsilon_i)}{\partial t} = -\frac{\partial(m_i n_i \epsilon_i v_i)}{\partial x} - P_i \frac{\partial v_i}{\partial x} + \frac{3}{2} \tau^{-1} k n_e (T_e \text{-} T_i) + \frac{\partial}{\partial x}(\kappa_i \frac{\partial T_i}{\partial x}) \tag{14}$$

and   Poisson's equation

$$\frac{\partial}{\partial t} E = 4\pi e \ (n_e v_e - Z n_i v_i) \tag{15}$$

In these equations (8) to (15), the wave equation for the external electric laser field E_L of radian frequency ω is solved using the temporally and spatially variable refractive index ñ, which depends on the plasma dynamics which determines the values of the electron density n_e and electron temperature T_e in the plasma frequency ω_p and collision frequency v

$$\nabla^2 E_L + ñ^2 (\omega/c)^2 \ E_L = 0 \tag{16}$$

$$ñ = 1 - \omega_p^2 / [\omega^2 (1 + iv/\omega)] \tag{17}$$

here v represents the Coulomb collision frequency [70] including the nonlinear dependence on the laser intensity I [27]

$$v = \left(\frac{m\pi}{2}\right) \ \frac{\omega_p^2 e^2 Z \ln\Lambda}{16(kT_e + I/n_c c)} \tag{18}$$

with the Coulomb logarithm

$$\ln\Lambda = \ln \ \{ \ (\tfrac{3}{2} kT)^{3/2} \ / \ [(\pi n_e)^{1/2} Z e^3] \} \tag{19}$$

and n_c is the critical density for electrons. The optical frequency of the laser field is ω and ω_p is the plasma frequency given by

$$\omega_p^2 = 4\pi e^2 n_e/m_e \tag{20}$$

When the laser irradiates the plasma, the electron heating power density W_L produced by the laser optical absorption in Eq. (13) is given by

$$W_L = (I \ 2\omega/c) \text{Im}(ñ) \tag{21}$$

while the time resolved nonlinear force in Eqs. (11) can be derived as [46][69] as

$$f_{NL} = \frac{\omega_p^2}{8\pi\omega^2} \sin^2(\omega t) \ \frac{\partial E_{Lx}^2}{\partial x} \tag{22}$$

This system determines the seven quantities n_e, v_e, T_e (the electron number density, velocity and temperature), n_i, v_i and T_i the same quantities for ions, and the internal longitudinal electric field E. The expressions ε_e, ε_i, P_e and P_i are internal energy densities and scalarly assumed pressures for electrons and ions (assumed scalar), viz.,

$$\varepsilon_e = 3kT_e/2m_e \tag{23}$$

$$\varepsilon_i = 3kT_i/2m_i \tag{24}$$

$$P_e = n_e kT_e \tag{25}$$

$$P_i = n_i kT_i \tag{26}$$

The following assumptions in the equations (13) to (20) have to be noted:

1)   the thermal conductivities of the electrons κ_e (Eq. 13) and ions κ_i (Eq. 14) are used in the classical form of Spitzer [70]. This is contrary to the knowledge that the thermal conductivity of the plasma may be decreased by a factor 1/f of 50 to 100, as seen from fitting the calculations of the thermal flow from the hot laser irradiated corona to the plasma interior in fusion pellets to experimental determinations. This has been discussed in numerous papers [69] and was simply explained by a double layer (DL) mechanism [67] showing that the electrons are mainly reflected between the hot and cold plasma and the ions only contribute to the thermal conductivity. Indeed we could have concluded similar mechanisms of spread double layers [69] in the following calculations of the density ripple and would have had to include the modified electronic thermal conductivity.

The complications of the thermal conduction, however, are not of importance in the short time steps. However, earlier numerical experiments with the numerical code [66] however have shown that, during the fast period of few picoseconds, a change of the thermal conductivity from the Spitzer value by a factor 70 does not greatly change the results within the corona. The processes between the hot corona and the cold plasma interior are not discussed here since this is a range of depth of plasma which goes far beyond the values of the following initial density profiles. A discussion of possible very fine variations would then be quite academic since they would be beyond experimental proof.

2)   In Eq. (8), the electron density and the ion density for the refractive index $\tilde{n}$ were assumed for each time step of the computations to solve the differential equation which depends only on space. This neglects that the time dependence of $\tilde{n}$ apart from its spatial dependence does not exactly permit the separation of the spatial and temporal wave equation in order to arrive at Eq. (8). Thus the approximation implies the approximation the following first order condition

$$(\partial\tilde{n}/\partial x) = (1/c)\,(\partial\tilde{n}/\partial t) \tag{27}$$

This is fulfilled even in extreme cases as seen from the following numerical results which justify the use of Eq. (16). Indeed this includes the very important slow time dependence of the refractive index. The stepwise solution of Eq. (16) uses the correct boundary conditions towards high plasma densities (decaying field) based on a specially economic matrix procedure as explained in [65][68].

3)   In the following computations the density and other plasma parameters are not fixed on the boundaries, but are allowed to vary. So that changes brought about by the laser may be initiated within the plasma and then develop in the direction towards the oncoming

incident laser light. Since we do not use herein an initially zero density with our code [66][68] for the first simulation step, our result of 1974 (see Fig. 7.7a of Ref. 57 or Fig. 10.10a of Ref. 69] with a much more simplified one fluid computation with a density beginning at the value zero at laser irradiation 0 cannot be reproduced directly. In the latter case it is possible to calculate how the laser intensity varies with depth into the plasma, initially ranging from the incident value (without any ripple) at the edge to a peak at the critical density where reflection occurs. 2 ps after the laser is switched on, the density ripple is generated and a rapid decay of the laser intensity is seen in the outer corona due to Laue-Bragg reflection. This effect has now also been observed experimentally [58]. This will be directly seen also in the following computation, which follow up the pulsations to 20 psec while the earlier computations [57] became numerically unstable after about 2 psec. It is to be mentioned that the density ripple and increased Laue-Bragg reflection are purely hydromechanical and electrodynamical mechanisms, and the parametric instability of stimulated Brillouin scattering (SBS) characterized by a frequency shift and generation of acoustic waves has not been included. While there is no doubt that SBS exists it was shown experimentally that it has little effect. Only less than 2% of the total energy transfer mechanism is due to SBS in the experiments [52] such that its neglect in the following hydrodynamic treatment is justified.

4)   The equipartition time τ for the thermal exchange between the electron and ion fluids is used in the standard formulation [65] [68] where the electron ion collision frequency ν is based on the classical value [70][69] and allowance for the anomalous collision frequency due to the quantum deviation [71] has been ignored at this stage.

The solutions of the wave equation (16) are expressed in the following form

$$E_L = E_{Lx}(x,t) \sin \omega t$$

where the amplitude is a complex value of a non-elementary function. This result takes into account the presence of a propagating and a partially standing wave. The time dependences of plasma density and temperature are slow compared to the wave period, satisfying the approximate condition of equation 27. After achieving the numerical solution with the boundary condition of outgoing waves only at high (supercritical) density [65][68] separation into the real part and imaginary part permits the identification of the incident laser intensity I by the relation

$$E^2_{Lx} = (8\pi/c)I \tag{28}$$

which is used in Eq. (21).

Eqs.(9)-(28) describe the successive motion of the plasma under the given initial and boundary conditions. An Eulerian difference scheme has been developed by a basically new numerical technique [64][68]. The Two-Step Lax-Wendroff Method was chosen to solve the continuity and momentum equations. The energy and Poisson's equations were solved using an implicit scheme. An explicit difference equation was used to compute the wave equation. In order to resolve the oscillations of the electric field E evolving from Poisson's equation (16) and the laser standing wave caused by the nonlinear force, one needs to adopt a time step and a space step as follows

$$\Delta t = 0.1/\omega_{pm} \tag{29}$$

$$\Delta x = 0.05\lambda \tag{30}$$

where ω_{pm} is the maximum plasma frequency and λ the vacuum wavelength of the laser field. With such time and space steps, the spurious growth of short wavelength oscillations from nonlinear interaction will emerge because the laser pulse length is usually longer than a few picoseconds. By using the smoothing method [72] a satisfactory hydrodynamic code has been obtained. Although this code has been used for many computations to explain several experimentally observed phenomena [64-66], these computations could only follow the plasma dynamics to 1~2 picoseconds for high laser intensities. To overcome this short time limitation in our computations, some numerical smoothing at the plasma boundary was introduced to filter spurious oscillations of the electron velocity, temperature and the longitudinal (electrostatic) field E. Only in this way can the computation follow the complex dynamic process up to 20 ps and more.

NUMERICAL EXAMPLE FOR DENSITY RIPPLES

In our simulation, the initial conditions at t=0 for the computation model use a fully ionized hydrogen plasma of electron temperatures T_e and T_i of 15 eV, zero velocities v_e and v_i for electrons and ions, equal electron and ion densities $n_e = n_i$ with a linear distribution of n_e and n_i along the depth x for a plasma layer of 15μm thickness and a maximum cutoff density of 10^{21} cm^{-3} at x = 15 μm and its half value at x=0. An overdense plasma region is assumed for x > 15 μm, on the side opposite to the incident of neodymium glass laser irradiation. For the first 0.5 ps no laser acts on the plasma so that the longitudinal field oscillations of the plasma are dampened by collisions [64] and the smooth hydro-expansion remains. After this relaxation process, at t = 0.5 ps a Nd glass laser with a characteristic wavelength of 1.06μm and a constant intensity of 10^{14} W/cm^2 is switched on at the vacuum-plasma boundary (x=0) from the left hand side (from -x). Computer experiments show that the time step and the space step should be chosen as 0.5 fsec and 0.1 μm, respectively, in this numerical example. All results are shown in Figs. 9-17.

The generation and relaxation of density ripples for electrons and ions can be seen in Figs. 9 and 10. As time develops, density ripples emerge on the initially smooth density profile. At t = 7.0 ps the ripples reach at a maximum amplitude of 10% above the

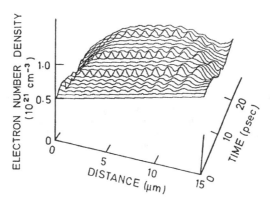

Fig. 9. At time t = 0, a fully ionized deuterium plasma of electrrons and ion temperatures of 15 eV, velocity zero and an electron density profile linearly increasing from 5 x 10^{20} cm^{-3} at x = 0 to 1.05 x 10^{21} cm^{-3} at a depth of 15 μm, is followed up by a genuine two fluid model [7][8] when 10^{14} W/cm^2 neodymium glass laser radiation is incident from -x. The Figure shows the temporal evolution of the electron density.

background densities. The distance between the ripples is in the range of 1.0 to 1.3 μm, close to the values calculated by the following formula

$$\Delta x = 1.5 \, [L(\lambda_o/2\pi)^2]^{1/3} \qquad\qquad (31)$$

where L is a plasma characteristic length. Formula (31) can be derived from Airy's Function determining the laser standing wave in plasmas [61]. This implies the ripples of plasma density are generated by the laser standing waves (see Fig.11) through the nonlinear force. These density ripples will relax with time and disappear at t = 9.0 ps. For later times of 12 ps and 18 ps, the second and the third maxima of density ripples appear, resulting in a pulsation period of 6 ps. It should be mentioned that the increase of the ripple amplitude grows nonlinearly with time as seen from the plots of Figs. 9 and 10.

The computed results for electric laser energy density $E^2/8\pi$ (see Fig. 11) shows the generation of the laser light standing wave in the plasma because the laser light is calculated as a complete solution of the wave equation (Eq. 16) including local reflection. Fig. 11 describes the penetration of the laser light into the plasma, directly determining the value of the nonlinear force term as well as the laser energy optically absorbed by the plasma. The laser standing wave also shows pulsation as do the dynamic behaviours of the ion and electron densities. The appearance of the strong standing waves with a maximum energy density of 3×10^{11} ergs/cm^3 and a minimum energy density of 10^9 ergs/cm^3 will produce a plasma ripple (see Figs. 9 and 10) by the nonlinear force [46], decreasing the laser light penetration in the plasma corona and resulting in much reflection at very low density at the ripple periodicity (Langmuir grating). When this reduction in wave propagation occurs, the density ripples will relax hydrodynamically and disappear so that more light then again into the plasma. The periodic of light penetration into plasmas will result in the pulsation of the laser light reflectivity. According to Fig. 11, a curve of the relative reflectivity of the laser light can be deduced since laser intensity I is proportional to the energy density. Obviously,

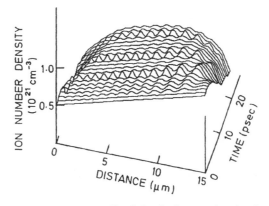

Fig. 10. The same as Fig. 9 for the ion number density.

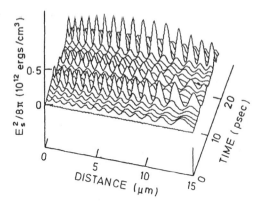

Fig. 11. The same as Fig. 9 for laser field energy density in plasmas where the units were based on a consequent use of cgs.

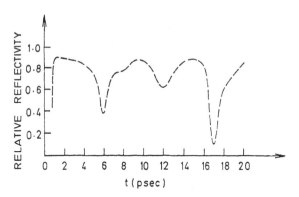

Fig. 12. The relative reflectivity corresponding to Fig. 11.

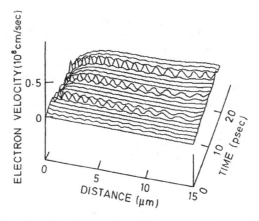

Fig. 13. The same as Fig. 9 for the electron velocity.

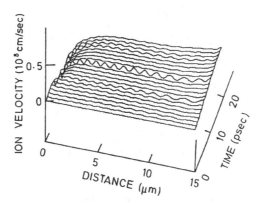

Fig. 14. The same as Fig. 9 for the ion velocity.

three minima of the reflectivity (see Fig. 12) appear at t = 6, 12 and 17 ps corresponding to the times when laser intensities become larger. Therefore, the computed result is in agreement with the observations of periodicity of the laser reflectivity in the range of 10 ps [5][73].

Similar to the pulsation of the density ripples, the developments in space and time of the ion and electron velocities (see Figs. 13 and 14) also display pulsations and ripples similar to the densities. When the ion velocity ripples emerge, their typical values are near 10^7 cm/s, which agrees with that estimated from the 4 Angstrom modulated spectra of the second harmonic emission in the following way: the diagram of the ion velocity (Fig. 14) near x = 0, shows an increase of the average velocity (averaged over the ripple) from 0 to 1 to 2 x 10^7 cm/s within 5 ps. Then it is nearly constant until 9 ps, raises then again to about 3 to 5 x 10^7 cm/s within 5 ps, is then constant etc.

This is exactly what Maddever [73] discovered as the explanation for the 4Å modulated spectra from the 10-15 ps pulsation of reflectivity: the plasma is accelerated to the velocities of 1 to 2 x 10^7 cm/s during the first pulsation corresponding to a Doppler shift of 4Å. Then the acceleration stops (as we understand due to ripples and reflection right on the edge of the corona as explicitly observed [58]. After the relaxation of the ripples, another block acceleration of the whole corona occurs with an <u>additional</u> 2 x 10^7cm/s velocity, resulting in the 8 Angstrom Doppler shift etc. Therefore, the 4Å periodic spectra should then be due to the pulsating added Doppler shift without needing parametric or other explanations for the broadening of the spectra. Our hydrodynamic model would thus support this conclusion by Maddever [73] as the pulsatingly increase of the corona velocity can be seen from Fig. 6 and can be understood by the alternation between ripple generation and relaxation.

In our computation, the "electrostatic" longitudinal field E is determined by the velocities and densities of both electrons and ions (see Eq. 15). The existence of pulsation ripples of density will cause an important effect on the distribution of the electrostatic field in the plasma. It has been found (see Fig. 15) that the electrostatic field demonstrates a clear

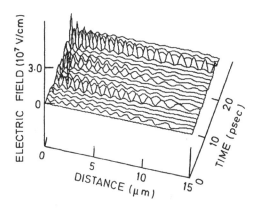

Fig. 15. The same as Fig. 9 for the electrostatic field in cgs units.

ripple behaviour with amplitudes of ± 3 x 10^7 V/cm and with a 6 ps period of build up and decay of pulsations. It is important to note that the amplitudes of these laser driven longitudinal (electrostatic) Langmuir oscillations are about 10 times less than the transverse laser electric field amplitudes of 10^{14} W/cm^2. The energy density of the Langmuir waves is only a few percent of the laser field energy. This result was also seen in preceding computations [64] to [69], and agrees with the measurements of Drake [51], showing that simulated Raman scattering (SRS), given by transfer of optical laser energy into Langmuir waves, does exist ,but is with respect to the energy transfer, being only a few percent, as seen in Fig. 15.

In Figs. 16 and 17, the spatial distributions of the electron and ion temperatures with their temporal evolutions are shown. During the first few picoseconds after the application of the laser field, the electron temperature shows a non-uniform distribution because the electrons are heated by optical absorption of the laser light. After a few picoseconds, the temperature reaches 300 eV and assumes a uniform distribution due to the high conductivity of the electrons at that temperature. Although ions are heated by thermal conductivity and collisions with electrons, their temperature doesn't show much change since the relaxation

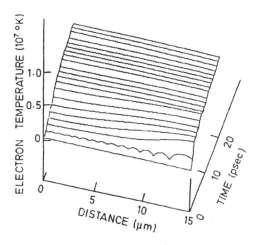

Fig. 16. The same as Fig. 9 for the electron temperature.

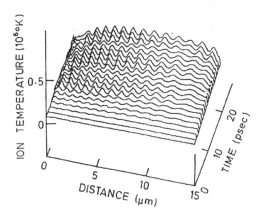

Fig. 17. The same as Fig. 9 for the ion temperature.

process of ions is very slow. Some small ripples in the temperature distribution are produced by adiabatic heating through compression by the nonlinear force.

The above numerical results clearly show that the plasma pulsations and ripples are produced by the nonlinear force [46][69] when the partly standing waves emerge due to the local reflection of the plasma. The large amplitude laser field values correspond to times when the reflectivity is low (less than few %) and the light penetrates the corona up to the critical density. This causes the density ripple within a few picoseconds and then light is cut off at the outermost corona (high reflectivity) because of the Bragg-Laue reflection by the ripple. Within 7 ps, the ripple relaxes hydrodynamically, the laser light can propagate again (low reflection), generates ripples again, cutting the light off again and so on.

The 10 to 15 ps pulsation of the light near 700 nm emitted by SRS from a laser-produced plasma [63] may be of a similar nature to the 10 ps pulsations, though other explanations have been discussed before. What is of special importance is that the pulsation was suppressed in this case and a very smooth, temporally constant SRS signal registered when Kato's [60] random phase plate RPP was used. It is the established opinion that the random phase plate suppresses filamentation and hot spots and the pulsation in the experiment [63] arises because the filamentation is suppressed by the RPP.

Based on our computer result we propose the following alternative explanation: the random phase plate produces many micro-beams in the laser beam, each shifted phase. If the one dimensional plane wave mechanism, discussed above, produces the standing waves and the density ripple, one may assume that this happens in each phase plate filament such that the nodes of the standing wave are spatially shifted on account of their phase difference and the ripple washes out immediately by radial plasma relaxation in the laser beam.

It may be possible that filamentation and hot spots are involved too, but if the phase differences act to suppress pulsating ripple generation, this induced spatial incoherence (ISI) [62] or broadband laser irradiation [61] produces a smooth laser-plasma interaction which is so important for direct drive of fusion pellets. We draw the conclusion that further experiments should check whether RPP, ISI etc. work by suppression of filamentation or by the phase controlled washing out of the pulsation of the ripple mechanism.

If the smooth and pulsation-free direct drive laser-plasma interaction can be realized experimentally [60 to 62] based on the above understanding of suppression of the density ripple, then the x-ray indirect drive [15] may no longer be needed as a basic process. Nevertheless knowledge of the x-ray mechanisms is of eminent value, as seen from the separation of the cut-off density from the ablation due to x-ray transport, e.g., measured by Gupta et al [73]. We emphasise again that direct drive is three times more efficient than indirect drive.

CONCLUSIONS

Laser fusion for large scale energy production offers a complete physical solution for a fusion power plant 10 to 15 years for energy production at a similar cost to light water reactors. Magnetic confinement using tokamaks is at least ten times more expensive as are solar energy methods. Furthermore there is a principle, due to Niu [3], that solar energy can never be as economic as light water reactors.

The physical solution for laser fusion is comparable to the situation in 1960 for manned flights to the moon: the physics is solved and the development of technology depends on a decision about the political urgency for a crash program. The climatic changes arising from the greenhouse effect, due to further burning of coal and oil for energy production, may trigger the need for a far-sighted decision.

The present day physical solution of laser fusion is based on the advances of the technology in the USA, including the Centurion-Halite underground nuclear experiments to confirm the high gain ignition of fusion capsules and includes projects such as the Athena-laser for producing 10 MJ laser pulses of several nsec duration and the CASCADE reactor. The physics of the gain of the laser driven capsules to produce 1000 MJ energy per irradiation is based on spark ignition and on indirect drive.

While these developments should in no way be delayed and should be taken as the first possible solution, the simultaneous development of laser techniques and studies of physical processes do promise further improvements so that a concerted effort may achieve in production of energy at one fifth of the cost of light water reactors in 50 to 80 years. This will herald the advent of a golden age due to safe and inexhaustible nuclear energy.

Looking further into the future (about 100 years from now) a reactor with a radioactivity level less than for burning coal using hydrogen and boron-11 offers clean low cost production of energy. This can be achieved if the technology of laser fusion with the present fuel of hydrogen isotopes, deuterium and tritium, were merged in a broadly ranging research program.

As an example of how the next generation of laser fusion reactors might be based on further improved physics leading to reduced costs, this paper has presented the results of volume ignition of the laser irradiated nuclear fuel capsules. The high efficiency arises from the very low entropy generation and by a very well understood volume ignition process, using the self heating of the reaction products in the plasma like in a Diesel engine. Fuel exhaustion of 80% and more can be expected. The experiments in Japan are pioneering the experimental verification of volume compression. We have presented here our recent numerical results on volume ignition including a view of the futuristic case of hydrogen boron-11.

The history of the direct drive of capsules by lasers has shown a variety of difficulties. The present best solution is Nuckolls' indirect drive where the laser light is converted into x-rays which then drive the capsule. After it was shown by R.P. Drake and C. Labaune et al that the parametric instabilities SRS and SBS are not the cause of the pulsations which appear in the laser-plasma interaction, the experiments by Maddever and Luther-Davies turned back to observations by Lubin and Soures 1975 which showed that other mechanisms are the reason for the difficulties. Combining results from earlier research, we have shown here how the complications are due to a pulsating density ripple and hydrodynamic relaxation mechanism. We have also demonstrated how this can be suppressed with Kato's random phase plate or Obenschain's induced spatial incoherence or the Shanghai broad band laser irradiation. In this way a smooth direct drive scheme becomes possible.

REFERENCES

[1]   Pres. George H.W. Bush, Press Conference, White House, 7 Nov. 1989.
[2]   Queen Elizabeth II, Christmas Message to Children 1989.
[3]   K. Niu, Fusion Energy, Cambridge University Press 1989.
[4]   C.-J. Winter, Physik. Blatter $\underline{45}$, 264 (1989).
[5]   James E. Quinn et al, General Electric, San Jose, Cf., Prospectus for Fission Reaction; W.J. Cook, U.S. News and World Rept., May 29, 1989, p.52.
[6]   H. Hora, G.H. Miley, and W. Mirbach, US Pat. Appl. No. 07/298159; German Patent Discl. No. 3, 733, 750 of 6 Oct., 1987.
[7]   A.H. Glaser, J. Hammel, H.R. Lewis, I.R. Lindermuth, G.H. McCall, R.A. Nebel, D. Scudder, J. Schlachter, R. Loveberg, P. Rosseau and P. Sherby, 12th Int. Conf. on Plasma Physics and Contr. Nucl. Fusion Res. Nice Oct. 1988, Synopses p.124; V.V. Yankov, Collaps of Z-pinch Necks for Inertial Fusion, Kurchatoo Inst. Atomic Energy, Moscow, Preprint IAE-490917 (1989).
[8]   J. Rand McNally Jr. Fusion Technology $\underline{7}$, 331 (1985).; H. Hora, L. Cicchitelli, G.H. Miley, M. Kagheb, A. Scharmann, and W. Scheid, Nuovo Cimento, $\underline{12D}$, 393 (1990).
[9]   P.K. Iyengar, Fifth Int. Conference on Emerging Nuclear Energy Systems, U. von Möllendorff ed. (World Scientific, Singapore 1990) p.
[10] M.L.E. Oliphant, P. Harteck, and Lord Rutherford, Proc. Roy. Soc. $\underline{A144}$, 692 (1934).
[11] M.L.E. Oliphant, and Lord Rutherford, Proc. Roy. Soc. $\underline{A141}$, 259 (1933).
[12] T.A. Weaver, Lawrence Livermore Laboratory, Laser Report 9 (No. 12), 1(1973); R.W.B. Best, Nucl. Instr. Meth. $\underline{144}$, 210 (1977).
[13] J.D. Lawson, Proc. Phys. Soc. $\underline{B70}$, 6 (1957).
[14] D. Pfirsch and K.H. Schmitter, Fusion Technology $\underline{15}$, 1471 (1989).
[15] J.H. Nuckolls, Phys. Today $\underline{35}$ (No. 9) 24 (1982).
[16] S.C. Kahalas, In Laser Interaction with Matter, G. Velarde, E. Minguez, and Y.M. Perlado eds. (World Scientific, Singapore, 1989) p.79.

[17] W.J. Broad, New York Times 137, No. 47, 451, March 21 (1988).

[18] Admiral J. Watkins, LMF Laboratory Microfusion Capability, Phase I Summary, Rept. DOE/DP-0069, DP-234 (April 1989).

[19] E. Storm, J.D. Lindl, E.M. Campbell, T.P. Bernat, L.W. Coleman, J.L. Emmett, W.J. Hogan, Y.T. Horst, W.F. Krupke, and W.H. Lowdervilke "Progress in Laboratory High Gain ICF: Prospects for the Future" (LNL, Livermore) Aug. 1988, Rept. No. 47312.

[20] General Atomic San Diego, Technologies Dept (GA-A17842 (Oct. 1985).

[21] R. Bock, Intern. Res. Facilities, Europ. Phys. Soc. 4th Seminar, Zagreb, March 1989 (EOS, Geneva) p.211.

[22] C. Rubbia, Nucl. Instr. Methods A278, 253 (1989).

[23] P. Vandevender et al, Laser and Particle Beams 3, 93 (1985).

[24] N.G. Basov and O.N. Krokhin, Sov. Phys. JETP 19, 123 (1964); A. Kastler, Compt. Rend. Ac. Sc. Paris 258, 489 (1964); J.M. Dawson, Phys. Fluids 7, 981 (1964).

[25] H. Hora, Inst. Plasma Phys. Garching, Rept. 6/23 (July 1964).

[26] R.F. Schmalz, Phys. Fl. 29, 1369 (1986).

[27] H. Hora, Physics of Laser Driven Plasmas (Wiley, New York 1981).

[28] H. Hora, and D. Pfirsch, 6th Int. Quantum Electr. Conf., Kyoto 1970, Conf. Digest, p.10.

[29] R.E. Kidder, Nucl. Fusion 14, 797 (1974).

[30] J.H. Nuckolls, Laser Interaction and Related Plasma Phenomena, H. Schwarz and H. Hora eds. (Plenum, New York, 1974) Vol. 63B, p.399.

[31] L. Cicchitelli et al, Laser and Particle Beams 6, 163 (1988).

[32] H. Hora and P.S. Ray, Z. Naturforsch. 33A, 890 (1978).

[33] H. Hora, Z. Naturforsch. 42A, 1239 (1987).

[34] G. Kasotakis et al, Laser and Particle Beams 7, 511 (1989).

[35] S. Atzeni, A. Caruso and V.A. Pais, Laser and Particle Beams 4, 393 (1986).

[36] S. Glasstone, and R.H. Loveberg, Controlled thermonuclear Reactions (NRC, New York (1960).

[37] G. Kasotakis et al, Nucl. Instr. Meth. A278, 110 (1989).

[38] C. Yamanaka and S. Nakai, Nature 319, 757 (1986); C. Yamanaka, S. Nakai et al, Phys. Rev. Lett. 56, 1575 (1986).

[39] G. Velarde, J.M. Aragones, J.A. Gago, L. Guanez et al, Laser and Particle Beams, 4, 349 (1986); M. Andre, D. Babonneau, E. Berthier, J.L. Bocher, J.C. Bourgade, M. Busquet, P. Combis, A. Soudeville, J. Coutaset, R. Dautray, M. Decroisette et al, Laser Interaction and Related Plasma Phenomena, H. Hora and G.H. Miley eds. (Plenum, New York, 1988) Vol. 8, p.503.

[40] K. Mima, H. Takabe, and S. Nakai, Laser and Particle Beams 7, 487 (1989); S. Nakai, Laser and Particle Beams 7, 475 (1989); see also: M.M. Basko, On the Spark and Volume Ignition of DT and D_2 Targets, Inst. Theor. and Experim. Physics, Moscow, Preprint 16-90 (1990).

[41] R. Castillo et al, Laser Interaction and Related Plasma Phenomena, H. Schwarz et al eds. (Plenum New York, 1981) Vol. 5, p.417; G.H. Miley, ibid, p.313.

[42] H. Hora et al, Plasma Physics and Contr. Nucl. Fusion (IAEA, Vienna, 1979) Vol. 3, p.237.

[43] S. Nakai et al, Laser Interaction and Related Plasma Phenomena, H. Hora and G.H. Miley eds. (Plenum, New York, 1990) Vol. 9, p.

[44] H. Hora, J.-C. Wang, P.J. Clark, and R.J. Stening, Laser and Particle Beams 4, 83 (1986); E.W. Becker, Laser and Particle Beams, 7, 743 (1989).

[45] A.G. Engelhardt, T.V. George, H. Hora and J.L. Pack, Phys. Fluids 13, 212 (1970).

[46] H. Hora, Phys. Fluids 12, 182 (1969).

[47] S. Jaeckel, B. Perry, and M. Lubin, Phys. Rev. Lett. 37, 95 (1976).

[48] L.D. Landau, J. Phys. USSR, 10, 25 (1946).

[49] V.N. Oraevski and R.Z. Sagdeev, Sov. Phys.-Tech. Phys. 7, 955 (1963).

[50] F.F. Chen, Laser Interaction and Related Plasma Phenomena, H. Schwarz and H. Hora eds. (Plenum, New York 1974) Vol. 3A p.291.

[51] R.P. Drake, Laser and Particle Beams 6, 235 (1988).

[52] C. Labaune et al, Phys. Rev. A32, 577 (1985).

[53] S.P. Obenschain, C.J. Pawley, A.M. Mostovych, J.A. Stamper, and S.E. Bodner, Phys. Rev. Lett 62, 768 (1989).

[54] H. Hora, Z. Phys. 226, (1969).

[55] M.S. Sodha, A.K. Ghatak, and V.K. Tripathi, Progression Optics, E.V. Wolf ed. (Academic Press, New York 1976) Vol. 13, p.171.

[56] H. Hora, J. Opt. Soc. Am. 65, 882 (1975).

[57] H. Hora, Laser Plasmas and Nuclear Energy (Plenum, New York, 1975).

[58] R.A.M. Maddever, B. Luther-Davies and R. Dragila, Phys. Rev. A (in print).

[59] Gu Min and H. Hora, Chinese Laser Journal 16, 656 (1989); Nucl. Fusion (in print).

[60] Y. Kato, K. Mima et al, Phys. Rev. Lett. 53, 1057 (1984).

[61] Gu Min et al, Opt. Comm. 66, 35 (1988).

[62] R.H. Lehmberg, and S.P. Obenschain, Opt. Comm. 46, 27 (1983); S.P. Obenschain et al, Phys. Rev. Lett 56, 2807 (1986).

[63] A. Giulietti, T. Afshar-Rad, S. Coe, O. Willi, Z.Q. Lin, W. Yu, ECLIM, Madrid 1988, Conference Proceedings, G. Velarde, E. Minguez, and Y.M. Perlado eds. (World Scientific Books, Singapore), p.208.

[64] P. Lalousis and H. Hora, Laser and Particle Beams 1, 283 (1983).

[65] H. Hora, P. Lalousis and S. Eliezer, Phys. Rev. Lett 53, 1650 (1984).

[66] H. Szichman, Phys. Fluids 31, 1702 (1988).

[67] L. Cicchitelli et al, Laser and Particle Beams 2, 467 (1984).

[68] P. Lalousis, Direct Electron and Ion Fluid Computation of High Electrostatic Fields in Dense Inhomogeneous Plasmas with Subsequent Nonlinear Optical and Dynamical Laser Interaction, PhD Thesis, Univ. New South Wales, 1983.

[69] H. Hora, Physics of Laser Produced Plasmas (Wiley, New York 1981).

[70] L. Sptizer, jr., Physics of Fully Ionized Gases (Wiley, New York, 1962).

[71] H. Hora, Opt. Comm. 41, 268 (1982).

[72] R. Shapiro, Rev. Geophys. and Space Phys. 8, 359 (1970).

[73] P.D. Gupta and S.R. Kumbhare, Phys. Rev. A40, 3265 (1989); T.S. Shirsat, H.D. Parab, and H.C. Pant, Laser and Particle Beams 7, 795 (1989).

[73] R.A.M. Maddever, Temporal and Spectral Characteristics of the Fundamental and Second Harmonic Emission from Laser-Produced Plasmas, Ph.D. Thesis, Aust. Nation. Univ. 1988.

LASER INDUCED TRANSITIONS IN MUON CATALYZED FUSION

S. Eliezer and Z. Henis

Plasma Physics Department
Soreq Nuclear Research Center, Yavne 70600, Israel

ABSTRACT

Nuclear fusion reactions of deuterium-tritium can be catalyzed by muons. When an energetic negative muon (μ) enters a deuterium-tritium mixture the following chain of reactions occurs: a) slowing down and atomic capture of μ, b) muon transfer to higher isotopes, c) μ-molecular formation d) nuclear fusion of dt. The muon is either released or captured by the nuclear fusion products (sticking). In analyzing the relevance of muon catalyzed fusion (mcf) for energy production, the problem is to maximize the number X_μ of fusion processes catalyzed by a single μ. X_μ is determined by the lifetime of μ and the frequency of the fusion events described above.

In this work it is suggested that powerful electromagnetic fields can increase the rates of the muon transfer and μ-molecular formation and thus enhance the value of X_μ.

The fusion cycle is inhibited by the slow transfer of muons between the ground states of deuterium and tritium. The fraction of muons reaching the deuterium ground state Q_{1S} can be decreased (and thus increasing the number of muons transferred from $d\mu$ to the tritium ground state) by changing the route of the muon deexitation cascade under laser irradiation. The muon cascade is determined by a competition between radiative transitions, external Auger transitions, quenching of the muon levels and transfer processes. The muon transfer cross section from the 2S level of deuterium is larger by a factor of about 5 than the transfer cross section from the 2P level. Therefore the following two step process is proposed in order to enhance the rate of the muon transfer (1): (a) a laser induced transition between the levels 2P and 2S of the deuterium (b) muon transfer by collisions to the tritium 2S level. The results show that a 0.2 eV laser (the Lamb shift in deuterium) with an intensity of about 10^9 W/cm^2 can decrease the value of Q_{1S} by a factor of three.

The rate of formation of the muonic molecule can also be enhanced by strong electromagnetic fields. We consider a three level system of the dtμ molecule a) a state in the continuum of tμ+d (b) the bound state of the molecule dtμ $(J,V)=(1,1)$ and (c) the bound state $(J,V)=(0,1)$. Under the influence of an external field a Stokes transition from the level (a) to level (b) takes place while the transition from (b) to (c) is occurring through an Auger process. It is shown that for a resonant laser frequency and intensities of 6×10^5 W/cm^2 the Stokes efficiency (defined as the ratio between Stokes induced transitions and spontaneous decay) is about 50. However due to the stringent resonance conditions for practical lasers one uses the off resonance regime yielding an intensity of about 10^9 W/cm^2.

I. INTRODUCTION

The atomic, molecular and nuclear phenomena associated with processes occurring when negative muons (μ) are absorbed in matter have been extensively investigated[1]. The electroneutrality and the small dimensions of muon atoms enable it to penetrate freely through the electron shells of the atoms. When a muonic atom approaches a deuterium or a tritium molecule a μ-molecule can be created. During this process the mesic atom is attached to another hydrogen isotope to form one of the centers of the electronic molecule. At this center of the molecule the dd or dt are kept close together by the negative muon and spontaneous fusion reactions occur in a very short time. The size of the muon atoms are of the order of the muon Bohr radius a_μ given by

$$a_\mu = a_o m_e / m_\mu \approx 2.6 \times 10^{-11} \text{ cm} \qquad (1)$$

where a_o is the Bohr radius for the electron atoms and $m_\mu / m_e \approx 207$ is the ratio of the muon and electron masses. If the lifetime of the muon ($\tau_o \approx 2.2 \times 10^{-6}$ sec) is long compared to the other processes, then many fusion reactions can occur during the lifetime of a muon so that in this case the muons have the role of a catalyzer.

When a negative muon enters a dense deuterium (D) - tritium (T) target it initiates a chain of reactions: (we denote by capital letters D and T the deuterium and tritium atoms while the small letter denotes the nuclei).

(a) Stopping and capture of μ$^-$ by d or t and the cascade to the ground state of the atom,

$$\mu^- + D \rightarrow (d\mu) + e^-$$
$$\qquad\qquad\qquad\qquad\qquad\qquad\qquad (2)$$
$$\mu^- + T \rightarrow (t\mu) + e^-$$

where e$^-$ denotes an electron. In general, the rate λ(sec^{-1}) is given by

$$\lambda = n\sigma v \text{ [sec}^{-1}] \qquad (3)$$

where n is the density of the target, σ is the cross section describing the process under consideration and v is the velocity of the projectile. The rate of the muonic atomic formation is $4 \times 10^{12} \phi$ sec^{-1}, where φ is the density in units of liquid hydrogen density $n_o = 4.25 \times 10^{22}$ cm^{-3}.

(b) μ$^-$ transfer from the deuterium to the tritium

$$(d\mu) + t \rightarrow (t\mu) + d \quad . \qquad (4)$$

This process can occur both from excited states at rates of about 10^{11} Sec^{-1} [2] and from the ground state of dμ at the lower rate 2.8×10^8 sec^{-1} [3].

(c) μ-molecular formation in an excited state, such as dtμ at a rate of about 3×10^8 sec^{-1} [2] and deexcitation to the ground state

(d) the nuclear fusion of dt

$$d + t \rightarrow {}^4He + n + 17.6 \text{ MeV} \qquad (5)$$

The muon is either released or captured by the nuclear fusion products (sticking). A scheme of the muon-catalyzed fusion cycle, presenting the chain of reactions occurring when a negative muon is stopped in a liquid deuterium-tritium mixture of equal proportions is shown in fig. 1.

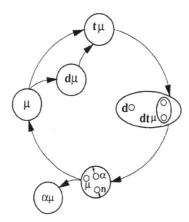

Fig. 1.   Muon-catalyzed fusion cycle in a liquid deuterium-tritium mixture of equal proportions.

The "energy cost" for the creation of a muon is one of the most important practical parameters in analyzing the relevance of the muon catalyzed fusion (mcf) for energy production. The muons are produced during the decay of pions which are created in nucleon-nucleon collisions. An optimistic estimation[4] requires 5 GeV of energy to produce one negative muon. Therefore if one muon catalyzes about 150 dt fusions (the present experimental value) the energy output is about 2.5 GeV (one dt fusion gives 17.6 MeV) per muon, so that no energy gain seems possible from "pure" fusion reactions. This difficulty may be solved in two ways: using mcf in a hybrid fusion - fission reactor[5] or by increasing the number X_μ of dt fusions per one muon. In order to achieve a meaningful energy gain in mcf (taking into account thermal to electricity, recirculated energy and accelerator efficiency) X_μ must exceed few thousands.

For high density mixture of DT and neglecting small effects the value of X_μ^{-1} is given by:

$$X_\mu^{-1} = \frac{\lambda_o}{\lambda_c} + W \qquad (6)$$

where λ_o is the muon-decay rate (0.455×10^5 sec^{-1}), λ_c is the mcf cycling rate (1/time between fusion neutrons) and W is the probability of muon loss per catalysis cycle. The cycling rate can be broken down into component terms according to:

$$\lambda_c^{-1} = \left[\frac{Q_{1S} \, C_d}{\lambda_{dt} \, C_t} + \frac{1}{\lambda_{dt\mu} \, C_d} \right] \phi^{-1} \tag{7}$$

C_d and C_t are atomic concentrations of deuterium and tritium in the mixture, $\lambda_{dt\mu}$ is the rate of the dtµ molecule formation, λ_{dt} is the rate of the muon transfer between the ground states of d and t, Q_{1S} is the probability that the dµ atom will reach the ground state before the muon is transferred to t.

The probability of muon loss W can be approximated by:

$$W = \omega_s + \frac{0.58 \, Q_{1S} \, C_d \, \lambda_{dd\mu} \, C_d \, \omega_d}{\lambda_{dt} \, C_t + \lambda_{dd\mu} \, C_d} + \frac{\lambda_{tt\mu} \, C_t \, w_t}{\lambda_{dt\mu} \, C_d} \tag{8}$$

$\lambda_{dd\mu}$ and $\lambda_{tt\mu}$ are the rates of the ddµ and ttµ molecular formation, ω_s, ω_d and ω_t are the sticking probabilities after fusion in the molecules dtµ, ddµ and ttµ respectively.

From fig. 1 and eq. (6), (7) one can see that there are three crucial bottlenecks for mcf energy production:
(1) the muon transfer process (2) the µ-molecular formation of dtµ and (3) the muon sticking. It was suggested in the past that powerful electromagnetic fields can effect the rates of µ-molecular formation and sticking and thus enhance the values of X_μ [6-11]. This work presents the possibility of increasing the rate of the muon transfer process and µ-molecular formation under laser irradiation.

II. THE DEUTERIUM ATOM CASCADE

The atomic capture and deexcitation cascade of muons in a deuterium – tritium mixture is determined by the competition between several processes [12]. The muon with an energy $E_\mu \sim 3$ keV is captured in a highly excited state [13] ($n \sim 14$) of the atoms dµ and tµ in a time $\sim 0.8 \times 10^{-12} \, \phi^{-1}$ sec. The Bohr radius of these excited atoms is close to the Bohr radius of the deuterium atom. Then the muons reach the muonic atom states with $n = 6-7$ in a time $\sim 0.5 \times 10^{-12} \, \phi^{-1}$ sec and finally, in a time $\sim 1.4 \times 10^{-11} \, \phi^{-1}$ sec they reach the ground state of dµ atoms.

The excitation occurs through the following processes:
(1) the chemical reaction

$$(\mu d)_i + D_2 \rightarrow (\mu d)_f + D + D \tag{9}$$

(2) the external Auger effect

$$(\mu d)_i + D_2 \rightarrow (\mu d)_f + D_2^+ + e^- \tag{10}$$

The Auger transitions have a dipole nature ($\Delta l = l_i - l_f \pm 1$) and transitions with the minimum change in the principal quantum number n permitted for the ionization of the molecule D_2 are predominant. The rate of the Auger transitions is density dependent since they are caused by collisions of excited muon atoms with atoms. The Auger transition rate is maximal at n=7, transitions with $\Delta n=1$ becoming possible from this level. With decreasing n the Auger transition rate falls rapidly.
(3) radiative transitions

$$(d\mu)_i \rightarrow (d\mu)_f + \gamma \qquad (11)$$

at a rate[13] given by:

$$\lambda_{if} = 4/3 \; \alpha \; |D_{if}|^2 \; \omega_{if}^3 \qquad (12)$$

where α is the fine-structure constant, D_{if} amd ω_{if} are the dipole moment and the energy of the transition. The radiative transitions are important for small values of n.

(4) Stark mixing of the different states nl with a given n of the dμ atom due to the electric field of a neighbour deuterium atom. The Stark transitions are important at high densities (or order of the hydrogen liquid density). For example deexcitation of the state 2S of the dμ atom can occur through radiative transitions to the state 1S in a Stark collision; the rate of this process was estimated [14] to be:

$$\lambda_{2S \rightarrow 2P \rightarrow 1S} = 4 \times 10^9 \; \phi \; (sec^{-1}) \qquad (13)$$

(5) μ - transfer to tritium which depend both on the density ϕ and on the tritium concentration c_t.

The relevant level structure and decay scheme of the dμ atom is shown in fig. 2. The strength of the radiative and external Auger transitions and of the muon transfer rates taken from Ref. (15) are shown in units of 10^{11} sec^{-1}.

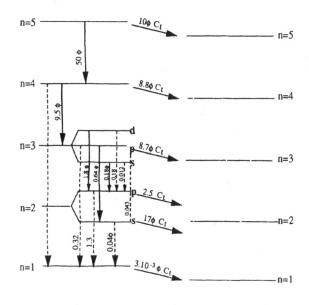

Deuterium   Tritium

Fig. 2. Cascade scheme of muonic deuterium

III. THE GROUND STATE POPULATION OF dμ IN A DEUTERIUM - TRITIUM MIXTURE

The kinetics of the muon cascade in a DT mixture can be described by the following rate equations:

$$\frac{dQ_i}{dt} = -\lambda_i Q_i + \sum_{j>i}^{N} \lambda_{ji} Q_j \qquad (14)$$

where Q_i is the population fraction of the level i of the, dμ atom λ_i is the rate of deexcitation of the level i including muon transfer and λ_{ji} is the rate of the transition j → i. The ground state population Q_{1S} of dμ is of considerable interest. $Q_{1S}C_d$ is the probability for a muon, initially captured by a deutron to reach the dμ ground state. If the muon reaches this 1S state the dtμ fusion cycle is significantly delayed due to the very low transfer rate $(3 \times 10^8 \phi c_t)$ from the ground dμ state to the ground tμ state. The value of Q_{1S} from muon catalyzed fusion experiments[16] is rather large; for example for $\phi = 0.72$, $c_t = 0.5$, $Q_{1S} = 0.72 \pm 0.2$ and for $\phi = 0.12$, $c_t = 0.5$, Q_{1S} 0.74 ± 0.27.

Several theoretical calculations of Q_{1S} have been carried out. The predicted value of Q_{1S} of ref (15) (assuming uninhibition of the muon transfer for the states n≥3 and statistical distribution between the sublevels of the n=3 level) is $Q_{1S} = 0.08$, much smaller than the experimental value.

A suppression in the muon transfer process for the levels n≥3 and a Boltzmann distribution of the sublevels of the n=3 level is assumed in[17]. The rates for the transfer process from excited states have been calculated[15,18] for collisions of dμ atoms with tritium atoms, not molecules. For atom-molecule collisions the transfer process can be strongly influenced by a substantial change in the final state density. For the levels n=1 and n=2 the muon transfer can be easily accompanied by dissociation of the target molecule, whereas those for n=3 and higher levels cannot. Thus the muon transfer rates of[15] for the levels n=1, n=2 probably apply but for states with n≥3 a molecular suppression may be active. The populations of the levels 3S, 3P and 3D are given by the Bolzman distribution:

$$Q_{3D} = 5/9 \; Q_{3S} \; \exp \; (-\Delta_{3D}/T)$$
$$Q_{3P} = 3/9 \; Q_{3S} \; \exp \; (-\Delta_{3P}/T) \qquad (15)$$

where $\Delta_{3D} = 72.1$ meV, $\Delta_{3P} = 66.5$ meV are the energy shifts between the 3S and 3D and 3S and 3P states and T is the temperature. The predicted value of Q_{1S} assuming muon transfer supression from the levels n≥3 and Bolzman distribution is $Q_{1S} = 0.5$ (for $\phi=1$, $c_t=0.5$). This value is in agreement with the experiments.

IV. DECREASE OF THE 1S POPULATION OF dμ UNDER LASER IRRADIATION

As described before the muon cascade is determined by the competition between radiative transitions, density dependent external Auger transitions, density dependent quenching of the muonic levels, and transfer processes which depend both on density ϕ and tritium concentration c_t. The cascade can take very different routes depending on the actual populations of these states. Therefore it is suggested to decrease the fraction of muons reaching the ground state of dμ by changing the route of the cascade under laser irradiation. One can see from fig. 2 that the cross section of the muon transfer process from state 2S state is larger by a factor of 5 than that transfer cross section from the 2P level. The energy difference between the 2S and 2P states (The Lamb Shift in the deuterium atom) is 0.2 eV and powerful lasers might be used for population change

between these two states. Therefore the following scheme is proposed
to enhance the rate of the muon transfer: a laser induced transition
between the levels 2P and 2S and then muon transfer by collisions to
the tritium 2S level.

The rate equations describing the kinetics of the muon cascade in
a DT-mixture under laser irradiation are given by:

$$\frac{dQ_i}{dt} = - \lambda_i Q_i + \sum_{j>i}^{N} \lambda_{ji} Q_j + (\delta_{2P,i} - \delta_{2S,i}) [B\rho Q_{2S} - (A+B)Q_{2P}], \quad (16)$$

where Q_i is the population of the level i, A and B are the Einstein
coefficients:

$$A = 4/3 \; \omega^3 \; \alpha/c^2 \; |D_{2P, \, 2S}|^2 \qquad [sec^{-1}]$$

$$ \tag{17}$$

$$B = 4\pi^2 e^2 |D_{2P, \, 2S}|^2/3\hbar^2 \qquad [cm^3 \; sec^{-2} \; erg^{-1}]$$

ρ is the energy density related to the laser intensity I_L by:

$$\rho = 2I_L/(\pi c \Delta v) \qquad [cm^{-3} \; sec \; erg], \qquad (18)$$

$\hbar\omega$ is the energy difference between the 2P and 2S levels and $|D_{2P, \, 2S}|$
is the dipole moment for this transition. The line width Δv is given
by the largest contribution which in this case is:

$$\Delta v_{Doppler} = v \frac{v}{c} \sim 1.4 \times 10^9 \; sec^{-1} \qquad (19)$$

at temperature of $\sim$ 300 K°.

We solved numerically the rate equations (16) under laser
irradiation, with initial conditions $Q_5(0)=0$, $Q_i(0)=0$, i=1,4 for two
cases: (a) assuming muon transfer suppression for the levels n≥3, and
(b) without muon suppression. The value of Q_{1S} is calculated for
different values of the density ϕ, tritium concentration c_t,
temperature T and laser intensity I_L.

The time dependence of Q_{1S} for the theoretical models described
above and the value of Q_{1S} under laser irradiation for a laser
intensity $I_o = 5 \times 10^8$ Watt/cm^2 is shown in fig. 3. Fig. 4 shows the
laser intensity dependence of the relative decrease in the asymptotic
value of Q_{1S}, $q = Q_{1S}(I_L)/Q_{1S}(I_L=0)$ for different values of density ϕ,
tritium concentration C_t and temperature T. One can see that for
moderate laser intensities of 10^9 W/cm^2, Q_{1S} may be decreased by a
factor of three. The results show that the decrease in Q_{1S} is larger
for (a) higher densities (b) larger tritium concentration and (c)
lower temperatures.

The number X_μ of fusions per muon with and without laser
irradiation (calculated from eq. (6-7)) as a function of the tritium
concentration is illustrated in fig. 5. It can be seen that the value
of X_μ increases under laser irradiation and shifts toward lower values
of c_t.

131

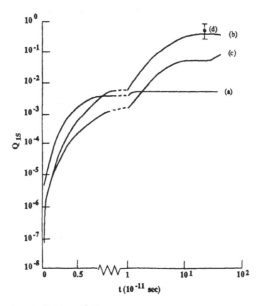

Fig. 3.   Time dependence of Q_{1S}:
         (a)   assuming muon transfer from dμ to tμ for dμ level
              $5 \geq n \geq 4$
         (b)   as (a) with $2 \geq n \geq 1$.
         (c)   for model (b) with a irradiance of $I_o = 5 \times 10^8$ W/cm^2
         I describes the asymptotic experimental value without laser
         irradiance.

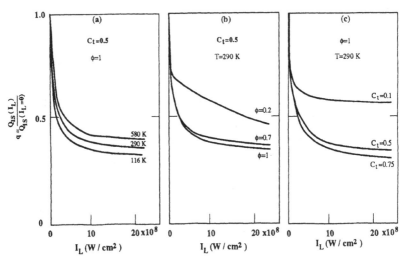

Fig. 4.   Laser intensity dependence of the relative decrease in Q_{1S}
         for different values of density ϕ, tritium concentration C_t
         and temperature T.

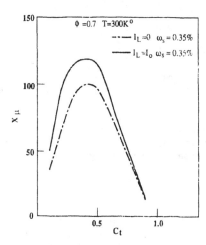

Fig. 5. The number of fusions per muon X_μ with and without laser irradiance as a function of the tritium concentration c_t.

V. LASER INDUCED u-MOLECULAR FORMATION

It was suggested in the past [8-10] that μ-molecular formation may be enhanced by strong electromagnetic fields. Our model of ref. (8) is briefly reviewed.

A three level system for the dtμ molecule is considered: (a) a state in the continuum of tμ + d (b) the bound state $(J,v) = (1,1)$ of the molecule dtμ and (c) the bound state $(J,v) = (0,1)$ of dtμ. The appropriate energies and lifetimes satisfy the conditions $E_a>E_b>E_c$ and $\gamma_a \gtrless \gamma_b$. Under the influence of an external field a Stokes transition from level (a) to level (b) takes place while the transition from (b) to (c) is occurring through an Auger process. The spontaneous decay rate γ_a from (a) to (b) and the decay of level (b) γ_b are included phenomenologically. The probabilities of the levels (a) and (b) ψ_a and ψ_b satisfy the equations: (within the rotating wave approximation):

$$\psi_a = -1/2\ \gamma_a\ \psi_a + 1/2\ i/\hbar\ V_{ba}\ e^{-i\Delta t}\ \psi_b$$

$$\psi_b = -1/2\ \gamma_c\ \psi_b + 1/2\ i/\hbar\ V_{ab}\ e^{-i\Delta t}\ \psi_a$$

(20)

where V_{ba} is the matrix element for the transition and Δ is the detunning given by:

$$\Delta = \frac{E_a - E_b}{\hbar} - \omega$$

(21)

ω is the laser frequency. The energy difference $(E_a - E_b)$ for dtμ is 0.66eV.

Assuming a dipole transition, the efficiency (defined as the ratio between Stokes induced transitions and spontaneous decay) is given by:

$$\eta = \frac{|V_{ab}|^2 \, \gamma_b}{4\hbar^2 \, \gamma_a \, (\Delta^2 + \gamma_b^2/4)} \qquad (22)$$

The results show that for a resonant laser frequency ($\Delta < \gamma_b/2$) and intensities of 6×10^5 W/cm^2 the efficiency for the dtμ formation is about 50.

However it must be emphasized that the resonance condition is extremely important for the estimation of η. For practical lasers one uses the off resonance regime yielding an intensity of about 10^9 W/cm^2.

VI. CONCLUSIONS

This work presents the possibility that strong electromagnetic fields could effect the rate of the muon catalyzed fusion cycle in a deuterium - tritium mixture. The fusion cycle is inhibited by the slow transfer of muons between the ground states of deuterium and tritium, the dtμ molecular formation and the muon loss due to sticking. We showed that a laser of moderate intensity of about 10^9 W/cm^2 with photon energy of 0.2 eV can reduce the value of the ground state population of the dμ atom Q_{1S} by a factor of three. Moreover, Q_{1S} can be reduced to zero by pulsed X-ray radiation of an intensity of the order of 10^{13} W/cm^3 (for broadening $\Delta\nu/\nu \approx 10^3$) by the two step process: (a) radiation induced transition from the 1S to 2P level of deuterium (b) muon transfer to the 2P level of tritium. The dtμ molecular formation rate can be increased by a laser of an intensity of the order of 10^9 W/cm^2 and a photon energy of 0.66 eV.

Acknowledgement

We would like to thank Dr. Y. Paiss for many fruitful discussions.

References

[1] See for example the following review articles:
   (a) S.S. Gershtein and L.I. Ponomarev, "Muon Physics," Vol. III, V.W. Hughes and C.S. Wu, Editors, Academic Press, N.Y. (1975), p. 141;
   (b) L.I. Ponomarev, Proc. Sixth Int. Conf. on Atomic Physics, Plenum, N.Y. (1978), p. 182;
   (c) J. Rafelski, "Exotic Atoms," K. Crowe and E. Duclos, Editors, Plenum, N.Y. (1979), p. 177;
   (d) L. Bracci and G. Fiorentini, Phys. Rep. 86, 169 (1982).
   (e) S. Eliezer, Laser and Particle Beams 6, 63 (1988).

[2] (a) L.I. Menshikov and L.I. Ponomarev, JETP Lett. 39, 663 (1984).
   (b) L.I. Menshikov and L.I. Ponomarev, Z. Phys. D2, 1 (1986).

[3] (a) S. Jones, Phys. Rev. Lett. 56, 588 (1986).
   (b) W.H. Breunlich et al., Phys. Rev. Lett. 58, 329 (1987).
   (c) K. Kabayashi, T. Ishihara and N. Toshima, Muon Cat. Fusion 2, 191 (1988).

[4] Yu. V. Petrov and Yu. M. Shabelskii, Sov. S. Nucl. Phys. 30, 66 (1979).

[5] (a) Yu,V. Petrov, Nature 285, 466 (1980).

    (b) S. Eliezer, T. Tajima, M.N. Rosenbluth, Nuclear Fusion 27, 527 (1987).

[6] L. Bracci & Fiorentini, Nature 297, 134 (1982).

[7] H. Takahashi and A. Moats, Atomkerenergie 43, 188 (1983).

[8] S. Eliezer and Z. Henis, Phys. Lett. 131A, 361 (1988).

[9] H. Takahashi, in "Muon Catalyzed Fusion", Ed. S. Jones, J. Rafelski, G. Hendrik, J. Monkhorst, AIP New York (1989).

[10] S. Barnet and A.M. Lane, in "Muon Catalyzed Fusion", Ed. S.E. Jones, J. Rafelski, G. Hendrik, J. Monkhorst, AIP New York (1989).

[11] T. Tajima, S. Eliezer and R.M. Kulsrud, in "Muon Catalyzed Fusion" Ed. S.E. Jones, J. Rafelski, G. Hendrik, J. Monkhorst, AIP New York (1989).

[12] (a) M. Leon and H.A. Bethe, Phys. Rev. 127, 637 (1962).

     (b) V.E. Markushin, Sov. Phys. JETP 53, 16 (1981).

[13] M. Leon, Phys. Lett. 35B, 413 (1971).

[14] G. Kodoski and M. Leon, Nuovo Cimento 1B, 41 (1971).

[15] L.I. Menshikov and L.I. Ponomarev, JETP Lett. 39, 663, (1984).

[16] S.E. Jones et al. Phys. Rev. Lett. 56, 558 (1986).

[17] B. Müller, S. Rafelski, M. Jandel, S.E. Jones, in "Muon Catalyzed Fusion" Ed. S.E. Jones, J. Rafelski, G. Hendrik, J. Monkhorst, AIP New York (1989).

[18] L.I. Menshikov and L.I. Ponomarev, Z. Phys. D2, 1 (1986).

[19] H.E. Rafelski, B. Miller, J. Rafelski, D. Trautman and R.D. Viollier preprint AZPH-TH/88-12 (1988) to appear in "Progress in Particle and Nuclear Physics".

LIGHT ION BEAM DRIVERS FOR INERTIAL CONFINEMENT FUSION

Juan J. Ramirez

Sandia National Laboratories
Albuquerque, New Mexico 87185

ABSTRACT

Intense beams of light ions are being developed at Sandia National Laboratories as a promising driver option for Inertial Confinement Fusion (ICF) implosions. The Particle Beam Fusion Accelerator II (PBFA II) will provide the physics basis for light-ion-beam driven ICF targets. Recent progress made in ion beam generation focusing on PBFA II has led to a record 5.4 TW/cm^2 peak focal intensity with >80 kJ proton energy delivered to a 6-mm diameter sphere. The driver-development program on PBFA II is reviewed. A design concept for a light ion beam driver for the Laboratory Microfusion Facility is also presented.

INTRODUCTION

Intense beams of light ions are being investigated at Sandia National Laboratories as a driver for Inertial Confinement Fusion targets[1]. High-gain ICF implosions using light ion beams (LIB) require ions with optimal range (30-40 mg/cm^2) that deliver >10 MJ to the target in ~15 ns. The required ion-beam power on target is ~1000 TW with an ion beam intensity >100 TW/cm^2. Light ion beams offer an attractive driver option for ICF implosions because they can be generated efficiently and at a low cost relative to lasers and heavy ion beams.

The Particle Beam Fusion Accelerator II[2] (PBFA II) is being developed to provide the physics basis for light-ion-beam driven ICF targets. At peak power, it will produce a 100-TW, 15-ns, 1.5-MJ pulse to drive a 24-MV, Li^+ ion diode. Recent progress made in the understanding of ion-diode performance has led to a record 5.4 TW/cm^2 peak focal intensity averaged over a target equivalent to a 6-mm diameter sphere. Achievement of higher power densities will require a high purity Li^+ source and operation at higher voltages.

The US Department of Energy has initiated planning for a Laboratory Microfusion Facility (LMF) as the next major ICF facility. It is expected to produce a 1000-megajoule yield with a 10-megajoule input to the target. Our design concept for the LMF consists of 36 accelerator modules driving independent Li+ extraction ion diodes located just outside the containment chamber. The 36 beams are transported and focused unto the target located at the center of the chamber. Twelve of these accelerator modules are similar to Hermes III[3]. The remaining twenty four modules are ~3 times more powerful.

This paper gives a brief review of the status of the ICF driver development effort at Sandia. The PBFA-II accelerator design and performance are discussed in Section II. Results of ion diode research on PBFA II are presented in Section III. The driver concept for the LMF is described in Section IV. Summary and conclusions are presented in Section V.

THE PBFA-II ACCELERATOR

An artist's drawing of PBFA II is shown in Fig. 1. It consists of 36 independent high-voltage, pulse-forming modules. The prime energy store consists of 36 Marx generators that store a total of 13 MJ energy at full charge[4]. Each Marx generator feeds a set of water-dielectric pulse forming transmission lines (PFLs). At peak power each PFL module delivers a 370-kJ, 3.2-MV, 55-ns output pulse[5]. In this design, currents from the individual modules are added at each of the eight layers of a common, 3.6-m diameter, vacuum insulator stack shown in Fig. 2. Each layer adds half of the current from nine separate PFL modules. A set of magnetically insulated vacuum transmission lines (MITLs) within the insulator stack add the voltages of the four top and bottom layers to generate 10-MV to 15-MV pulses. These two summed outputs feed a common ion diode located at the midplane of the insulator stack. Plasma opening switches (POS) located top and bottom just upstream of the ion diode provide the final stage of compression and voltage amplification to generate the 24-MV pulse required for the Li+ ion diode[6]. The top and bottom switches are synchronized to maintain symmetric power flow into the ion diode.

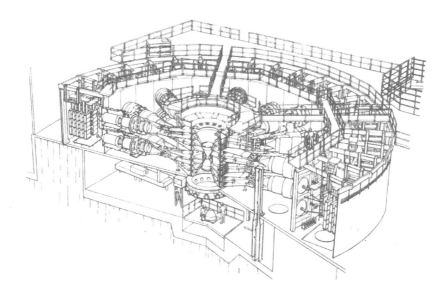

Figure 1. Drawing of PBFA II.

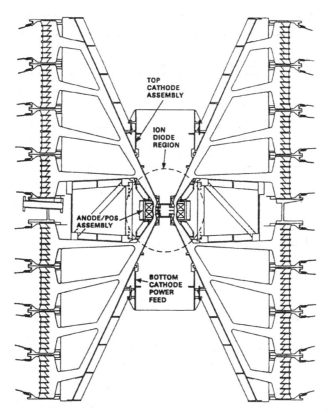

Figure 2. PBFA II center section showing MITL structure that adds the voltages from various levels of the insulator stack and deliver the two summed output pulses to a common ion diode.

PBFA II was first fired in December 1985. Significant progress has been made in accelerator performance, reliability, synchronization and shot rate[7]. At the 3/4-energy level we are able to field and sustain one fully diagnosed ion diode shot per day. The module synchronization is now routinely better than the original specifications as is shown in Table I.

PBFA II was designed to use the vacuum feed region as part of an inductive energy store system. The energy stored in the vacuum-feed inductance is delivered to the ion-diode load by the plasma opening switch. During the past year, three POS configurations have been tested at the 1/2-energy level into a 5-ohm electron beam load[8]. These configurations make use of self-generated and applied magnetic fields to control the movement of the plasma in the switch region. All three switches demonstrated fast opening times into the 5-ohm load. Excellent agreement was obtained between the measured and calculated current waveforms upstream and downstream of the switch. The most flexible switch configuration uses an externally applied field to inject plasma into the switch region in combination with a fast self-generated field to enhance the magnetic pressure that drives the switch open[9]. This switch will be implemented in the near future to drive an ion-diode load.

Table I. Timing Performance of PBFA-II Subsystems

	Marx Generators		Gas Switches		Pulse Forming Lines	
	Spread	Std. Dev.	Spread	Std. Dev.	Spread	Std. Dev.
All Shots	30.7 ns	7.1 ns	10.6 ns	2.3 ns	15.1 ns	3.4 ns
Last 50	26.5 ns	6.3 ns	8.3 ns	1.8 ns	13.0 ns	3.0 ns
Orig.Spec.	<30 ns		<10 ns		<15 ns	

ION DIODE RESEARCH

The ion diode used on PBFA II is shown in Fig. 3. It uses an applied magnetic field to insulate the electrons and prevent them from reaching the anode. Substantial progress has been made on ion-beam generation and focusing[10-15]. This progress has been aided by advances in the theory of magnetically insulated ion diodes and by the development of an expanded set of intense ion-beam diagnostics.

The high-power pulse generated by PBFA II is fed to the diode from the top and bottom as shown in Fig. 3. The electric fields generated at the anode-cathode gap are high enough (>1 MV/cm) to generate copious emission from the cathode surfaces. The cathode becomes a space-charge-limited emitter of electrons which are accelerated towards the anode. When there is a source of ions at the anode (from an active ion source or by the formation of a plasma at this electrode), it becomes a space-charge-limited emitter of ions that are accelerated towards the cathode. Under these conditions, most of the current is carried by the electrons. The electron current reaching the anode must thus be minimized for efficient ion-diode operation. In an applied-B ion diode, a transverse magnetic field is applied in the region between the anode and cathode to constrain the electron flow. The electrons flow in a sheath along a magnetic flux surface in the azimuthal (ExB) direction. The sheath establishes an equipotential surface called the "virtual cathode." The ions, being more massive, can easily cross the magnetic field and are accelerated in the gap between the anode and the virtual cathode.

An analytic theory of applied-B ion diodes has been recently developed that explains many of the features seen in the operation of these diodes[16]. This theory follows the motion of the virtual cathode due to the applied and self-generated fields. One key result of this theory is the existence of a critical voltage, V_*, at which the ion current diverges. This voltage is proportional to the product of the applied magnetic field and the anode-cathode gap. The theory provides an explanation for a known scaling relationship that was obtained previously

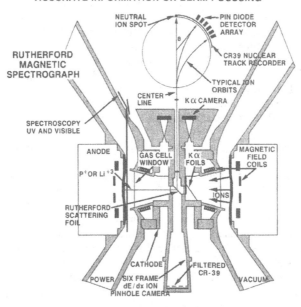

Figure 3. Applied-B ion diode used on PBFA II showing various intense ion beam diagnostic packages.

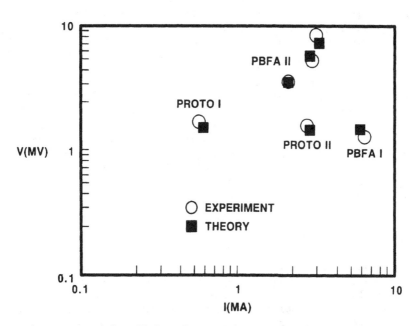

Figure 4. Measured ion diode voltage and current, at peak power compared with values predicted by analytic theory.

from analysis of ion-diode data from various accelerators and which suggested the existence of this limiting voltage[17]. Fig. 4 shows the measured ion diode voltage and current, at peak power, for various accelerators compared with values predicted by the analytic theory. The theory also provides some insights into a possible explanation for the falling diode impedance observed to occur past peak power[18]. The virtual cathode location, and thus the effective anode-cathode gap, depends on the amount of flux between the anode and the virtual cathode. A decrease in flux in this region due to flux penetration into the anode plasma would result in a smaller effective gap and an increased ion current.

Development of a high-purity Li^+ source is essential for the production of higher intensity beams. Several candidate sources are being developed. The electrohydrodynamically (EHD)[19] driven source produces Li^+ ions by field emission from closely spaced cusps on the surface of a layer of liquid lithium. The cusps are formed by an electrohydrodynamic instability that is driven by the applied voltage pulse. These sources have been used to generate high brightness, low current density beams but have not been tested at the power levels required for PBFA II. A second source (BOLVAPS/LIBORS)[20] uses a lithium-bearing film that is heated rapidly by a current pulse to produce a thin layer of lithium vapor that is then ionized by a laser pulse. Small scale experiments have demonstrated the production of a lithium plasma of adequate density and low neutral particle content. A third source being investigated (LEVIS)[21] uses a laser beam to form the lithium vapor layer and to ionize it. Plans call for these sources to be tested by next spring, and a decision will be made on which source to emphasize for future ion-diode experiments on PBFA II.

The rapid progress being made in understanding ion-diode behavior has been strongly influenced by development of an array of intense ion-beam diagnostic packages and calculational tools for use in PBFA II. An x-ray pinhole camera provides a measure of the ion fluence by imaging K_α radiation produced when ions strike a metal target near the diode axis[22]. A six-frame ion pinhole camera records two dimensional profiles of the ion beam on axis as a function of the ion kinetic energy[23]. A magnetic spectrograph provides time-integrated and time-resolved data on the ion kinetic energy and ion species. It also provides time-resolved information on the ion current density and power density[24]. Nuclear activation is used to measure the time-integrated ion current[25]. A complete description of these diagnostics has been presented elsewhere by R. J. Leeper[26]. A number of new computer programs have been developed as part of the PBFA II ion-diode research[27-28]. These codes have been very useful in developing new diagnostic capabilities and in interpreting data from ion-beam focusing experiments.

LIGHT ION BEAM LMF DRIVER CONCEPT

An artist's drawing of the LMF driver concept is shown in Fig. 5. It consists of 36 accelerator modules driving independent Li^+ extraction ion diodes[29]. The individual ion beam is extracted from the coaxial ion diode and transported to a solenoidal lens located near the inner wall of the target chamber. This magnetic lens focuses the beam on the pellet located at the center of the target chamber as shown in Fig. 6. The shape of the

Figure 5. Drawing of the light ion beam LMF driver concept.

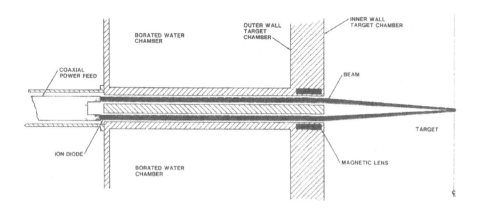

Figure 6. Drawing of LMF target chamber region showing ion beam generation and transport geometry for one beam line.

power pulse delivered to the target is controlled by the synchronized firing of various module groups. Twelve of the accelerator modules are similar to Hermes III, and the remaining 24 modules are ~3 times more powerful. This design would deliver ~20 megajoules ion energy to a 1-cm radius pellet with the proper pulse shape for high-gain ICF implosions.

Hermes III is a new generation electron beam accelerator recently completed at Sandia to provide enhanced capabilities for the above-ground weapons effects simulation program. A drawing of Hermes III is shown in Fig. 7. This accelerator combines high-power linear induction accelerator cavities with MITLs to add eighty 1-MV pulses in a parallel/series combination and thus generate the 22-MV, 730-kA, 40-ns output pulse that powers the electron beam diode. A cross section of the Hermes III MITL system is shown in Fig. 8. It consists of an adder section where the inputs from the 20 cavities are added and an extension section that delivers the output of the adder to the electron-beam diode. The outer conductor of the coaxial adder section is formed by the surface of the cavity bores. This conductor is interrupted at regular intervals by the radial gaps in the cavities through which power is fed to the adder. The inner conductor of the MITL is formed by a long cantilevered cathode shank. This cathode shank decreases in radius at each cavity feed thus increasing the impedance of each MITL section. The impedance of the nth section is made n-times the impedance of the first. The voltage at any point along the adder is thus given by the sum of the voltage pulses delivered by all the cavities up to that point. The extension section is a constant impedance coaxial MITL that provides a matched impedance to the adder. Hermes III has proven to be a very efficient, highly reliable accelerator that exceeded performance specifications and became operational one year ahead of schedule[30].

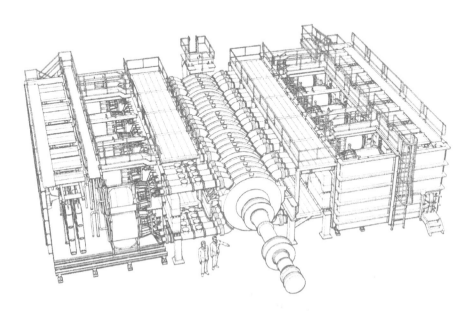

Figure 7.   The Hermes III accelerator.

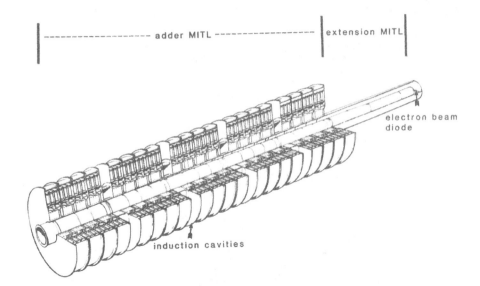

adder MITL ————————————— extension MITL

electron beam diode

induction cavities

Figure 8. Cross section drawing of the Hermes III MITL system.

Hermes III produces a negative output pulse to power an electron beam diode. The polarity of the output pulse will be reversed so that the LMF module can power an extraction ion diode. Reversing the output polarity has a significant effect on the performance of the MITL. Analytic calculations and computer simulations show significantly more complex magnetically insulated flow for the positive polarity adder. An initial experiment with a positive polarity adder has been performed on Hermes III. It demonstrated efficient current and power transport, through the adder and extension sections, to an electron beam diode load[31]. Additional positive polarity experiments are currently under way. These experiments will include a study of the coupling efficiency to an extraction ion diode.

The applied-B extraction ion diode is shown in Fig. 9. It is similar to the one tested previously by Slutz[32]. The major diode issues include beam divergence, impedance history, source uniformity and purity, ion generation efficiency, and the large magnetic fields required. The beam microdivergence required (~6 mrad) is the most stressing point in the LMF design.

The baseline LMF transport and focusing system is a first order achromatic magnetic lens system that consists of two lenses: (1) an ion diode which acts as a self-field lens, and (2) a B_z solenoidal lens[33]. The region from the diode to the target is filled with ~1 Torr helium gas to provide good charge and current neutralization for ballistic transport. An added feature of this design involves the temporal bunching of the ion pulse as it is transported from the diode to the focusing lens and then to the target. This is accomplished by applying a ramped voltage pulse to the diode which results in the proper velocity ramp for beam bunching.

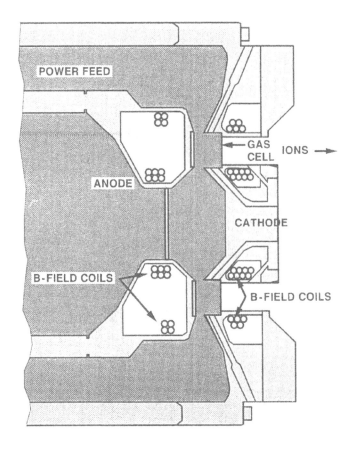

Figure 9. The applied-B extraction ion diode.

The LMF target explosions will be contained within a chamber which, for radiation-shielding purposes, will be immersed in a tank of borated water. The baseline transport concept uses a 1.5-m radius chamber constructed of 5-cm thick aluminum walls[34]. X-rays from the thermonuclear explosions deposit their energy near the surface of the first wall, vaporizing some material. The mass of the vaporized material gives rise to peak pressures of several gigapascals. These peak pressures are high enough to generate shocks that move through the first wall, possibly damaging the wall. Methods of mitigating these damaging shocks are being investigated. We are also investigating a 3-m radius target chamber to be used in case an acceptable 1.5-m radius chamber is not possible. Several back-up transport schemes are being investigated for use with the larger target chamber[33].

SUMMARY AND CONCLUSIONS

Intense beams of light ions are being developed at Sandia for application to ICF. The mainline program on PBFA II will provide the physics basis for light ion-beam driven ICF targets. Significant progress has been made in accelerator reliability, synchronization and shot rate. Steady progress in ion-beam generation and focusing has led to a record 5.4 TW/cm^2 peak focal intensity averaged over a target equivalent to a 6-mm diameter sphere. Achieving higher focal intensities will require a high purity Li$^+$ source and operation at higher voltages. One of three promising lithium sources will be chosen for future ion-diode experiments on PBFA II. The higher voltage required for these experiments will be provided by one of three POS configurations presently under test.

A design concept for a light ion-beam LMF driver has been formulated. This design uses the pulsed power accelerator technology recently demonstrated on Hermes III. This design would deliver ~20 megajoules of ion energy to a 1-cm radius pellet with the pulse shape required for high gain ICF implosions. Critical issues for this design have been identified, and a research program to address these issues has been initiated. Our goal is a technology validation experiment on Hermes III that includes a demonstration of the beam generation, transport, bunching and focusing at the performance levels required for the LMF.

ACKNOWLEDGEMENTS

I wish to acknowledge the contributions of my colleagues at Sandia National Laboratories, private industry, universities and other national laboratories. Although they are too numerous to list here, the vitality of the LIB ICF program is built around the strength of their individual contributions.

REFERENCES

1. J. P. VanDevender and D. L. Cook, Science, 232, 831 (1986).
2. B. N. Turman, et al., in Proc. 5th IEEE Pulsed Power Conf.,
    Arlington, VA, June 10-12, 1985, pp. 155-161.
3. J. J. Ramirez, et al., in Proc. 7th Int'l Conf. on High-Power
    Particle Beams, Karlsruhe, Germany, July 1988, pp. 148-157.
4. J. M. Wilson, in Proc. 4th IEEE Pulsed Power Conf.,
    Albuquerque, NM, June 6-8, 1983, pp. 64-67.
5. L. X. Schneider, et al., in Proc. 7th IEEE Pulsed Power Conf.,
    Monterey, CA, June 11-14, 1989.
6. G. E. Rochau, et al., in Proc. 6th IEEE Pulsed Power Conf.,
    Arlington, VA, June 29 – July 1, 1987, pp. 77-80.
7. D. L. Cook, et al., in Proc. 7th Int'l Conf. on High-Power
    Particle Beams, Karlsruhe, Germany, July 1988, pp. 35-45.
8. G. E. Rochau, in Proc. 7th IEEE Pulsed Power Conf., Monterey, CA,
    June 11-14, 1989.
9. C. W. Mendel, et al., in Proc. 15th IEEE Conf. on Plasma Sci.,
    p. 128 (1988).
10. D. J. Johnson, et al., J. Appl. Phys., 58, 12 (1985).
11. P. L. Drieke, et al., J. Appl. Phys., 60, 878 (1986).
12. J. E. Maenchen, et al., in Proc. 6th Int'l Conf. on High-Power
    Particle Beams, Kobe, Japan, June 9-13, 1986.
13. S. Humphries, Jr., Nucl. Fusion, 20, 1549 (1980).

14. C. L. Olson, J. Fusion Energy, **1**, 309 (1982).

15. J. E. Maenchen, et al., J. Appl. Phys., **65**, 448 (1989).

16. M. P. Desjarlais, Phys. Rev. Lett., **59**, 2295 (1987).

17. P. A. Miller, J. Appl. Phys., **57**, 1473 (1985)

18. M. P. Desjarlais, Phys. Fluids B, **1**, 8 (1989).

19. A. L. Pregenzer, J. Appl. Phys., **58**, 4509 (1985).

20. P. L. Drieke and G. C. Tisone, J. Appl. Phys., **59**, 371 (1986).

21. K. W. Bieg and G. C. Tisone, (private communication).

22. J. Maenchen, et al., Rev. Sci. Instrum., **59**, 1706 (1988).

23. W. A. Stygar, et al., Rev. Sci. Instrum., **59**, 1703 (1988).

24. R. J. Leeper, et al., Rev. Sci. Instrum., **59**, 1700 (1988).

25. R. J. Leeper, et al., Rev. Sci. Instrum., **59**, 1860 (1988).

26. R. J. Leeper, et al., in Proc. 7th Int'l Conf. on High-Power
    Particle Beams, Karlsruhe, Germany, July 1988, pp. 258-266.

27. T. A. Mehlhorn, et al., Rev. Sci. Instrum., **59**, 1709 (1988).

28. J. P. Quintenz, et al., Bull. Am. Phys. Soc., **30**, 1603 (1985).

29. J. J. Ramirez, et al., in Proc. 8th Topical Meeting on the Technology
    of Fusion Energy, Salt Lake City, Utah, October 9-13, 1988.

30. J. J. Ramirez, et al., in Proc. 7th IEEE Pulsed Power Conf.,
    Monterey, CA, June 11-14, 1989.

31. D. L. Johnson, et al., in Proc. 7th IEEE Pulsed Power Conf.,
    Monterey, CA, June 11-14, 1989.

32. S. A. Slutz, et al., Bull. Am. Phys. Soc., 1803 (1987).

33. C. L. Olson, in Proc. 1989 Particle Accelerator Conf.,
    Chicago, IL, March 20-23, 1989.

34. R. R. Peterson, et al., Univ. of Wisconsin Report No. UWFDM-768,
    August, 1988 (unpublished).

DEVELOPMENT OF HIGH POWER KrF LASER FOR ICF LASER DRIVER

AND LASER INTERACTION EXPERIMENTS

H. Yoneda, H. Nishioka, A. Sasaki, K. Ueda, and T. Takuma

Institute for Laser Science
University of Electro-Communications
Chofuga-oka, Chofushi, Tokyo 182, Japan

ABSTRACT

We briefly report high power KrF laser system and the research for laser plasma interaction in ILS.

The system has three staged amplifiers of 600J output energy. Angular multiplexed 5ns beamlets will be brought to the target with maximum averaged intensity up to $10^{15} W/cm^2$.

Target experiments have been started with one beam of our system. The laser plasma interaction of short wavelength laser was investigated, which includes the absorption and radiation processes. The target experiments were carried out using ns and ps pulses for varying scale length of plasma. The irradiance was $10^{13-14} W/cm^2$ for 20ns pulses and $10^{14-15} W/cm^2$ for 2ps pulses. In the ps laser system, undesirable prepulse was controlled by a saturable absorber (power ratio; 10^{-7}). Good absorption properties were observed even for the long scale length plasma. The ps pulse interaction suggested that the absorption was largely sensitive to the prepulse for small scale length plasma.

INTRODUCTION

A KrF laser has optimized wavelength for inertial confinement fusion.[1] An electron beam pumped KrF laser has been studied in recent several years.[2] These investigations were concentrated to the study of the laser medium. These results indicated that it has a high intrinsic efficiency (up to 10%) and the extraction efficiency does not changed in the wide range of pumping rate. This means that it has the possibility of scale up to a 500kJ machine with keeping the high efficiency.[3]

As a next step, we have to achieve the development of efficient pumping technology and the efficient pulse compression technique. In this paper, we will report some new technology for improvement of efficiency and some results of demonstration.

Though many implosion experiments have been studied with various glass lasers, they are limited in the condition where laser wavelength is longer than .35μm and pulse duration is shorter than 1ns. There is small data for quarter micron laser interaction and long scale length plasma. Interaction phenomena are supposed to be changed in the different wavelength and scale length of plasma. The interaction physics with 250nm light and several ns pulses have to be investigated. A high peak power laser is required for such experiments because of the scaling under wide range irradiation intensity. For this reason, ultra short pulse front-end oscillator was developed. In this paper, we will also report some results of target interaction experiment with the above lasers.

Pulse power technology

In a reactor size laser system, the dimension of laser medium is supposed to be 2x3x10m.[3] For the large volume pumping, large current and large aperture of electron beam will be required. For avoiding the self pinching effect and the difficulty of low impedance pulse line, the electron beam diode will be divided into many smaller modules, one of which is electro-magnetically isolated to each other. In this sense, an edge relaxed Blumlein line was developed. Generally, the energy density of a pulse line is determined by water breakdown. In coaxial line, inner surface of cylindrical electrode is most severe for this breakdown, while, in parallel line, there is no particular point except for edge. To remove this edge effect, acryl plates (Fig.1), of which the breakdown strength is much larger than water dielectric, was inserted in this point. The most edge electric field was introduced into the Acryl plate because of its low dielectric constant and an edge free electric field was performed in parallel line. With this technology, the modular-type multi-lines can be easily constructed very close to each other. This technology are demonstrated in the final amplifier (Fig.2).

The electron diode have to generate a spatially uniform and temporally constant electron beam. Many cathode materials were tested; e.g. Ti, Ta, Sus, Carbon and velvet with various configuration. The result denotes that a velvet cathode can generate uniform electron beam better than any other materials (Fig.3) and it also has a fast turn-on time and constant impedance in 100ns operation (Fig.4).

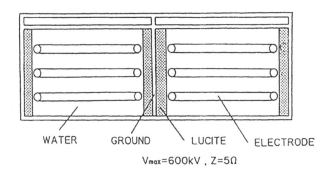

WATER　　GROUND　　LUCITE　　ELECTRODE

$V_{max}=600kV$, $Z=5\Omega$

Fig.1 Schematic drawing of edge relaxed Blumlein line

Fig.2 Photograph of our final amplifier

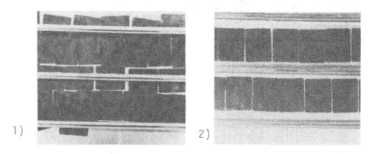

1)   2)

Fig.3 The e-beam profiles on anode foil
1)a cheese grater cathode
2)a velvet cathode

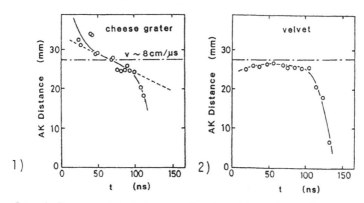

1)   2)

Fig.4 Temporal history of effective A-K distance
1)a cheese grater cathode
2)a velvet cathode

KrF Laser Technology

KrF laser medium has an extremely high small signal gain and easily suffers a large ASE loss without sufficient laser field. A 3-d ASE code was developed for the estimation of the energy depletion under realistic amplifier condition.[4] Extraction efficiency was calculated as a function of aspect ratio (L/D), which means the ratio of the gain length to diameter. The calculated result is shown in fig.5. Optimum amplifier parameter can be determined (L/D=3) with keeping the high stage gain and the high extraction efficiency.

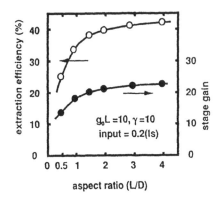

Fig.5

Calculated extraction efficiency as a function of amplifier aspect ratio (L/D)

KrF laser medium does not storage the pumping energy for 100ns excitation. It is necessary to demonstrate the efficient and cost-effective pulse compression. An angular multiplexing system contained 9 beam optics was designed with our three stage amplifier train (Fig.6).

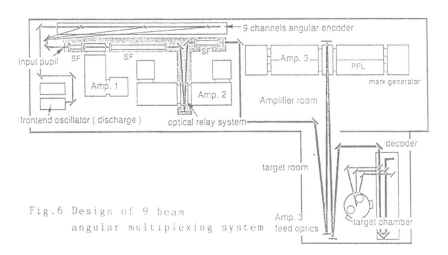

Fig.6 Design of 9 beam angular multiplexing system

Amplifier No.1, No.2 and No.3 were tested with oscillator mode. The maximum output energy were 25, 80 and 600J, respectively. In the hatching region, relay optical system was chosen for decreasing the number of optical components and the cost. The quality of our optical system was estimated with our ray-tracing code. Some calculated results are shown in fig.7. Total aberration was calculated with initial field angle as a parameter, which denotes the incident angle into the relay optics This figure showed that the distortion of irradiation pattern on the target plane was kept small unless the field angle is larger than 0.5 degree.

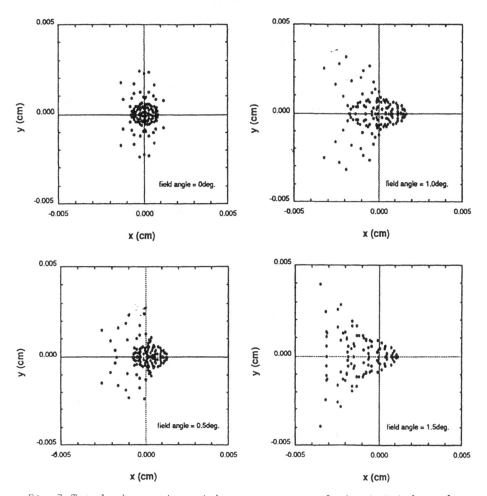

Fig.7 Total aberration with a parameter of the initial angle

A temporal domain multiplexing method was also developed for additional increasing of irradiation power (Fig.8). An unpolarized beam was divided into a pair of polarized beams and the reflected beam returned to the optical axis after the adjustable delay section. A pair of pulses propagate in the amplifier chain as if these are a 'single' beam (Fig.9a) and easily separated into two beams (Fig.9b) by means of a thin film polarizer[5]. After the amplification, the delay between a pair of pulses can be removed with the previous method. Finally, a pair of pulses can make one pulse of twice power. With this technique, the irradiation power will be doubled without changing main optical components and cost.

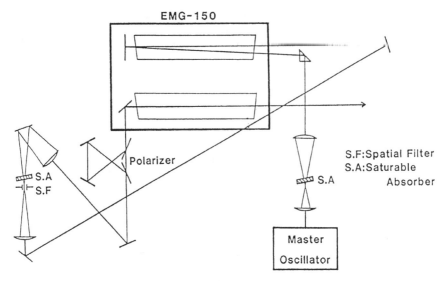

Fig.8 Schematic drawing of a temporal multiplexing system

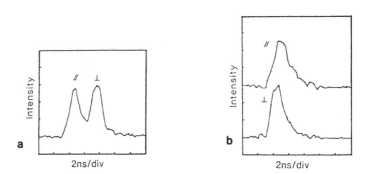

Fig.9 Waveform of a pair of pulses
    a) in the amplifier chain
    b) after separation by means of a thin film polarizer

Irradiation uniformity is also important for the direct drive fusion scheme. For this purpose broadband ASE oscillator[6] was developed. While increasing incoherency of laser light is useful for uniform irradiation, it also degrades the beam focusability. For this reason optimum incoherency has to be carefully decided. We controlled the coherency of laser beam carefully and good irradiation uniformity was obtained with moderate focusability (Fig.10).

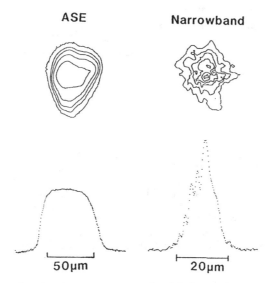

ASE   **Narrowband**

50μm   20μm

Fig.10 The farfield pattern of ASE broadband oscillator light and narrowband coherent laser.

For studying target interaction physics, a wide variety of irradiation power is required. To obtain high peak power laser, an ultra short pulse oscillator was developed (Fig.11). In contrast of the ns operation, the KrF laser medium operates in a storage mode because the pulse duration is much shorter than the effective life time. This means that high peak power pulses can be easily produced with a ps amplification. On the other hand, ASE control is extremely difficult for the short pulse amplification because the long pulse ASE light is easily amplified with small signal gain and that results in decreasing laser-ASE contrast ratio. A good saturable absorber for KrF laser was developed[7]. ASE light was well controlled and high peak power laser was easily obtained by the aid of acridine in ethanol.

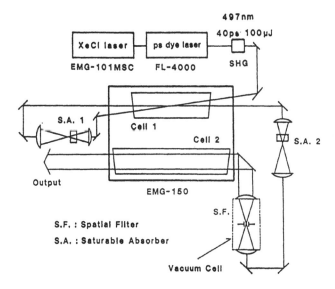

Fig.11 Schematic drawing of ultra short pulse oscillator

Target experiments

The laser plasma interaction with short wavelength laser light is very important for ICF research. Though many target experiments have been studied with the first, second and third harmonics of glass lasers, there is a small amount of data for 250nm laser light. Especially for scaling to reactor size target, it is important to obtain the data for interaction properties with various plasma size. By using one beam of our laser system, target shooting experiments was started under the condition of various plasma scale-length and irradiated power density.

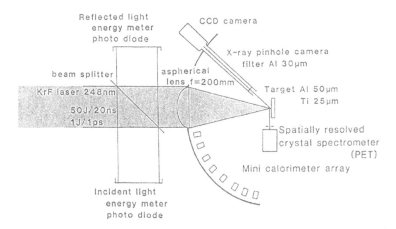

Fig.12 Experimental setup of target irradiation

In the first experiment, the irradiation condition was selected to be 20ns and 1×10^{13}W/cm^2. Al was selected for a target material. A long scale length plasma was created with this long pulse laser. A schematic diagram of the experimental setup is shown in fig.12. The dimension of created plasma was measured with a x-ray pinhole camera and length of plasma was determined to be 400μm (Fig.13). The lines of H-like and He-like ion was observed with a spatially resolved x-ray crystal spectrometer. The electron temperature profiles was estimated by the line ratio of 1s-2p/1s^2-1s3p and 1s-3p/1s^2-1s5p. These results denoted that the plasma was expanded isothermally and Te=400-500eV (Fig.14). To investigate absorption property, the scattered light was measured with several sets of mini-calorimeters. Figure 15 shows that most scattered light was re-entered the final focus lens (0.3%) and total absorbance was approximately 99%.

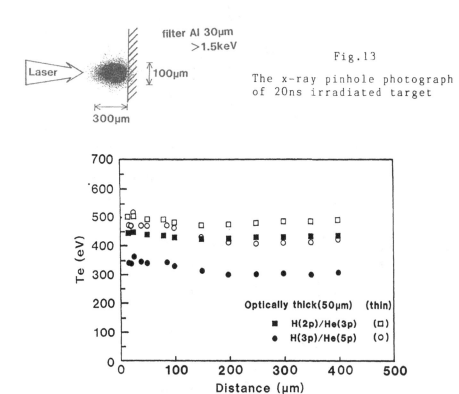

filter Al 30μm
>1.5keV

Fig.13

The x-ray pinhole photograph of 20ns irradiated target

100μm

300μm

Fig.14 Electron temperature versus distance from initial target position.

For studying the interaction in high density and short scale length plasma, the irradiation condition was changed to 2ps 1×10^{15}W/cm^2. Initial target condition was changed with prepulse laser light. If energy of prepulse light is large enough for producing preplasma on the target surface, the main laser interacts with the preplasma. In our system,

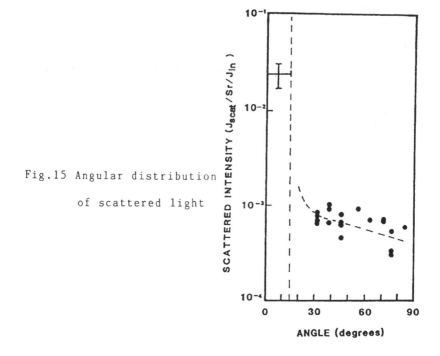

Fig.15 Angular distribution

of scattered light

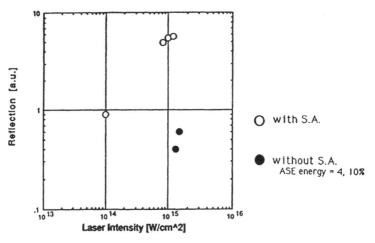

Fig.16 Reflectivity as a function of irradiation intensity
and prepulse energy

prepulse energy was controlled by newly developed saturable absorber; acridine in ethanol, with energy ratio of 10^{-4} to 1. The typical duration of prepulse was about 10ns. Backscattered intensity was measured as a function of irradiation intensity and prepulse energy. Typical results are shown in figure 16. The absorption changed drastically by a small amount of prepulse energy (4%). To understand this phenomena, reflectivity was calculated as a function of plasma scale length. In this calculation, only collisional absorption was considered and collision frequency of plasma was assumed to be a liner function of electron density. A calculated result is shown in fig.17. With normal expansion velocity(10^7cm/s), the plasma length can be estimated to be 0.2μm in the case of no preplasma condition. The reflectivity, therefore, is supposed to be 0.4 from the calculated results. The experimental results said that the reflectivity of preplasma condition decreased with an order of magnitude. These results denoted that very small prepulse light (4% energy) produced tens of scale length plasma before the main laser irradiation.

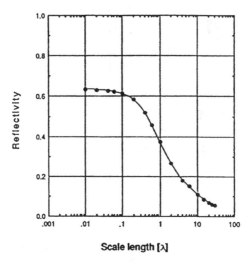

Fig.17 Calculated reflectivity versus scale length

Table 1. Comparison of detectable line (6-8Å)
between 2ps and 20ns irradiation

	2 p s	2 0 n s
He-like	$1s^2 - 1s3p$ $1s^2 - 1s4p$	$1s^2 - 1s3p$ $1s^2 - 1s4p$ $1s^2 - 1s5p$ $1s^2 - 1s6p$
H-like	———	$1s - 2p$ $1s - 3p$ $1s - 4p$ $1s - 5p$

To compare x-ray emission property of plasma, the Al targets were irradiated with the same intensity(5×10^{13}W/cm^2) but different pulse length (2ps & 20ns). Line emission between 6-8A were observed and the detected lines were summarized in table 1. Only He like lines were detected in the case of 2ps laser while H like lines also detected, of which intensity was the nearly same as He like lines, in that of 20ns . In 2ps irradiation, the emission length of Heβ line is much larger than that of thermal expansion during the laser pulse. In addition, the spectral broadening were the same value for both cases of the irradiation (Fig.18). These results suggested that this line emission was occurred after laser irradiation and some expansion in the case of 2ps irradiation.

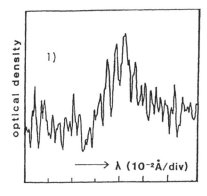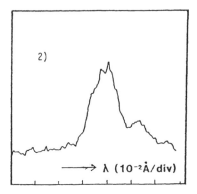

Fig.18 Heβ spectral line shape of
1)2 ps pulse irradiation
2)20 ns pulse irradiation

ACKNOWLEDGMENTS

The authors wish to acknowledge the excellent technical assistance of T. Arai, T. Kimura and H. Kuranishi. This work was supported by a Grant-in Aid for Scientific Research from the Ministry of Education, Science and Culture in Japan.

REFERENCES

1) Richardson, M.C. et.al. Phys. Rev. Lett., 56, 2048 (1986)
   Rosocha, L.A. et.al. Fusion Technol. 11, 497 (1987)
2) Salesky, E.T and W.D. Kimura, IEEE J. Quantum Electro., QE-21, 1761 (1985)
   Kannari, F. Appl. Phys. Lett. 45, 305 (1986)
   Ueda, K. et.al. Proc. of SPIE vol.710, pp.7 (1986)
3) Ueda, K. and H. Takuma, "Shortwavelength Lasers and their Applications "(Springer-Verleg) pp.178 (1988)
4) Sasaki, A., K. Ueda and H.Takuma, J. Appl. Phys. 65, 231 (1989)
5) Ueda, K. et. al. Technical Digest of CLEO'86 (1986)
6) Lehmberg, R.H. et.al. Fusion Technol. 11, 532 (1987)
7) Nishioka, H. et.al. Opt. lett. 14, 692 (1989)

THE GAMMA-RAY LASER - STATUS AND ISSUES IN 1989

C. B. Collins, J. J. Carroll, M. J. Byrd, K. N. Taylor,
T. W. Sinor, F. Davanloo, J. J. Coogan, and C. Hong

Center for Quantum Electronics
University of Texas at Dallas
P.O. Box 830688
Richardson, TX 75083-0688

ABSTRACT

Efforts to demonstrate the feasibility of a gamma-ray laser scored major advances in 1988 and 1989. Culminating with the discovery of giant pumping resonances below 4 MeV for dumping isomeric populations, priority issues were brought into better focus by the lessons learned from a wealth of new results. Perceptions were advanced so greatly that we must reassess the critical issues for the 1990's. Only the bottom line remains the same. *A gamma-ray laser is feasible if the right combination of energy levels occurs in some real material.* The likelihood of this favorable arrangement has been markedly increased by the experimental results to date.

INTRODUCTION

From the beginning of research into the feasibility of a gamma-ray laser, it had been realized that levels of nuclear excitation which might be efficiently stimulated would be very difficult to pump directly. To have sharply-peaked cross sections for stimulated emission, such levels must have very narrow widths for interaction with the radiation field. This is a fundamental attribute that has led to the frequent misconception that "absorption widths in nuclei are too narrow to permit effective pumping with x-rays."

The same concerns had been voiced in atomic physics before Maiman's great discovery, and it has proven very useful to pursue this analogy between ruby and gamma-ray lasers. The identification and exploitation of a bandwidth funnel in ruby were the critical keys in the development of the first laser. There was a broad absorption band exciting a state of Cr^{3+} which quickly decayed by cascading its population into levels of lower energy. A reasonably favorable pattern of branching insured that much of the cascading populated the narrow level. At the core of the simplest proposal for pumping a gamma-ray laser is the use of the analog of this effect at the nuclear level as shown in Fig. 1. A detailed analysis of this mechanism was reviewed[1] as early as 1982 while the first evidence for its existence was reported[2,3] as a breakthrough in 1987. Yields of gamma-ray fluorescence in ^{77}Se and ^{79}Br were enhanced by eleven orders of magnitude by this effect[2,3] when isotopic samples were irradiated with bremsstrahlung pulses from the DNA/PITHON nuclear simulator.

Also shown in Fig. 1 is a further refinement of the incoherent pumping scheme benefiting from upconversion. As has been discussed before, upconversion as shown in Fig. 1c has many advantages. While it clearly enhances performance and efficiencies; upconversion also makes threshold itself much more accessible. Higher energy isomers need less pump energy to reach the broad states that would optimize bandwidth funneling, and the required pump energies can fall in the range where strong x-ray lines may be found to concentrate the spectral intensities.

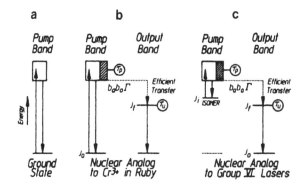

Figure 1. Schematic representation of the energetics of the priority schemes for pumping a gamma-ray laser with flash x-rays. The large width of the level defining the pump band is implied by the height of the rectangle representing the level and the shaded portion indicates that fraction, b_0, which is attributed to the transition to the upper laser level. Angular momenta of the ground, isomeric, and fluorescent levels are denoted by J_0, J_i, and J_f, respectively.
a) Traditional two-level approach.
b) Three-level analog of the ruby laser serving to illustrate the important concept of bandwidth funneling.
c) Refinement of the three-level scheme which incorporates upconversion in order to lessen the energy per photon which must be supplied in the pumping step.

Whether or not the initial state being pumped is isomeric, the principal figure of merit for bandwidth funneling is the partial width for the transfer, $b_a b_0 \Gamma$. Constituent parameters are identified in Fig. 2 where it can be seen that the branching ratios b_a and b_0 specify the probabilities that a population pumped by absorption into the i-th broad level will decay back into the initial or fluorescent levels, respectively. It is not often that the sum of branching ratios is unity, as channels of decay to other levels are likely. However, the maximum value of partial width for a particular level i occurs when $b_a = b_0 = 0.5$.

EARLY PERCEPTIONS

In 1986 one of the strongest tenets of theoretical dogma insisted that for processes of optical pumping involving long-lived isomers,

$$b_a b_0 \Gamma \leq 1.0 \ \mu eV \qquad , \qquad \qquad (1)$$

so that the efficacy of bandwidth funneling would be seriously limited in all important cases. The first successes[2-4] in demonstrating bandwidth funneling at the nuclear level showed partial widths of 39, 5, and 94 μeV for the excitation with bremsstrahlung of isomers of [77]Se, [79]Br, and [115]In, respectively, from ground state populations. While providing a "moral victory" by breaking the absolute limits of Eq. (1), these results still left an aura of credibility to the rule-of-thumb that partial widths for isomers would be limited to the order of magnitude of μeV.

The actual measurement of partial widths involves the correlation of fluorescence yields excited by a pulse of continuous x-rays in the scheme of Fig. 2 with those expected from the expression,[2-5]

$$N_f = N_0 \sum_i \xi_i \frac{\varphi_i}{A} \qquad , \qquad \qquad (2a)$$

where N_0 and N_f are the numbers of initial and fluorescent nuclei respectively, (φ_i/A) is the spectral intensity of the bremsstrahlung in keV/keV/cm^2 at the energy E_i of the i-th pump band, and the summation is taken over all of the possible pump bands capable of cascading to the same fluorescence level of interest. The ξ_i is a combination of nuclear parameters including the partial width $b_a b_o \Gamma$ in keV,

$$\xi_i \simeq \frac{(\pi b_a b_o \Gamma \sigma_o/2)_i}{E_i} \qquad , \qquad (2b)$$

where σ_o is the peak of the Breit-Wigner cross section for the absorption step. The Breit-Wigner cross section peaks at

$$\sigma_o = \frac{\lambda^2}{2\pi} \frac{2I_e+1}{2I_g+1} \frac{1}{\alpha_p+1} \qquad , \qquad (2c)$$

where λ is the wavelength in cm of the gamma ray at the resonant energy, E_i; I_e and I_g are the nuclear spins of the excited and ground states, respectively; and α_p is the total internal conversion coefficient for the two-level system shown in Fig. 1a. The combination of parameters in the numerator of Eq. (2b) is termed the integrated cross section for the transfer of population according to the scheme of Fig. 2.

Tempering expectations that these successes might be readily extended to the pumping of actual isomeric candidates for a gamma-ray laser was a concern for the conservation of various projections of the angular momenta of the nuclei. Many of the interesting isomers belong to the class of nuclei deformed from the normally spherical shape. For those systems there is a quantum number of dominant importance, K, which is the projection of individual nucleonic angular momenta upon the axis of elongation. To this is added the collective rotation of the nucleus to obtain the total angular momentum J. The resulting system of energy levels resembles those of a diatomic molecule for which

$$E_x(K,J) = E_x(K) + B_x J(J + 1) \qquad , \qquad (3)$$

where $J \geq K \geq 0$ and J takes values $|K|$, $|K| + 1$, $|K| + 2,\ldots$. In this expression B_x is a rotational constant and $E_x(K)$ is the lowest value for any level in the resulting "band" of energies identified by other quantum numbers x. In such systems the selection rules for electromagnetic transitions require both $|\Delta J| \leq M$ and $|\Delta K| \leq M$, where M is the multipolarity of the transition.

In most cases of interest, the lifetime of the isomeric state is large because it has a value of K differing considerably from those of lower levels to which it would, otherwise, be radiatively connected. As a consequence, bandwidth funneling processes such as shown in Fig. 1c must span substantial changes in ΔK and component transitions have been expected to have large, and hence unlikely, multipolarities.

Attempts to confirm these rather negative expectations in an actual experiment have been confounded by the rarity of the 29 candidate isomers of interest for a gamma-ray laser. Experiments[5] in which the simpler cycle of Fig. 1b was pumped through a change of $\Delta J = 4$ or 5 with a pulsed source of continua, at first confirmed these reservations, showing an integrated cross section of only 10^{-25} cm²-eV. Such values implied that one of the constituent transitions was significantly hindered as was expected for nuclei in which K and J remain good quantum numbers at all energies of relevance. The corresponding partial width was only 37 μeV, again tending to confirm the order of magnitude for the rule-of-thumb, Eq. (1). Dogma would insist that partial widths decrease further as the values of ΔK needed for transfer would be increased.

GIANT PUMPING RESONANCES

The candidate isomer $^{180}Ta^m$ was the one of the 29 possibilities that was the most initially unattractive as it had the largest change of angular momentum between isomer and ground state, 8ℏ. However, because it was the only isomer for which a macroscopic sample was readily available, $^{180}Ta^m$ became the first isomeric material to be optically pumped to a fluorescent level.

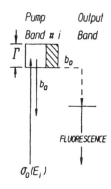

Figure 2. Schematic representation of the decay modes of a gateway state of width Γ sufficiently large to promote bandwidth funneling. The initial state from which population is excited with an absorption cross section σ_0 can be either ground or isomeric.

This particular one of the 29 candidates for a gamma-ray laser, $^{180}Ta^m$, carries a dual distinction. It is the rarest stable isotope occurring in nature[6] and it the only naturally occurring isomer.[7] The actual ground state of ^{180}Ta is 1^+ with a half-life of 8.1 hours while the tantalum nucleus of mass 180 occurring with 0.012% natural abundance is the 9^- isomer, $^{180}Ta^m$. It has an adopted excitation energy of 75.3 keV and half-life in excess of 1.2×10^{15} years.[7] Deexcitation of the isomer is most readily affirmed by the detection of the x-rays from the $_{72}Hf$ daughter resulting from decay of the ^{180}Ta ground state with an 8.1 hour half-life.

The target used in experiments[8] conducted at the end of 1987 was enriched to contain 1.2 mg of $^{180}Ta^m$ in 30 mg of ^{181}Ta. Deposited as a dusting of oxide near the center of the surface of a 5 cm disk of Al and overcoated with a 0.25 mm layer of Kapton, this sample was believed free from self-absorption of the x-rays from the daughter Hf.

Figure 3 shows the spectra of the enriched target before and after 4 hours' irradiation with the bremsstrahlung from a linac having a 6 MeV end point energy. Figure 4 shows the dependence upon time of the counting rate observed in the $Hf(K_\alpha)$ peaks after irradiation. Data points were plotted at the particular times at which the instantaneous counting rate equaled the average counting rate measured over the finite time interval shown. The figure shows the close agreement of the measured rates to the decay expected for a half-life assumed to be 8.1 hours.

From these data and from the calibrated doses of the pump radiation, the integrated cross section for the deexcitation of the isomer could be readily calculated if the reaction were assumed to occur through a gateway state narrow in comparison to the range of energies spanned by the irradiation. A *minimum value* of $\sigma\Gamma = 4.8 \times 10^{-22}$ cm^2-eV was obtained for the integrated cross section if the gateway energy was assumed to be near 2.0 MeV. Even larger cross sections would result from the assumption that the gateway lies at higher energies where the pumping flux was decreased. This is an enormous value exceeding anything reported for any interband transfer by two orders of magnitude. In fact, it is 10,000 times larger than the values measured for nuclei usually studied in our work.

The relatively straightforward analysis needed to extract the partial width for dumping the isomer gives an astonishing value for the partial width of,

$$b_a b_0 \Gamma(^{180}Ta^m) \simeq 0.5 \text{ eV} \tag{4}$$

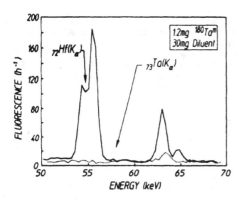 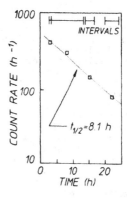

Figure 3. (Left) Dotted and solid curves show, respectively, the spectra obtained before and after dumping some of the isomeric ^{180}Ta^m contained in a target sample enriched to 5%. An HPGe detector was used to obtain the dotted spectrum before irradiation. The feature at 63 keV is from traces of natural activity in the counting shield. The solid curve shows activity resulting from the transmutation of the pumped ^{180}Ta measured in the same sample and counting system after irradiation. The prominent additions are the K_α and K_β hafnium x-ray lines resulting from electron capture in the ^{180}Ta.

Figure 4. (Right) Plot of the counting rates measured for the Hf(K_α) fluorescence from the target as functions of the time elapsed from the end of the irradiation. The vertical dimensions of the data points are consistent with 1σ deviations of the measured number of counts accumulated during the finite counting intervals shown at the top of the graph. The dotted line shows the rate expected for a half-life of 8.1 hours.

With this result of Eq.(4) the guideline of Eq.(1) is completely destroyed as a meaningful rule. The tenet of faith limiting to 1 μeV the partial widths for pumping isomers to radiating states has been proven to be nearly a million times too pessimistic. More recently[9] the reaction ^{176}Lu(γ,γ')^{176}Lu^m was reported to proceed with integrated cross sections and effective widths of comparable magnitude.

CALIBRATION OF PUMPING RESONANCES

It was the purpose of our most recent work to examine the systematics of the giant pumping resonances for the photoexcitation of isomeric states to determine whether such results were curiosities associated only with ^{176}Lu and ^{180}Ta^m or whether they were part of a more generally prevalent phenomena. To the extent possible it was intended to try to learn whether the large integrated cross sections were the results of a large number of reaction channels of more conventional size or rather, a few of unprecedented magnitude.

Unfortunately, the type of continuously variable electron accelerators that provided bremsstrahlung spectra with endpoint energies that could be selected in classical investigations[10] of (γ,γ') reactions is no longer available. For the experiments conducted in 1988 and 1989 we could arrange access only to a combination of accelerators, some with limited variability of endpoint energies such as DNA/PITHON at Physics International and DNA/Aurora at Harry Diamond Laboratories. Additional capabilities were contributed by two medical linacs having fixed endpoint energies of 4 and 6 MeV. This enabled us to examine the photoexcitation of 19 isomeric nuclei, most of them over the full range from 0.5 to 11 MeV. The diversity of accelerators and locations contributed an operational benefit by tending to minimize the possibility of inadvertently introducing any systematic bias which might have been associated with any particular machine or its environment.

The intensity of the bremsstrahlung produced by these different accelerators at the position of the irradiation target was most conveniently expressed by a normalized spectral intensity function F(E) multiplied by the total fluence deposited during the experiment. Then, when the resulting excitation was corrected for the finite duration of the irradiation by standard nuclear counting techniques, the activation yield could be readily determined. That quantity of activation is the ratio (N_i/N_0) which would be obtained from Eq. (2a) by dividing both sides by N_0.

A plot of the activation per unit x-ray dose can be useful as a sensitive indication for the opening of (γ,γ') channels whenever photons of the requisite energies, E_i become available. A change of the endpoint energy, E_{end} of the bremsstrahlung spectrum modulates the spectral intensity function $F(E_i)$ at all of the important gateway energies, E_i. The largest effect in the excitation function occurs when E_{end} is increased from a value just below some gateway at $E_i = E_k$ to one exceeding it so that $F(E_k)$ varies from zero to some finite value. In earlier work[10] plots of such quantities as functions of the endpoint energies of the irradiating spectra showed very pronounced activation edges which appeared as sharp increases at the energies, E_i, corresponding to excitation of new gateways. In our most recent work the relatively coarse mesh onto which endpoint energies could be set prevented any precise definition of the energies at which activation edges occurred for most nuclei. However, their existence was very clearly indicated in the data.

Normalized activation measurements obtained from all four accelerators for the excitation of $^{87}Sr^m$ are shown in Fig. 5, which also indicates the threshold energy for (γ,n) reactions, E_n, at 8.428 MeV. The datum at 1.2 MeV represents an upper limit since no fluorescence photons were observed above the background level. Although lacking in resolution in the critical range from 1.2 to 6 MeV, the data allow several conclusions to be drawn about the photoexcitation process. Most important is that, even though the amount of normalized activation above 1.2 MeV is suprising, there is no evidence to support the significant participation of any non-resonant processes. This type of mechanism, if present, would be heavily dependent upon the density of nuclear states, which rises sharply at energies approaching E_n. The slow increase in the excitation function above 6 MeV relative to the change seen below 4 MeV would seem to preclude this as the dominant means of photoexcitation.

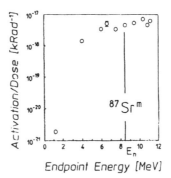

Figure 5. Normalized activation obtained from irradiations with all four accelerators for $^{87}Sr^m$. The size of the symbols is comparable to one standard deviation except where error bars are explicitly shown. The point at 1.2 MeV determined from a DNA/PITHON exposure is an upper bound on the excitation since no fluorescence photons were observed above the level of background. The vertical line indicates the neutron evaporation threshold at $E_n = 8.4$ MeV.

The large increase in normalized activation from 1.2 to 4 MeV indicates that at least one resonant gateway of significant magnitude lies in that range. The experimental resolution, however, does not allow a clear observation of the activation edges so the details of these states cannot be directly determined.

Fortunately, however, the isomer $^{87}Sr^m$ is distinguished by the degree to which its photoexcitation has been characterized in the literature. An early work,[10] which used an x-ray source of the type no longer available, examined the production of this isomer by bremsstrahlung with endpoints which could be continuously varied up to 3 MeV. The tunability of that device allowed three distinct gateways to be identified at 1.22, 1.88, and 2.66 MeV, and their integrated cross sections to be measured. In the usual units of 10^{-29} cm²-keV, these were reported to be (8.5 + 4 - 3), (16 + 8 - 5) and (380 + 200 - 100), respectively.

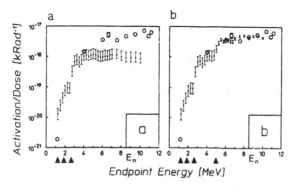

Figure 6. *Normalized activation obtained from irradiations with all four accelerators for $^{87}Sr^m$ as previously shown in Fig. 5. Also displayed are excitation functions calculated from Eq. (5) for specific gateways indicated by the triangular markers:*
a) Expected photoexcitation through only the three known gateways of Ref. 10. The error bars indicate the uncertainties in the measurements of that work.
b) Expected photoexcitation through the three known gateways plus that of a hypothetical state near 5 MeV with integrated cross sections on the order of 4000×10^{-29} cm²-keV. The error bars indicate only the uncertainties in the measurements of Ref. 10.

Figure 6a shows the ^{87}Sr data together with the normalized activations that should have been excited through the known gateways by photons with energies below 4 MeV. These values were calculated from Eq. (2a) for many endpoint energies by using typical bremsstrahlung spectra scaled to the various endpoints. The composite graph produced in this way exhibits the required activation edges at 1.22 and 2.66 MeV in agreement with the actual measurements. No edge is apparent at 1.88 MeV, but this is due to the comparable magnitude and proximity of this level to the one at 1.22 MeV.

The correlation of the expected values near 4 MeV with the datum there indicates that no new states are required to explain all of the normalized activation obtained with the 4 MeV linac.

Conversely, the data above 6 MeV in Figs. 6a significantly exceed the photoexcitation which could have been produced through the three known gateways. This extra activation must have therefore represented (γ, γ') reactions which proceeded through one or more unidentified levels. The simplest picture which matches the data is that of a single gateway near 5 MeV with an integrated cross section of the order of 4000×10^{-29} cm²-keV. The normalized activation expected from this state, as well as those previously identified, is shown in relation to the experimental data in Fig. 6b.

The nuclide ^{87}Sr provides a benchmark for other (γ, γ') studies since this is the only instance in which the current work can be compared with earlier experiments over a significant range of energies. It is rewarding that the present measurements of the photoexcitation of ^{87}Sr^m below 4 MeV are completely explained by resonant absorption of photons through gateways already reported in the literature.[10] The previous identification of these states has provided a means of validating the methology described above. Clearly, data such as shown in Figs. 6a and 6b can give useful indications of the locations and integrated cross sections of participating resonances even when insufficient experimental resolution is available for a precise determination of these quantities.

SYSTEMATICS OF GIANT PUMPING RESONANCE

While the measurements of isomeric fluorescence from ^{87}Sr^m were critically important in calibrating our experimental procedures, they did not reveal a giant resonance for pumping that nuclide. Of the 19 species examined in the past two years, the giant resonances for pumping have been most prevalent in the mass region from ^{167}Er to ^{191}Ir. The activation of ^{167}Er^m was typical of these nuclides and the relevant rates are shown in Fig. 7.

An examination of Fig. 7 indicates some interesting behavior. Although the normalized activation achieved with ^{167}Er nuclei is nearly two orders of magnitude larger than that of the ^{87}Sr benchmark, both isomers display similarly slow increases in activation above 6 MeV. This is surprising since non-resonant processes might be expected to be more significant for the photoexcitation of a nucleus as massive as that of ^{167}Er. Nevertheless, resonant absorption appears to be the dominant means of isomeric production for this nucleus as it was for the benchmark nuclide.

The data of Fig. 7 has a compelling resemblance to that for the excitation of one of the gateways in ^{87}Sr. The same sharp jump in activation with increasing x-ray endpoint is followed by relatively level yield up to 11 MeV. It is possible to fit the data of Fig. 7 with a single gateway near 4 MeV but more detailed considerations suggest there are two gateways in ^{167}Er. On the other hand, the excitation function for the reaction ^{179}Hf$(\gamma, \gamma')^{179}$Hf^m shown in Fig. 8 seems compelling in suggesting a single jump in activation at energies below 4 MeV. An integrated cross section of 60,000 ($\times 10^{-29}$ cm²-keV) at an excitation energy of 2.6 MeV would explain the data, but that is a minimal as opposed to a unique possibility. What is most significant is that this is the largest cross section ever observed for pumping nuclei with x-rays to obtain γ-ray fluorescence.

CURRENT ISSUES

It is an interesting speculation that at certain energies of excitation collective oscillations of the core nucleons could break some of the symmetries upon which rest the identification of the pure single particle states. This might occur as a result of couplings to states built upon cores of non-fissioning shape isomers.[11] Such states show double minima in energy as functions of elongation, even at low values of spin. At some values of excitation energy the shape of such a nucleus would be unstable and projections upon its principal axes would no longer be conserved. In this way the transition from a K value characteristic of the ground state to one consistent with the isomer might occur by mixing with such a state.

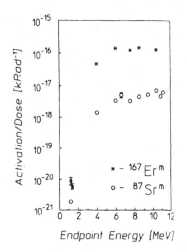

Figure 7. *Normalized activation obtained from irradiations with all four accelerators for $^{167}Er^m$. Also included for the purpose of comparison are those values for $^{87}Sr^m$. The size of the symbols is comparable to one standard deviation except where error bars are explicitly shown. The point at 1.2 MeV determined from a DNA/PITHON exposure is an upper bound on the excitation since no fluorescence photons were observed above the level of background. The neutron evaporation thresholds are not indicated for ^{167}Er and ^{87}Sr, but are 6.4 and 8.4 MeV, respectively.*

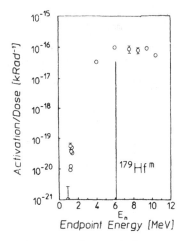

Figure 8. *Normalized activation obtained from irradiations with all four accelerators for $^{179}Hf^m$ ($T_{1/2}$ = 18.68 sec). The size of the symbols is comparable to one standard deviation except where error bars are explicitly shown. The vertical line indicates the neutron evaporation threshold at E_n = 6.1 MeV.*

In any case the pervasiveness found for the giant values of integrated cross sections for photoexcitation in the mass-180 region argues for some type of core property varying slowly with increasing nuclear size. Further work is needed to resolve the precise cause of this phenomenon.

Whatever the mechanisms, the experimental fact remains that interband transfer processes reaching isomeric levels can be pumped through enormous partial widths approaching 1 eV, even when the transfer of angular momentum must be as great as $\Delta J = 8$. Elucidation of the process, together with identification of the gateways, has been propelled into a place of importance for the 1990's.

The most available of the isomeric candidates for a gamma-ray laser, $^{180}Ta^m$, was shown to benefit greatly from this facility for bandwidth funneling. Successfully pumped with bremsstrahlung pulses having peak intensity of only $4W/cm^2$, the great width for the transfer in $^{180}Ta^m$ provided for adequate fluorescence signals from a milligram of isomer. This fixes a pragmatic scale for the evaluation of the other 28 candidates whenever samples become available in milligram quantities.

ACKNOWLEDGEMENT

This work was supported by IST/SDIO and directed by NRL.

REFERENCES

1. C. B. Collins, F. W. Lee, D. M. Shemwell, B. D. DePaola, S. Olariu, and I. I. Popescu, The Coherent and Incoherent Pumping of a Gamma Ray Laser with Intense Optical Radiation, J. Appl. Phys. 53:4645 (1982).
2. J. A. Anderson and C. B. Collins, Calibration of Pulsed Bremsstrahlung Spectra with Photonuclear Reactions of ^{77}Se and ^{79}Br, Rev. Sci. Instrum. 58:2157 (1987).
3. J. A. Anderson and C. B. Collins, Calibration of Pulsed X-Ray Spectra, Rev. Sci. Instrum. 59:414 (1987).
4. C. B. Collins, J. A. Anderson, Y. Paiss, C. D. Eberhard, R. J. Peterson, and W. L. Hodge, Activation of $^{115}In^m$ by Single Pulses of Intense Bremsstrahlung, Phys. Rev. C 38:1852 (1988).
5. J. A. Anderson, M. J. Byrd, and C. B. Collins, Activation of $^{111}Cd^m$ by Single Pulses of Intense Bremsstrahlung, Phys. Rev. C 38:2838 (1988).
6. A. B. W. Cameron, "Essays in Nuclear Astrophysics," C. A. Barnes, D. D. Clayton, and D. N. Schramm, eds., Cambridge Univ. Press, Cambridge (1982).
7. E. Browne, Nucl. Data Sheets 52:127 (1987).
8. C. B. Collins, C. D. Eberhard, J. W. Glesener, and J. A. Anderson, Depopulation of the Isomeric State $^{180}Ta^m$ by the Reaction $^{180}Ta^m(\gamma,\gamma')^{180}Ta$, Phys. Rev. C 37:2267 (1988).
9. J. J. Carroll, J. A. Anderson, J. W. Glesener, C. D. Eberhard, and C. B. Collins, Accelerated Decay of $^{180}Ta^m$ and ^{176}Lu in Stellar Interiors through (γ,γ') Reactions, Astrophys. J. 344:454 (1989).
10. E. C. Booth and J. Brownson, Electron and Photon Excitation of Nuclear Isomers, Nucl. Phys. A98:529 (1967).
11. M. Girod, J. P. Delaroche, D. Gogny, and J. F. Berger, Hartree-Foch-Bogoliubov Predictions for Shape Isomerism in Nonfissile Even-Even Nuclei, Phys. Rev. Lett. 62:2452 (1989).

MODELLING OF INTENSE LINE RADIATION FROM LASER-PRODUCED PLASMAS

Yim T. Lee and M. Gee

Lawrence Livermore National Laboratory
Livermore, CA, U.S.A.

I INTRODUCTION

Resonance line radiation has been observed in many laser-produced plasma experiments. These lines which are emitted by a dipole transition from an excited state to ground state are usually the brightest in the spectra. Measured intensity ratios have been used to estimate plasma temperature and density[1-11]. However, for these measurements to be useful, the plasma must be optically thin to the lines. Micro-dot targets[4-5] have been used to produce plasmas with small enough optical depths in the transverse direction to allow the lines to escape.

Re-absorption (i.e. trapping) of resonance lines in a plasma will not just make the interpretation of an emission spectrum difficult, but it can affect the plasma condition. For an example, the plasma can become hotter and more ionized. However, for low - Z plasmas such as aluminum and chlorine which are of interest here, the effect of trapping of optically thick lines on the plasma temperature is usually small. This is because radiative loss in these plasmas is only a small fraction of the total energy. However, trapping will enhance populations of ions in excited states. This usually results in a more ionized plasma because ionization from excited states dominates over other collisional and radiative processes at the density of laser-produced plasmas.

Trapping of optically thick resonance lines can also have an important effect on the performance of a x-ray laser. It populates the lower state of a laser transition, thereby resulting in a laser with a lower or no gain. The effect of trapping of optically thick lines on the gain coefficients of both recombination and collisional excitation pumped soft x-ray lasers has been investigated recently[12-13]. It was found that large ion velocity-gradients usually generated in laser-produced plasmas play an important roll in the transfers of optically thick lines. As a result of Doppler shifts, line absorption occurs only over the distance at which the shifts equal to the width of a line. The importance of such an effect on transfers of optically thick lines in astrophysical plasmas was pointed out by V. V. Sobolve in 1957[14]. In the laser-produced plasmas which have been used as amplifying media, the distance over which absorption occurs is also much less than the plasma size. This means absorption occurs only locally. As a result, escape probability should give good estimates of the line-transfers in these plasmas. This was pointed out several years ago[15-16] and was recently shown to be true by comparing the results from escape probability calculations to that from detailed line-transfer calculations in both planar[12] and cylindrical[17] geometries.

In this paper, we discuss modelling of Lyman-α (i.e. Ly-α) radiation emitted from laser-produced plasmas. We are interested in the application of one of these line radiations to pump a transition of an ion in a different plasma spatially separated from the emitting source. The interest is in perturbing the plasma rather than just probing it as in some backlighting experiments. As a result of pumping, the populations of certain excited levels are inverted. The resulting gain coefficients depend strongly on the population inversion density which in turn depends on the brightness of the pump radiation. As a result, we must produce an intense bright radiation source. In addition, to pump a transition effectively, we also need a pump line with a width larger than the mismatch of the resonance since the widths of the pumped transitions are rather narrow.

As an example on the modelling of resonance line radiation, we consider the Ly-α lines of aluminum emitted from laser-produced plasmas. Recently, one of these lines has been proposed[18-19] as a pump line for the 2s-5p transitions in lithium-like iron. As a result of pumping, population inversions and hence gains are possible for the transitions between the n=5 and n=4 levels. Recently, we have performed gain calculations of lithium-like iron ions in a steady-state plasma or an exploding-foil target[18]. These results show that gain coefficients of 3-5 cm^{-1} are achievable in an exploding-foil plasma where the lithium-like ions are pumped by a line with brightness greater than 10^{12} Wcm^{-2}.

To produce such an intense line radiation, we consider a source to be a massive slab target which is irradiated by high-power optical laser beams. In particular, we investigate the following two types of targets: 10μm aluminum slab and 1μm aluminum foil on top of 15 μm CH slab. These slab targets have been used in many laser-interaction experiments which measure shockwaves[20], x-ray conversion[21], and intense K- and L-shell line emissions[22]. In a 10μm slab target, the Ly-α lines are mostly emitted from the front side of the slab which is irradiated directly by the optical laser. In a 1 μm aluminum foil target, the line radiation is mainly emitted from the back side of the foil which is between the aluminum and CH. In these targets, we find that in order to produce an intense bright line the plasma condition has to be such that the source function of the lines would peak inside the critical density.

In next section, we discuss the lithium-like iron laser scheme in which the 5p levels are pumped by aluminum Ly-α radiation emitted from a laser-produced plasma. The line brightness which is needed for pumping to achieve population inversions and gains are considered. Section III presents calculations of the Ly-α lines emitted from an aluminum slab target. We also show how these line brightness depends on the optical laser irradiance on a target.

II APPLICATION OF INTENSE BRIGHT RESONANCE LINE RADIATION

In this section, we discuss application of resonance line radiation in the pumping of a x-ray laser. Here, as an example, we discuss pumping of the 5p levels in lithium-like iron ions by a Ly-α line ($2p_{3/2}$-$1s_{1/2}$ transition) of aluminum. This resonance was first discovered by Lee, et al. in 1988[18]. Several other line pumped laser

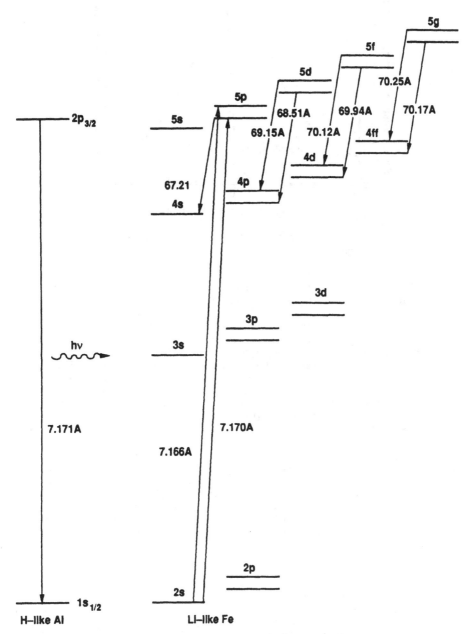

Fig. 1. A laser scheme based on pumping of lithium-like iron ions by using a Ly-α line of aluminum.

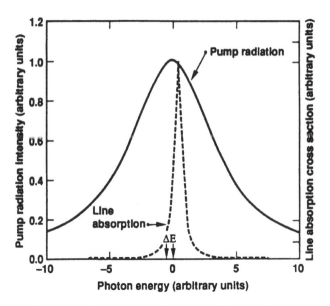

Fig. 2.   Overlapping between the pump line radiation and the
          pumped transition.

schemes in which lithium-like ions are photoexcited by either Ly-α or
He-α radiation have been proposed recently[19].

    A simplified energy level diagram for lithium-like iron is given in
Figure 1. Also shown in the figure are the wavelengths of the pumping and
pumped transitions. These wavelengths match better than a few parts in
ten thousands. In this laser scheme, pumping of the 5p levels will result
in populating the other n=5 levels such as 5d, 5f, and 5g because of
strong collisional mixing among them. As a result of pumping, the
populations of the n=5 and n=4 levels are inverted. The transitions
between these levels in lithium-like iron have wavelengths of
approximately 70 Å.

    To achieve lasing on these transitions, exploding foil targets have
been proposed as possible laser amplifying media[18]. These plasmas are
usually produced by irradiating two opposite high-power optical laser
beams in line focus on a thin foil with thickness of a 1000 Å.
Previous calculations[18] have shown that the 5g-4f transitions in
lithium-like iron have gains of 2-3 cm^{-1} if an ion is pumped by a line
with intensity greater than 10^{12} Wcm^{-2}. Among the dipole allowed 5-4
transitions, the 5g-4f transitions have the highest gains because of their
large oscillator strengths.

    In these gain calculations, it is assumed that the pumped and pumping
transitions are overlapped as shown in Figure 2. The widths of the pumped
transitions are determined by thermal Doppler broadening because Stark and
opacity broadenings are negligible at the density of the exploding foil
targets. Assuming an ion temperature of 500 eV, we calculate the width of
the $2s_{1/2}$-$5p_{1/2}$ transition in lithium-like iron to be 0.34 eV. This
approximately equals the mismatch of the resonance shown in Figure 1.
Therefore, the width of the pump line has to be larger that. In next
section, we show that it is possible for laser-produced plasmas to emit
Ly-α lines with widths which are 4-5 times larger than their thermal
Doppler widths.

III CALCULATION OF LY-α RADIATION FROM ALUMINUM SLAB TARGETS

To produce intense bright Ly-α lines, we consider a possible source to be a slab target which is irradiated by a high-power optical laser beam. In our calculations, we have used both a 0.35 μm wavelength laser beam in a gaussian pulse of 450 ps. FWHM and a 0.53 μm wavelength laser beam in a flat top pulse of 1 ns. Two types of targets are used: a 10 μm aluminum slab (target A) and a 15 μm CH slab (target B) with 1 μm of aluminum foil on top. In target A, line radiation is mostly emitted from the front surface of a slab which is heated directly by an optical laser beam. In target B, line radiation is mostly emitted from the back surface of a CH slab. The intensity of the line radiation at the front side of the aluminum foil is much smaller and the line profile is significantly Doppler shifted.

Our calculations of Ly-α line radiation emitted from these targets consist of two steps. First, we model absorption of an optical laser beam and heating of a target by using the hydrodynamics code LASNEX[23]. In these one-dimensional calculations, the plasmas are assumed to expand into a spherical cone to account for their diverging flows. Calculations done this way have been found to give reasonable results when they are compared to two-dimensional simulations[24]. In the second step of the calculations, hydrodynamics output such as temperature, density, and continuum radiation are input together with an atomic model to the kinetics code XRASER[25] to compute detailed transfers of optically thick resonance lines. The line-transfers are calculated by using a multi-angles and multi-frequencies numerical method[12]. We have also assumed complete re-distribution of the line profile in the calculations. In these calculations, the intrinsic line width is assumed to be the thermal Doppler width. Other contributions such as Stark broadening and electron broadening are in general very small at the density of the slab target. However, as given later, the widths of the emission lines from these targets are much wider than the intrinsic widths of the lines because of broadening by opacity and ion velocity gradients in a plasma.

As an example of a typical LASNEX calculation, we plot in Figure 3 the temperatures at different zones of a 10 μm aluminum slab target at a time near the peak laser irradiance. The target is assumed to be irradiated by a 0.35 μm wavelength laser beam in a gaussian pulse of 450 ps. FWHM and at a peak intensity of 8×10^{14} Wcm^{-2}. The charge-state distribution at different zones of the target is shown in Figure 4. We see that hydrogen-like population is less than 10% throughout most of the underdense plasma and reaches a maximum value of 40% inside the overdense plasma. However, near the front surface of the slab, hydrogen-like population starts to increase with the decrease in electron density. This is because the contribution of ionization from excited states decreases very fast with density, approximately as a square of density. The recombination rates, on the other hand, decrease only linearly with density. Therefore, the plasma should recombine with the decrease in density as shown in Figure 4.

To estimate the importance of trapping of resonance lines in a slab target, we have calculated the optical depths at different locations in the target. The optical depth just inside the critical density are about 20 at a time near the peak irradiance of a laser pulse. This shows that most of the lines emitted inside the critical density are trapped. In addition, as shown later, the transfers of these lines are very important in determining the line emission from the target.

In Figures 5a-5b, we plot the source function of the Ly-α lines

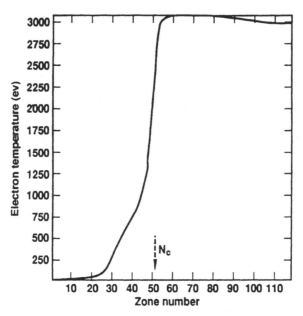

Fig. 3.   Electron temperature inside a 10μm aluminum slab at a
time near the peak intensity of a laser beam.   The laser
beam which has a wavelength of 0.35 μm is in a gaussian
pulse of 450 ps FWHM and at an intensity of 8×10^{14}
Wcm^{-2}.   The arrow points at the zone where the critical
density is located.

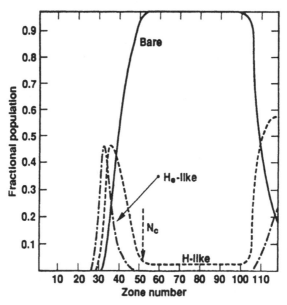

Fig. 4.   Fractional populations of aluminum ions inside a 10μm
slab at a time near the peak intensity of a laser beam.
The laser parameters are the same as in Figure 1.   The
arrow points at the zone where the critical density is
located.

inside an aluminum slab target at a time near the peak laser irradiance. In these figures, we express brightness in an unit of photons /mode, $N(h\nu)$. Radiation intensity in an unit of Wcm^{-2} can be obtained from $N(h\nu)$ as,

$$I_{h\nu} = [8\pi(h\nu)^3/c^2] N(h\nu)$$
$$\approx 6.332 \times 10^4 (h\nu)^3 N(h\nu) \quad W/(cm^2\text{-}eV)$$

where $h\nu$ is the photon energy in electron-volt. As an example, a resonance line with an energy of 1729 eV, a brightness of 0.005 photons/mode and a line width of 1 eV is equivalent to an intensity of approximately 10^{12} Wcm^{-2}. Figure 5a shows the results for a 10 μm aluminum slab which is irradiated by a laser pulse with an intensity of 8×10^{14} Wcm^{-2} and Figure 5b shows the results corresponding to the laser intensity of 1.6×10^{14} Wcm^{-2}. Comparing these results, we find that irradiating the target with a higher laser irradiance would produce a plasma such that the source function of the lines would have larger values. In addition, these values are located at a higher density part of the overdense plasma.

In the slab targets, the source function of the Ly-α lines depends on the populations of the 2p levels of hydrogen-like aluminum ions. Since these levels are mainly populated by electron collisional excitation from the ground state, their populations are approximately proportional to the collisional excitation rates and to the ground state population of hydrogen-like aluminum. The collisional rates depend on both the plasma temperature and density. Because of the linear dependence of the collisional rates on the electron density, the plasma should be as dense as possible. The dependence of the collisional rates on the temperature is more complicated. For dipole allowed transitions such as the $1s_{1/2}$-$2p_{1/2}$ and $1s_{1/2}$-$2p_{3/2}$, the rate coefficients usually are the largest near the temperatures which equal to the excitation energy. At a temperature below one half of the excitation energy, the rate coefficients decrease exponentially with temperature. Therefore, the plasma should have a temperature between one half of the excitation energy and the excitation energy.

The dependence of the ground state populations of hydrogen-like aluminum ions on the condition of a plasma is also complicated. This is because the plasma is at non-Local Thermal Equilibrium. In addition, trapping of optically thick resonance lines can deplete the ground state populations. According to our calculations, an aluminum plasma with a temperature in a range of 800-1200 eV and at a density of a few times 10^{22} cm^{-3} would have hydrogen-like populations at least of 10%. At a temperature above 1200 eV, the plasma will be over ionized, containing mostly fully stripped ions. For an example, hydrogen-like aluminum populations in a plasma with a temperature greater than 1500 eV and an electron density of 10^{22} cm^{-3} are less than a few %.

The density consideration of a plasma suggests that, as a source of bright Ly-α lines, massive slab targets are better than thin foil targets. These foil targets which are burned through by the optical laser beams produce rather uniform but low density plasmas. Typical densities in a foil target which is irradiated by a laser driver with 0.35μm wavelength are a few times 10^{20}-10^{21} cm^{-3}. These densities are more

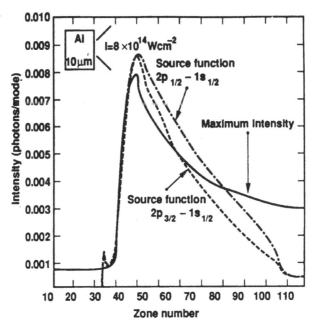

Fig. 5a. Maximum intensity and source function of aluminum Ly-α lines at a time near the peak intensity of a laser beam.

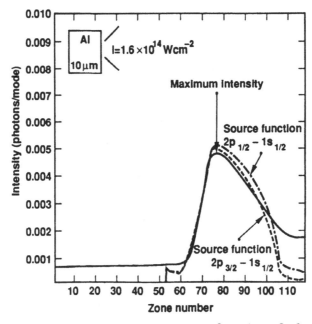

Fig. 5b. Maximum intensity and source function of aluminum Ly-α lines at a time near the peak intensity of a laser beam.

suitable as a laser amplifying medium than as a source of line radiation.
In addition, thin foil targets also emit line radiation which is
significantly Doppler shifted   Such resonance lines have been observed in
both experiments and computer simulations[26].

Figure 6 shows the source functions of the Ly-α lines in a 1μm
aluminum foil of target B.   In the calculations, the target was assumed to
be irradiated by a 0.53 μm wavelength laser beam in a 1 ns flat top
pulse with an intensity of 4.0×10^{14} Wcm^{-2}.   To model the
radiation at the boundary between the aluminum and the CH, we had to fine
zone the aluminum at the boundary.   Typically, the first few zones at the
surface were choosen so they will have optical depth less than 0.1 at a
time near the peak irradiance of a laser pulse.

Also shown in Figures 5a-5b and 6 are the maximum intensity of the
Ly-α line radiation at each zone of the slab target.   The intensity is
averaged over solid angles.   In target A, the maximum intensity at the
front surface of the slab is 6-7 times larger than the local source
function.   This is because the intensity is characteristic of the values
of the source function which is within several optical depths from the
surface of the slab target.   Near the surface, most of the line radiation
are transfered here from the dense part of the target where the lines are
first emitted.   In target B, the maximum intensity at the front surface of
the foil is approximately 10 times larger than the local source function.
However, at the back surface of the foil, the maximum intensity is smaller
than the local source function by as much as 70%.   We can understand this
result as follows.   Since there is no radiation entering the aluminum foil
at the boundary between the aluminum and the CH, the angle averaged
intensity at the first zone of the boundary is at least a factor of two
less than the intensity at an angle perpendicular to the target surface.
If the line radiation is emitted isotropically, the angle averaged
intensity would be exactly half of the intensity at any angle.   Taking
this into consideration, we see that the intensity at an angle

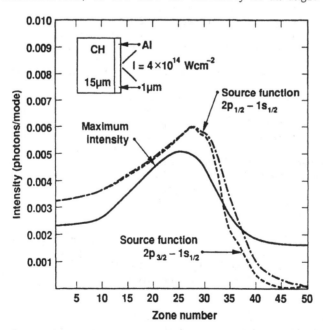

Fig. 6.   Maximum intensity and source function of aluminum
Ly-α lines at a time near the peak intensity
of a laser beam.

179

perpendicular to the target surface is actually larger than the local
source function. Again, as in the front surface of the foil, part of the
intensity is characteristic of the values of the source function inside
the target surface.

Figures 7a-7b present the angle averaged intensity profile of the
Ly-α lines at two zones 40 and 118 of a 10 μm aluminum slab
target. The laser irradiance on the target is the same as in Figure 5a.
Zone 40 is inside the critical density, while zone 118 is the first zone
at the surface of the slab. In our calculations, the first zone has been
choosen so it will have a small enough optical depth to ensure the line
emission intensity from the surface of the slab would be the same as the
line intensity in the first zone. Therefore, Figure 7b also represents
the line emission profile from the target.

The Ly-α lines consist of two fine-structure components, namely
the $2p_{1/2}$-$1s_{1/2}$ and $2p_{3/2}$-$1s_{1/2}$ transitions. These two
transitions are separated in energy of approximately 1.3 eV. In Figures
7a-7b, zero photon energy corresponds to the energy of the $2p_{1/2}$-$1s_{1/2}$
transition, while the arrow points at the energy of the $2p_{3/2}$-$1s_{1/2}$
transition.

Inside the critical density, the Ly-α lines are red shifted by an
amount of approximately 0.2 eV. This is due to the contribution of these
lines which are first emitted inside the ablation front. These lines are
red shifted because the ions are moving away from the front surface of the
slab. According to LASNEX calculations, ions inside the ablation surface
have velocity which can Doppler shift the line to a lower energy by 0.2
eV. The intensity profile at the surface of the slab is also red shifted,
but by an amount of 0.3-0.4 eV.

Figures 7a-7b show that the width of the Ly-α lines is
approximately 3.7 eV inside the critical density and approximately 3.5 eV
at the surface of the slab. This suggests that the width of each
fine-structure component is at least of 2.0 eV. Since the largest ion
temperature inside the critical density is approximately 1200 eV according
to LASNEX calculations, this means that the thermal Doppler width should
be no more than 0.80 eV. Therefore, the results in these figures suggest
that the lines are broadened by opacity and ion velocity gradients.

At the surface of the slab, the values of the intensity profile near
the line-center of the $2p_{1/2}$-$1s_{1/2}$ transition are larger than that
near the line-center of the $2p_{3/2}$-$1s_{1/2}$ transition. Since the $2p_{3/2}$
level population is a factor of two greater than the $2p_{1/2}$ level
population and the radiative rates for the two transitions are the same,
one might expect that the line emission intensity of the $2p_{3/2}$-$1s_{1/2}$
transition would be higher than the intensity of the $2p_{1/2}$-$1s_{1/2}$
transition. However, as discussed previously, most of the line emission
come from the high density part of the plasma. Therefore, line absorption
plays a very important roll in determining the line intensity profile at
the surface. The absorption cross section of the $2p_{3/2}$-$1s_{1/2}$
transition is a factor of two larger than that of the $2p_{1/2}$-$1s_{1/2}$
transition. As a result, there will be more absorption of the high energy
component of the Ly-α lines. This explains why the maximum value of
the line emission profile is near the low energy component of the Ly-α
lines.

Figure 7b shows that at the high energy side of the profile there is
a small peak at approximately 2 eV above the line-center of the
$2p_{3/2}$-$1s_{1/2}$ transition. In our calculations, we have found that the
line emission profile would have that structure if the laser irradiance on

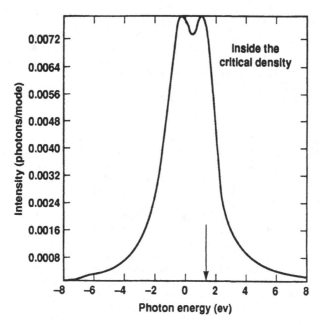

Fig. 7a. Intensity profile of aluminum Ly-α lines at
a zone number 40 which is inside the critical
density surface of the slab. The arrow locates
the line-center of the $2p_{3/2}$-$1s_{1/2}$ transition.

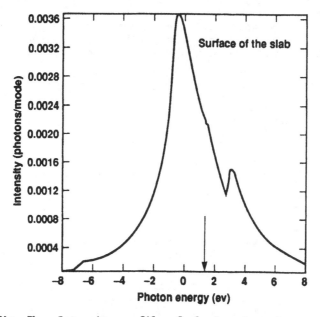

Fig. 7b. Intensity profile of aluminum Ly-α lines at
the surface of the slab target. The arrow
locates the line-center of the $2p_{3/2}$-$1s_{1/2}$
transition.

the target is greater than 4×10^{14} Wcm^{-2}. As the laser irradiance
on the target increases, the location of the peak also shifts toward
higher energy. Based on these results, we see that the peak represents
the contribution from the source of the Ly-α lines which are emitted
outside the critical density. At the laser irradiance of 4×10^{14}
Wcm^{-2}, the critical density surface would be expanding at the velocity
of a few times 10^7 cm/sec. Ions moving at such a velocity would emit
lines which are blue shifted by about 2 eV. This agrees with the results
in Figure 7b.

In Figure 8, we show how the brightness of the Ly-α emission from
a 10μm aluminum target would depend on the optical laser irradiance.
To show the emission intensity profile, we plot in Figure 8 the maximum
intensity and the line-center intensity at the first zone of the surface
of the slab. Here, by line-center intensity, we mean the line brightness
at the photon energy of the $2p_{3/2}$-$1s_{1/2}$ transition. In these
calculations, we assume a 0.35 μm wavelength laser driver in a
gaussian pulse of 450 ps. FWHM. Also shown in the figure is the maximum
intensity of the Ly-α lines in the slab target. This intensity is
several times brighter than the maximum intensity at the surface of the
slab. This is because most of the radiation emitted inside the slab are
trapped, but only a small fraction of it can escape from the target. In
addition, the maximum intensity of the Ly-α lines inside the target
increases very fast with the irradiance of the optical laser driver.

As a function of the optical laser irradiance on a target, the
intensity profile at the surface increases much slower than the maximum
intensity inside the target does. At the irradiance above 1.2×10^{15}
Wcm^{-2}, the line-center intensity also starts to decrease. At these high
irradiances, the plasma inside the ablation surface becomes hot enough to
emit Ly-α line radiation. These lines, however, are red shifted
because of the large ion velocity moving in the direction away from the
surface. For an example, at a laser irradiance of 1.2×10^{15} Wcm^{-2},
the temperature and ion velocity inside the ablation front could reach 900
eV and 2.0×10^7 cm/sec, respectively. Ions moving with such velocity

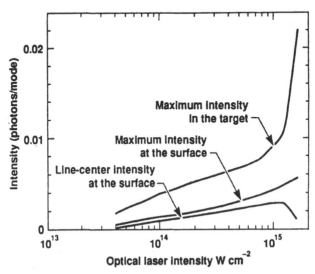

Fig. 8. Maximum intensity and emission intensity of the
aluminum Ly-α lines vs. optical laser driver
irradiance.

will emit Ly-α lines which are red shifted by as much as 2.5 eV. Such a shift is about the same as the line width of the $2p_{3/2}$-$1s_{1/2}$ transition. This explains why the line-center intensity decreases with laser irradiance above 1.2×10^{15} Wcm^{-2} as shown in Figure 8.

The results in Figure 8 suggest that pump lines with brightness of about .003 photons/mode are achievable using laser-produced plasmas. To produce a much brighter pump line would require more sophisticated target designs. That is beyond the scope of the present paper. However, a pump line with brightness of .003 photons/mode, according to previous calculations, can be used to pump the 5p levels in lithium-like iron to achieve gains of 2-3 cm^{-1} in the 5g-4f transitions. These values for the gains are certainly measurable in an experiment.

Acknowledgments

I am grateful to the following colleagues for many helpful discussions on several aspects of the work presented here: E. Alley, M. Howard, and H. Scott.

This work was performed under the auspices of the U. S. Department of Energy by the Lawrence Livermore National Laboratory under Contract W-7405-ENG-48.

REFERENCES

1 B. K. F. Young et al., Phys. Rev. Lett. 61, 2851 (1988).
2 J. P. Apruzese, D. Duston, and J. Davis, J. Q. S. R. T. 36, 339 (1986).
3 C. DeMichelis and M. Mattioli, Nucl. Fusion 21, 677 (1981).
4 M. J. Herbst et al., Rev. Sci. Instrum. 53, 1418 (1982).
5 P. G. Burkhalter et al., Phys. Fluids 26, 3850 (1983).
6 V. A. Boiko, I. Yu. Sobelev and A. Ya. Faenov, Fiz. Plazmy 10, 143 (1984) [Sov. J. Plasma Phys. 10, 82 (1984)]. 7 J. C. Gauthier et al., J. Phys. D 16, 1929 (1983).
8 P. Alaterre et al., Opt. Commun. 49, 140 (1984).
9 M. H. Key et al., Phys. Rev. Lett. 44, 1669 (1980).
10 R. L. Kauffman, R. W. Lee, and K. Estabrook, Phys. Rev. A 35, 4286 (1987).
11 R. W. Lee, J. Q. S. R. T. 27, 87 (1982).
12 Y. T. Lee, et al., "Application of Escape Probability to Line-Transfer in Laser-Produced Plasmas", Lawrence Livermore National Laboratory Report, UCRL-102219 (1989).
13 D. C. Eder, Phys. Fluids: B 1 (12), 2462 (1989).
14 V. V. Sobolev, Soviet Astr.-AJ 1, 131 (1987).
15 R. A. London, "1986 Laser Program Annual Report," Lawrence Livermore National Laboratory, CA, UCRL-50021-86 (1987).
16 G. Pert, J. Phys. B: Atom. Molec. Phys. 9, 3301 (1976).
17 D. C. Eder, private communication, Lawrence Livermore National Laboratory, CA.
18 Y. T. Lee, et al., J. Q. S. R. T. (1990, in press).
19 Y. T. Lee, et al., "Resonant Photopumping of Lithium-like Ions in Laser-Produced Plasmas", NATO-ASI volume on Non-Equilibrium Processes in Partially Ionized Gases (1990, in press).
20 For example, A. Ng, et al., Phys. Rev. Lett. 24, 2604 (1985).
21 For example, P. Alaterre, et al., Phys. Rev. A 34, 4184 (1986).
22 For example, D. L. Matthews, et al., J. Appl. Phys. 54, 4260 (1983).
23 G. B. Zimmerman and W. L. Kruer Comments Plasma Phys. Controlled Fusion 11, 51 (1975).

24 W. C. Mead, et al., Phys. Rev. Lett. <u>47</u>, 1289 (1981).

25 P. L. Hagelstein, Ph. D. thesis, Massachusetts Institute of
    Technology, (1981).

26 M. S. Maxon, et al., "1985 Laser Program Annual Report," Lawrence
    Livermore National Laboratory, CA, UCRL-50021-85 (1986).

OBSERVATION OF GAIN ON XUV TRANSITIONS IN Ne-LIKE AND Li-LIKE IONS

T. Boehly, D. McCoy, M. Russotto, J. Wang, and B. Yaakobi

Laboratory for Laser Energetics
University of Rochester
250 East River Road, Rochester, NY 14623-1299

I.   INTRODUCTION

Population inversion and amplified spontaneous emission (ASE) at soft x-ray wavelengths has been achieved using two methods of pumping: collisional and recombinational. In the collisionally excited x-ray laser, thermal electrons in the plasma excite the upper laser state. The inversion is a result of the faster deexcitation rate of the lower level. The first collisionally pumped laser was demonstrated by Lawrence Livermore National Laboratory (LLNL)[1] in Ne-like Se. This scheme has been extensively studied in various other elements and extended to several Ni-like species.[2] These experiments used thin exploding-foil targets to ensure that refraction effects are minimized. The pump laser burns through the target foil near the peak of the pulse, causing the foil to explode symmetrically. The resulting density profile is relatively flat in the central region and falls off in both directions along the incident laser axis. This allows x rays to propagate along the axis of the plasma and be significantly amplified before leaving the plasma.

Recent results indicate that slab targets can also be used to create collisionally excited x-ray lasing.[3] These targets are thick enough that the pump laser does not burn through the target mass during the pulse. The slab targets give rise to a steep density gradient in the direction of the incident laser. Since there is no axis of symmetry in the density profile, all rays suffer refraction away from the target. This behavior was originally thought to be a severe detriment to the attainment of gain. The results of Ref. 3 are thus unexpected and merit further study.

We present results of experiments which utilize slab targets and produce lasing in Ne-like Ge, using lower laser energy than in previous experiments. The wavelengths of the five lasing lines range from 196 Å to 287 Å. To produce sufficient excitation of the 3p state in the Ne-like Ge ion (of excitation energy 1480 eV), a temperature in the range of 600–1000 eV is required. Such temperatures are produced using on-target intensities of ~2 x 10^{13} W/cm^2 as evidenced by the observation of Ne-like lines in the x-ray and XUV spectra. We present length scaling of the output and also show measurements of the divergence of this x-ray laser.

In a recombination-pumped laser the mean ionization of the plasma is higher than that of an equilibrium plasma at the same temperature. At this nonequilibrium condition, which occurs following a rapid cooling, three-body recombination and radiative cascading dominate over collisional ionization and excitation processes. This tends to populate the upper levels (e.g., 4f and 5f in Li-like ions) more than the lower excited states, thus creating an inversion. Furthermore, the lower level (3d) has a short lifetime due to a strong radiative transition to the ground state.[4]

This nonequilibrium situation is typically created by heating the plasma with a short laser pulse and then allowing it to cool in a time which is short compared to the recombination time for the ions involved. Since the recombination time for ions of interest (i.e., lasing at wavelengths <200 Å) is of the order of tens of picoseconds, these types of experiments usually require laser pulses which are shorter than 100 ps. Such experiments have succeeded in producing various types of recombination lasing.[5] In long-pulse experiments, any lasing occurs while the pump laser is heating the plasma; it must therefore occur far from the target surface where the plasma absorbs very little laser energy and is thus able to cool quickly. At this point, collisional excitation and ionization (dominant in the region heated by the laser) is insignificant compared to recombination.

Recent results by Hara et al.[6] have shown that the intensity of the 3d-4f and 3d-5f lines of Li-like Al increase exponentially with length in experiments using very low power, 5 ns laser pulses. Further studies of this type of experiment are required. We present results of experiments with Li-like Al performed at somewhat higher intensities than reported in Ref. 6. We show apparent ASE of the lines at 105 Å and 154 Å. We also present results which extend the Li-like scheme to shorter wavelengths, i.e., a Li-like Ti transition at 47 Å. The Ti experiments also showed a lasing line at 326.5 Å. A tentative identification of this line is a $2p_{1/2}^5 3p_{1/2}$ (J = 0) to $2p_{1/2}^5 3s_{1/2}$ (J = 1) transition in Ne-like Ti which has been obtained by extrapolating the wavelength values of this Ne-like transition in various elements.[7] Finally, we present measurements of the angular distribution of the output of these x-ray lasers.

II.   EXPERIMENTAL CONFIGURATION

Experiments were performed using the Glass Development Laser (GDL)[8] at the Laboratory for Laser Energetics at the University of Rochester. The targets were irradiated with 650 ps pulses of 1054 nm light at various intensities. The line focus was produced by the addition of a cylindrical corrector lens to the f/3 spherical focus lens normally used in laser plasma interaction experiments. A grazing incidence grating spectrometer (GIGS) is used to view the plasma end-on and detect the output of the x-ray laser. The spectrometer has an acceptance angle of ~20 mrad in the direction along the entrance slit. Perpendicular to the slit, a focusing mirror provides collection angle of ~1 mrad. A crystal spectrometer and an x-ray pinhole camera are used to measure the x-ray spectra and the intensity distribution of the optical laser at line focus, respectively. All diagnostics are time integrated and use film for recording. Kodak 2497 film is used in the pinhole and crystal spectrometer and Kodak 101 film is used in the GIGS spectrometer. An advantage of using film in the GIGS rather than a one-dimensional microchannel plate is that a record of the angular output intensity perpendicular to the dispersion direction is obtained on film.

The targets consist of a 125-μm Mylar backing which is coated with ≥1-μm thickness of Ge. This thickness of Ge is much larger than the heat-front penetration depth during laser irradiation of Ge. The Mylar backing serves to provide a resilient support which ensures the flatness of the Ge surface. Segments of 40-μm glass fibers are used as alignment reference pins to be imaged in the target alignment microscope. Using these pins and the edge of the target we are able to align the target, spectrometer, and line-focus axes to better than 5 mrad accuracy.

Measurements of gain are performed by comparing the output intensity from targets of different length. The target length is varied by using different target slab lengths while maintaining the laser line-focus length. This ensures that variations due to laser irradiance differences are kept to a minimum since the targets sample only the central region of the beam. For example, the longest targets (22 mm) are overfilled by the line focus by ~2 mm. The x-ray and XUV film densities are converted to intensity using x-ray calibration data.[9] The reflectivity of the grating and collection mirror have not been absolutely calibrated; all intensities are therefore in relative units.

III.   RESULTS

(A)   Ne-Like Germanium

The presence of gain and the effect of laser length on x-ray laser output is demonstrated in Figs. 1 and 2. Figure 1 shows a small portion of the XUV spectra (from GIGS) which include the J = 2 - 1 lines from two shots of different target lengths. Both spectra are shown after conversion to relative intensity. Note that for the 10-mm long targets the lasing lines are barely discernible, whereas for the 15 mm length the lines begin to be the dominant feature in the spectrum. For focus lines shorter than 10 mm, the lasing lines could hardly be seen in the spectrum.

If the line focus is extended to 22 mm, the lasing lines become the brightest lines in the spectrum. Figure 2 shows the Ge spectrum (from GIGS) in the range of ~20–380 Å from a 22-mm long target irradiated at an intensity of 1.5×10^{13} W/cm^2. The lasing lines of Ne-like Ge are labeled a to e. The nonlasing lines have about the same intensity as that in the shots of Fig. 1. The carbon K absorption edge is at 43.6 Å and results from the carbon in the film emulsion and impurity coating on the GIGS optics. The lasing line designations are shown in Table 1.

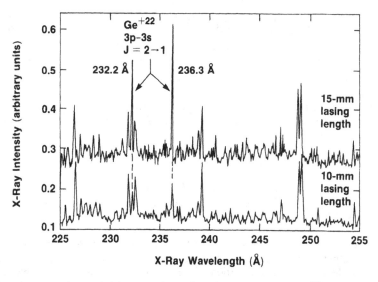

Fig. 1   A comparison of the output intensity of the J = 2 - 1 Ge^{+22} lines (232 Å and 236 Å) for two x-ray laser lengths. The lower spectrum is for a 10-mm plasma length and the upper spectrum is for a 15-mm length.

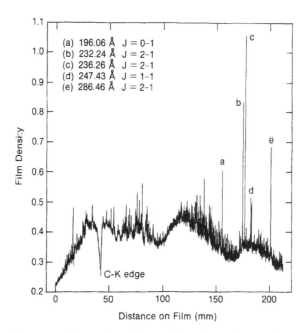

Fig. 2   A wide range of the spectrum from a 22-mm long Ge x-ray laser target. The lines
designated a – e are the various 3p-3s lasing lines observed in Ne-like Ge.

Table 1

Designation	Transition	Wavelength (Å)	Gain (cm⁻¹)
(a)	$(2p^5_{1/2}3p_{1/2})_0 - (2p^5_{1/2}3s_{1/2})_1$	196.06	1.5
(b)	$(2p^5_{3/2}3p_{3/2})_2 - (2p^5_{3/2}3s_{1/2})_1$	232.24	2.5
(c)	$(2p^5_{1/2}3p_{3/2})_2 - (2p^5_{1/2}3s_{1/2})_1$	236.26	2.5
(d)	$(2p^5_{3/2}3p_{3/2})_1 - (2p^5_{3/2}3s_{1/2})_1$	247.32	1.0
(e)	$(2p^5_{3/2}3p_{1/2})_2 - (2p^5_{3/2}3s_{1/2})_1$	286.46	1.8

Figure 3 is a plot of the relative intensity of the J = 2 - 1 lasing line at 236 Å as a
function of the Ge target length ℓ. The data points are obtained from different shots which
had nominally the same optical laser conditions (±10%). The solid curve is a best fit of the
analytic expression:

$$I \propto [\exp(g\,\ell) - 1]^{3/2}/[(g\,\ell)\exp(g\,\ell)]^{1/2} \qquad (1)$$

to the data. From this we derive the gain coefficient, g. This method was likewise applied
to the other lasing lines. The results are shown in Table 1 which lists the gain for the
various lines. The labels a – e correspond to the labeling in Fig. 2.

The angular width of the x-ray laser output can be measured using a spectrometer.
Since the output beam of the x-ray laser is of limited extent (~200 μm) and narrow
(~10 mrad), the rays fill only a portion of the entrance slit of the spectrometer and are
relayed to the film plane as a narrow beam. Hence, depending on the acceptance angle of
the spectrometer, only a limited length of the film is exposed by the x-ray laser. In
contrast, spontaneously emitted x rays are isotropic and they fill the slit and form a spectral
line which extends all across the film.

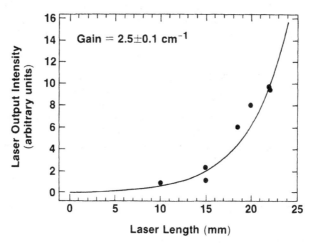

Fig. 3   Intensity of the 236 Å line of Ge^{+22} versus plasma length. The gain is obtained from the solid curve which is a fit of Eq. (1) to the experimental points.

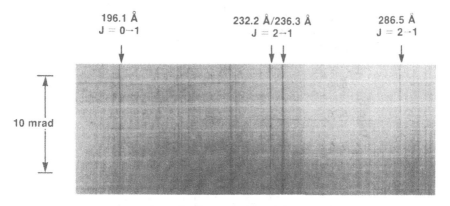

Fig. 4   A print of the spectrometer film in the region 180–300 Å. The variation of intensity along the lasing lines (vertical in photo) represents the angular distribution of the output of the x-ray laser. Note that the nonlasing lines uniformly fill the spectrometer film.

Figure 4 is a print of the spectrometer film in the range ~180–300 Å. The limited angular extent of the four lasing transitions can be seen in comparison to the nonlasing transitions. These non-lasing lines fill the spectrometer slit and show no dependence on angle. A measure of the x-ray laser beam divergence can be obtained by plotting the intensity along the spectral line.

The angular peak of the narrow x-ray laser beams can be seen to be very near the top edge of the spectrometer film. This offset position is a result of the refraction of the x-ray laser beam by the plasma density profile. This deflection angle is ~10–15 mrad away from the target surface.

The inherent narrowness of the x-ray laser beam makes the measurement of its total intensity difficult. This is a result of the uncertainty as to what portion of the x-ray beam the spectrometer is detecting in the direction perpendicular to the slit. We noted significant (20%–30%) variation in output intensity for shots of the same length. The scatter in the data is a result of two effects. The first is shot to shot variations in the experimental conditions, e.g., the optical laser conditions. The second is variations in the alignment of the GIGS spectrometer just mentioned.

(B)   Li-Like Laser in Aluminum and Titanium

For the Li-like Al experiments we used the same 650-ps pulse and line focus as described in Sec. II, except that we reduced the laser energy to about 25 J so that the intensity was 1.9×10^{12} W/cm^2. The experimental configuration was likewise the same.

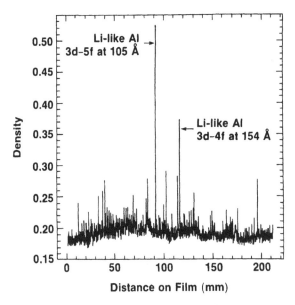

Fig. 5   The spectrum from a 14-mm long Al target and an irradiance of 1.9×10^{12} W/cm^2. Note that the Li-like Al lines at 105 Å and 154 Å dominate the output spectrum for this target.

Figure 5 shows the XUV spectrum from a 14-mm long Al target. Note that the 3d-5f and 3d-4f lines at 105 Å and 154 Å dominate the spectrum. Figure 6 shows the intensity (as recorded by GIGS) of the 3d-5f line of Li-like Al versus the target length. This figure shows that for lengths up to about 14 mm, the output of the laser increases exponentially with target length. At lengths longer than 14 mm, the output intensity remains about constant. This holds true for lengths up to the maximum measured, 22 mm. This is possibly the result of refraction which limits the useful length of the plasma. By this we imply that the curvature of rays due to refraction is such that only a 14 mm length of the plasma contributes to the gain. For lengths greater than 14 mm, the x-rays are refracted out of the plasma column before they can sample additional length of plasma.

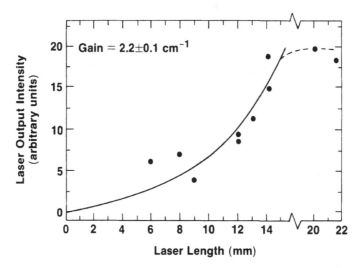

Fig. 6 The output intensity of the 3d-5f Li-like Al line at 105 Å for various target lengths. The points are the experimental data and the solid line is a fit of Eq. (1). Note that the line shows exponential growth for lengths less than 14 mm but is nearly constant for longer lengths (dashed curve).

Similar experiments were performed using Ti targets instead of Al. In order to obtain the proper ionization (He-like Ti), the laser energy was increased to 200 J corresponding to an intensity of 1.25×10^{13} W/cm^2. We observed an exponential growth of the laser output of the 3d-4f line at 47 Å for increasing length, as is shown in Fig. 7.

It should be noted that neither the Li-like Al nor the Li-like Ti laser exhibited the narrow cone of emission that was observed in Fig. 4 for the Ne-like Ge lasers. This may be a result of the fact that the lasing occurs at a relatively low density where the gradient is also low (as discussed in Sec. I). A broader region of gain will yield a beam which is broader in angle. The resulting output of the Li-like x-ray lasers may, therefore, be broad enough to uniformly fill the spectrometer slit.

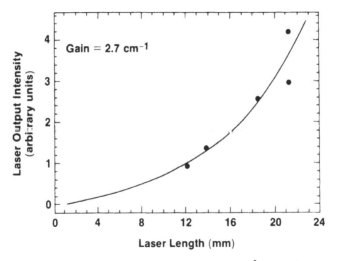

Fig. 7 The output intensity of the 3d-4f Li-like Ti line at 47 Å for various target lengths. The points are the experimental data and the solid line is the fit to Eq. (1).

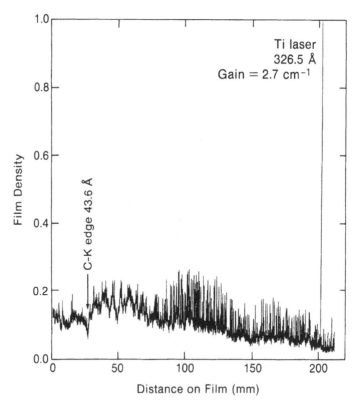

Fig. 8 The spectrum from a 22-mm long Ti target which shows the lasing line at 326.5 Å. This line has been tentatively identified as a J = 0 - 1, 3p-3s transition in Ne-like Ti. The J = 2 - 1 lines which normally also lase in Ne-like ions are beyond the range of this spectrometer.

(C)   Lasing in Ti at 326.5 Å

An additional line was observed to lase in Ti at 326.5 Å. The output of this laser is shown in Fig. 8 which is the GIGS spectrum from a Ti target which was 22 mm long. The line at 326.5 Å is the most intense line in the Ti spectrum, far brighter than the line at 47 Å. This laser has been tentatively identified as a Ne-like Ti transition. Extrapolation of known Ne-like transitions indicate that the J = 0 - 1 line of Ne-like Ti should correspond to this observed wavelength.[7] The other Ne-like laser lines (J = 2 - 1, J = 0 - 1, etc.) lie beyond the current wavelength range of our spectrometer. Figure 9 is the spectrum of a similar shot in the region 10–40 Å which has the n = 3 to 2 lines of Ne-like and F-like Ti labeled. A group of lines which contains higher ionization states is also indicated. The presence of both Ne-like and Li-like Ti suggests that at peak intensity, we have ionized the Ti past the Ne-like state and that lasing must occur later as the plasma recombines back into Ne-like. It is still unclear what the pumping mechanism is for the inversion. Besides collisional excitation, it has been suggested that Ne-like lasers can be also pumped by recombination from the F-like ions.[2] There have also been line coincidences identified by Nilsen[10] which could also contribute to the inversion by photo-pumping.

Figure 10, which is a print of the spectrometer film in the range 300–340 Å, shows the limited transverse extent of the lasing line at 326.5 Å. This narrowness allows us to determine the gain in the plasma in a different manner than that in Secs. (A) and (B). By comparing the radiation intensity along the axis of the x-ray laser to the emission slightly off-axis, we obtain the ratio of the ASE (on-axis) to the spontaneous emission (off-axis) from which gain on axis can be determined. The narrow beam in Fig. 10 allows us to determine both of these intensities using the same spectrometer. The ratio of these two intensities as a function of the gain-length of the laser is given by:

$$I/I_0 = [\exp(g\ell) - 1]/(g\ell) \qquad (2)$$

From this relation we determine the gain coefficient of the 326.5 Å Ti laser to be 2.7 cm^{-1}.

IV.   SUMMARY AND FUTURE WORK

We have demonstrated several types of lasing using slab targets. The high densities present in slab targets where the gain is higher, apparently compensate for the refraction losses resulting from density gradients which are steeper than in exploding-foil targets. We have developed both analytical[11] and numerical[12] models which treat the refraction of x-ray laser beams in various types of targets. We expect to elucidate this question by using these models. The numerical code solves the three-dimensional ray trajectory equation through the two-dimensional density profiles generated by a separate hydrocode. The local gain (as a function of temperature and density) has been calculated using a detailed atomic physics code.[13] By following various rays through the plasma, applying the calculated gain, and summing their contributions we can compare the predictions to the experiments and comment on the relative importance of refraction in various target types.

A comparison of the angular intensity distribution of one of the Ne-like Ge lasers (236 Å) and the Ti laser (326 Å) is shown in Fig. 11. The Ti laser is about half as divergent as the Ge laser. This difference in divergence may be related to the region where the conditions for lasing are optimal in each case. The Ti laser, having a lower atomic number, will have an optimal density for lasing which is lower than that of Ge. Different optimal conditions will cause each laser to propagate in different regions of the plasma, thus resulting in different output profiles. Moreover, if the Ti laser is driven by other excitation mechanisms, the time at which the two x-ray lasers peak will also be different.

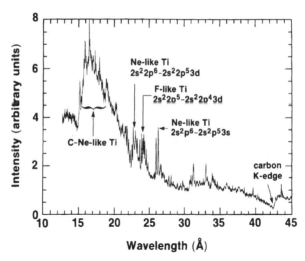

Fig. 9   The spectrum from a 22-mm long Ti target. This spectral region (10–40 Å) contains the n = 3 to n = 2 lines of Ne-like and F-like Ti. The presence of Ne-like emission supports the claim that the lasing line at 326.5 Å is Ne-like Ti.

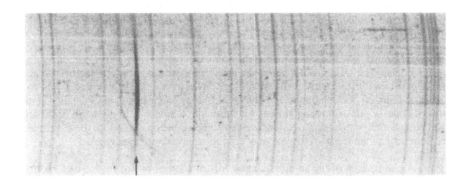

Fig. 10   A print of the spectrometer film in the region of the Ti laser at 326.5 Å. The narrow angular extent of the x-ray laser output can be readily seen (see also Fig. 4). Both the ASE (on-axis) and the unamplified spontaneous emission can be measured from this spectrum.

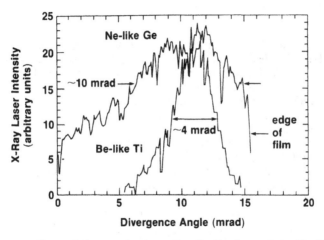

Fig. 11 A comparison of the angular intensity distributions of the Ne-like Ge laser (236 Å) and the Ti laser (326 Å).

Further analysis of the output characteristics of these lasers should reveal the basic manner in which they operate.

We will attempt to identify the transition of the Ti laser at 326.5 Å by extending the range of the GIGS to include the other Ne-like lasing transitions. If it is also a Ne-like laser there are a number of additional questions to be resolved. For example, the gain is higher than predicted by the scaling of gain in Ne-like ions with atomic number as demonstrated by Lee et al.[3] An extrapolation of that data to Ti results in a gain of less than 1 cm^{-1}. Is this evidence of a different pumping mechanism? Finally, in contrast to the Ge laser, the Ne-like Ti laser is predicted to favor lower densities. How can this be reconciled with the assumption that slab targets succeed because of higher densities?

Finally, the Li-like lasers (in Al and Ti) appear to behave very differently than the Ne-like lasers in that no angular distribution peak could be discerned in the output intensity. A highly directional beam results when only a small portion of the available rays are preferentially amplified in a narrow region of the plasma. The lack of measured directionality plus the fact that the output of the Li-like Al emission does not continue to grow exponentially with lengths greater than ~13 mm may be a result of a limited propagation length of the rays within the plasma. The path length of rays in the plasma may be reduced as a result of the position where inversion by recombination is created. On the other hand, the lack of directionality in the Li-like lines casts doubt on the claim of gain on these lines.

ACKNOWLEDGMENTS

The authors thank W. Beich, C. Frantz, and S. Noyes for technical assistance in performing the experiments. This work was supported by the Naval Research Laboratory under contract N00014-86-C-2281 and by the Laser Fusion Feasibility Project at the Laboratory for Laser Energetics which has the following sponsors: Empire State Electric Energy Research Corporation, New York State Energy Research and Development Authority, Ontario Hydro, and the University of Rochester.

REFERENCES

1. D. L. Matthews, P. L. Hagelstein, M. D. Rosen, M. J. Eckart, N. M. Ceglio, A. U. Hazi, H. Medecki, B. J. MacGowan, J. E. Trebes, B. L. Whitten, E. M. Campbell, C. W. Hatcher, A. M. Hawryluk, R. L. Kaufman, L. D. Pleasance, G. Rambach, J. Scofield, G. Stone, and T. A. Weaver, "Demonstration of a Soft X-Ray Amplifier," Phys. Rev. Lett. 54:110 (1985).

2. C. J. Keane, N. M. Ceglio, B. J. MacGowan, D. L. Matthews, D. G. Nilson, J. E. Trebes, and D. A. Whelan, "Soft X-Ray Laser Source Development and Applications Experiments at Lawrence Livermore National Laboratory," J. Phys. B: At. Mol. Opt. Phys. 22:3343 (1989); R. A. London, M. D. Rosen, M. S. Maxon, D. C. Eder, and P. L. Hagelstein, "Theory and Design of Soft X-Ray Laser Experiments at the Lawrence Livermore National Laboratory," J. Phys. B: At. Mol. Opt. Phys. 22:3363 (1989).

3. T. N. Lee, E. A. Mclean, and R. C. Elton, "Soft X-Ray Lasing in Neonlike Germanium and Copper Plasmas," Phys. Rev. Lett. 59:1185 (1987).

4. D. Kim, C. H. Skinner, A. Wouters, E. Valeo, D. Voorhees, and S. Suckewer, "Soft X-Ray Amplification in Lithiumlike Al XI (154 Å) and Si XII (129 Å)," J. Opt. Soc. Am. B 6:115 (1989).

5. P. Jaegle, G. Jamelot, A. Carillon, A. Klisnick, A. Sureau, and H. Guennou, "Soft-X-Ray Amplification by Lithiumlike Ions in Recombining Hot Plasmas," J. Opt. Soc. Am. B 4:563 (1987); G. J. Pert, "Recombination Lasers in the XUV Spectral Region," Plasma Phys. & Controlled Fusion 27:1427 (1985); S. Suckewer, C. H. Skinner, H. Milchberg, C. Keane, and D. Voorhees, "Amplification of Stimulated Soft-X-Ray Emission in a Confined Plasma Column," Phys. Rev. Lett. 55:1753 (1985); J. F. Seely, C. M. Brown, U. Feldman, M. Richardson, B. Yaakobi, and W. E. Behring, "Evidence for Gain on the C VI 182 Å Transition in a Radiation-Cooled Selenium/Formvar Plasma," Opt. Commun. 54:289 (1985); P. R. Herman, T. Tachi, K. Shihoyama, H. Shiraga, and Y. Kato, "Soft X-Ray Recombination Laser Research at the Institute of Laser Engineering," IEEE Trans. Plasma Sci. 16:520 (1988).

6. T. Hara, K. Ando, N. Kusakabe, H. Yashiro, and Y. Aoyagi, "Soft X-Ray Lasing in Al Plasma Produced by a Low Power Laser," Proc. Japan. Acad. 65:60 (1989).

7. J. H. Scofield, private communication.

8. W. Seka, J. M. Soures, S. D. Jacobs, L. D. Lund, and R. S. Craxton, "GDL: A High-Power 0.35 μm Laser Irradiation Facility," IEEE J. Quantum Electron. QE-17:1689 (1981).

9. B. L. Henke, S. L. Kwok, J. Y. Uejio, H. T. Yamada, and G. C. Young, "Low-Energy X-Ray Response of Photographic Films. I. Mathematical Models," J. Opt. Soc. Am. B 1:818 (1984); B. L. Henke, F. G. Fujiwara, M. A. Tester, C. H. Dittmore, and M. A. Palmer, "Low-Energy X-Ray Response of Photographic Films. II. Experimental Characterization," J. Opt. Soc. Am B 1:828 (1984).

10. J. Nilsen, private communication.

11. B. Boswell, D. Shvarts, T. Boehly, and B. Yaakobi, "X-Ray Laser Beam Propagation in Double-Foil Targets," to appear in Phys. Fluids B.

12. R. S. Craxton, "X-Ray Laser Emission from Long Line-Focus Plasmas," 10th Annual Anomalous Absorption Conference, Durango, CO, 19–23 June 1989; also R. S. Craxton, "CASER – A New Code for Calculating the X-Ray Laser Emission from Line-Focus Plasmas," LLE Review 38:88 (1989).

13. M. Klapisch, "A Program for Atomic Wavefunction Computations by the Parametric Potential Method," Comput. Phys. Commun. 2:239 (1971).

NUCLEAR-DRIVEN SOLID-STATE LASERS
FOR INERTIAL CONFINEMENT FUSION

Mark A. Prelas and Frederick P. Boody

Fusion Studies Laboratory, University of Missouri-Columbia
323 Electrical Engineering, Columbia, MO 65211 USA

Abstract

A new class of visible excimers are potential pumping sources for solid-state lasers such as Nd:Glass, GGG, GSGG, Alexandrite, and Emerald. These fluorescence sources are based on alkali excimers. The alkali metals have very low ionization potential and the excimer lines are in the visible. Hence, the alkali excimers have high intrinsic efficiencies (i.e., $\eta_{int} = h\nu/W^*$). Calculations of potential laser efficiencies indicate that nuclear-driven alkali excimer flashlamps are competitive with semiconductor laser pump sources. Some experimental evidence for remotely pumping solid-state lasers will be presented to support the concepts described in this paper.

Introduction

Nuclear-Pumped Lasers (NPLs) directly interface with the ions created in nuclear reactions such as fusion and fission. Experimental investigations have resulted in the discovery of 18 NPLs [Prelas, Boody, Miley, & Kunze, 1988].

The major limitation of directly interfacing ions from nuclear reactions with lasers is the maximum power density available [Prelas, Boody, Miley, & Kunze, 1988]. One method of overcoming the power density limitations of directly interfacing the ions from nuclear reactions to lasers is to directly interface the ions to a medium which can store molecular energy [Wessol, Prelas, Merrill, & Speziale, 1988] or which can produce narrow band fluorescence [Prelas, Boody, Miley, & Kunze, 1988]. The advantage of either method is that resonance transitions can be excited, thereby lowering the power density threshold of a potential laser medium. Miley, 1984, reported on a technique which utilizes $O_2(^1\Delta)$ as a molecule for energy storage and Boody and Prelas, 1983, reported on a technique which uses Xe_2^* excimer fluorescence to photolytically drive the XeF(B-X) laser.

This paper focuses on the use of nuclear generated visible narrow band fluorescence sources which can be used to drive solid-state lasers [Prelas, et. al., 1988] and the applicability of such lasers for inertial confinement fusion drivers.

Nuclear-Driven Solid-State Lasers

The keys to achieving solid-state lasers with multi-megawatt average power output are the pumping geometry and the pumping source. To ensure beam quality, the pumping geometry must be one that produces thermal gradients only in the direction of propagation, i.e. a large diameter thin disk, insulated on the edge and uniformly illuminated and cooled on the face. The then unavoidable axial gradient can be minimized by keeping the disk very thin and pumping steady state, so as to produce the lowest maximum temperature possible. The effects of the minimized axial gradient are themselves minimized by double passing through the laser medium. The pump source must be efficient and affordable and able to uniformly illuminate the laser medium under the above geometrical and temporal restraints.

Geometry has been a problem because large diameter disks have been available only in Nd:glass, which has a high saturation intensity, narrow absorption bands, and poor thermal conductivity. Nd:YAG has much better laser and thermal properties, but even narrower absorption bands and can not be grown in large diameters. Pump sources able to meet the demands of large saturation intensity and narrow absorption bands have been inefficient, pulsed, expensive, and of limited area, making uniform illumination difficult.

New Nd^{+3},Cr^{+3} co-doped solid-state laser materials that can be grown in large diameters [Laser Focus, 1989], and which boast both relatively low saturation intensities and broad absorption spectra, which make them easier to pump, and which have the high thermal conductivity of crystals (relative to glasses), such as GSGG [Krupke, et. al., 1986], promise to change this by making possible the use of pump sources that were not previously feasible. One of those pump sources may be nuclear-driven fluorescers.

Nuclear-driven fluorescers (NDFs) utilize charged particles from nuclear reactions, such as fission, to excite fluorescer gases. While limited to charged particle power densities of <10 kW/cm^3, nuclear excitation can efficiently and uniformly excite large volumes without additional power conversion hardware [Prelas, Boody, Miley & Kunze, 1988]. NDFs are inherently high average power long pulse steady-state devices. However, the deleterious effects of nuclear radiation, as well as low peak fluorescence intensities, have made NDFs impractical for pumping solid-state lasers. We have developed a new geometry for NDFs that solves the radiation problem by shielding the laser from radiation and using a lightpipe to transport the fluorescence around the shield. This geometry also substantially concentrates the fluorescence which, combined with the lower saturation intensity requirements, brings the required pumping power density into the range of NDFs.

The remotely-located nuclear-driven fluorescer concept is illustrated in Fig. 1. Key features are: 1) the use of a visible nuclear-driven excimer fluorescer that will efficiently produce photons that can be absorbed by solid-state laser media and 2) a large diameter to length ratio hollow lightpipe that efficiently transports the fluorescence around the radiation shield to the laser medium. While similar in concept, this approach varies significantly in detail from one we proposed previously for pumping photodissociation excimer gas lasers [Boody, Prelas, 1983], in which the fluorescer and laser media were separated by only a thin window. That geometry was dictated by the facts that the fluorescence was VUV and could not be transported long distances (one or two reflections), that the gas laser medium was not damaged by radiation, and that the larger saturation intensity of the gaseous medium required higher power densities. Even though the laser medium was not subject to radiation damage, optics -- even thin windows -- have proved to be quite sensitive to radiation-induced absorption, particularly in the UV. Radiation-induced absorption is much less of a problem for visible fluorescence. The desire to remove the optics from the radiation field and to switch to the visible, combined with the recent advances in solid-state laser materials, led us to consider solid-state lasers.

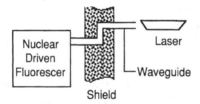

Fig.1. Nuclear-pumping using a remotely located nuclear-driven fluorescer. Isolation of the laser from radiation enables the use of a solid-state laser medium

Visible Nuclear-Driven Fluorescers

The requirements for an NDF laser pump are 1) efficient conversion of fission fragment energy to light, 2) wavelength of the fluorescence within the laser medium's absorption band and long enough to be efficiently reflected by the walls of the fluorescence cell and lightpipe, and 3) transparency to its own output. At the pressures of hundreds of torr needed to stop fission fragments before they hit the wall, resonance line radiators are highly self-absorbed so that, for the long path lengths in the fluorescer inherent with NDFs, excimer fluorescers are preferred. With excimers the ground state of the molecule is unstable so that the molecule dissociates almost immediately after radiation and only a small amount of the radiation gets reabsorbed. Yet, because they radiate to the ground state from the first excited level or higher, and because there is usually only one radiation channel, their fluorescence efficiency is high, as high as 50% for rare-gas excimers. Most excimers, however, tend to be efficient fluorescers only at short wavelengths because the initial excited state is formed in rare-gas buffers which have large W* values. Thus only a few of the well known rare-gas, rare-gas halide, and metal halide excimers fall within the visible, mostly the metal halides, and the efficiencies tend to be in the 10-15% range.

TABLE 1. Alkali Metal Properties

Alkali Metal	λ_{ex} (nm)	I_1 (eV)	I_2 (eV)	W^*_{est} =1.2 I_1	$\dfrac{h\nu_{ex}}{W^*_{est}}$	T_{100} (C)
Li	459	5.36	75.3	6.4	.42	1097
Na	437	5.12	47.1	6.1	.46	701
K	575	4.32	31.7	5.2	.42	586
Rb	605	4.16	27.4	5.0	.41	514
Cs	713	3.87	23.4	4.6	.37	509

Since the photon energies of low saturation intensity solid-state lasers are 1-2 eV, to achieve a reasonable laser system efficiency, the energy cost or W* value of the fluorescence photon precursor must be minimized. This means the use of an atom with low excited state energies and first ionization energy but, to avoid wasteful higher ionizations, a high second ionization energy. The materials that best fulfill these criteria are the alkali-metals. Table 1 lists first and second ionization potentials, I_1 and I_2, estimated W* value, excimer radiation wavelength, λ_{ex} [Luh, Bahns, Lyyra, et. al., 1988], and the temperature required to provide a vapor pressure of 100 torr, T_{100} [Handbook of Chemistry & Physics, 1964], for each of the alkali-metals. Since published W* values, $W^*=W/(1+N_{ex}/N_i)$, for alkali-metals do not exist, they are estimated as 1.2 times I_1, in analogy to the W^*/I_1 values we calculate from published [Platzman, 1961] W and N_{ex}/N_i values for rare and molecular gases. Additionally, the D-line radiation efficiency of low pressure sodium lamps is about 34% of the power deposited in the vapor, which translates to an effective energy absorbed per photon emitted of 1.2 I_1, in agreement with the W* estimates. Dimers exist at significant densities in electrically excited alkali-metals, but only in the bound singlet ground state. However they are formed in both **singlet** and triplet excited states.

At the high power densities of lamp and most other electrically pumped systems, the triplet states are rapidly converted to singlet states by collisions with electrons. The singlet radiation is then absorbed by the high density of ground states. While the excimer radiation from sodium lamps, either at low or high pressure, is normally quite small, the excitation conditions typical of lamps — high power and electron density — are the opposite of those required for producing excimer radiation. The excimer bands were first clearly observed in potassium by Rebbeck and Vaughn, 1971, in an unusual discharge with very low power and electron density, similar to that produced by nuclear excitation. The strongest radiation under these conditions was the excimer band.

Coupling

Fig. 2 traces a visible fluorescence photon from its emission in the NDF to its absorption in the solid-state laser medium. In the tapered fluorescence cell, fission fragments from reflectively coated fissioning source material, present as an aerosol or a "glass wool", excite an alkali-metal excimer fluorescer to emit visible photons which reflect off the source material and the cell walls and are transmitted through the fluorescer/fuel mixture and the cell window into the hollow lightpipe. The lightpipe, which has high transmission due to its high diameter-to-length ratio, transmits >90% of the fluorescence to an active mirror laser amplifier, similar to one developed at the U. of Rochester [Kelly, Smith, Lee, et. al., 1987]. Since the fluorescence profile in the lightpipe is uniform, so is the absorbed power in the active mirror, minimizing radial temperature gradients. Both pump and laser are double passed, minimizing the effects of the unavoidable axial temperature gradients.

A number of loss mechanisms make the coupling of the fluorescence to the laser less than perfect. They include 1) absorption of the fluorescence within the fluorescence cell by either the fluorescer or by the fuel, with which is mixed; 2) absorption by the reflecting surfaces of the fluorescence cell due to an absorbing film, which may be present, and to less than perfect reflectivity; 3) absorption in the output window due to radiation-induced absorption or the deposition of a film of fuel or fluorescer; 4) absorption due to the less than perfect reflectivity of the lightpipe walls, especially at angles near the principle angle for the p-polarized component of the fluorescence; 5) gaps in the lightpipe waveguide or the fluorescence cell wall to allow for cooling; and 6) imperfect absorption in the laser medium.

Reabsorption in the fluorescer is minimized by using an excimer. By design the reactor core fuel loadings do not exceed a few mg/cm^3 so that absorption by the fuel is small [Prelas, Boody, Miley, Kunze, 1988]. Film formation on cell walls is minimized by maintaining them at elevated temperatures. In the visible, radiation-induced absorption in the thin output window is drastically reduced by thermal annealing [Lyons, 1985], at the high temperatures characteristic of both alkali-metal vapors and NDFs, and by photobleaching [Lyons, 1985 and Schneider, Babst, Henschel, et. al., 1986].

Although the total internal reflection of solid lightpipes is attractive, they have problems with radiation-induced absorption, acceptance angle, and excessive mass. Hollow lightpipes minimize radiation-induced absorption, can be quite light and their acceptance angle is essentially ±90°. However, they do not reflect perfectly and, at large angles with respect to the direction of propagation, a large number of reflections is required even for propagating short distances. Thus the number of reflections must be kept to a minimum by using a large diameter to length ratio. This ratio is maximized by combining the fluorescence, from as many cells as are required to produce a critical reactor, to feed only one large diameter lightpipe at each end of the core as shown in Fig. 3. At the active mirror end, the lightpipe would split to match the transverse dimensions of the active mirrors. The reflecting material will probably be silver because it has the highest reflectivity with the least angle dependence at the wavelengths of interest. We calculate a transmission of 0.9, at 550nm, for a 5 m long silver coated hollow lightpipe with a diameter of 1.2 m and two 90° bends, such as is shown in Fig. 3.

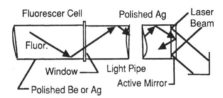

Fig. 2. Coupling of visible nuclear-driven fluorescence to an active mirror solid-state laser amplifier.

Due to the large angular spread of the fluorescence, the laser disk must be located very close to the end of the waveguide to minimize losses. This probably precludes the use of brewster angle arrangements and pumping from both both faces, but is compatible with an active mirror configuration as shown in Figs. 2 and 3. A small cooling gap for cooling between the end of the lightpipe and the disk will not increase losses greatly because most of the fluorescence at angles near 90° is highly attenuated by the large number of reflections.

Performance Estimate

We have estimated the performance of a large-scale remote NDF pumped solid-state laser such as shown in Fig. 3., with a fission driver made up of 24 conical fluorescence cell pairs with a length of 1.2 m and an inside diameter of 0.20 m for a total active core volume of 1 m^3. The cells have a sapphire wall, that is second surface coated with silver, and sapphire output windows. The charged particle power density in the fluorescer is 100 W/cm^3. The efficiency of each of the conversion steps is listed in Table 2. The total reactor power would be 170 MW and, if a fluorescence efficiency of 20% is assumed (about half of those estimated in Table 1), the total fluorescence power would be 20 MW. The fluorescence intensity at the surface of the Nd,Cr:GSGG active mirror would be 2 kW/cm^2. With a spectral matching efficiency of 50%, an active mirror thickness of 0.5 cm, a Cr doping such that the absorbed power fraction is 80%, and assuming a fluorescence lifetime of 250 µs [Krupke, Shinn, Marian, et. al., 1986], the average upper state density is 1.3x10^{18} cm^{-3} and stored energy is 0.25 J/cm^3. Using an effective σ_e for Nd,Cr:GSGG of 1.3x10^{-19} cm^2 [Krupke, Shinn, Marian, et. al., 1986], small signal gain is 0.17 cm^{-1} and, assuming an extraction efficiency of 80%, extracted laser power is 3 MW. Total system efficiency, nuclear power to light power, is 1.7%. Had higher fluorescence or spectral matching efficiency been assumed, the total system efficiency would be even higher, Table 2.

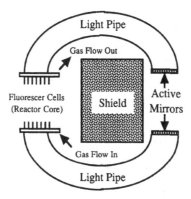

Fig. 3. Layout of Remote Nuclear-Driven Fluorescer Pumped Solid-State Laser showing locations of the fluorescence cells (which also make up the reactor core), radiation shield, lightpipes, and active mirrors.

Applications to ICF

There is an advantage in using Indirectly Driven Nuclear-Pumped Lasers (IDNPLs) for Inertial Confinement Fusion (ICF) which is well known to researchers in the field but has not

TABLE 2. Potential Efficiency of Remotely Located Nuclear-Driven Fluorescer Pumped Solid-State Laser (Fission and Fusion Reactions).

Category	Fission Driver	Fusion Driver
Nuclear Pumping Efficiency	0.6	1.0
Alkali Fluorescence Efficiency	0.2 - 0.4	0.2 - 0.4
Cell Extraction Efficiency	0.9	0.9
Lightpipe Transmission Efficiency	0.9	0.9
Spectral Matching Efficiency	0.5 - 1.0	0.5 - 1.0
Active Mirror Absorption Efficiency	0.8	0.8
$\lambda_{pump} / \lambda_{laser}$	0.55	0.55
Laser Extraction Efficiency	0.7 - 0.9	0.7 - 0.9
η_l (Wall plug efficiency)	0.125 - 0.32	0.125 - 0.32
Total Efficiency	0.015 - 0.076	0.025 - 0.127

been formally published. In an IDNPL, charged particles excite an intermediate media which produces either a long-lived molecular excited state [Wessol, Prelas, Merrill, Speziale, 1988] or narrow band fluorescence [Prelas, Boody, Miley, Kunze, 1988]. The long lived molecular excited state or the narrow band fluorescence can then be transported out of the reactor to a laser media and the upper laser level is populated by resonance transitions with the intermediate species.

The advantage of using an IDNDF is that the charged particles produce the intermediate species plus heat. The intermediate species can be used to drive a laser and the heat can then be used by the thermal energy conversion system. It is evident from Figures 4 to 7 that IDNPLs have a better energy flow than electrical driven drivers.

The energy flow equations for the electrically pumped ICF driver (Fig. 4) are,

$$E_{th} = E_f \ (E_f \text{ is fusion energy} \& \ E_{th} \text{ is thermal energy})$$

$$E_e = E_{th} \ \eta_e \ (E_e \text{ is electrical energy, } \eta_e \text{ is } \frac{E_e}{E_{th}})$$

$$E_d = f_e \ E_e \ (E_d \text{ is energy deposited in driver } \& \ f_e \text{ fraction of electrical}$$
$$\text{energy recirculated})$$

$$E_l = (1-f_e) \ E_e \ (\text{load energy})$$

$$E_f = G \ \eta_l \ E_d \ (\eta_l \text{ is laser energy} / E_d \ \& \ G \text{ is pellet gain.})$$

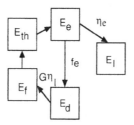

Fig. 4. Energy flow diagram for an electrically pumped ICF driver. (E_f is fusion energy, E_{th} thermal energy, E_d energy deposited in driver, η_l laser energy / energy deposited in laser, η_e electrical energy / thermal energy into converter, f_e fraction of electrical energy recirculated, and G pellet gain.)

The energy flow equations for an ideal NPL ICF driver (Fig. 5) are,

$$E_{th} = E_f(1-f_{cp}) \ (f_{cp} \text{ is fraction of fusion energy in charged particles.})$$
$$E_d = f_{cp}E_f$$
$$E_e = \eta_c E_{th}$$

$$E_l = E_e, \text{ and } \eta_{th} = (1-\frac{1}{G})\eta_c.$$

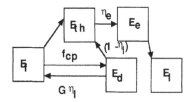

Fig. 5. Energy flow in an Ideal IDNPL driven ICF reactor. (f_{cp} is fraction of fusion energy in charged particles.)

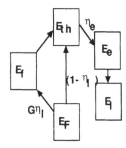

Fig. 6. Energy flow for a fission pumped ICF driver. (E_F is fission power.)

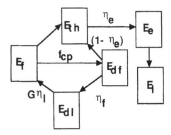

Fig. 7. Energy flow for a dual media type nuclear-pumped laser driven inertial confinement fusion plant. (E_{df} is the energy deposited in first media, E_{dl} energy deposited in driver media, and η_f production efficiency of intermediate transport species--photons or excited states.)

The energy flow equations for a fission pumped ICF driver (Fig. 6) are,

$$E_{th} = E_f + (1-\eta_l)E_F$$
$$R = E_f/E_F \text{ (fusion/fission energy ratio)}$$
$$E_l = E_{th}\eta_e$$
$$E_f = \eta_l G E_F$$
$$\eta_{th} = (1+R-\eta_l)\eta_e/(1+R).$$

The energy flow equations for dual media fusion pumped ICF drivers (Fig. 7) are,

$$E_{th} = (1-f_{cp}) E_f$$
$$E_{df} = f_{cp} E_f$$
$$E_{dl} = \eta_f E_{df}$$
$$E_l = (1-\eta_f)\eta_e E_f$$
$$E_f = G \eta_l E_{dl}$$
$$\eta_{th} = (1-\frac{1}{G\eta_l})\eta_e.$$

The thermal efficiency of the ICF power plant scales as $(1-\frac{1}{\eta_e\eta_l G})\eta_e$ for an electrical powered driver and $(1-\frac{1}{G})\eta_e$ for an ideal driver using fusion reactions. Use of a fission reactor to pump the ICF driver approaches the ideal IDNPL thermal efficiency as shown in Table 3 (Wessol, Prelas, Merrill, Speziale, 1988). An analysis of IDNPL's driven by fusion products and using molecular energy storage or fluorescence sources have a slightly modified thermal efficiency of $(1 - \frac{1}{\eta_l G})\eta_e$.

Parameters for an ICF power plant using either an electrically powered driver, a fission powered driver, or a fusion powered driver are given in Table 3. The analysis shows some interesting differences in the various driver approaches. If the driver power and plant power are held constant, then the fission driver has better performance at lower laser efficiencies ($\eta_l=0.01$--$E_L=7.5$ MJ, $\eta_{th}=0.333$) than at higher laser efficiencies ($\eta_l = 0.1$ -- $E_L = 14$, $\eta_{th} = 0.332$). Basically, this is a result of minimizing the size of the fission driver.

In order to maintain a 1 GW electrical output the power contributed by fusion must increase as fission power decreases. The overall performance of a fission driven driver is better than the electrically driven driver or the fusion driven driver. A 1 to 2 percent fission driven laser efficiency (defined as laser energy/energy deposited) has already been obtained in the laboratory and as shown in Table 2, appears to be within easy reach of remotely driven solid-state laser technology (with predicted η_l between 12 and 32%). Optimization of the fission driven ICF plant could take better advantage of the high efficiency alkali excimer driven solid-state laser. The fusion driven driver has better performance at low laser

TABLE 3. Comparison of Various Types of ICF Drivers (Electric, Fission, and Fusion).

η_l	Electric Driver			Fission Driver			Fusion Driver		
	f_e	E_L(MJ)	η_{th}	E_f/E_F	E_L(MJ)	η_{th}	f_{cp}	E_L(MJ)	η_{th}
0.005	--------	--------	---------	1	7.5	0.333	--------	--------	--------
0.010	--------	--------	---------	2	10	0.332	1.000	30	0.167
0.020	0.750	60	0.083	4	12	0.332	0.500	20	0.250
0.050	0.300	21	0.233	10	14	0.332	0.200	17	0.300
0.080	0.187	19	0.271	16	14	0.332	0.125	16	0.313
0.100	0.150	18	0.283	20	14	0.332	0.100	16	0.317
0.120	0.125	18	0.292	24	14	0.332	0.083	16	0.319
0.140	0.107	18	0.298	28	15	0.332	0.071	16	0.321
0.160	0.094	17	0.302	32	15	0.332	0.062	15	0.323
0.180	0.083	17	0.306	36	15	0.332	0.056	15	0.324
0.200	0.075	17	0.308	40	15	0.332	0.050	15	0.325

E_f/E_F is the fusion energy to fission energy ratio, E_L is the driver energy, f_e is fraction of electrical energy recirculated to the driver, f_{cp} is the necessary fraction of fusion energy released as charged particles, η_l for electric driver is driver energy/electrical energy into driver system, η_l for fission or fusion driver is driver energy/energy deposited in driver, and η_{th} is the thermal efficiency of the ICF plant. This analysis assumes that the plant uses a 33% efficient steam cycle, that the fusion driver uses a 50% efficient fluorescer, that the pellet gain G is 200, and that the electrical power output of the plant is 1 GW.

efficiencies than does the electrical driven driver but the advantage evaporates as laser efficiency increases. However, we argue that a solid-state laser pumped with fluorescence generated by nuclear reaction products is easier to scale to 10 MJ than an electrically pumped solid-state laser. As shown in Table 2, η_L values greater than 0.1 are feasible for large scale fusion driven systems.

Our analysis indicates that an efficient direct energy conversion cycle which significantly improves η_e, provided that one can be developed, would favor electrical drivers. One such direct energy conversion cycle, based on photovoltaic conversion of narrow band fluorescence from the interaction of fusion products with excimer forming gases, has been reported by Prelas and Charlson, 1989.

Conclusion

We have presented a promising new concept, the remote nuclear-driven fluorescer, for pumping very high average power solid-state lasers. We have presented an example system with a steady-state output power of 3 MW and a total system efficiency of 1.7%.

Interfacing of nuclear-pumped lasers to ICF power plants is very promising. Analysis shows that fusion/fission hybrids, where the fission hybrid is an NPL, demonstrates favorable characteristics at low laser conversion efficiencies. Fusion product driven NPLs show favorable scaling as well at low to intermediate driver efficiencies. In order for electrically driven drivers to compare well to the fission driven and fusion driven NPLs, the wall-plug efficiency of the driver must be 10% or better or the electrical energy conversion efficiency must be substantially improved. Additionally, nuclear-pumped lasers scale to high energies more readily than electrically pumped lasers which would give NPLs an advantage over other laser concepts.

Acknowledgement

Partial support from Idaho National Engineering Laboratory and the National Science Foundation's Presidential Young Investigator Program is gratefully acknowledged.

References

F.P Boody and M.A. Prelas, *Excimer Lasers-83*, C.H. Rhodes, H. Egger, & H. Pummer, eds., (AIP,1983), 349.

Handbook of Chemistry and Physics 45th Ed., p. D-95, (Chemical Rubber, 1964).

J.H. Kelly, D.L. Smith, J.C. Lee, et al., Opt. Lett. **12**, 996 (1987).

see for example, W.F. Krupke, M.D. Shinn, J.E. Marion, et al., J. Opt. Soc. Am. B **3**, 102 (1986).

Laser Focus World, p. 24, March 1989.

Wei-Tzou Luh, John T. Bahns, A. Marjatta Lyyra, et al., J. Chem. Phys. **88**, 2235 (1988).

Peter B. Lyons, SPIE Vol. 541, 89 (1985).

R.L. Platzman, Int. J. Appl. Radiat. Isotopes 10, 116 (1961).

M.A. Prelas, F.P. Boody, J.F. Kunze, and G.H. Miley, Laser and Particle Beams **6**, 25,(1988).

M. A. Prelas, F. P. Boody, G. H. Miley, D. Woodall, T. Speziale, and E. Wills, "Remote Nuclear-driven Solid-state Lasers" Int. Workshop on Laser Interactions & Related Plasma Phenomena, Monterey, CA, Oct., 1987

M.M. Rebbeck and J.M. Vaughn, J. Phys. B: Atom. Molec. Phys. **4**, 258 (1971).

W. Schneider, U. Babst, H. Henschel, et al., SPIE Vol. 721, (1986).

D. Wessol, M. Prelas, B. Merrill, T. Speziale, "Feasibility Study of a Nuclear-driven $O_2(^1\Delta)$ Generator to Drive a 18 MW Average Power Iodine Laser for Inertial Confinement Fusion", Laser Interactions and Related Plasma Phenomena, Vol. 8, H. Hora and G. Miley, editors, (1988)

LASING WITHOUT INVERSION

Heidi Fearn and Marlan O. Scully

Center for Advanced Studies
Department of Physics and Astronomy
University of New Mexico, Albuquerque, New Mexico 87131
and
Max-Planck Institute für Quantenoptik
D-8046 Garching bei München, West Germany

INTRODUCTION

We shall investigate both the degenerate and nondegenerate Λ quantum beat laser. This involves a three-level atomic system with an upper excited state and two closely spaced lower levels. The lower levels are degenerate when there is no external electromagnetic field. When the levels are split, by an applied field, electric dipole transitions between them are forbidden. We find that the Λ-type arrangement of the energy levels allows for gain without population inversion, i.e. can display gain even when only a small fraction of the atoms are in the upper excited state.

The degenerate Λ quantum beat laser can be realized in several ways, e.g., we will look at using microwaves or coherent picosecond excitation to establish coherence between the lower levels or a Raman type interaction. When coherence is established between the split states, (lower levels) the transition probabilities between these states and the excited state are modified by quantum interference effects. Quantum interference may result in a zero transition probability for photon absorption, which is the hallmark of the noninversion lasing phenomenon. As an introduction we examine the experimental arrangement of Fig. 1, which illustrates the physics in its simplest form. The figure shows a microwave cavity containing three-level atoms whose lower two levels are driven by an intense microwave field. There also exists a weak, incoherent, pumping mechanism which populates the upper level a. The lasing

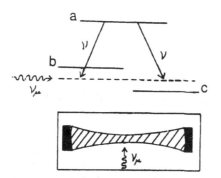

Fig. 1. Lasing without inversion.

field only operates between the mirrors as indicated. The gas of atoms is taken to be dilute so that we consider the collisionless regime and the levels are long lived in relation to the transit time for an atom to cross the lasing region.

In this model the atoms are injected into the lasing region at a rate r where they spend a time duration τ. The energies from the excited levels to the ground states are given by $\hbar\omega_a, \hbar\omega_b$ and $\hbar\omega_c$. The dipole forbidden transition from level b to c is induced by a microwave field of frequency ν_μ and the corresponding Rabi factor is $\Omega \exp(-i\phi)$, where Ω is the Rabi frequency and ϕ is the phase of the microwave field. The interaction Hamiltonian for the degenerate system in the interactive picture is

$$V_I = \hbar \sum_j \left[N(t,t_j)(g_1 a^\dagger \sigma_1^j \exp(-i\Delta t) + g_2 a^\dagger \sigma_2^j \exp(i\Delta t)) + \frac{1}{2}\Omega \exp(i\phi)\sigma_\mu^j \right] + h.a. \quad (1)$$

where a and $a^\dagger$ are the annihilation and creation operators for the light with frequency ν, which induces the transitions from $|a\rangle \to |c\rangle$ and $|a\rangle \to |b\rangle$. The dipole coupling constants for these transitions are g_1 ans g_2 respectively. The atomic lowering operators for the jth atom are given by $\sigma_1^j = |c^j\rangle\langle a^j|$, $\sigma_2^j = |b^j\rangle\langle a^j|$, and $\sigma_\mu^j = |c^j\rangle\langle b^j|$ and the summation is over all atoms. $N(t,t_j)$ represents a step function which is unity in the limits $t_j \leq N(t,t_j) \leq t_j + \tau$ and zero otherwise. In Eq. (1), $\omega_{bc} = \nu_\mu$ and $\omega_{ac} - \omega_{ab} = 2\nu$ (where $\omega_{ij} = \omega_i - \omega_j$) have been assumed, also $\Delta = \nu_\mu/2$. The Schrödinger equation is transformed via a unitary transformation $\Psi = U\Psi'$ which allows us to treat the microwave field to all orders. If the coupling constants $g_1 = g_2 = |g|$, the results show that the linear gain equation for the average photon number, $\bar{n}$, of the laser field is given by [1]

$$\frac{d}{dt}\bar{n} = \alpha\left(\rho_{aa}^0(\bar{n}+1) - \frac{1}{2}[\rho_{bb}^0(1+\cos\phi) + \rho_{cc}^0(1-\cos\phi)]\bar{n}\right) \quad (2)$$

where the linear gain coefficient $\alpha = rg^2\tau^2$. In Eq. (2) the first term represents stimulated and spontaneous emission whilst the second term represents absorption from the lower levels. Specifically, when $\phi = 0$ and $\rho_{bb}^0 = 0$ the second term vanishes and there is no absorption. In this case any amount of population in the upper level $|a\rangle$ will lead to net gain and therefore lasing. This lack of absorption in the Λ quantum beat laser is a result of quantum interference effects.[2] When an excited atom relaxes, the total transition probability is the sum of the $a \to b$ and $a \to c$ probabilities. The transition probabilities from the two lower levels to the upper level are obtained by squaring the sum of the two probability amplitudes. The coherence between the two lower levels may lead to interference terms which result in a zero transition probability for photon absorption. If the microwave field were turned off, it is clear from Eq. (2) that there would be no net gain when $2\rho_{aa}^0 < \rho_{bb}^0 + \rho_{cc}^0$. Hence, we have shown that for no absorption we have lasing even if only a small fraction of the atoms are excited into the upper state.

NONDEGENERATE MICROWAVE SCHEME

For the nondegenerate case we begin with the interaction Hamiltonian,

$$V_I = \hbar \sum_j \left[N(t,t_j)(g_1 a_1^\dagger \sigma_1^j \exp(-i\Delta t) + g_2 a_2^\dagger \sigma_2^j \exp(i\Delta t)) + \frac{1}{2}\Omega \exp(i\phi)\sigma_\mu^j \right] + h.a. \quad (3)$$

where a_k and $a_k^\dagger$, represent the annihilation and creation operators for the light with frequency ν_k, and we assume $\omega_{ac} = \omega_a - \omega_c = \nu_1$ and $\omega_{ab} = \nu_2$ (see Fig. 2). The unitary transformation which allows us to treat the microwave field to all orders is given by,

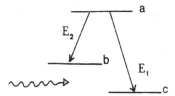

Fig. 2. Lasing without inversion using microwaves.

$$U = \exp\left[-i\lambda\left(\sigma_\mu^j e^{i\phi} + \sigma_\mu^{j\dagger} e^{-i\phi}\right)\right] \tag{4}$$

where $\lambda = \Omega t/2$. This can be expanded in a series to obtain

$$U = \prod_j \left[1 - i\lambda\left(\sigma_\mu^j e^{i\phi} + \sigma_\mu^{j\dagger} e^{-i\phi}\right) - \frac{\lambda^2}{2}\left(\sigma_\mu^j e^{i\phi} + \sigma_\mu^{j\dagger} e^{-i\phi}\right)^2 + \ldots\right]$$

$$= \prod_j \left[|a^j\rangle\langle a^j| + \left(|b^j\rangle\langle b^j| + |c^j\rangle\langle c^j|\right)\cos\lambda - i\left(\sigma_\mu^j e^{i\phi} + \sigma_\mu^{j\dagger} e^{-i\phi}\right)\sin\lambda\right]. \tag{5}$$

We then have the equation of motion for Ψ' given by

$$i\hbar\dot{\Psi}' = V_I'\Psi' \tag{6}$$

where

$$V_I' = \sum_j U^\dagger N(t,t_j)\left(\hbar g_1 a_1^\dagger \sigma_1^j e^{-i\Delta t} + \hbar g_2 a_2^\dagger \sigma_2^j e^{i\Delta t} + adj\right)U. \tag{7}$$

Using $\Omega = 2\Delta$, $\sigma_\mu^{j\dagger}\sigma_1^j = \sigma_2^j$ and $\sigma_\mu^j\sigma_2^j = \sigma_1^j$ we find that,

$$V_I' = \frac{1}{2}\hbar \sum_j N(t,t_j)\left[g_1 a_1^\dagger\left(\sigma_1^j + \sigma_2^j e^{-i\phi}\right) + g_2 a_2^\dagger\left(\sigma_2^j - \sigma_1^j e^{i\phi}\right)\right] + h.a. \tag{8}$$

We may now find the equation of motion for the photon number as follows;

$$\dot{n}_k = \frac{i}{\hbar}\left[V_I', n_k\right] \tag{9}$$

where $n_k = a_k^\dagger a_k$ is the photon number, for $k = 1$ or 2. We obtain;

$$\dot{n}_1 = \frac{i}{2}\sum_j N(t,t_j)\left[g_1^*\left(\sigma_1^{j\dagger} + \sigma_2^{j\dagger} e^{i\phi}\right)a_1 - a_1^\dagger g_1\left(\sigma_1^j + \sigma_2^j e^{-i\phi}\right)\right] \tag{10a}$$

$$\dot{n}_2 = \frac{i}{2}\sum_j N(t,t_j)\left[g_2^*\left(\sigma_2^{j\dagger} - \sigma_1^{j\dagger} e^{-i\phi}\right)a_2 - a_2^\dagger g_2\left(\sigma_2^j - \sigma_1^j e^{i\phi}\right)\right] \tag{10b}$$

213

Similarly, we can obtain equations of motion for $\dot{a}_1$, $\dot{a}_2$, $\dot{\sigma}_1^j$ and $\dot{\sigma}_2^j$ and their adjoints.

$$\dot{a}_1 = -\frac{i}{2}\sum_j N(t,t_j)g_1[\sigma_1^j + \sigma_2^j e^{-i\phi}] \tag{11a}$$

$$\dot{a}_2 = -\frac{i}{2}\sum_j N(t,t_j)g_2[\sigma_2^j - \sigma_1^j e^{i\phi}] \tag{11b}$$

$$\dot{\sigma}_1^j = \frac{i}{2}N(t,t_j)\left[\left(g_1^* a_1 - g_2^* a_2 e^{-i\phi}\right)\left(\sigma_a^j - \sigma_c^j\right) - \sigma_\mu^j\left(g_2^* a_2 + g_1^* a_1 e^{i\phi}\right)\right] \tag{11c}$$

$$\dot{\sigma}_2^j = \frac{i}{2}N(t,t_j)\left[\left(g_2^* a_2 + g_1^* a_1 e^{i\phi}\right)\left(\sigma_a^j - \sigma_b^j\right) - \sigma_\mu^{jt}\left(g_1^* a_1 - g_2^* a_2 e^{-i\phi}\right)\right] \tag{11d}$$

We now integrate Eqs. (11) and substitute the results into Eqs. (10). Note that n is slowly varying so that $n(t_j) \simeq n(t)$ and the summation over all atoms has been transformed into an integral as follows,

$$\sum_j \rightarrow r\int_{-\infty}^t dt_j. \tag{12}$$

The equations of motion for n_1 and n_2 become;

$$\langle \dot{n}_1 \rangle = \frac{1}{4}\tau^2 r\left[2|g_1|^2\rho_{aa}(1+\bar{n}_1) - |g_1|^2(\rho_{bb}+\rho_{cc})\bar{n}_1 \right.$$
$$+ (\rho_{cc}-\rho_{bb})|g_1 g_2|\sqrt{\bar{n}_1\bar{n}_2}\,e^{-i(\phi-\psi)}$$
$$- 2|\rho_{bc}||g_1|^2\bar{n}_1\cos(\phi+\theta)$$
$$\left. + |\rho_{bc}||g_1 g_2|\sqrt{\bar{n}_1\bar{n}_2}\left(e^{i(\psi-2\phi-\theta)} - e^{i(\theta+\psi)}\right)\right] \tag{13}$$

$$\langle \dot{n}_2 \rangle = \frac{1}{4}\tau^2 r\left[2|g_2|^2\rho_{aa}(1+\bar{n}_2) - |g_2|^2(\rho_{bb}+\rho_{cc})\bar{n}_2 \right.$$
$$+ (\rho_{cc}-\rho_{bb})|g_1 g_2|\sqrt{\bar{n}_1\bar{n}_2}\,e^{i(\phi-\psi)}$$
$$+ 2|\rho_{bc}||g_1|^2\bar{n}_2\cos(\phi+\theta)$$
$$\left. + |\rho_{bc}||g_1 g_2|\sqrt{\bar{n}_1\bar{n}_2}\left(e^{-i(\psi-2\phi-\theta)} - e^{-i(\theta+\psi)}\right)\right] \tag{14}$$

where $\rho_{bc} = |\rho_{bc}|\exp(-i\theta)$ and $g_1 g_2^* = |g_1 g_2|\exp(i\psi)$ and ρ_{ij} is the expectation of the atomic operator $|i\rangle\langle j|$.

If we consider the case when $\bar{n}_1 = \bar{n}_2$ then we obtain,

$$\langle \dot{n} \rangle = \frac{1}{4}\tau^2 r\left[2(|g_1|^2 + |g_2|^2)\rho_{aa}(1+\bar{n}) \right.$$
$$- (|g_1|^2 + |g_2|^2)(\rho_{bb}+\rho_{cc})\bar{n}$$
$$+ 2(\rho_{cc}-\rho_{bb})|g_1 g_2|\bar{n}\cos(\phi-\psi)$$
$$+ 2|\rho_{bc}|(|g_2|^2 - |g_1|^2)\bar{n}\cos(\phi+\theta)$$
$$\left. + 2|\rho_{bc}||g_1 g_2|\bar{n}\left(\cos(2\phi+\theta-\psi) - \cos(\theta+\psi)\right)\right]. \tag{15}$$

We see that it is possible to choose the atomic and microwave phases so that the absorption terms in Eq. (15) vanish. When $\phi = \psi = 0$ and $g_1 = g_2 = |g|$ Eq. (15) reduces to the degenerate case treated in the introduction.

Quantum interference provides insight into the physical mechanism which allows for lasing without inversion. An example which does not depend on microwaves to establish coherence between the two lower levels, is given by atomic state preparation via picosecond pulse excitation.

ATOMIC STATE PREPARATION VIA PICOSECOND PULSES

It has recently been shown that under steady state conditions the way that a train of short pulses propagates through a medium depends resonantly on the relationship between the frequency representing the splitting of the ground state ω_{bc} and the pulse repetition frequency[4] ω_p. Under resonant conditions such that $\omega_{bc} = n\omega_p$ for $n = 1, 2 \ldots$, a strong series of pulses may result in the formation of a superposition state of the lower two levels.

Consider a three-level atom interacting with a laser field, Fig. 3. We take levels a, b and c with the energies as defined previously. The decay rates from these levels are γ_a, γ_b and γ_c respectively. Atomic transitions $a \to c$ and $a \to b$ are induced by fields 1 and 2 with frequencies ν_1 and ν_2 as before.

The equation of motion for the atomic density matrix is given by

$$\dot{\rho} = -\frac{i}{\hbar}\Big[H,\,\rho\Big] - \frac{1}{2}\Big(\Gamma\rho + \rho\Gamma\Big) \tag{16a}$$

where

$$H = H_0 + V \tag{16b}$$

$$H_0 = \hbar\omega_a |a\rangle\langle a| + \hbar\omega_b |b\rangle\langle b| + \hbar\omega_c |c\rangle\langle c| \tag{16c}$$

$$V = -\frac{1}{2}g_1 E_1(t)e^{-i\nu_1 t}|a\rangle\langle c| - \frac{1}{2}g_2 E_2(t)e^{-i\nu_2 t}|a\rangle\langle b| + h.a. \tag{16d}$$

$$\Gamma = \gamma_a |a\rangle\langle a| + \gamma_b |b\rangle\langle b| + \gamma_c |c\rangle\langle c| \tag{16e}$$

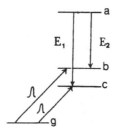

Fig. 3. Lasing without inversion using picosecond pulses.

Hence for the nondegenerate case the equations of motion for the jth atom are given by,[3]

$$\dot{\rho}_{ac}^j(t) = -\left(i\omega_{ac} + \gamma_1\right)\rho_{ac}^j - \frac{i}{2}g_1 E_1(t)e^{-i\nu_1 t}\left(\rho_{aa}^j - \rho_{cc}^j\right)$$
$$+ \frac{i}{2}g_2 E_2(t)e^{-i\nu_2 t}\rho_{bc}^j \tag{17a}$$

$$\dot{\rho}_{ab}^j(t) = -\left(i\omega_{ab} + \gamma_2\right)\rho_{ab}^j - \frac{i}{2}g_2 E_2(t)e^{-i\nu_2 t}\left(\rho_{aa}^j - \rho_{bb}^j\right)$$
$$+ \frac{i}{2}g_1 E_1(t)e^{-i\nu_1 t}\rho_{cb}^j \tag{17b}$$

where $\gamma_1 = (\gamma_a + \gamma_c)/2$ and $\gamma_2 = (\gamma_a + \gamma_b)/2$.

The diagonal elements and the $b \rightarrow c$ density matrix element are given by,

$$\rho_{\lambda\lambda}^j(t) = \rho_{\lambda\lambda}^0 e^{-\gamma_\lambda(t-t_j)} \quad \text{for} \quad \lambda = a, b, c \tag{18}$$
$$\rho_{bc}^j(t) = \rho_{bc}^0 e^{-(i\omega_{bc} + \gamma_{bc})(t-t_j)} \tag{19}$$

where $\gamma_{bc} = (\gamma_b + \gamma_c)/2$.

The total density matrix elements, for all the atoms, are given by,

$$\rho_{a\lambda}(t) = \sum_j \rho_{a\lambda}^j(t) \tag{20}$$

where $\lambda = b$ or c. The equations of motion for the two fields are then,

$$\dot{E}_1(t) = -\frac{\nu_1}{\epsilon_0} Im\left[e^{i\nu_1 t}g_1^* \rho_{ac}(t)\right] \tag{21}$$
$$\dot{E}_2(t) = -\frac{\nu_2}{\epsilon_0} Im\left[e^{i\nu_2 t}g_2^* \rho_{ab}(t)\right] \tag{22}$$

We assume that the atoms are injected into the cavity in a fashion such that the phases of all the atoms are equal as they enter the cavity. The atoms are injected at a regular rate requiring that

$$\exp\left[i\omega_{bc}(t_j - t_{j-1})\right] = 1 \tag{23}$$

This could be realized through an interaction between the atom and a regular picosecond pulse series.

Integrating Eqs. (17) by using Eqs. (12), (18), and (19) and substituting into Eq. (20) gives for the off-diagonal elements,

$$\rho_{ac}(t) = -\frac{i}{2}re^{-i\nu_1 t}\left[\frac{\left(\frac{g_1}{\gamma_a}\rho_{aa}^0 - \frac{g_1}{\gamma_c}\rho_{cc}^0\right)E_1}{\gamma_1 + i(\omega_{ac} - \nu_1)} - \frac{\frac{g_2}{\gamma_{bc}}\rho_{bc}^0 E_2}{\gamma_2 + i(\omega_{ab} - \nu_2)}\right] \tag{24}$$

$$\rho_{ab}(t) = -\frac{i}{2}re^{-i\nu_2 t}\left[\frac{\left(\frac{g_2}{\gamma_a}\rho_{aa}^0 - \frac{g_2}{\gamma_b}\rho_{bb}^0\right)E_2}{\gamma_2 + i(\omega_{ab} - \nu_2)} - \frac{\frac{g_1}{\gamma_{bc}}\rho_{cb}^0 E_1}{\gamma_1 + i(\omega_{ac} - \nu_1)}\right] \tag{25}$$

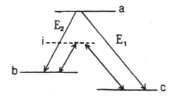

Fig. 4. Lasing without inversion using a Raman interaction

Substituting the total density matrix elements into Eqs. (21) and (22) for the equations of motion of the fields gives,

$$\dot{E}_1 = \frac{\nu_1 r}{2\epsilon_0} \left\{ \frac{\gamma_1 |g_1|^2}{\gamma_1^2 + (\omega_{ac} - \nu_1)^2} \left(\frac{\rho_{aa}^0}{\gamma_a} - \frac{\rho_{cc}^0}{\gamma_c} \right) E_1 \right.$$

$$\left. - \frac{1}{\gamma_{bc}} \left(\gamma_2 \cos\psi - (\omega_{ab} - \nu_2)\sin\psi \right) \frac{|g_1 g_2| \rho_{bc}^0 E_2}{\gamma_2^2 + (\omega_{ab} - \nu_2)^2} \right\} \qquad (26)$$

$$\dot{E}_2 = \frac{\nu_2 r}{2\epsilon_0} \left\{ \frac{\gamma_2 |g_2|^2}{\gamma_2^2 + (\omega_{ab} - \nu_2)^2} \left(\frac{\rho_{aa}^0}{\gamma_a} - \frac{\rho_{bb}^0}{\gamma_b} \right) E_2 \right.$$

$$\left. - \frac{1}{\gamma_{bc}} \left(\gamma_1 \cos\psi + (\omega_{ac} - \nu_1)\sin\psi \right) \frac{|g_1 g_2| \rho_{cb}^0 E_1}{\gamma_1^2 + (\omega_{ac} - \nu_1)^2} \right\} \qquad (27)$$

where we have used our previously defined phase ψ.

Consider the special case of resonance $(\omega_{ac} - \nu_1) = (\omega_{ab} - \nu_2) = 0$, where we have $E_1 = E_2$ and all decay and coupling constants are equal.[3] Then, it is clear that the absorption (second) terms in Eqs. (26) and (27), cancel each other and we again have lasing without inversion.

Experimental high-resolution coherence spectroscopy using pulse trains has clearly demonstrated the formation of a coherent superposition state of the lower atomic levels.

The final example we shall consider is that of a Raman interaction which establishes the coherence in the lower two atomic levels.

RAMAN MIXING OF THE LOWER LEVELS

A further mechanism for generating a coherent superpostion of the lower two levels is by Raman interaction (see Fig. 4). The mixing of the lower levels is accomplished via an interaction with an intermediary state. The steady state population of the intermediate state is however zero. Theory and experiment by the Pisa group[5,6] and theoretical work by Walls and Zoller[7] have shown that such a coherently prepared ground state would lead to vanishing absorption as we have discussed above.

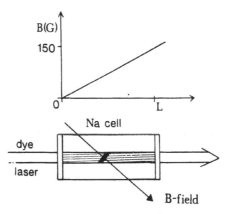

Fig. 5. Experiment to show nonabsorption in sodium vapour. Resonance fluorescence appears as the shaded region. Dark line appears at right angles to the magnetic field gradient.

Rather than to present more equations, we shall instead describe the physics which is beautifully illustrated by the experiment of Alzetta et. al[6] (see Fig. 5).

A cell containing sodium vapour was excited by a multimode dye-laser so as to observe reonance fluorescence along the beam. A magnetic field gradient (of up to 150 Gauss) was then applied, at an angle to the direction of the laser beam, in order to Zeeman split the sodium D lines. Dark lines then appeared across the fluorescent path of the beam, their width being of the same order as the Zeeman (or hyperfine) RF line.

This strong decrease of vapour fluorescence has been interpreted by Alzetta et. al. as due to a resonance effect which reduces the atomic absorption. The atoms are essentially in a nonabsorbing state, which is the trade mark of the quantum interference phenomenon discussed above.

CONCLUSIONS

The most obvious application of the Λ quantum beat laser would be to realize laser operation for those wavelengths where population inversion is difficult. It is interesting to notice that the noninversion laser will also experience reduced spontaneous emission noise. The ratio of the phase diffusion coefficients for the noninversion laser (Λ superscript) and a normal inverted laser (n superscript for normal laser) is given by,[1]

$$\frac{D_{\phi\phi}^{(\Lambda)}}{D_{\phi\phi}^{(n)}} = \frac{\rho_{aa}^{(\Lambda)}}{\rho_{aa}^{(n)}} = \frac{\rho_{aa}^{(n)} - \rho_{cc}^{(n)}}{\rho_{aa}^{(n)}}. \tag{28}$$

We have also treated a far more complicated problem using a (closed) three-level system with nondegenerate levels, going to third order in the optical fields. This treatment is rather lengthy and shall be presented elsewhere.

To conclude this brief summary we note that it is possible to have lasing even when only a small fraction of the atomic population is in the upper level using the Λ-type quantum beat laser. It can also be shown that when the upper two levels are closely spaced and coherence is established in these levels rather than the lower levels we obtain inversion without lasing. The atom is effectively trapped in the excited state and unable to radiate. In this case (V-type system) we may arrange for the absorption of light without stimulated emission.

ACKNOWLEDGEMENTS

HF would like to thank the sponsors of the 9th International Workshop on Laser Interaction and Related Plasma Phenomena for a scholarship to attend the conference. This work was partially supported by the Office of Naval Research.

REFERENCES

1. M. O. Scully, Shi-Yao. Zhu, Phys. Rev. Letts. **62**, 2813 (1989); M. O. Scully, "Lasing Without Inversion via the Lambda Quantum Beat Laser," Proc. NATO Advanced Research Workshop on, *Noise and Chaos in Nonlinear Dynamical Systems*, Torino, March 7-11 (1989).
2. U. Fano, Phys. Rev. **124**, 1866 (1961).
3. Shi-Yao Zhu, M. O. Scully and E. E. Fill, "Lasing without inversion due to initial coherence between lower atomic states," Proc. Rochester conference (1989).
4. J. Mlynek, W. Lange, H. Harde and H. Burggraf, Phys. Rev. **A24**, 1099 (1981); H. Harde and H. Burggraf, Optics Commun. **40**, 441 (1982); Y. Fukada and T. Hashi, Optics Commun. **44**, 297 (1983).
5. G. Orriols, Nuovo Cimento **53**, 1 (1979).
6. G. Alzetta, L. Moi and G. Orriols, Nuovo Cimento **52**, 209 (1979).
7. D. F. Walls and P. Zoller, Optics Commun. **34**, 260 (1980).

Ion Acoustic Parametric Decay Instabilities in Laser-Plasma Interactions

K. Mizuno, W. Seka†, R. Bahr†, R.P. Drake††, P.E. Young††, J.S. De Groot, and K.G. Estabrook††

Department of Applied Science
University of California
Davis, California 95616

† Laboratory for Laser Energetics
University of Rochester
Rochester, New York 14633

†† Lawrence Livermore National Laboratory
University of California
Livermore, California 94550

I.   INTRODUCTION

One of the critical issues in laser driven fusion pellet is the degradation in fusion yield due to plasma physics effects. Microwave simulation experiments[1-3], and computer simulations[4,5] have shown that Ion Acoustic Parametric Decay Instability[6] (IADI), in which an electromagnetic wave decays into an electron plasma wave, and an ion acoustic wave near the critical density n_c (where electromagnetic wave frequency equals plasma frequency), is an important process. The IADI can produce a significant number of hot electrons[1-5] in a large scale plasma. These hot electrons are a concern in proposed inertial confinement fusion (ICF) studies because they can preheat the target and degrade compression. The ion wave turbulence excited by IADI will also be a source of anomalous resistivity[7], so that the thermal electrons are strongly heated due to anomalous Joule heating, and heat flow is reduced. In spite of these results, it has been hoped that IADI would not constrain laser fusion target designs, for two reasons. First, once the plasma has formed, the laser intensity will have been significantly attenuated by collisional absorption when it reaches the densities near the critical surface where IADI occurs. Second, the threshold for IADI may be substantially increased by convective stabilization in the steepened plasma density profile near the critical density.

The second harmonic emissions relevant to IADI have been studied by many authors[8-14] in laser produced plasma. We have made extensive studies of the IADI by measuring second harmonic emissions in laser produced plasma. We have shown that the IADI threshold is quite low in a large scale length plasma. The threshold decreased as the laser spot size increased. The measured threshold is an order of magnitude lower than previously reported values in small spot experiments. The threshold values for IR and short wavelength Green lasers are quite low, and reached homogeneous-plasma collisional values in a planar plasma produced by large spot size laser irradiations. The results are in agreement with LASNEX calculations[15] for a flux limit of f = 0.1. These low threshold values indicate that IADI is potentially important in a large scale plasma, even in short-wavelength laser-pellet interactions which are applicable to laser fusion research. We have also studied the IADI vs. target material such as CH, Al, Cu, Mo, and Au. The threshold value of the IADI was low even for a high Z target.

II. Experimental Arrangement

The experiments were conducted on the GDL laser system (λ_L = 1 μm and 1/2 μm, τ_L = 1 nsec, E_L < 200 J, I_L = 10^{13} to 3 x 10^{15} W/cm^2), Omega Laser System (λ_L = 0.35 μm, E_L = 1 kJ), at the Laboratory for Laser Energetics, University of Rochester, and Phoenix (Janus) laser system (λ_L = 1.06 μm, τ_L = 1 nsec, E_L < 100 J, $I_L \approx$ 3 x $10^{12} \sim$ 1 x 10^{15} W/cm^2), at the Lawrence Livermore National Laboratory (Fig. 1). In the experiments, the laser normally irradiates a planar target with a 1 ns FWHM Gaussian pulse and a maximum energy of 200J. The laser light was focused through a f/3 lens onto the target. The spot size was varied from 85 μm to 900 μm. Experiments were also performed with various target materials; CH, Al, Cu, MO and Au. The target was thick enough (50 μm thickness) so that no laser burnthrough was observed.

We monitored the emission spectrum near the second harmonic (2ω) of the incident laser frequency (ω), that was collected at 45^0 to the incident laser direction and in the plane of polarization. A focusing mirror (GDL experiments) or a focusing lens (Phoenix experiments) was used to collect the emission near the second harmonic. The signal is fed into a spectrometer. The spectrum is streaked using a streak camera (resolution of 1 Å and 30 ps typical). The data are recorded on Kodak film. The data are digitized and processed using computers in LLE and LLNL. Experimental results are interpreted with the help of LASNEX computer calculations.

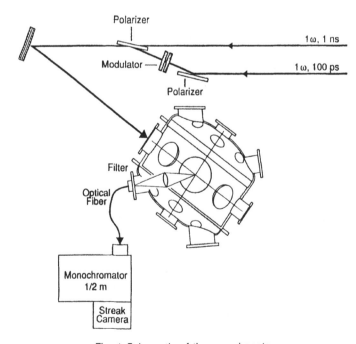

Fig. 1 Schematic of the experiments

III. Experiments

(A) The IADI threshold

The second harmonic spectrum[14] is measured to study the IADI. Figure 2 displays a three dimensional plot of the IADI measured with a MO target. A well-defined Stokes structure (2nd and 3rd peaks) is clearly shown. We see the time history of the signal. The 2ω signal, the first Stokes mode, and then the second Stokes mode are excited. Occasionally, we observe a third Stokes mode although the signal is weak. Figure 3 shows a typical spectrum near the second harmonic from a CH target for various laser intensities. Emission is observed only at 2ω for low power laser irradiation. This signal is probably due to electron plasma waves generated by resonance absorption. Above a well-defined threshold, the Stokes sidebands (first Stokes mode with $\Delta\lambda \approx 23$ Å) are clearly observed. This Stokes mode is an electron plasma wave excited by the IADI. The value of $\Delta\lambda$ is proportional to the frequency of the ion acoustic wave. We have observed a sharper Stokes peak for a high Z target ((MO, AU) as shown in the right hand side in Fig. 3. For high intensity laser irradiation, the Stokes mode becomes broad and no peaks are observed.

We have found that the IADI threshold is quite low for irradiation with a large laser spot size. Figure 4(A) gives Stokes intensity vs. vacuum laser intensity I_L with laser spot size as a parameter. The IADI intensity increases strongly with I_L near the threshold. For the large laser spot (600 μm), the measured threshold is as low as 5×10^{12} W/cm^2. The minimum laser power for which we have seen the IADI is 5×10^{12} (shown by the arrow). The threshold values increase with decreasing laser spot size. For the 350 μm diameter spot, the threshold is about 2×10^{13} (curve (b)). The threshold is high, about 5×10^{13} (curve (c)), for a small spot size (100 μm). The 2ω signal can be observed for laser intensity well below the IADI threshold as is shown in Fig. 3.

The IADI intensity increases with increasing laser spot size. The intensity of the Stokes mode normalized by the intensity of the 2ω signal, $I_{sto}/I_{2\omega}$, is plotted vs. laser spot diameter in Fig. 4(B). For a small spot, no Stokes mode is observed. Near the threshold,

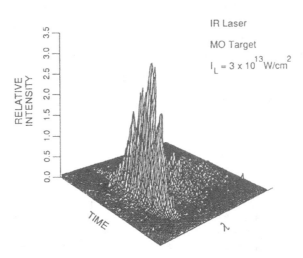

Fig. 2 Three dimensional plot of the IADI. The relative intensity vs. time and wavelength is shown. The vacuum laser intensity, $I_L = 3 \times 10^{13}$ W/cm^2, wavelength, 1 μm, and the laser spot diameter, 350 μm (MO target)

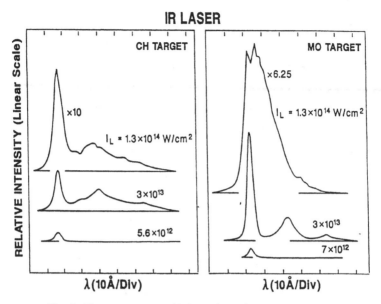

Fig. 3 Wave spectrum with laser intensity as a parameter

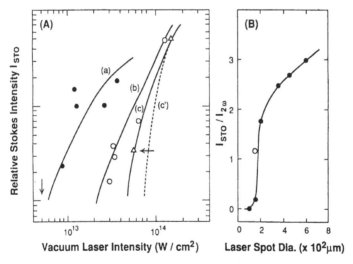

Fig. 4 (A)Stokes mode intensity, I_{sto}, vs. vacuum laser intensity, I_L, with laser spot diameter as a parameter: (a) solid circles, 600 μm; (b) empty circles, 350 μm; and (c) triangles, 100 μm (CH target); (B) Stokes intensity normalized by the 2ω intensity, $I_{sto}/I_{2\omega}$ vs. laser spot diameter. $I_L = 4 \times 10^{13}$ W/cm²

the value $I_{sto}/I_{2\omega}$ sharply increases with spot size. The ratio $I_{sto}/I_{2\omega}$ also increases with the spot size above the threshold. The two Stokes peaks are observed for a spot diameter of 150 μm. One of them is due to normal IADI, and the other is due to IADI-R, which will be discussed below. The empty circle shows the intensity sum of the two different kind of Stokes modes in Fig. 4(B).

For smaller laser spot sizes (100 and 150 μm), we have observed a Stokes mode with a small $\Delta\lambda$ below the threshold intensity of IADI. The data is highlighted by an arrow in Fig. 4(A), and also an empty circle in Fig. 4(B). We call this sideband IADI-R to distinguish it from the normal IADI. From the value of $\Delta\lambda$, we can estimate that the Stokes mode is excited at a plasma density n ≈ $0.98\ n_c$, (using the Bohm-Gross dispersion relation), very close to the critical surface. Therefore, this Stokes mode can be considered as IADI-R excited by Resonance absorption. We have observed only the Stokes mode (no anti-Stokes), so that this is not due to amplitude (or phase) modulation of the laser. For a large spot size laser, IADI-R is not clearly seen because a strong IADI Stokes mode is excited so that IADI-R may be masked. For higher laser intensity, the normal Stokes mode is observed for the 100 μm laser spot size as is seen in the larger size laser experiments. The threshold of the normal Stokes mode was higher. The estimated curve is shown in Fig. 4 (dotted). The threshold was about $(7-9) \times 10^{13}$. No IADI Stokes mode was observed for I_L = 6×10^{13}.

We have also performed a series of green laser (1/2 μm) experiments. We have again observed a very low threshold $((2-3) \times 10^{13})$ for the large spot diameter of 900 μm. The $\Delta\lambda$ of the Stokes mode is about 12 Å which scales very well from the IR laser result of 23 Å.

(B) Ion Acoustic Decay Instability vs. Target Material

We have made detailed experiments vs. target materials (CH, MO, Au). Typical results for high Z targets (MO and Au) are shown with two dimensional contour plots (time and wavelength), and the time integrated spectrum in Fig. 5. For high Z targets, the Stokes spectrum is narrow, and the contour line is simple and steady with time. For a low Z target, the Stokes spectrum is rather broad, and the contour line is rather complicated and unsteady with time. Figure 6 gives the relative intensity of the Stokes mode vs. vacuum laser intensity for the various target materials (CH, MO, AU). We obtained almost the same value of the threshold. However, a large amplitude for the Stokes mode is observed for a low Z target. The 2ω signal does not have a clear threshold, and it can be detected even at a laser intensity below the threshold of IADI, as is expected. The intensity ratios of the Stokes mode to 2ω signal ($I_{sto}/I_{2\omega}$) vs. vacuum laser intensity are plotted as a parameter of target material in Fig. 7. At slightly above the threshold value, the ratio $I_{sto}/I_{2\omega}$ depends strongly on the target material. Its value decreases strongly with increasing Z targets. The IADI is weakly excited for high Z targets, because the growth rate is small. For high intensity laser irradiations, however, the ratio $I_{sto}/I_{2\omega}$ depends weakly on the target material because the IADI saturates at a low level for a low Z target.

IV.   Theory and Discussion

We can analytically estimate the self-consistent underdense plasma density and plasma scale length using one dimensional theory. The plasma density is self-consistently modified due to the ponderomotive force of the incident laser, and a step-plateau like density profile (with a steep density at the critical surface and underdense plateau) will be produced. The plasma expands approximately with the sound speed. The plasma expansion distance $(C_{st}L)$ is much smaller than the large laser spot diameters so that a one dimensional assumption is valid. The average underdense plasma density is larger than 0.9 n_c for the threshold laser intensities of 5×10^{12} (IR) and 3×10^{13} (Green). On the other

MO TARGET

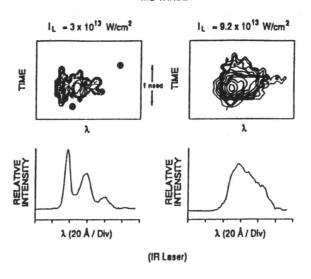

$I_L = 3 \times 10^{13}$ W/cm^2 $I_L = 9.2 \times 10^{13}$ W/cm^2

TIME   λ   1 nsec   TIME   λ

RELATIVE INTENSITY   λ (20 Å / Div)   RELATIVE INTENSITY   λ (20 Å / Div)

(IR Laser)

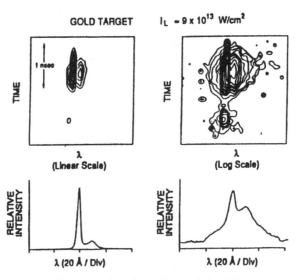

GOLD TARGET $I_L = 9 \times 10^{13}$ W/cm^2

1 nsec   TIME   0   λ (Linear Scale)   TIME   λ (Log Scale)

RELATIVE INTENSITY   λ (20 Å / Div)   RELATIVE INTENSITY   λ (20 Å / Div)

Fig. 5 Two dimensional intensity plot of the IADI and the IADI spectrum for an IR laser

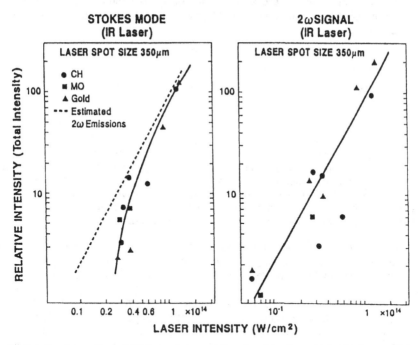

Fig. 6 Relative intensities of Stokes mode and 2ω signal vs. laser intensity for an IR laser

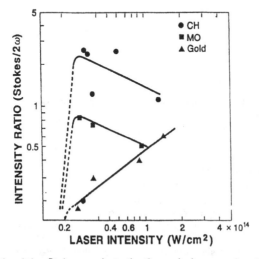

Fig. 7 Intensity ratio of the Stokes mode to the 2ω emission as a function of laser intensity. Laser spot diameter is 350 μm and CH target for an IR laser

hand, since we have measured the $\Delta\lambda \approx 23$ Å and 12 Å of the Stokes modes for IR and Green lasers, we know that the IADI is excited at 0.84 n_c. Using the Bohm-Gross dispersion relation for the electron plasma waves, the plasma density of the instability region is given by the relation,

$$\frac{n_e}{n_c} = 1 - 3 \times (M/m) \, (\Delta\lambda/\lambda_{2\omega_0})^2 /(Z + 3 \, T_i/T_e) \tag{1}$$

This value depends weakly on the temperature ratio for a low Z target, and is nearly independent of the temperature ratio for a high Z target. Therefore, the IADI is excited on the long scale length underdense shelf rather than at the critical density with a sharp density gradient. The convection loss of the electron plasma waves is small on the underdense shelf, and the IADI threshold is essentially determined by collisional damping of the electron plasma waves. Therefore, the IADI threshold is quite low.

We have calculated the IADI threshold vs. $k\lambda_{De}$ using Eq. (4), where k is the wave number of the plasma waves, and λ_{De} is Debye length. This calculation includes collisional and Landau damping of the electron plasma wave, and Landau damping of ion acoustic wave. The threshold values are plotted in Fig. 8 (solid curves). When $k\lambda_{De}$ is larger than 0.24, the threshold intensity increases strongly due to Landau damping of the electron plasma waves. On the other hand, the growth rate of the IADI is proportional to k^2 as is shown in Eq. 5. Therefore, the IADI amplitude profile has a peak near the turning point of the threshold curve as shown in curves (c) and (e). We have estimated these values from the linear growth rate after 5 e-foldings of the unstable waves. No saturation mechanism is taken into account in this calculation so that only the peak value should be compared with the experimental results in the weak laser power regime (slightly above the threshold). In fact, the wave spectrum becomes broad and turbulent-like, so that the experimental results do not agree with the calculations in the high laser power regime. We can estimate the value of $(k\lambda_{De})_{peak}$ using the following equation obtained from the ion acoustic wave dispersion relation,

$$(k\lambda_{De})_{peak} = \frac{\Delta\lambda}{\lambda_{2\omega}} \times \frac{\omega}{\omega_{pe}} \times \left(\frac{m}{M} \times \left((Z + 3 \, \frac{T_i}{T_e}) \right) \right)^{-1/2} \tag{2}$$

We obtain $k\lambda_{De} \approx 0.23$ by substituting the measured value of $\Delta\lambda = 23$ Å (IR exp), and 12 Å (Green exp.), the electron to ion temperature ratio $T_e/T_i = 1.6$ and Z of CH is 3.5 (from LASNEX calculations), and $\lambda_{2\omega} = 0.53$ μm (IR exp.), and 0.265 (Green exp.) into the equation. Notice this value is independent of T_e itself. The measured values are shown (solid circles) for both IR and Green lasers in Fig. 8. The laser intensities in the instability region are estimated from LASNEX calculations in both cases (IR laser and Green laser experiments). The threshold values and the most unstable values of $k\lambda_{De}$ agree well with collision dominated theory. The threshold power is so low that ablative steepening is more important than ponderomotive force in determining the underdense plasma density profile. We have obtained a plasma scale length L = 34 μm in the instability region for a flux limiter f = 0.1 from LASNEX calculations (the inset of Fig. 8). The scale length is still fairly large and the underdense shelf is shallow. The threshold values increase only slightly (curve (b)), when we include convection loss effects.

We have found that the threshold increases significantly with decreasing laser spot size (Fig. 4). The instability regime may shift from a collisional dominated regime to a convective loss dominated regime, so that the threshold increases if the plasma density profile becomes steep. There are two possible explanations: (1) For a small spot size, the plasma expansion distance ($C_s\tau_L$) is smaller than the spot diameter, so that the geometry should be nonplanar, and the scale length of the underdense plasma will be steeper and the shelf will be deeper. (2) LASNEX results indicate that the plasma density scale length strongly depends on the flux limiter (small scale length (4 ~ 5 μm), and deep underdense shelf ($n/n_c \lesssim 0.7$) for a small flux limit of f = 0.03), as shown in Fig. 8 (the inset). If the

flux limit decreases for small laser spot size, the IADI threshold will increase. Notice that the convective threshold intensity will increase nonlinearly as the plasma scale length decreases, since it is proportional to T_e (T_e increases with I_L).

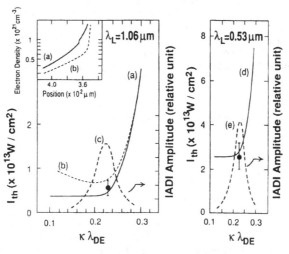

Fig. 8 The threshold of $k\lambda_{De}$ (a) and (d); collisional and Landau damping, and (b) convective damping + (a), and Stokes amplitude (c) and (e) vs. $k\lambda_{De}$. The inset is from LASNEX calculations of the plasma density profile, (a) flux limiter f = 0.1 and (b) f = 0.03. $I_L = 5 \times 10^{12}$ W/cm^2.

Our experimental results have shown that the threshold of the IADI increases slowly for the higher Z targets. The IADI is written in the following equation with the dipole approximation[6]:

$$(\gamma + \Gamma_e)(\gamma + \Gamma_i) = \alpha k^2 E_0^2 \tag{3}$$

Here, γ, Γ_e, Γ_i, E_0, and α are the IADI growth rate, the damping coefficients of electron plasma wave and ion acoustic wave, the electric field of the incident laser, and a constant. The IADI threshold is obtained for $\gamma = 0$, so that the threshold is

$$\alpha k^2 E_0^2)_{th} = \Gamma_e \Gamma_i \tag{4}$$

while the IADI growth rate is

$$\gamma \approx \frac{\alpha k^2 E_0^2}{\Gamma_e} - \Gamma_i \tag{5}$$

in the moderate intensity regime ($\Gamma_e \gg \gamma \gg \Gamma_i$). For high Z targets, Γ_e increases because of the increasing collision frequency. However, the Γ_i decreases for a high Z target because ZT_e/T_i increases so that the Landau damping of the ion acoustic wave decreases strongly. As a result, the threshold values determined from the product of Γ_e and Γ_i does not change significantly with Z. However, the growth rate is proportional inversely to Γ_e as is given in Eq. 5. Therefore, the IADI is expected to be weakly excited for high Z targets even when the IADI threshold is low. Our experiments agree with this theoretical prediction.

V.   SUMMARY

We have made extensive studies of the Ion Acoustic Decay Instabilities (IADI) in laser-pellet interactions. A well-defined IADI is observed. The result scales very well from an IR laser (1 μm) to a short wavelength Green laser (1/2 μm). The threshold decreased as the laser spot size increased. The measured threshold is an order of magnitude lower than the values in small spot experiments. The threshold values for IR and short wavelength Green lasers are quite low, and reached homogeneous-plasma collisional values in a planar plasma produced by a large spot size laser irradiations. The results are in agreement with LASNEX calculations for a flux limit of f = 0.1. The plasma density profile obtained from LASNEX calculation with a flux limit of f = 0.03 did not agree with our low threshold results. We have also shown that the IADI threshold only slowly increases for high Z targets.

These low threshold values indicate that IADI is potentially important in a large scale plasma, and even in short wavelength laser-pellet interactions which are applicable to laser fusion research.

ACKNOWLEDGEMENTS

We acknowledge fruitful and stimulating discussions with W.L. Kruer and E.M. Campbell. The research and materials incorporated in this work were partially developed at the National Laser Users Facility at the Laboratory for Laser Energetics, University of Rochester, with financial support from the USDOE under Cooperative Agreement DE-AS08-88DP10779. The work was partially supported by the Plasma Physics Research Institute, UC Davis and LLNL, and the Lawrence Livermore National Laboratory under the auspices of the USDOE under contract W-7405-ENG-48.

REFERENCES

1.   K. Mizuno, J.S. De Groot, K.G. Estabrook, "Electron Heating due to S-Polarized Microwave (Laser) Driven Parametric Instabilities", Phys. Rev. Lett. 52, 271 (1984) and "Electron Heating Caused by Parametrically Driven Turbulence Near the Critical Density", Phys. Fluids 29, 568 (1986).

2.   K. Mizuno, J.S. De Groot, Wee Woo, P.W. Rambo, K.G. Estabrook, "Hot-Electron Production Due to the Ion Acoustic Decay Instability in a Long Underdense Plasma", Phys. Rev. A 38, 433 (1988).

3.   J.S De Groot, K. Mizuno et al. "Recent Advances in Microwave Simulation of Laser-Plasma Interactions", Laser Interaction and Related Plasma Phenomena, H. Hora and G. Miley eds. (Plenum, New York) V6, 561 (1984) and "Microwave Modeling of Moderate Power Laser-Plasma Interactions", V7, 167 (1986).

4.   P.W. Rambo, W. Woo, J.S. De Groot, K. Mizuno, "Electron Heating Caused by the Ion Acoustic Decay Instability in a Finite-Length System", Phys. Fluids 27, 2234 (1984).

5. K. Estabrook, W.L. Kruer, "Parametric Instabilities near the Critical Density in Steepened Density Profiles", Phys. Fluids 26, 1888 (1983).

6. K. Nishikawa, "Parametric Excitation of Coupled Waves", J. Phys. Soc. Jpn. 24, 916; 1152 (1968); W.L. Kruer, "Parametric Excitation of Coupled Waves", The Physics of Laser Plasma Interactions (Addison-Wesley Publishing Company, CA, 1988).

7. K. Mizuno and J.S. De Groot, "Microwave Simulation Experiments of Electron Heating and Heat Transport Relevant to Laser Fusion", in 1987 International Conference on Plasma Physics, Kiev, USSR, volume 1, 654 (1987).

8. J. L. Bobin, M. Decroisette, B. Meyer, and Y. Vitel, "Harmonic Generation and Parametric Excitation of Waves in a Laser-Created Plasma", Phys. Rev. Lett. 30, 594 (1973).

9. K. Eidman and R. Sigel, "Second-Harmonic Generation in an Inhomogeneous Laser-Produced Plasma", Phys.. Rev. Lett. 34, 799 (1975).

10. N. G. Vasov, V. Yu, Bychenko, O.N. Krokhin, M.V. Osipov, A.A. Rupasov, V.P. Silin, G.V. Sklizkou, A.N. Starodub, V.T. Tikhonchuk, and A.S. Shikanov, "Investigation of $2\omega_0$-harmonic Generation in a Laser Plasma", Sov. Phys. JETP 49, 1059 (1979).

11. X. Zhihan, Xu Yuguang, Yin Guangyu, Zhang Yanzhen, Yu Jiajin, and P.H. Lee, "Second-harmonic emission from laser-plasma interactions", J. Appl. Phys. 54, 4902 (1983).

12. Y. Takada, N. Nakano, and H. Kuroda, "Second-harmonic emission from a picosecond laser-produced plasma", Phys. Fluids 31, 692 (1988).

13. P. D. Carter, S.M.L. Sim, and T. P. Hughes, "Time Resolved Spectroscopy of Second Harmonic Emission from Laser Irradiation Microballoons", Optics Communications 27, 423 (1987).

14. K. Tanaka, W. Seka, L.M. Goldman, M.C. Richardson, R.W. Short, J.M. Soures, and E.A. Willians, "Evidence of Parametric Instabilities in Second-harmonic Spectra from 1054 nm Laser-Produced Plasma", Phys. Fluids 27, 2187 (1984).

15. G.B. Zimmerman, and W.L. Kruer, "Numerical Simulation of Laser-Initiated Fusion", Comments Plas. Phys. Contr. Fusion 2, 51 (1975).

SPECTRAL AND TEMPORAL PROPERTIES OF $1\omega_0, 2\omega_0$ AND $3/2\omega_0$ EMISSION FROM LASER-PRODUCED PLASMAS

R. Dragila, B. Luther-Davies and R.A. Maddever

Laser Physics Centre, Research School of Physical Sciences
The Australian National University, PO Box 4, Canberra ACT 2601
Australia

INTRODUCTION

This paper summarises the results of an extensive study of the spectral and temporal properties of emission from laser-produced plasmas at the fundamental ($1\omega_0$), second harmonic ($2\omega_0$), and three-halves harmonic ($3/2\omega_0$). Some of the results have been presented in detail in recent publications[1,2]. In the experiments short pulses (20-400psec) of 1.053μm radiation from a Nd:glass laser were focussed to high intensity ($10^{14} - 10^{17}$W/cm²) using an F=1 lens onto planar targets of glass, aluminium or lead. The observation of two common features of the spectra at all frequencies motivated this work. Firstly it was found that all time integrated spectra contain apparently random spectral modulations with the spectral width of these modulations simply reflecting the resolution of the recording system. Fourier analysis of the spectral envelope demonstrated that the modulations had the characteristic of "pink noise", that is the power spectrum of the modulations decayed smoothly to high modulation frequencies. A consequence of this characteristic is that if the recording system has poor spectral resolution, there will be a distinct impression that the spectra consists of a few well defined spectral peaks. However, as the resolution is improved, it will be found that these apparently well defined peaks develop more and more substructure and eventually become unrecognisable.

The second "universal" characteristic comes from the time resolved measurements. It has been found that all emissions pulsate with the duration of the pulsations being typically 10-20psec and their separation around 30-50psec although as in the case of the spectral modulations a great degree of randomness exists in this behaviour. An important observation is that temporal modulations always accompany spectral modulations and it is only at very low intensities ($\approx 10^{14}$W/cm²) that both disappear. This clearly suggests a link must exist between these two phenomena. Additionally the basic characteristic of the spectral modulation and temporal pulsations in essentially independent of the experimental variables (emission frequency – the same behaviour occurs for $1\omega_0$, $2\omega_0$, and $3/2\omega_0$ emission; laser intensity for intensities $>10^{14}$W/cm^2; pulse duration in the range 100-400psec; and target mass). This observation is important in the context of explaining the observations since the relative insensitivity to such parameters allows many mechanisms to be eliminated.

To link temporal and spectral characteristics we have shown[1] that it is necessary to postulate that the emission within a relatively narrow band undergoes several rapid phase-modulations (equivalent to a non-linear Doppler shift) during the laser pulse which causes the emission to sweep out to one side of the harmonic and back again. The extent of such a sweep defines the extent of any red or blue shift in the time integrated spectrum whilst the spectral fine structure has a spacing inversely related to the duration of the phase event. By using a 1-D plasma code we were able to show numerically that the action of the ponderomotive force, which makes the density profile a strong function of laser intensity, was one mechanism that could cause the required phase modulation of the reflected laser light and produce spectra very similar to those observed experimentally. The phase modulations could also be accompanied by intensity modulations although strictly speaking this is a secondary feature of the mechanism.

Since frequency sweeps with a narrow "instantaneous" bandwidth are an inherent feature of this model, the relevance of the time resolved spectral measurements becomes obvious since they should reveal the predicted frequency sweeping. The very short duration of the bursts of emission make this measurement difficult although we have unequivocally observed frequency sweeping of the second harmonic spectra in exact agreement with our expectations. Only samples of our extensive results are presented here.

EXPERIMENTAL RESULTS

The emission was collected by an appropriate optical system (either a lens or an optical fibre) and analysed using either a 1m grating spectrograph fitted with a 1200l/mm grating and having a best resolution resolution of $\approx$0.02Å for the time integrated spectra; or an Imacon 500 streak camera for time resolved measurements of the emission intensity with a temporal resolution of $\approx$2psec; or the combination of the streak camera and spectrograph for time resolved spectral measurements in which case the spectral resolution was $\approx$1Å and the temporal resolution $\approx$10psec. The time integrated spectra were recorded photographically in early work although all the data presented here was obtained using a Reticon array with known noise and fixed pattern performance. The $3/2\omega_0$ spectra were too broad at high intensities to fit on the 25mm long Reticon array and hence some studies were made with reduced resolution using a 300l/mm grating. Since with very short duration pulses ($\approx$20psec) we have observed that the spatial distribution of the second harmonic contains a characteristic speckle pattern (indicating the emitting surface is rough) care was taken in constructing the optical system to ensure that any motion of that speckle pattern did not introduce apparent temporal modulations into the emission.

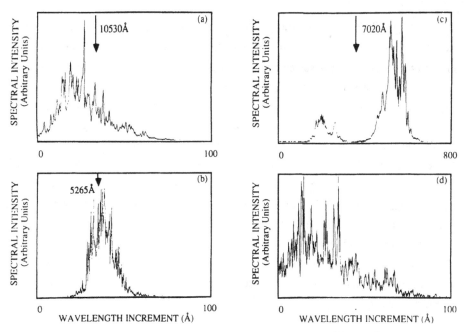

Figure 1. Time integrated spectra. (a): $1\omega_0$, laser intensity $7\times10^{16}W/cm^2$; (b): $2\omega_0$, laser intensity $\approx10^{16}W/cm^2$; (c): $3/2\omega_0$ low resolution – 300l/mm grating, laser intensity $3\times10^{16}W/cm^2$; (d): $3/2\omega_0$ high resolution – 1200l/mm grating, laser intensity $\approx10^{16}W/cm^2$.

Examples of time integrated $1\omega_0$, $2\omega_0$ and $3/2\omega_0$ spectra are shown in figure 1. In the case of the $3/2\omega_0$ spectrum both low resolution and high resolution examples are presented. The former illustrates that at low spectral resolution the spectrum has the characteristic red and blue peaks and an apparent regular low frequency modulation. In the latter only a portion of the red-peak is shown but now the modulation apparent in the low resolution case has become unrecognisable having split into many sub-peaks. Fourier analysis of the envelope of such spectra demonstrates that the modulations have the character of pink noise. The spacing between the red and blue peaks in the $3/2\omega_0$ spectra increases monotonically with increasing laser intensity. At the highest intensities ($\approx10^{17}W/cm^2$) the separation indicates a temperature in the quarter critical region of around 8keV which is appropriate when the corona is dominated by superthermal electrons as would be expected to be the case for the conditions of these experiments. For the purpose of this paper, however, we will focus only on the spectral modulations. From figure 1 it is clear that these modulations are present on each emission and have a very similar character in spite of the fact that very different physical mechanisms in the plasma create the $1\omega_0$, $2\omega_0$ and $3/2\omega_0$ emission. Any mechanism leading to modulations must, therefore, be independent of the generating mechanism.

In figure 2 we present samples of our time resolved data. In figure 2a we time resolve the $2\omega_0$ emission intensity (no spectral resolution) where the light was simply collected by a lens and then scattered off a screen onto the streak camera. The strong pulsations are quite apparent and again very similar behaviour was recorded at both the $1\omega_0$ and $3/2\omega_0$. The $1\omega_0$ emission had to be frequency doubled using a KD*P crystal to shift its frequency into the band detectable by the S20 photocathode in the Imacon streak camera. In figure 2b we show an example of a time resolved $2\omega_0$ spectrum. There is some complication in interpreting this data since there is an uncorrected pin-cushion distortion at the output of the streak camera which provides an underlying curvature. However, careful examination of the figure shows that the bursts are not parallel as it the case in figure 2a and hence even without this distortion being corrected shows it is quite clear that the emission frequency is sweeping during the pulsations. We have observed similar behaviour for $1\omega_0$ emission, whilst in the case of the $3/2\omega_0$ emission the burst duration tends to be shorter ($\approx$10psec) and hence if frequency sweeping exists it cannot be detected due to the limited resolution of the spectrograph/streak camera combination.

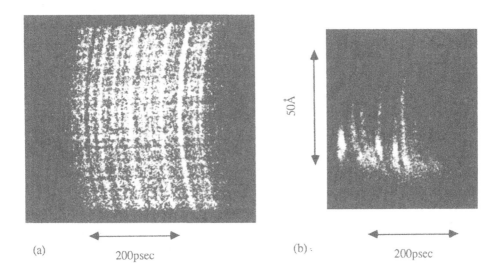

(a)   200psec   (b)   200psec

50Å

Figure 2. Time resolved emission. (a) $2\omega_0$ no spectral resolution, laser intensity 6×10^{14}W/cm^2, pulse duration 400psec; (b) $2\omega_0$ spectrally resolved, laser intensity 4.3×10^{15}W/cm^2, pulse duration 220psec.

DISCUSSION

Although only sample data have been presented in this paper, more than a hundred time resolved spectra and several hundred time integrated spectra have been recorded as part of this study providing us with a wide data base for our conclusions. Some of the time resolved spectra show very clear frequency sweeping although in much of the data the limited temporal resolution and the very short burst duration makes detection of frequency sweeping difficult. On the body of evidence, however, we conclude that frequency sweeping (or frequency [FM] modulation) is the dominant mechanism determining the spectra and occurs whether we could detect it or not. As a result, the instantaneous emitted bandwidth must be relatively narrow (compared to the overall spectral width), and consistent, for example, with resonance absorption as the main second harmonic generating process. We can justify our conclusion in part on the basis that frequency sweeping occurs in some of the data (figure 2b), but it could be argued that other data fails to support this view or is at best equivocal and, therefore, we must explain why we can rule out alternative explanations.

The most likely alternative scenario is that the emission occurs simultaneously across a wide spectral band for at least some of the bursts (which is equivalent to amplitude [AM] modulation of a broad band signal). To test this proposition one has to consider as we did previously[1] the result on the time integrated spectra. Basically we have shown that time integrated spectra are broadened, and red (or blue) shifted relative to the laser harmonics. Most importantly, however, they <u>always</u> display random spectral modulations. Spectra such as those in figure 1 are, however, generated by time integrating data such as that shown in figure 2b. We can, therefore, use this link to argue the case for the dominance of frequency sweeping during bursts even if it was not detectable in the streak records. To construct equivalent time integrated spectra from a postulated time resolved spectrum in the FM case one multiplies the Fourier transform of the frequency spectrum of the phase event by the Fourier transform of the time event. In the AM case multiplication is replaced by convolution. Only for FM modulation will the fine spectral modulations, therefore, appear since the convolution operation involving a broad band of frequencies with a fine structure introduced by the temporal modulations smears out that fine structure.

The basic characteristics of the spectral modulations and temporal bursts have been defined from these experiments. Although there is a shot to shot variation in the characteristics of the spectra even for nominally identical experimental conditions, the average character of the spectral modulations and of the bursts were remarkably constant over a very wide range of different target and laser parameters. It is essential that any explanation of these phenomena can account for this insensitivity.

THEORY

The fact that plasma emissions of different physical character and originating from different regions of the plasma all basically display the same features leads us to conclude that there is one common mechanism leading to both the spectral modulations and the pulsations. The basic property of the mechanism is that it leads directly or indirectly to rapid phase modulation of the emitted light. The mechanism must also occur primarily in the underdense region where the plasma density is below $n_c/4$ to affect $3/2\omega_0$ emission. Many models have been proposed which lead to pulsation of the plasma emission. These incorporated various mechanisms such as cavitons[3]; stimulated Brillouin scattering (**SBS**) with feedback[4]; double stimulated Brillouin scattering[5]; and others[6,7]. However, none of the existing models is satisfactory and even if it fitted some particular data, it would not fit other experimental data. For example, most of the mechanisms proposed result in a time scale for the pulsations which depends on the scalelength of the underdense plasma which contradicts the observation that the pulsation characteristics are independent of the pulse duration.

We, therefore, invoke a new model where the reflection is generated by SBS possibly seeded either by reflection from the critical surface or by long wavelength ion waves which propagate down the density profile in the supersonically expanding plasma. The growth of the ion density fluctuations results in the appearance of a non-linear phase mismatch[8] term that forces the instability to saturate. The hydrodynamic properties of the plasma which results in a steady increase in the extent of the underdense corona with time, and hence an increase in the size of the scattering region, in combination with this non-linear phase mismatch causes the reflectivity to pulsate. The pulsations are driven not only by the constructive and destructive interference of the backscattered wave due to changes in the interaction length but also by changes in the ion wave amplitude that respond to the relative phasing of the incident and reflected waves.

Although we have now introduced SBS into the problem and hence pre-suppose the existence of ion waves resulting from that instability, we point out that the frequency of these waves is too small for them to directly modulate the time integrated spectra in the manner that is observed. In our model, therefore, SBS leads primarily to the temporal modulation of the intensity of the light reaching the high density regions by providing the second reflection point for the incident wave. Note that since the ion acoustic waves are of such low frequency, little broadening of the spectra would occur due to the SBS itself.

Let us consider a laser produced plasma expanding into vacuum. We assume that the heating radiation undergoes stimulated Brillouin scattering within some region $x_1(t) < x < x_2(t)$ in the underdense plasma and that the plasma flow through this region is supersonic. For the sake of simplicity we begin with several additional assumptions: (i) The electron density within $x_1 < x_2$ is homogeneous and has a value $n_0 < n_c$; (ii) The scattering region flows with constant velocity $u > c_s$: (iii) The linear size of this region $L_0 = x_2 - x_1$ is constant; (iv) Within the scattering region the frequencies and the linear phases of the interacting heating (ω_0, k_0), backscattered (ω_1, k_1) and ion acoustic $(\omega_2 = k_2 u - k_2 c_s, k_2)$ waves are perfectly matched, i .e. $\omega_0 + \omega_2 = \omega_1$, with the linear phase mismatch given by $\delta k_L = k_0 + k_2 - k_1 = 0$; (v) We assume $k_2^2 \lambda_D^2 \ll 1$. Such an approximation is acceptable for high intensity heating radiation when $\Gamma(L_{HD}/k_2)^{1/2} \gg 1$ where

$$L_{HD} = \max[n(dn/dx)^{-1}, u(du/dx)^{-1}]$$

characterizes the actual inhomogeneity of the underdense plasma and $\Gamma^{-1}$is the critical interaction length of the convective instability[7]; (vi) We finally assume that $\delta k_L \gg \Gamma$ outside the plasma layer (x_1, x_2).

In a steady state the dimensionless amplitudes of the three interacting waves are then described by the following set of equations[7]:

$$\frac{da_0}{d\xi} = -a_1 a_2$$

$$\frac{da_1}{d\xi} = -a_0 a_2^*$$

$$(1 - M)\frac{da_2}{d\xi} = a_0 a_1^* + i\frac{\delta k_{NL}}{\Gamma}a_2 \qquad (1)$$

where

$$E = eE_0\left[a_0 \exp(i\omega_0 t + ik_0 x) + \left(\frac{c_0\omega_1}{c_1\omega_0}\right)^{1/2} a_1 \exp(i\omega_1 t + ik_0 x) + c.c \right]$$

is the net electric field of the heating and backscattered waves, E_0 is the amplitude of the heating wave c_0 and c_1 are the corresponding group velocities,

$$\delta n = 2in_0\frac{E_0}{E_p}[a_2 \exp(i\omega_2 t + ik_3 x) + c.c.]$$

is the density perturbation associated with the ion acoustic wave, $\xi = \Gamma x$ is the dimensionless spatial variable, $M = u/c_s$ is the Mach number of the flow, and δk_{NL} is the non-linear wavenumber shift of an arbitrary origin. Now, we will replace the three complex amplitudes a_0, a_1, a_2 by a set of six real functions $A_{0,1,2}$ and $\phi_{0,1,2}$ by the following transformation

$$a_{0,1} = (M - 1)^{1/2} A_{0,1}(\xi)e^{i[\Delta\Omega_{0,1}\tau + \phi_{0,1}(\xi)]}$$

$$a_2 = A_2(\xi)e^{i[\Delta\Omega_2\tau + \phi_2(\xi)]} \qquad (2)$$

where $\tau = c_s\Gamma t$ is the dimensionless temporal variable and

$$\Delta\Omega_0 = 0, \ \Delta\Omega_1 + \Delta\Omega_2 = 0.$$

Inserting eq. (2) in o eqs. (1) one obtains

$$\frac{dA_m}{d\xi} = -A_j A_k \cos \phi \qquad m,j,k = 0,1,2; \; m \neq j \neq k \tag{3}$$

and

$$\frac{d\phi}{d\xi} + \tan \phi \frac{d}{d\xi} \ln(A_0 A_1 A_2) = \frac{\delta k_{NL}}{\Gamma(M-1)} + \Psi \tag{4}$$

where $\phi = \phi_1 + \phi_2$ and Ψ is an arbitrary constant which is defined by the requirement that $\sin\phi \rightarrow 0$ when $A_{1,2} \rightarrow 0$ (see Ref. 9). The boundary conditions read as follows:

$$A_0(\xi = \Gamma L_0) = (M-1)^{-1/2}$$

$$A_0^2(\xi = 0) - Q_1 = A_{10}^2 = A_1^2(\xi = 0)$$

$$A_0^2(\xi = 0) - Q_2 = A_{20}^2 = A_2^2(\xi = 0) \tag{5}$$

where A_{10} and A_{20} are the "seed" amplitudes of the secondary waves and $Q_{1,2}$ are the two integrals of the system (3)

$$Q_1 = A_0^2(\xi) - A_1^2(\xi) = const$$

$$Q_2 = A_0^2(\xi) - A_2^2(\xi) = const \tag{6}$$

In what follows we will treat eqs. (3) and (4) as a system describing parametric generation in a weak coupling limit ($\delta k_{NL} > \Gamma$). Assuming that the heating wave is intense and that $A_0(\xi) \cong const$ and following the standard procedure (see e.g. Ref. 10), one obtains the following expression for the gain of the output wave which has been normalized to A_{20}, the amplitude of the seed ion-acoustic wave

$$g_1 = G \left(\frac{\Gamma}{\delta k_{NL}} \right)^2 \sin^2 \delta k_{NL} L_0 \tag{7}$$

where G is a constant dependent on the phase relation beween the pump (heating) wave and the "seeds". For example, if $A_{20}^2 \gg A_{10}^2$ then $G \rightarrow 1$ (independently of the phase relation between the "seed" and the pump since the generated field A_1 will adjust its own phase to provide maximum gain $g_1 \equiv A_1^2(\xi = \Gamma L_0)/A_{20}^2)$. In such a situation, the coefficient of reflectivity, R, for **SBS** backscattering, behaves like:

$$R \propto \left(\frac{k_0}{\delta k_{NL} L_0} \right)^2 \sin^2 \delta k_{NL} L_0. \tag{8}$$

Let us now drop assumptions (i) and (iii), which restrict the above equation to a homogeneous plasma having a scattering region which is constant in length, and consider a inhomogeneous plasma where the scattering takes place in a bounded region of size L_{HD} Since, L_{HD} is in reality a slow function of time compared to the time-scale of the ion-acoustic oscillations, and is given by $L_{HD} \simeq c_s t$, then one can replace L_0 in eq. (8) by $c_s \tau$. This results in the temporal oscillations of the coefficient reflectivity $R \propto \sin^2(\delta k_{NL} c_s t)$ with an oscillation period, τ, given by

$$\tau = \frac{\pi}{\delta k_{NL} c_s} . \tag{9}$$

Finally, it remains to evaluate the non-linear wavenumber shift, δk_{NL}. According to Kurin and Permitin[8] and Kurin[7], δk_{NL} can be expressed as

$$\delta k_{NL} = 4 \, \alpha \, A_2^2 \Gamma \tag{10}$$

where either $\alpha = \alpha_0$ if produced by displacement of the plasma by the ponderomotive force associated with the heating and backscattered waves, or $\alpha = \alpha_2 >> \alpha_0$ if δk_{NL} results from generation of higher harmonics of the ion acoustic wave. There is, however, a competitive process of ion trapping that also causes a non-linear wavenumber shift[11]:

$$\frac{\delta k_{NL}^{(t)}}{k_2} \cong 1.6 \left(\frac{e\Phi_0}{T_e} \right)^{1/2} F \left(\frac{\omega_2}{k_2 v_i}, \frac{T_e}{T_i} \right) \tag{11}$$

where e is the electron charge, Φ_0 is the electrostatic potential associated with the density perturbations ($e\Phi_0/T_e = \delta n/n_0$), T_i is the ion temperature and v_i is the ion thermal velocity. Assuming that the ions have a Maxwellian velocity distribution and that the ion waves are weakly damped, the introduced function becomes[11]:

$$F\left(z, \frac{T_e}{T_i} \right) = \frac{c_s z (z^2 - 1) \exp \left(\frac{-z^2}{2} \right)}{2\pi \, [(z^2 - 1)(1 + k_2^2 \lambda_D^2) \frac{T_e}{T_i} - 1] v_g} \tag{12}$$

where v_g is the group velocity of the ion wave. For comparison, if determined by generation of harmonics, δk_{NL} becomes [from eq. (10) after setting $\alpha = \alpha_2$],

$$\frac{\delta k_{NL}^{(h)}}{k_2} = -\frac{1}{12 k_0^2 \lambda_D^2} \left(\frac{\delta n}{n_0} \right)^2 \tag{13}$$

Let us now evaluate δk_{NL} for some realistic parameters. If we assume that the non-linear phase-shift is due to particle trapping then from figure 2 of Ref. 11 one obtains $\delta k_{NL}^{(t)} \cong 6 \times 10^{-2} k_2$ under the assumptions of $k_2 \cong 2k_0$, $T_e/T_i \approx 10$ and $e\Phi_0/T_e = 0.05$. Assuming, further that $c_s = 1 \times 10^7$ cm/sec, one obtains from the separation between the bursts of the backscattered light, $\tau^{(t)} \approx 40$ psec, which means that the corresponding half-width of the bursts themselves would be ≈ 20 psec. On the other hand, if one assumes that generation of harmonics is the cause of δk_{NL} then, for the same plasma parameters, eq. (13) gives,

$$\frac{\delta k_{NL}^{(h)}}{k_2} \approx \frac{2.5 \times 10^4}{k_0^2 \lambda_D^2}$$

Therefore, for $T_e \approx 1$ keV and $n_0/n_c = 0.1$ one obtains $\delta k_{NL}^{(h)}/k_2 \approx 1.6 \times 10^{-3}$ which is well below the corresponding value due to ion trapping. As can be seen by comparing equations (11) and (13), $\delta k_{NL}^{(h)}$ is much more sensitive to the amplitude of the density fluctuations associated with the ion acoustic wave. However, since it is the ion trapping, in particular that imposes the upper limit on the amplitude of the ion waves $\delta n/n_0$, such that $\delta n/n_0 < 0.2$ (e.g. see Ref. 12), then for the considered set of realistic plasma parameters, we always have $\delta k_{NL}^{(t)}$ $> \delta k_{NL}^{(h)}$. If the de-phasing length $\cong \delta k_{NL}^{-1}$ is short relative to the average value of L_{HD} during the laser pulse then the backscattered radiation is emitted in a series of many short bursts. We note that for the modulation effect to occur the size of the "parametric resonator" has to be $L_0 \cong$ $n/\delta k_{NL}$ (at the end of the laser pulse) where n is the total number of the bursts. Since $\delta k_{NL} <<$ k_2 and since one also has to satisfy the linear wavenumber matching ($\delta k_L=0$) within the "resonator", such an effect can occur only in that region of the underdense plasma where the plasma is either weakly inhomogeneous or rare, i.e. well below the critical where $k_2 \cong 2k_0$ (k_0 is the vacuum wavenumber of the incident radiation). However, it does not mean that SBS does not occur at higher densities as well.. It only means that the modulation of the backscattered radiation originates from the low density plasma.

Finally, the separation between the bursts, $\tau^{(t)}$, and consequently the duration of the bursts, $\tau_b^{(t)}$ both scale as

$$\tau^{(t)}, \tau_b^{(t)} \propto \lambda_0 T_e^{-1/2} \tag{14}$$

i.e. both would be 10 times longer for a CO_2 laser-produced plasma than for a Nd laser-produced plasma, for otherwise identical conditions. Also, most importantly, both are only very weakly dependent on the intensity of the heating radiation through T_e (for our plasmas[13] $T_e \propto I_0^{0.1}$). This is in agreement with our observations, since variation of the laser radiation by two orders of magnitude had no significant effect on the burst duration and separation. Notice, that since the intensity of the scattered wave behaves as $\sin^2(\delta k_{NL}L_0)$ the ratio of the burst duration to the burst separation is 1:2. Dependence on the target material is also very weak and is brought in by the speed of sound, $c_s \propto (Z/A)^{1/2}$ where Z is the charge and A is the atomic weight. As a result of these weak dependences of the plasma and laser parameters, the mechanism appears able to explain the observed temporal modulations of the back-scattered heating radiation.

CONCLUSIONS

We have suggested a mechanism that can lead to temporal modulation of the phase and the amplitude of plasma emission at a variety of frequencies. Its basis is the existence of two reflection points for the fundamental wave in the plasma one corresponding to the normal critical density surface and the other located in the low density region. The reflectivity in the latter region is produced by SBS which saturates due to the growth of a non-linear phase mismatch term proportional to the ion wave amplitude. Plasma expansion then leads to pulsation of the reflectivity on a time scale $\tau \approx \pi/(\delta k_{NL}c_s)$ where δk_{NL} is the non-linear wavenumber mismatch. It has been shown that the most dominant source of this mismatch is ion trapping and that this results in a reflectivity time scale that scales only very weakly with most experimental parameters, in accordance with the experimental observations. Its major sensitivity is to laser wavelength where a linear dependence is expected. This is in reasonable agreement with the observation that the pulsations are about ten times slower for CO_2 radiation[14,15] than is observed using Nd lasers and observation of faster pulsations for $0.5\mu m$ heating radiation[16]. However, the data obtained in the $0.35\mu m$ experiment[17] under very similar experimental conditions as in Ref. 16 reveal pulsations that are slower than prediction following the linear scaling law.

The presence of **SBS** can, however, have further consequences and result in temporal modulation of the other harmonics. If **SBS** occurs at densities below $n_c/4$ then all harmonics of the heating radiation starting from $\omega_0/2$ have to reveal corresponding temporal modulation driven by the modulation of the pump. Since harmonic generation is a non-linear process, one may expect that the duration of the bursts will be shorter than those of the fundamental, while their *separation* remains unchanged. These conclusions are consistent with observations of a wide range of harmonics, from ω_0 to $11\omega_0$ (Ref. 14), in agreement with our scaling law (14).

The mechanism leads directly to phase modulation of the reflected light as the reflection point moves back and forward between the low density region and the critical surface. This accounts for the frequency sweeping observed in our experiments. Additionally, the reflectivity pulsations from the low density region result in the intensity of the laser light at higher densities being modulated in time, thereby imprinting the same modulation characteristics on all processes occurring at those densities (resonance absorption, harmonic generation, two plasmon decay instability, Raman scattering, etc). This explains the reason for similar bursts in emission produced by a wide range of different physical processes each of which, however, is sensitive to the amplitude of the pump.

We consider that the model presented here provides a very good explanation of most of the observations of burst phenomena reported in the literature. To fully simulate all the physics involved would take a quite massive computational effort which is outside the scope of our resources. However, such a simulation might be essential if the physics of short pulse laser produced plasmas is to be fully understood.

References

1. R. Dragila, R.A.M. Maddever, and B. Luther-Davies, Phys. Rev. A, **36**, 5292 (1987).

2. R.A.M. Maddever, B. Luther-Davies, and R. Dragila, accepted in Phys. Rev. A.

3. N.E. Andreev, V.P. Silin, and G.L. Stenchikov, Zh. Eksp. Teor. Fiz. **78**, 1396 (1980); [Sov. Phys. JETP **51**, 703 (1980)]; N.E. Andreev, G. Auer, K. Baumgartel, and K. Sauer, Phys. Fluids **24**, 1492 (1981).

4. K. Baumgartel and K. Sauer, Phys. Rev. A **26**, 3031 (1982); C.J. Randall and J.R. Albritton, in Laser Program Annual Report, Lawrence Livermore National Laboratory UCRL-50021-82, 3-45 (1982).

5. T.J.H. Pattikangas and R.R.E. Salomaa, unpublished.

6. N.S. Erokhin, S.S. Moiseev, and V.V. Mukhin, Nucl. Fus. **14**, 333 (1974); L.M. Gorbunov and A.N. Polyanchev, Zh. Eksp. Theor. Fiz. **74**, 552 (1978); [Sov. Phys. JETP **47**, 290 (1978)].

7. V.V. Kurin, Fiz. Plazmy **10**, 418 (1984); [Sov. J. Plasma Phys. **10**, 245 (1984)].

8. V.V. Kurin and G.V Permitin, Fiz. Plazmy **8**, 365 (1982); [Sov. J. Plasma Phys. **8**, 207 (1982)].

9. V.V. Kurin and G.V. Permitin, Fiz. Plazmy **5**, 1084 (1979); [Sov. J. Plasma Phys. **5**, 607 (1979)].

10. R. L. Bryer, in Nonlinear Optics, (P. G. Harper and B. S. Wherrett, Eds.) Academic Press (London 1977), p.47.

11. H. Sugai, R. Hatakeyama, K. Saeki, and M. Inutake, Phys. Fluids **19**, 1753 (1976).

12. W.M. Manheimer, Phys. Fluids **20**, 265 (1977).

13. M.D.J. Burgess, R. Dragila, B. Luther-Davies, K.A. Nugent, and G.J. Tallents, Laser Interaction and Related Plasma Phenomena, (H.J. Schwarz and H. Hora, Eds.) Plenum, (N.Y. and London 1984), Vol.6, p. 461.

14. P.A. Jaanimagi, G.D. Enright, and M.C. Richardson, IEEE Trans. Plasma Sci. **PS-7**, 166 (1979).

15. V.Yu. Baranov, M.F. Kanevskiy, S.M. Kozochkin, D.D. Malyuta, Yu.A. Satov, and A.P. Streltsov, unpublished.

16. Zunqi Lin, O. Willi, R.G. Evans, and J. Szechi, Rutherford Laboratory Report **RL-82-106** (1982).

17. E. McGoldrick and S.M.L. Sim, Opt. Comm. **40**, 433 (1982).

A survey of Raman spectra from laser-produced plasmas

R. Paul Drake[a] and R.E. Turner

Lawrence Livermore National Laboratory
Livermore, CA USA 94550
(a) *also University of California, Davis*

ABSTRACT

The observed spectra of Raman light from laser-produced plasmas, most often attributed to stimulated Raman scattering, have been sufficiently complex to confound simple expectations. This has led to an interest in any trends that may exist in the observed spectral structure. In the present article, we survey the Raman spectra that have been observed in experiments using short-wavelength lasers (< 1.1 μm) and relatively large laser energies (>~ 1 kJ) to produce copious Raman scattering. In most but not all cases, the spectra exhibit both steady and time-dependent structure that is more complex than the simplest theoretical considerations would suggest. However, trends in this structure are difficult to discern. The observed spectra thus provide an interesting challenge for theories of nonlinear mechanisms that may affect SRS.

INTRODUCTION

When a laser-light wave produces and irradiates a plasma, a number of mechanisms may act to scatter or absorb the laser light. Scattered light produced by scattering of the laser-light wave from electron-plasma waves in the plasma is known as Raman light. Such light is most often attributed to the stimulated Raman scattering (SRS) instability[1], in which the laser-light wave also beats with the scattered-light wave to resonantly drive the plasma wave. SRS has received considerable study because of its importance to laser fusion. The scattered-light wave carries energy away from the laser-fusion target, reducing its efficiency. In addition, the electron-plasma wave produces hot electrons[2,3] by Landau damping, and these can penetrate the fuel capsule, preheat the fusion fuel, and reduce the gain of the laser-fusion target. The present article concerns the spectra of Raman light that have been observed in certain experiments.

For reasons to be discussed shortly, the SRS instability would be naively expected to produce smoothly varying spectra up to wavelengths of $2\lambda_o$, where λ_o is the wavelength of the incident laser light. However, measurements of the SRS spectrum have frequently revealed surprising structure. In particular, the long-wavelength limit of the observed spectrum has often[4-7] been less than $2\lambda_o$. In addition, the observed emission has often shown local structure in time and spectrum as well.[5,8] Such observations have stimulated theoretical work concerning possible, more complex explanations of the observed spectral structure.[9-11] Such work in turn has led to an interest in any systematic trends to be found in such structure, and in particular in the long-wavelength limit of the Raman spectrum. The present article surveys the data from a range of experiments, and may prove to be a useful summary. The experiments chosen are those using more than ~1 kJ of laser energy and having $\lambda_o \leq 1.1$ µm. Such experiments have produced relatively planar plasmas with long enough scale lengths to produce copious Raman scattering.

BASICS OF SRS

To set the context within which to interpret the observations, we first review the basic properties of SRS and ask what they imply. In the absence of damping, the SRS instability can occur at all densities up to $n_c/4$, where n_c is the critical density of the laser light. The dispersion relations for the laser-light wave, the scattered-light wave, and the electron-plasma wave, having frequencies and wavenumbers ω_o, ω_s, ω_{epw}, and k_o, k_s, k_{epw}, respectively, are

$$\omega_o^2 = \omega_{pe}^2 + c^2 k_o^2 , \tag{1}$$

$$\omega_s^2 = \omega_{pe}^2 + c^2 k_s^2 , \quad \text{and} \tag{2}$$

$$\omega_{epw}^2 = \omega_{pe}^2 + 3 v_{th}^2 k_{epw}^2 . \tag{3}$$

Here the speed of light is c, the plasma frequency is ω_{pe}, and the thermal velocity of the electrons, given by the square root of the ratio of electron temperature to electron mass, $(T_e/m)^{1/2}$, is v_{th}. The waves involved in the instability must satisfy the matching conditions for resonance,

$$\omega_o = \omega_s + \omega_{epw} \quad \text{and} \tag{4}$$

$$k_o = k_s + k_{epw} , \tag{5}$$

as well. Recognizing from (2) and (3) that $\omega_s \geq \omega_{pe}$ and that $\omega_{epw} \geq \omega_{pe}$, eq. 4 implies that $2\omega_{pe} \leq \omega_o$, or that the density, n, is $\leq n_c/4$. In addition, because the second term in eq. 3 is always small, lower densities correspond to smaller ω_{epw}, larger ω_s, and smaller scattered-light wavelengths, λ_s.

The growth rate of the SRS instability in a homogeneous plasma, γ_o, is a key factor in the threshold or gain for SRS in any plasma. It is given by

$$\gamma_o = (1/4) v_o k_{epw} (\omega_{pe}^2 / \omega_s \omega_{epw})^{1/2}, \tag{6}$$

in which v_o is the oscillating velocity of the electron in the laser-light wave. As fig. 1 shows, γ_o varies gradually with density. Note that γ_o also varies little, in absolute value, with laser wavelength. The principal dependances on laser wavelength in eq. 6 cancel one another. From consideration of the growth rate alone, one would expect that an adequately homogeneous plasma would produce a continuous spectrum having its maximum below $n_c/4$. In real plasmas, the structure of the observed spectrum will depend largely upon the mechanism(s) that limit the growth or amplification of the scattered light wave and the electron plasma wave. Some such mechanisms are discussed next. Other such mechanisms include density or temperature gradients (which introduce phase mismatch),[1,12-14] the impact of turbulence[15] or of ion waves,[16] and saturation by hot electron production[17] or other means.

Landau damping imposes a generic limit on SRS in all plasmas having a population of thermal electrons, whether or not suprathermal electrons are present. Landau damping limits the growth of any electron plasma wave having a phase velocity close enough to v_{th}. The phase velocity of the plasma waves driven by SRS decreases toward v_{th} as the plasma density decreases. This happens because ω_{epw} decreases and k_{epw} increases with decreasing density. One specific way to model this is to assume that Landau damping becomes very large when $k_{epw}\lambda_D$ reaches some value ~ 0.3, where λ_D is the electron Debye length. The impact of Landau damping is to limit the observed spectrum at short wavelengths.[7, 18, 19]

A number of mechanisms may limit the SRS spectrum at long wavelength. Three of these are straightforward to account for and will not be a primary focus

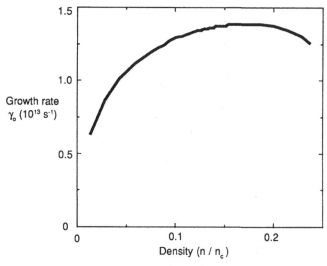

Fig. 1. The growth rate of SRS in a homogenous plasma of a given density is shown as a function of density, assuming no damping to be present. The growth rate alone produces no significant dependence of the observed SRS upon density. The curve shown is for $\lambda_L = 0.35$ μm, $I_L = 10^{15}$ W/cm², and $T_e = 1.5$ keV.

of discussion later. First, a density maximum that is less than $n_c/4$ limits the SRS spectrum. Specifically, the SRS spectrum is limited to wavelengths below that corresponding to the density maximum, as has been observed in numerous experiments.[19-24] Second, collisional damping can quench SRS at densities near $n_c/4$ while weak SRS is observed from lower densities.[25] This only occurs if the plasma is sufficiently collisional, relative to γ_o, which is not the case for most of the reported observations of SRS. Third, if the plasma becomes large enough, then the scattered light from the highest densities can be reabsorbed, producing a long-wavelength limit to the observed spectrum. This effect may have been present in certain experiments with large plasmas,[25, 26] but has yet to be clearly identified. In the following, the spectral features of interest are those to which none of these limitations apply.

On the basis of the discussion thus far, one can construct the generic, time-resolved SRS spectrum that would be observed if more complex mechanisms did not interfere. Figure 2 shows a contour plot of such a spectrum. A laser beam is assumed to strike an initially solid target, producing a plasma and driving SRS within it. This configuration is common to most of the reported experiments. The spectrum extends in wavelength from a lower limit, imposed by Landau damping, to $2\lambda_o$. It shows an onset when some threshold or gain condition is satisfied, but note that the observed onset may or may not be simultaneous at all wavelengths. This depends upon the specific plasma properties and saturation mechanisms that are present. The emission grows brighter as time advances, because the plasma size and density-gradient scale length increase. Note that the specific functional dependence of the increase with time again depends upon the specific plasma properties and saturation mechanisms that are present.[27] The end of the emission depends in addition upon the laser pulse shape and the target thickness. We can now proceed, after a brief aside, to examine and evaluate data with reference to the example of fig. 2.

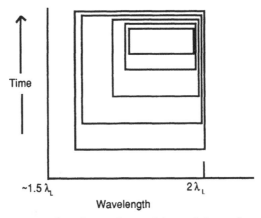

Fig. 2. A contour plot shows the anticipated dependence of SRS intensity on time and wavelength. The plot is derived from generic considerations of SRS growth and Landau damping, discussed in the text.

As an aside, we note that the discussion just completed is somewhat different than that implied by some of the early experimental papers on SRS. If one
assumes the electron-density profile to be strictly linear (or exponential) and the
laser beam to be uniform, then one's theoretical expectations for SRS are rather
different.[1] However, with one possible exception,[28] the laser-target interaction is
too complex, particularly when the plasma is formed by the temporally coherent
but spatially modulated laser beams used in most experiments, to guarantee that
the experiment will adequately satisfy such restrictive assumptions. This is the
grounds for taking the broader approach used here.

RESULTS OF EXPERIMENTS

Data Meeting The Expectations

The Raman spectra observed in experiments are not always complex and
mysterious. Indeed, under some conditions, the observed spectra are quite similar
to the generic example of fig. 2. One such condition is when sufficient laser intensity and energy are used to irradiate gold targets with 0.35 μm laser light. One
must use enough intensity to overcome collisional damping[29,30] and enough energy
to produce an adequate density-gradient scale length.[31] An example of such data
is shown in fig. 3. The details of this experiment, and of those producing data for
subsequent figures, are given in Table 1. The Raman spectrum extends to $2\lambda_o$ and
the brightest emission is near the end of the laser pulse, as anticipated. The total
amount of Raman light measured in this experiment was >0.7 % of the laser
energy, as shown in the table. In this and the other cases the total Raman light
energy was determined using an array of photodiodes.[7] In the context of other
observations of SRS that have produced more complex spectra, the 0.35 μm, gold-
target experiments are the most collisional; that is, they have the largest rate of
electron-ion collisions for a given value of γ_o.

In one other reported case, the *time-integrated* Raman spectrum is close to
expectations. This example is from the irradiation of a gold target with 1 μm
laser light in a 0.6 ns Gaussian pulse having a peak intensity of 3.3 x 10^{15} W/cm².
The targets was irradiated with 3.25 kJ at normal incidence, and the spectrum
was observed at 20°. In this case, the spectrum did extend to $2\lambda_o$. Limited data
are available under such conditions; an example may be found in fig. 3 of ref. 32.
This particular experiment is the most strongly driven for which data are available, in the sense that v_o/c and v_o/v_{th} are the largest.

In all other reported cases, the observed spectra are more complex than
those just discussed. Two aspects of this complexity will be our focus here. First,
as discussed in the introduction, there is often a spectral region having weak or
undetected emission, and known as the "gap", in between a region of strong
Raman emission and the emission near $2\lambda_o$. Second, the observed spectra often
show significant and interesting time-dependent structure.

Data showing a gap below $2\lambda_o$

The gap may be illustrated by data obtained using CH exploding-foil targets. Such targets are thin enough that they explode and decay to densities well
below $n_c/4$ prior to the end of the laser pulse. Data from three similar experi-

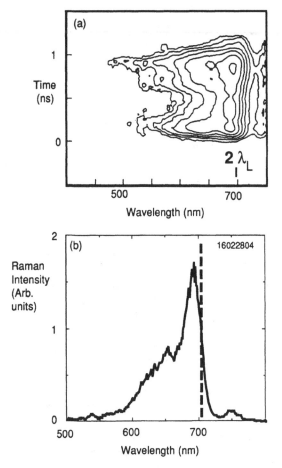

Fig. 3. The Raman spectrum observed at 27° from the normal of a gold disk target irradiated with 0.35 μm light. (a) A contour plot shows the time-resolved spectrum. The contours indicate factor of two changes in the spectral intensity of the Raman emission. These data are quite similar to the expected data shown in fig. 2. (b) The observed spectrum at the time of maximum emission.

Intensity	36	13	36	~15	32
(10^{14} W/cm^2)					
Target:					
Material	Au	CH	CH	CH	Be
Type	Thick	|———— Exploding ————|		Thick
Experiment shown	16022804	93112105	16081202	94040208	16082602
Raman Fraction (%)	> 0.7	8	3	2	2

(a) The square pulses have 100 ps rise and fall times.

ments using three laser wavelengths are shown in figure 4. These spectra show the following features: First, each spectrum shows a short-wavelength limit that we attribute to Landau damping. Second, each spectrum shows a decreasing maximum wavelength late in time that we attribute to the decay of the plasma density. Third, the spectra show various onset behavior. The detailed differences are not understood, but may results from differences in λ_o, in average laser intensity, and in detailed structure of the laser beams in the three cases. Fourth, each spectrum shows a gap that is between 10% and 25% of λ_o. For low-Z targets, any dependence of the gap on laser wavelength is smaller than the dependence on time or laser intensity. The gap is often time-dependent, as seen in fig. 4 and discussed further below. In addition, in other experiments similar to that shown in fig. 4b, which irradiated CH targets with 0.35 μm laser light,[19] the gap has been observed to decrease with increasing laser intensity.

Other experiments reported in the literature show a similar gap.[4-7] Indeed, the data discussed above showing no gap are rather unique. Slight changes of experimental conditions have produced a gap. For example, a gap was observed in experiments using a Gaussian pulse of 0.53 μm light rather than a square pulse of 0.35 μm light but otherwise identical to that of fig. 3. (See fig. 6 of ref. 7.) In addition, the experiment at 0.53 μm produced much more total scattered light, presumably by stronger stimulated Brillouin scattering (SBS).[33] As another example, the spectrum shown in fig. 3 of ref. 32, mentioned above, shows no gap. In contrast, a similar experiment, having slightly lower laser intensity and some geometric differences, produced a spectrum with a gap, as shown in fig. 4 of the same ref.

One would hope that a survey of such data would reveal obvious and systematic trends in the size of the gap. Alas, this is not the case. With the exception of the dependence on laser intensity in the specific experiments just discussed (and in those of Tanaka et al, ref. 4), any trends are not obvious. This suggests that complex, nonlinear effects are responsible for the gap. Possible

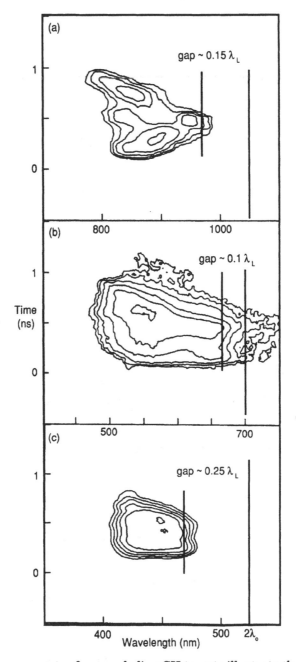

Fig. 4. Raman spectra from exploding CH targets illustrate the common gap below $2\lambda_o$. The laser wavelength and observation angle were (a) 0.53 µm, 135°, (b) 0.35 µm, 153°. (c) 0.26 µm, 135°.

important variables include laser wavelength, target Z, laser intensity, laser pulse shape, and detailed structure of the laser beam. The data discussed here may help those who develop models of such mechanisms to evaluate their plausibility.

Complex, time-dependent spectral structure

As was clear above, the structure of the spectrum near $2\lambda_o$ is often time-dependent. We will use the data of figs. 5 and 6 to illustrate some time-dependent features that are frequently observed. The spectra shown in fig. 6 correspond to the three features indicated in fig. 5, and will be discussed in turn.

In experiments using flat-topped (square) laser pulses, which have short rise times (~100 ps), we often observe an initial burst of Raman light very early in the laser pulse. As is seen in fig. 6a, such a burst may well extend to $2\lambda_o$, which is 702 nm in this case. The feature just above $2\lambda_o$ is the usual feature near $\omega_o/2$, attributed to the two-plasmon decay instability. Emission resulting from two-plasmon decay may also, of course, contribute to that observed just below $2\lambda_o$, as may be clearly seen in fig. 6b. The spectrum of the initial burst is often smoother that that shown in fig. 6a, which shows a dip at 600 nm and a spike near 650 nm.

After the initial burst, the Raman emission often weakens at some or all wavelengths. In the data of figs 5 and 6, the emission is seen to weaken at wavelengths above ~630 nm. During this period, the data show a distinct gap and Raman emission comparable to that of the doublet near $\omega_o/2$. The data are reminiscent of the Raman spectra of Tanaka et al.,[4] produced in plasmas having comparable density-gradient scale lengths. Later in time, as the scale length increases, the emission near $\omega_o/2$ does not weaken (see fig. 5), but the Raman

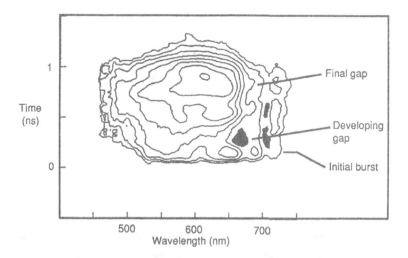

Fig. 5. The spectral power of Raman light, from an experiment using 0.35 μm light to irradiate a thick Be target, illustrates complex, time-dependent structure. The contours show factor of two changes in SRS intensity. The shaded areas show intensity minima.

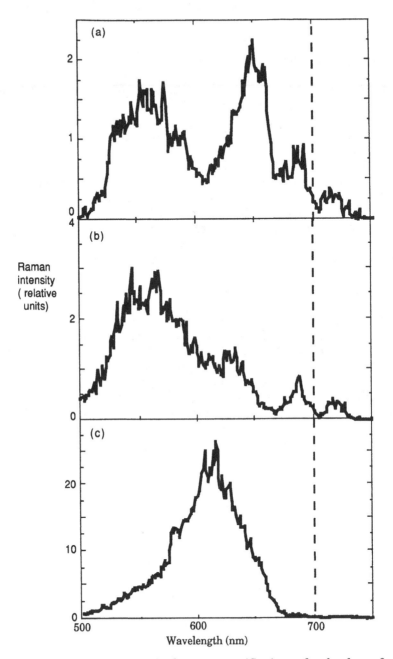

Fig. 6. The Raman spectrum is shown at specific times, for the data of
fig. 5. The units of spectral power are relative but are consitent in
the three plots. (a) at 150 ps; (b) at 285 ps; and (c) at 920 ps.

growth rate of SRS and SBS (or by altering the noise source for SBS).[9, 34,35] In contrast, the variations illustrated by the present data occur on hydrodynamic time scales rather than on the time scales of instability growth. This suggests that the development and competition of the saturated state of the instabilities will be relevant here. And in addition, a successful explanation of the present data will probably need to consider hydrodynamic effects as well.

DISCUSSION

The present article has been structured as a survey of available Raman data rather than as a presentation of data in support of some specific hypothesis. Its intent has been to demonstrate and clarify the features of the data that demand explanations beyond those of generic SRS theory. In this discussion, we offer a brief survey of three categories of explanation for the observed effects.

First, one can attempt to explain the structure in the observed spectra by invoking enhanced damping of some of the plasma waves driven by SRS. The point of view of such explanations is that SRS ought to be growing everywhere, as the discussion of figs. 1 and 2 suggests, and that the inherent problem is to turn it off. One source of enhanced damping is ion waves. Such waves act to damp electron-plasma waves,[36] and may be localized in density. One source of such localization is variations in seeding or growth rate for Stimulated Brillouin Scattering (SBS). By means of damping, such ion waves can restrict the Raman spectrum just as collisional or Landau damping can,[25] and there is circumstantial evidence that this effect is present in some of the data discussed above.[37] However, SBS is not the only feasible source of enhanced damping. Similar effects could be produced by ion turbulence,[10] driven by two-plasmon decay or other mechanisms, or by cavitation,[9] developing as a long-term response of the plasma to the saturated instabilities. In the opinion of the present authors, some mechanism involving enhanced damping is the most likely explanation of the observations.

Second, one can attempt to explain the observed structure by invoking enhanced growth of the SRS plasma waves. The point of view of such explanations is that SRS is too weak to be observed without assistance, as one often concludes when using linear-gradient theory[1] to evaluate experiments. One possible source of enhanced plasma waves is Landau growth, driven by an inverted electron distribution function, as first suggested by Simon and Short.[11] A specific model of such a mechanism, invoking the bump-on-tail instability, has been developed and presented by Simon and coworkers.[11, 38-40] In general, one can specify an electron distribution function that drives the growth of plasma waves so as to produce Raman light over any chosen range of frequencies. And given a saturation model for the plasma waves, one could use the Raman data to infer the time-dependent (though spatially averaged) electron distribution function.

Unfortunately, neither the theoretical understanding of electron heating nor the experimental measurements of the electron distribution function are sufficient to provide direct evidence for or against this model. Evaluations of the indirect evidence in the literature differ sharply in their conclusions[38-41].

Third, one can attempt to explain the observed structure by invoking local, convective stabilization of SRS via density profile steepening. Such an explanation would also have the point of view that SRS must be quenched rather than enhanced, but would do so by steepening rather than by damping. Such steepening might be caused, for example, by a localized instability such as two-plasmon decay or by ablative effects. The observed narrowing of the gap with increasing rather than decreasing laser intensity argues against such an effect. In a contrary sense, the strongest argument for this type of explanation may be that it has no serious advocates in the literature.

CONCLUSION

We have surveyed the Raman spectra observed in experiments using >~ 1 kJ of relatively short-wavelength laser light to produce and irradiate plasmas from planar targets. The general properties of the SRS instability lead one to expect the observed spectra to have certain features. Some of these, such as a short wavelength limit that may plausibly be attributed to Landau damping, are present in all cases. And in a few, rare circumstances, spectra have been observed that resemble in all respects one's naive expectations. More often, the spectra show a long-wavelength limit and/or time-dependent structure that demand more complex explanations. The long-wavelength limit has been observed to increase with laser intensity but otherwise this limit does not scale cleanly with laser wavelength or target properties. The time-dependent structure shows some reproducible qualitative features but varies in detail from experiment to experiment.

Such observations suggest that the observed spectral structure is produced by the nonlinear competition of two or more laser-plasma effects. Unfortunately, it is not trivial to identify these because a large number of effects might in principle contribute. These include but are not limited to stimulated Raman scattering, stimulated Brillouin scattering, the two-plasmon decay instability, cavitation, ablative steepening, and hot-electron generation. Some experiments have identified suggestive correlations but detailed confirmation of any specific hypothesis has not been obtained. Similarly, some theoretical work has developed models of the interaction of specific effects, but none of these theories are sufficiently complete to allow a prediction of spectral structure from the definition of an experiment. The present survey may be of value to those who seek to define more detailed experiments or to develop more complete theoretical models.

ACKNOWLEDGEMENTS

The authors acknowledge fruitful discussions with and technical contributions, found in the cited references, by many members of the laser-plasma community, including but by no means restricted to E.A.Williams, W.L.Kruer, T.W.Johnston, Kent Estabrook, D.W.Phillion, H.A.Baldis, R.L.Berger, W.Seka,

REFERENCES

1. C.S.Liu, M.N.Rosenbluth, R.B.White, Raman and Brillouin scattering of electromagnetic waves in inhomogeneous plasmas, Phys. Fluids 17:1211 (1974).

2. R.P.Drake, R.E.Turner, B.F.Lasinski, K.G.Estabrook, E.M.Campbell, C.L.Wang, D.W.Phillion, E.A.Williams, and W.L.Kruer, Efficient Raman Sidescatter and Hot-Electron Production in Laser-Plasma Interaction Experiments, Phys. Rev. Lett. 53:1739 (1984).

3. R.P.Drake, R.E.Turner, B.F.Lasinski, E.A.Williams, K.G.Estabrook, W.L.Kruer, E.M.Campbell, T.W.Johnston, X-ray emission caused by Raman scattering in long-scale-length plasmas, Phys. Rev. A, 40:3219 (1989).

4. K.Tanaka, L.M.Goldman, W.Seka, M.C.Richardson, J.M.Soures, E.A.Williams, Stimulated Raman scattering from uv-laser-produced plasmas, Phys. Rev. Lett. 48:1179 (1982)

5. R.E.Turner, D.W.Phillion, E.M.Campbell, K.G.Estabrook, Time-resolved observations of stimulated Raman scattering from laser-produced plasmas, Phys. Fluids 26:579 (1983).

6. H.Figueroa, C.Joshi, C.E.Clayton, Experimental studies of Raman scattering from foam targets using a 0.35-um laser beam, Phys. Fluids 30:586 (1987)

7. R.P.Drake, R.E.Turner, B.F.Lasinski, E.A.Williams, D.W.Phillion, Kent Estabrook, W.L.Kruer, E.M.Campbell, T.W.Johnston, K.R.Manes, J.S.Hildum, Studies of Raman scattering from disk targets irradiated by several kilojoules of 0.53-um light, Phys. Fluids 31:3130 (1988) and references therein.

8. Wolf Seka, University of Rochester, private communication, November 1987.

9. H.A.Rose, D.F.DuBois, B.Bezzerides, Nonlinear coupling of stimulated Raman and Brillouin scattering in laser-plasma interactions, Phys. Rev. Lett. 58:2547 (1987).

10. W.Rozmus, R.P.Sharma, J.C.Samson, W.Tighe, Nonlinear evolution of stimulated Raman scattering in homogeneous plasmas, Phys. Fluids 30:2181 (1987).

11. A.Simon, R.W.Short, A new model of Raman spectra in laser-produced plasmas, Phys. Rev. Lett. 53:1912 (1984).

12. M.N.Rosenbluth, Parametric Instabilities in Inhomogeneous Media, Phys. Rev. Lett. 29:565 (1972).

13. R.P.Drake, E.A.Williams, P.E.Young, Kent Estabrook, W.L.Kruer, H.A.Baldis, T.W.Johnston, Drake, Williams, Young, Estabrook, Kruer, Baldis, and Johnston Reply, Phys. Rev. Lett. 61:2387 (1988)

14. E.A.Williams and T.W.Johnston, Phase-inflection parametric instability behavior near threshold with application to laser-plasma stimulated Raman scattering (SRS) instabilities in exploding foils, Phys. Fluids B 1:188 (1989).

15. D.R.Nicholson, A.N.Kauffman, Parametric instabilities in turbulent, inhomogeneous plasma, Phys. Rev. Lett. 33, 1207 (1974); D.R.Nicholson, Parametric instabilities in plasma with sinusoidal density modulation, Phys. Fluids 19, 889 (1976).

16. H.C.Barr, F.F.Chen, SRS with SBS ion waves, Phys. Fluids 30:1180 (1987).

17. Kent Estabrook, W.L.Kruer, and B.F.Lasinski, Heating by Raman backscatter and forward scatter, Phys. Rev. Lett. 45:1399 (1980).

18. W.Seka, E.A.Williams, R.S.Craxton, L.M.Goldman, R.W.Short, K.Tanaka, Convective stimulated Raman scattering instability in UV laser plasmas, Phys. Fluids 27:2181(1984).

19. R.P.Drake, P.E.Young, E.A.Williams, Kent Estabrook, W.L.Kruer, B.F.Lasinski, C.B.Darrow, H.A.Baldis, T.W.Johnston, Laser-Intensity Scaling Experiments in Long-scalelength, Laser-produced plasmas, Phys. Fluids 31:1795 (1988).

20. H.Figueroa, C.Joshi, H.Azechi, N.A.Ebrahim, K.G.Estabrook, Stimulated Raman scattering, two-plasmon decay, and hot electron generation from underdense plasmas at 0.35 μm, Phys. Fluids 27:1887(1984).

21. R.P.Drake, D.W.Phillion, Kent Estabrook, R.E.Turner, R.L.Kauffman, E.M.Campbell, Hydrodynamic expansion of exploding-foil targets irradiated by 0.53 um laser light, Phys. Fluids B 1:1089 (1989).

22. J.A.Tarvin, Gar E.Busch, E.F.Gabl, R.J.Schroeder, and C.L.Shepard, Laser and plasma conditions at the onset of Raman scattering in an underdense plasma, Laser and Particle Beams 4:461 (1986).

23. C. Garban-Labaune, private communication (1987).

24. O.Willi, D.Bassett, A.Giulietti, S.J.Kartunnen, Nonlinear interaction of an intense laser beam with millimeter sized underdense plasmas, Opt. Comm. 70:487 (1989).

25. R.P.Drake, E.A.Williams, P.E.Young, Kent Estabrook, W.L.Kruer, D.S.Montgomery, H.A.Baldis, T.W.Johnston, Narrow Raman Spectra: The Comptetition between collisional and Landau damping, Phys. Fluids B 1:2217 (1989).

26. C.B.Darrow, R.P.Drake, D.S.Montgomery, P.E.Young, Kent Estabrook, W.L.Kruer, T.W.Johnston, Experimental studies of stimulated Raman scattering in reactor-size, laser-produced plasmas, submitted to Phys. Fluids B (1990).

27. R.P.Drake, E.A.Williams, P.E.Young, Kent Estabrook, W.L.Kruer, H.A.Baldis, T.W.Johnston, Evidence that stimulated Raman scattering from laser-produced plasmas is an absolute instability, Phys. Rev. Lett. 60:1018 (1988)

28. S.P.Obenschain, J.A.Stamper, C.J.Pawley, A.N.Mostovych, J.H.Gardner, A.J.Schmitt, and S.E.Bodner, Reduction of Raman Scattering in a Plasma to Convective Levels Using Induced Spatial Incoherence, Phys. Rev. Lett. 62:768 (1989).

29. R.E.Turner, K.G.Estabrook, R.L.Kauffman, D.R.Bach, R.P.Drake, D.W.Phillion, B.F.Lasinski, E.M.Campbell, W.L.Kruer, E.A.Williams, Evidence for collisional damping in high-energy Raman-scattering experiments at 0.26 um, Phys. Rev. Lett. 54:189 (1985).

30. R.P.Drake, E.A.Williams, P.E.Young, Kent Estabrook, W.L.Kruer, H.A.Baldis, T.W.Johnston, Observation of thermal emission in the stimulated Raman scattering frequency band from a laser produced plasma, Phys. Rev. A 39:3536 (1989).

31. R.P.Drake, The scaling of absolutely-unstable, stimulated Raman scattering from planar, laser-produced plasmas, Phys. Fluids B 1:1082, (1989).

32. D.W.Phillion, E.M.Campbell, K.G.Estabrook, G.E.Phillips, F.Ze, Stimulated Raman scattering in large plasmas, Phys. Fluids 25:1405 (1982)

33. R.P.Drake, R.E.Turner, B.F.Lasinski, E.M.Campbell, W.L.Kruer, E.A.Williams, and R.L.Kauffman, Measurements of absorption and Brillouin sidescattering from planar plasmas produced by 0.53 um laser light, Phys. Fluids B, 1:1295 (1989).

34. C.J.Walsh, D.M.Villeneuve, and H.A.Baldis, Electron Plasma-Wave Production by Stimulated Raman Scattering: Competition with Stimulated Brillouin Scattering, Phys. Rev. Lett. 53:1445 (1984).

35. D.M.Villeneuve, H.A.Baldis, J.E.Bernard, Suppression of Stimulated Raman Scattering by the Seeding of Stimulated Brillouin Scattering in a laser-produced plasma, Phys. Rev. Lett. 59:1585 (1987).

36. R.J.Faehl, W.L.Kruer, Laser light absorption by short wavelength ion turbulence, Phys. Fluids 20:55 (1977).

37. H.A.Baldis, P.E.Young, R.P.Drake, W.L.Kruer, Kent Estabrook, W.A.Williams, T.W.Johnston, Competition between the stimulated Raman and Brillouin scattering instabilities in 0.35-um irradiated CH foil targets, Phys. Rev. Lett. 62:2829 (1989).

38. A.Simon, W.Seka, L.M.Goldman, and R.W.Short, Raman Scattering in inhomogeneous laser-produced plasma, Phys. Fluids 29:1704 (1986).

39. A.Simon, R.W.Short, Alternative Analysis of CO2-laser-produced plasma waves, Phys. Fluids 31:3371 (1988).

40. A.Simon, R.W.Short, Energy and Nonlinearity Considerations for the Enhanced Plasma Wave Model of Raman Scattering, Phys. Fluids B 1:1073 (1989).

41. A. Simon, R.W.Short, Comments on "Studies of Raman scattering ... " Phys. Fluids B 1,:1341 (1989); R.P.Drake, R.E.Turner, B.F.Lasinski, E.A.Williams, D.W.Phillion, Kent Estabrook, W.L.Kruer, E.M.Campbell, and T.W.Johnston, Reply to the comments of Simon and Short, Phys. Fluids B, 1:1342 (1989); R.P.Drake, The effect of enhanced plasma waves on Thomson scattering with a high-frequency probe laser, Phys. Fluids B 1:2291 (1989); R.P. Drake, Comment on "Energy and Nonlinearity Considerations ...", Phys. Fluids B 2, Jan. (1990); A.Simon, Reply to the comments of Drake, Phys. Fluids B 2, Jan. (1990).

EXPERIMENTAL STUDY OF BEAM-PLASMA INSTABILITIES IN LONG

SCALELENGTH LASER PRODUCED PLASMAS

A. Giulietti*, S. Coe, T. Afshar-rad, M. Desselberger,
O. Willi, C. Danson[+] and D. Giulietti[#]

Imperial College, Blackett Laboratory, Prince Consort
Road, London SW7 2BZ, U.K.
* Istituto di Fisica Atomica e Molecolare, Pisa, Italy
+ Rutherford Appleton Laboratory, Chilton, Didcot, U.K.
# Dipartimento di Fisica dell' Universita' di Pisa, Italy

ABSTRACT

Plasmas with scalelegths of the order of 1 mm were
produced by focusing four green laser beams of the Rutherford
Appleton Laboratory Vulcan laser onto thin foil targets in a line
focus configuration. A delayed green laser beam was focused
axially into the preformed plasma up to irradiances of $2 \times 10^{15} W/cm^2$.
Delays were chosen that the preformed plasma was underdense and
the electron temperatures ranged from 0.3 KeV to 1 KeV. Several
beam plasma instabilities were investigated under those
experimental conditions including laser filamentation and
whole-beam self-focusing, stimulated Brillouin and Raman
scattering. The interacting laser beam was smoothed by using a
random phase plate and an induced spatial incoherece system.
It was found that both these methods give rise to a virtual
suppression of self-focusing and to a significant reduction in the
level of the Raman and Brillouin instabilities. In addition,
time resolved spectra of stimulated Raman and Brillouin
scattering were recorded showing distinct differences when random
phase plate and induced spatial incoherence system were used in
comparison with the normal laser beam.

I. INTRODUCTION

The most relevant non linear effects in laser plasma
interaction, such as filamentation, Stimulated Brillouin
Scattering (SBS) and Stimulated Raman Scattering (SRS) are
strongly dependent on the scalelength of the plasma. On the
other hand laser fusion targets are expected to be surrounded by
long scalelength, underdense plasmas and a laser-plasma
interaction path of several millimeters has to be considered.
For this reason the experimental study of the interaction of an
intense laser beam with long scalelength, quasi homogeneous
underdense plasmas is of considerable interest.

An experimental campaign is currently performed at the SERC

Central Laser Facility of Rutherford Appleton Laboratory (U.K.) in order to investigate the non linear effects induced by an intense laser beam on a preformed plasma of millimetre size. The main results up to now include direct observation of filamentation and whole beam self-focusing[1], measurements of backscattered laser energy due to SBS and SRS and time resolved spectroscopy of backscattered light[2,3].

The interaction was studied with both an ordinary, coherent laser beam and a laser beam smoothed to reduce its coherence and its non-uniformities. Two different smoothing techniques were used, namely Random Phasing Plate (RPP)[4] and Induced Spatial Incoherence (ISI)[5]. All the observed non linear effects were found strongly depressed when the laser beam was smoothed. The level of backscattering was reduced of orders of magnitude, both for SBS and SRS[6], filamentation was "frozen" with RPP and virtually suppressed with ISI[7].

Although the effect of beam smoothing is very promising for laser fusion, a deeper understanding of the nature of interaction in these beam conditions is necessary. In this respect we belive that time resolved spectroscopy of the laser light scattered unelastically from the plasma is a powerful diagnostics. This paper is also devoted to present a first analysis of some details of SBS and SRS spectra obtained in the experiment. Several interesting features can be explained as a consequence of filament formation in case of the unsmoothed laser beam.

In the next section we describe the experimental method. Sections III. and IV. are devoted to the analysis of SBS and SRS spectra respectively, as obtained from the interaction of the preformed plasma with the unsmoothed laser beam. In section V. we present the effect of beam smoothing on filamentation and backscattering; this latter was fund not only drastically reduced, but also spectrally changed. Finally section VI. is devoted to the conclusion.

II. EXPERIMENTAL METHOD

A general view of the experimental set-up is shown in Fig.1. A long-scalelength plasma was generated by irradiating a thin foil stripe target by four frequency doubled (λ_0 = 527 nm) beams of the Vulcan glass laser. The beams were focused onto target in two opposing pairs in a line focus geometry using an off-axis lens mirror scheme[8]. The length and the width of the focal spot for each beam were 2.0 mm and 0.3 mm, respectively. All the four beams were superimposed onto the target resulting in a typical irradiance of 10^{14} W/cm^2 in a 600 ps FWHM laser pulse. The targets consisted of an aluminium stripe 300 μm wide, 700 nm thick and 0.8 mm long supported on 100 nm formvar substrate. A delayed second harmonic, 600 ps FWHM, laser beam was then interacted axially with this preformed plasma, as shown in Fig.1. Consequently the interaction length was about 800 μm along which the plasma temperature and the density were approximately constant. Typically the interaction beam irradiance was in the range (10^{14} - 10^{15} W/cm^2). By varying the delay between the interaction beam and the heating beams it is possible to control the plasma conditions at the time of the interaction. For the results presented in this paper the interaction beam delay was 1.8 ns.

The backscattered Raman and Brillouin signals generated by the interaction beam were imaged out via the incident focusing lens. The scattered light was then split in two separate

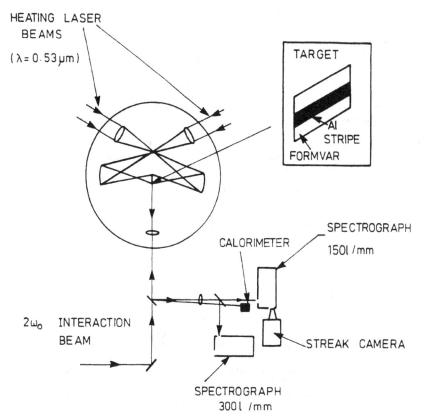

HEATING LASER
BEAMS

($\lambda = 0.53\,\mu m$)

TARGET

Al
STRIPE

FORMVAR

SPECTROGRAPH
$150\,\ell/mm$

CALORIMETER

STREAK CAMERA

$2\omega_o$ INTERACTION
BEAM

SPECTROGRAPH
$300\,\ell/mm$

Fig.1. Schematic of the experimental arrangement. The two pairs of heating beams, the interaction beam as well as the target geometry are shown. Also the main components of the Raman backscatter detection channel are shown, as designed to obtain time integrated/time resolved spectra and energy measurements. A similar detection channel was used for Brillouin backscatter, but with higher dispersion gratings.

channels and focused down onto calibrated photodiodes and calorimeters to measure the scattered energy levels. In addition time-resolved measurements were made of SRS and SBS spectra by spectrometers coupled to optical streak cameras. Measurements made of SRS and SBS spectra can be used to estimate the plasma conditions[2]. For the experiment considered here these spectra indicate an electron density of 8% of the critical electron density and a background temperature of 500 eV. This is consistent with the predictions of the plasma conditions made by the model of London and Rosen[9]. In fact spectra have also several different features indicating that special conditions are created locally at definite times. An equivalent plane monitor recorded the incident interaction beam focal spot distribution at planes equivalent to a 200 μm separation throughout the length of the preformed plasma.

Finally the interaction beam was smoothed by two different methods. One of them was a RPP inserted before the focusing lens. It breaks the incident laser beam up into a large number of individual beamlets which are then randomly shifted in phase. These beamlets are then overlapped onto the target by a focusing lens. The resulting interference pattern produced eliminate any gross long scalelength non-uniformities from the laser beam profile but introduces pronounced short-scalelength intensity fluctuations which remain frozen in the focal spot distribution throughout the laser pulse. The second device was an ISI system consisting of a pair of crossed echelons whose steps introduced a delay between two adjacent beamlets. In that case a broad band laser was used so that the coherence time of the laser, t_c, was shorter than the incremental delay between individual beamlets. This causes the instantaneous interference pattern produced to be averaged out over times long compared to t_c, leaving a smooth focal spot profile.

III. SBS SPECTRA

All the spectra of Brillouin backscattered light showed two or more components. The shift of each component from the laser light wavelength was different and usually the different components were seen at different time . The duration of each burst of SBS light was definitely shorter than the laser pulse duration. This behaviour was attributed to the influence of filamentation on the Brillouin instability[2]. Our configuration was specially convenient to study features of SBS spectra because the preformed plasma expands mainly in the transverse direction to the interaction beam without significant plasma flow towards the laser, resulting in a Doppler-free Brillouin shift which is given by

$$\Delta\lambda/\lambda_0 = 2(c_s/c)(1-N)^{1/2} / (1+4k_0^2 \lambda_D^2)^{1/2} \tag{1}$$

where k_0 is the local wavenumber of the laser light and λ_D is the Debye length. The relative density $N = n_e/n_c$ is normalised to the critical density. c_s and c are the ion sound velocity and the speed of light, respectively.

Most of our spectra were obtained in a range of delay between the heating and the interacting pulses for which $N \ll 1$ and $4k_0^2 \lambda_D^2 \ll 1$. In these conditions the temperature is easily evaluated from the red shift.

Several SBS spectra show components with different shift. If we consider them in order of increasing red shift, we can estimate different temperatures for different SBS sources. At a delay of 1.8 ns a first component agrees with $T_e \approx 500$ eV, which is a reasonable value for the background plasma[9]. Other components give temperatures from 1 to 1.5 KeV. We assume that those higher temperatures can be produced in some filamentary regions as already discussed[2,7].

However, for some spectra, the value of the maximum red shift is too high to be reasonably attributed to increased temperature by filamentation. Moreover in these spectra the evolution of the Brillouin scattering is peculiar, showing a first stage of increasing red shift, followed by a decreasing stage, as shown in Fig.2.

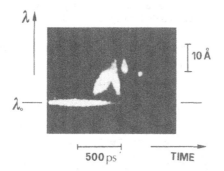

Fig.2. This image is representative of a series of anomalous SBS spectra, characterized by very large maximum red shift and peculiar evolution of the spectrum. Target:Al stripe 700 nm thick. Heating irradiance: 10^{14} W/cm². Interaction irradiance: 8×10^{14} W/cm². Heating-interaction pulse delay:1.8 ns. Emission at $\lambda_0 = 527$ nm is wavelength fiducial but not time fiducial.

We believe that all the features of such anomalous SBS spectra can be explained in terms of Self-Phase Modulation (SPM) of the laser light penetrating into filamentary regions which are suddenly evacuated by ponderomotive and/or thermal pressure[10]. Under this assumption the stage of the increasing shift is the evidence of the creation of very strong filamentary regions, while the stage of shift decreasing back to ordinary SBS value, is a mark of the saturation of the filamentation process. Also the short times of SBS, where no emission is observed, can be explained in the SPM context as due to the mismatch between the pumped ion acustic wave and the fast shifting laser light, both involved in the SBS process.

IV. SRS SPECTRA

Let briefly recall the main features expected from a Raman spectrum. The phase matching conditions give for the shift $\Delta\lambda = (\lambda_R - \lambda_0)$ of the Raman scattered light

$$\Delta\lambda/\lambda_0 = N^{1/2}(1 + 3k^2\lambda_D^2)^{1/2}[1 - N^{1/2}(1 + 3k^2\lambda_D^2)^{1/2}]^{-1}, \qquad (2)$$

where k is the wavenumber of the electron plasma wave. The equation (2) applies to Raman forward and backscattering, but the difference is in the wavenumber k which is given in the cold plasma limit [11],

$$k/k_0 = 1 \pm [(1-2N^{1/2})/(1-N)]^{1/2}, \qquad (3)$$

where "+" sign corresponds to Raman backscattering. At higher electron temperatures, eq.(3) is modified by a fairly complicated function depending on the relative density N and the electron thermal speed v_e [11]. Eq.(3) determines the density limit N<0.25 for the Raman process.

In the present experiment N < 0.1, and the largest possible λ_R corresponds to the highest electron density present into the interaction region, at a given temperature. Even though the density is expected to be uniform along the beam axis, different densities are present in the cross section of the interaction region, and a broad λ_R is expected. The density cut-off towards shorter wavelengths at a certain temperature is given by Landau damping of the plasmons.

All the SRS spectra we examined for this work were obtained with Al stripe targets and 1.8 ns delay between the heating and interaction pulses. As reported in section II, we expect in these conditions a temperature $T_e \approx 500$ eV, which is also confirmed by the low-shift component of SBS spectra. Electron density is expected $n_e \approx 0.08\, n_c$. These values can give us a rough estimation of both upper and lower wavelength limits for SRS spectra. These limits are always respected, but all the spectra show other striking features. First of all the usual trend of the spectrum towards the shorter wavelength (in agreement with an expanding plasma) is not always respected. Secondly each spectrum shows very fast and unreproducible time structures, as in a spiky emission. The actual duration of each spike is on the scale of a few tens of picoseconds and it is possibly limited by the temporal resolution given by the streak camera slit (20 ps).

The time slope of the spectrum of each spike varied shot by shot, but was generally consistent with a very fast evolution of the SRS. These two interesting features are visible in the time resolved spectrum of Fig.3. Generally speaking all the features are consistent with evidencing a Raman scattering occurring in a region far from homogeneity and whose parameters are rapidly changing in time. This picture is of course in agreement with an unstable interaction due to filamentation, but we need a self-consistent model to describe it. In what concerns particularly the spiky structure of the emission we cannot ignore that another interesting explanation was recently suggested[12].

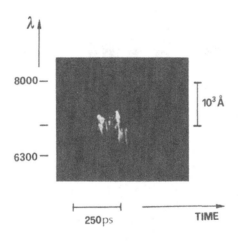

λ

8000 —

—

6300 —

10^3 Å

├─────────────┤ ─────────→
    250ps   TIME

Fig.3 A time resolved SRS spectrum shifting towards blue
due to plasma expansion. Moreover time structures
of duration of few tens of picoseconds are
apparent: they are attributed to an unstable
interaction due to filamentation. Target:Al stripe
700 nm thick. Heating irradiance: 8×10^{13} W/cm^2.
Interaction irradiance:6×10^{14}W/cm^2.
Heating-interaction pulse delay: 1.8 ns.

V. EFFECTS OF BEAM SMOOTHING

In the second part of the experiment the interaction beam was
smoothed with two devices, namely a RPP and an ISI system. The
aim was to study the influence of reduced coherence on the
instabilities. · The RPP was just put behind the f/2.5 focusing
optics. Owing to the beam diameter of about 8 cm, this
configuration produced in the focal region diffraction driven
structures[4] of a few microns cross section and 10-15 μm in depth.
On the other hand the use of ISI required an improved bandwidth[5]
of the Vulcan oscillator, to get a coherence time $t_c \approx 2$ ps. In
fact each echelon had 5 steps with a step to step optical delay
of 5 ps. In order to obtain an effective beam smoothing along the
whole plasma, a longer focal length lens (f/10) was used for ISI.
Figure 4 shows images taken from the equivalent plane monitor
for the far field of the narrow-band laser beam, narrow-band laser
beam with RPP and the broad-band ISI beam. For the unsmoothed
beam pronounced structure with a typical scalelength of 15-25 μm
and intensity ratios of up to 5:1 can be seen. This gross
structure is eliminated with the use of the RPP, but the
interference pattern produced by the RPP results in a fine scale
structure with pronounced intensity fluctuations on a 2-3 μm
scalelength. It can be seen that the far field image for the ISI
beam is observed to have very little structure compared with the
other two cases.
Both RPP and ISI devices had strong effect on the Brillouin
and Raman backscattered energy. In Fig.5 and 6 the backscattered
absolute levels of SBS and SRS with respect to irradiance are
presented for coherent, RPP and ISI interaction beams[6]. It
should be noted that in determining these levels an f/10 focusing
lens was used for the coherent and ISI interaction beams. This

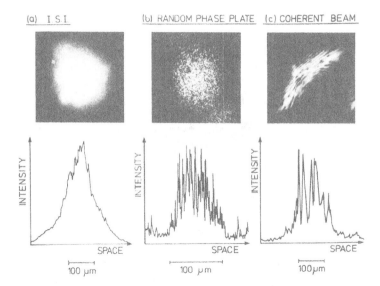

Fig.4. Typical focal spot distributions measured with
the equivalent plane monitor: a) broad-bandwidth
ISI beam (f/10 focussing lens), b) narrow-bandwidth
beam with RPP (f/2.5 focussing lens), and c)
standard narrow-bandwidth beam (f/10 focussing
lens). Note that the magnification for the RPP
image and densitometer trace is twice that of the
other two cases.

generated a focal spot of approximately 120 μm diameter along the
length of the preformed plasma. This was measured using a
multi-frame equivalent plane monitor setup for the incident
interaction beam. For the RPP interaction beam an f/2.5 focusing
lens was used, resulting in a FWHM focal spot varying between 70μm
and 140 μm along the plasma length. Consequently, since the
intensity variations along the interaction region for the
different focusing conditions did not vary significantly, a
comparative study between the different interaction beams could
be made. Fig.5 shows the variation in the SBS level with
irradiance for the three interaction beams considered above. It
can be seen that the RPP reduces the SBS level by a factor of
10-20, and with ISI the level is reduced to below the threshold of
detection. This corresponds to a reduction of a least a factor
of 500 for irradiances $I \approx 5 \times 10^{14}$ W/cm^2. For the coherent beam a
saturation level was obtained at about 1% and a definite threshold
value for the irradiance was observed at about 2×10^{14} W/cm^2. This
value was also corroborated using the spectral and the spatial
diagnostics used to monitor the SBS emission. A more detailed
discussion of the saturation levels of both SBS and SRS is given
elsewhere [3]. Fig.6 shows the variation of SRS level with

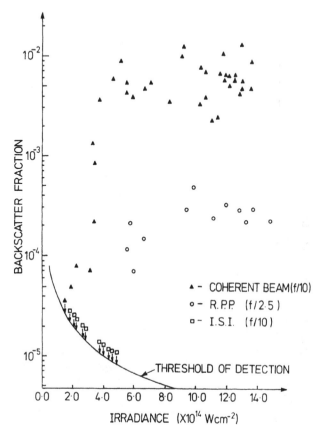

Fig.5 Variation of SBS backscattered fraction with
average laser irradiance. An arrow (↓)
indicates that the SBS signal was lower than the
threshold of the detection. ▲ coherent beam
(f/10), ○ RPP (f/2.5), □ ISI (f/10).

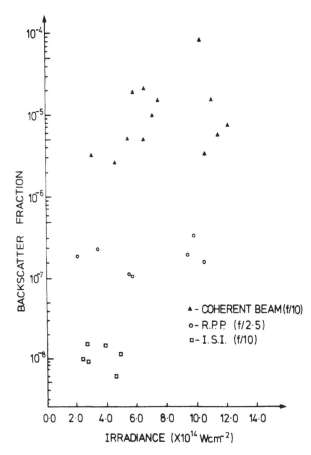

Fig.6 Variation of SRS backscatter fraction with average
laser irradiance. Symbols are the same as in Fig.5.

irradiance. Similar reductions are observed in the level of SRS emission with RPP and ISI as in SBS case. Further, a similar threshold value of $I \approx 1-2 \times 10^{14}$ W/cm^2 was determined for the backscattered SRS from the diagnostic channels, set up to monitor the spectral and the spatial profiles, since no detectable signal could be recorded below this irradiance. Although the observed thresholds for the SRS and SBS are characteristic of a stimulated process, we attribute this to the onset of filamentation instability, which has an expected threshold of $I \approx 10^{14}$ W/cm^2. This whould explain why we observe similar threshold values for both the backscattered SRS and SBS processes. With the previous observations already discussed, establishing the importance of filamentation in the generation of SRS and SBS, we must conclude therefore that these reductions are primarily due its suppression. Such a intermediate mechanism is necessary to account for these dramatic reductions.

This interpretation is fully consistent with the net reduction of filamentation which we observed by imaging the exit plane of the interaction region on a gated image intensifier[6] with 120 ps gate time . With RPP in use, very thin but stable structures are observed as shown from the equivalent plane picture (Fig.4b) in place of chaotically evolving filaments and whole beam self-focusing seen with the usual coherent beam [13]. With the ISI the beam was very smooth during the interaction.

Also the spectra of the residual SBS and SRS light changed significantly with smoothed beam [10]. SBS spectra had usually single component with a red shift consistent with the background plasma temperature. SRS spectra did not show structures in time as observed with the coherent beam.

VI. CONCLUSION

We designed a new method to study laser interactions with underdense plasmas, which was experimentally tested. The method allows to obtain a preformed plasma whose temperature, density and scalelength at the interaction time are determined by choice of target material and thickness, irradiance of heating laser light and delay between heating and interacting beams.

Stimulated instabilities as self-focusing and filamentation, Brillouin and Raman scattering were studied at nominal interaction irradiances of 10^{14} to 3×10^{15} w/cm^2 and 527 nm laser wavelength. The plasma was definitely underdense, about 0.5 KeV in temperature and several hundreds of micrometers both in longitudinal and transverse scalelength.

With an ordinary narrowband laser bean both Brillouin and Raman instabilities were found to be driven by filamentation occuring in the underdense plasma. The same threshold observed for the three phenomena was a first evidence for that; it was also confirmed by time resolved images of Brillouin and Raman backscattering sources, showing filamentary structures. Time resolved spectra of Brillouin and Raman scatterd light showed features consistent with enhanced scattering from filaments. Spectra also evidenced activation of nonlinear processes, including self-phase modulation of the laser light in the filamentary plasma.

The laser bean interacting with the preformed plasma was subsequently smoothed with two different devices, a random phase plate and a pair of echelons inducing spatial incoherence. In the last arrangement the laser oscillator was turned to broad-band operation. A distinct inhibition of filamentation was observed with both devices as well as a drastic reduction of Brillouin and Raman backscattering. Time resolved spectra of backscattered light did not show any more nonlinear feature when the plasma was interacted with the smoothed beam.

Many observation and measurements require further analysis. An important point to be clarified is the relatively low level of backscattering observed with the unsmoothed beams. The next experimental compaing will be also devoted to this problem. On one hand new configurations will be used to improve plasma homogeneity, including smoothing of the heathing laser beans. On the other hand several possible saturation mechanisms will be investigated. Harmonics of the laser light emitted during the interaction will provide a supplementary diagnostics. In particular second harmonic generation in underdense plasmas is very promising to study local deviation from homogeneity.[14,15]

REFERENCES

1) S.E. Coe, T. Afshar-Rad and O. Willi, Opt.Comm.,**73**,299,(1989).
2) O. Willi, D. Bassett, A. Giulietti and S.J. Karttunen, Opt. Comm.,**70**, 487,(1989).
3) T. Afshar-rad, S. Coe, O. Willi "Effect of filamentation on stimulated Raman and Brillonin scattering in long preferred plasmas",preprint, (1989).
4) Y. Kato, K. Mima, N. Miyanaga, S. Aringa, Y. Kitagawa, M. Nakatsuka and C. Yamanaka, Phys. Rev. Lett., **53**, 1057, (1984).
5) R.H. Lehmberg and S.P. Obenschain, Opt. Comm., **46**, 27, (1983).
6) S.E. Coe, T. Afshar-Rad, M. Desselberger, F. Khattak, O.Willi, A.Giulietti, Z.Q. Lin, W. Yu and C. Danson, Europhys.Lett., **10**, 31, (1989).
7) A. Giulietti, T. Afshar-Rad, S.E. Coe, O. Willi, Z.Q. Lin and W. Yu, Laser Interaction with Matter - Edit. by G.Verlarde (World Scientific, Singapore, 1989), pag. 208.
8) I.N. Ross, J. Bonn, R. Corbett, A. Damerell, P. Gottfeldt, C. Hooker, M. Key, G. Kiehn, C. Lewis and O. Willi, Appl. Opt., **216**, 1584, (1987).
9) R.A. London and M. Rosen, Phys. Fluids, **29**, 3813, (1986).
10) T. Afshar-Rad, S.E. Coe, C. Danson, M. Desselberger, A.Giulietti, D. Giulietti and O. Willi, Book of Abstracts, O-8, 20th ECLIM, Schliersee, (1990).
11) S.J. Karttunen, Laser and Particle Beams, **3**, 157, (1985).
12) H. Hora, L. Cicchitelli, S. Eliezer, Gu Min, R.S. Pease, W. Scheid and A. Scharmann, Book of Abstracts, O-9, 20th ECLIM, Schliersee, (1990).
13) S. Coe, Ph.D.Thesis, Imperial College of Science Thechnology and Medicine, London, (1989)
14) A. Giulietti, D.Giulietti, D. Batani, V. Biancalana, L. Gizzi, L. Nocera, and E. Schifano, Phys. Rev. Lett.,63, 524, (1989).
15) P. E. Young, H. A. Baldis, T. W. Johnston, W. L. Kruer and K. G. Estabrook Phys.Rev.Lett., 63, 2812 (1989).

ANALYSIS OF 2ω AND 3ω/2 SPECTRA FROM PLASMAS PRODUCED BY IRRADIATION OF THIN FOIL TARGETS

A. Giulietti, D. Batani*, V. Biancalana*,
D. Giulietti*, L. Gizzi@, L. Nocera and E. Schifano*

Istituto di Fisica Atomica e Molecolare
via del Giardino, 7, 56100, Pisa, Italy
* Dipartimento di Fisica, Universita' di Pisa, Italy
@ Imperial College, London, U.K.

ABSTRACT

Thin plastic foils have been irradiated at 1.064 μm laser wavelength and an intensity up to 5×10^{13} W/cm^2. The plasma produced became underdense during the laser pulse and electron temperatures of several hundred eV's were obtained. Second harmonic and three-halves harmonic emission have been studied perpendicularly to the laser beam axis. Time resolved imaging and spectroscopy gave us novel information on the physical mechanisms involved in these processes. The 2ω line was found to be red shifted consistently with the frequency sum between the incident laser light and the Brillouin backscattered radiation. Such an effect was already postulated to explain second harmonic side emission from filaments occurring in an underdense corona. However the occurrence of frequency sum in a plasma was demonstrated for the first time in our experiment. 3ω/2 spectra showed an intense broadband red-shifted component and a very faint blue component. The analysis of spectral features of 3ω/2 light, in terms of coupling between laser light and electron waves generated by Two Plasmon Decay (TPD) of the laser light itself, allowed us to conclude that the TPD plasmons propagate through the $n_c/4$ layer before they couple.

I. INTRODUCTION

In the last years a lot of work has been done on the study of parametric instabilities developing in long scale length undercritical plasmas similar to those the laser radiation have to pass through before reaching the target denser regions in ICF experiments.[1,2]

Among the numerous plasma diagnostics, the detection and analysis of harmonics of laser radiation (both integer and half-integer) have been some of the most powerful[3-5].

In particular $3\omega/2$ radiation, detected at different angles, has been utilized to follow the motion of the quarter-critical layer ($n_e \approx n_c/4$) and to measure electron temperature in the same region [6-13]. However the comparison between experimental results and theory was not always satisfactory. A complete understanding of the spectral features must take into account a number of possible contributions, including reflection of laser radiation inside the plasma and plasmon propagation[14] before they couple with laser radiation to produce $3\omega/2$ harmonic.

The second harmonic has been mainly studied in presence of the critical density layer where its generation is by far more probable [15,16]. Only a few works report on the observation of 2ω emission from undercritical plasmas[17-19]. The mechanisms of 2ω generation and the information on plasma parameters that can be obtained are completely different in the two cases.

In the present experiment time resolved spectra and images at 2ω and $3\omega/2$ were obtained at 90^0 to the beam axis from a plasma produced by laser irradiation of a thin plastic film. The plasma was underdense for most of the time of the laser pulse. To reduce sources of uncertainty in the analysis of the 2ω and $3\omega/2$ spectra, we performed the experiment with a small aperture of the focusing and detection optics .

The analysis of 2ω spectra allowed us to confirm the hypothesis proposed by other authors [20] for the emission of second harmonic from an underdense plasma at 90^0 to the laser beam axis. On the other hand the $3\omega/2$ spectra evidenced the relevance of the effect of propagation of plasmons created by Two Plasmon Decay (TPD), before they couple to generate $3\omega/2$ radiation.

In section II we recall elements of the theory of $3\omega/2$ and 2ω emission. The experimental apparatus is presented in section III. In section IV and V the experimental results are reported and discussed.

II. THEORY

Three-Halves Harmonic Generation

The $3\omega/2$ radiation can be produced from both the coupling of the incident laser radiation with $\omega/2$ plasma waves or the coalescence of three of those waves. However, as the three wave process is of higher order, coupling of laser radiation with plasma wave is expected to prevail in our experiment. Thus only the latter process will be considered. $\omega/2$ plasmons can be produced close to $n_c/4$ plasma density layer by TPD and Stimulated Raman Scattering (SRS); but, due to the lower threshold of the TPD as compared with SRS, only the former mechanism of plasmons production will be considered.

In TPD the laser ($\omega_o, \underline{K}_o$) and plasma ($\omega_{B,R}, \underline{K}_{B,R}$) waves satisfy the energy and momentum conditions:

$$\omega_o = \omega_B + \omega_R \qquad (1)$$

274

$$\underline{K}_o = \underline{K}_B + \underline{K}_R \qquad (2)$$

which, with the dispersion relations for photons and plasmons, give the plasmon frequencies:

$$\omega_B \approx \omega_o/2 \ (1+3/4 \ (K_B^2 - K_R^2) \ \lambda_D^2 \ ) \qquad (3)$$

$$\omega_R \approx \omega_o/2 \ (1-3/4 \ (K_B^2 - K_R^2) \ \lambda_D^2 \ ) \qquad (4)$$

where λ_D is the Debye length.

With $K_B > K_R$, the plasmons (ω_B, $\underline{K}_B$), travelling with a wavevector component in the direction of the pump wave i.e. $\underline{K}_o \cdot \underline{K}_B > 0$ are blue shifted, while the plasmons (ω_R, $\underline{K}_R$), for which $\underline{K}_o \cdot \underline{K}_R$ can be positive or negative, are red shifted from the exact $\omega_0/2$ frequency. In the following we will call them "blue" and "red" plasmons respectively.

In the most general case the plasmons from TPD propagate[14] through the inhomogeneous plasma before they couple with the pump photons. As a consequence the matching conditions for the $3\omega/2$ emission process can considerably change in space, because in the quarter critical region the $\underline{K}_{R,B}$ wavevectors dramatically depend on density. We consider here only density gradients parallel to $\underline{K}_o$, as those due to the target foil explosion. A perpendicular component of the density gradient is also usually present in laser produced plasmas due to the non-uniformity of illumination and eventually filamentation. This component can be shown to have a minor influence on $3\omega/2$ spectra. On the contrary a gradient component perpendicular to $\underline{K}_o$ is essential to generate second harmonic in underdense plasmas (see next subsection).

For laser radiation propagating along density gradient and coupling with TPD (red or blue) plasmons, the energy and momentum conservation for the process gives for the wavelength shift $\Delta\lambda_{3/2}$ from exact $2\lambda_0/3$ value:

$$\Delta\lambda_{3/2} =$$
$$\lambda_0/2 \ (v_e/c)^2 \ [1-(2\sigma/3^{1/2}) \ (11\alpha^2 \pm 4\cdot6^{1/2} \ \alpha^2 \cos\theta - 8\sin^2\theta)^{1/2}] \qquad (5)$$

Where v_e is the electron thermal velocity, θ the angle between $\underline{K}_o$ and $\underline{K}_{3/2}$, $\sigma = (\underline{K}_o \cdot \underline{K}_{B,R}) \ / \ | \ \underline{K}_o \cdot \underline{K}_{B,R} \ |$ and $\alpha = |\underline{K}_{B,R}|/|\underline{K}'_{B,R}|$, $\underline{K}_{B,R}$ and $\underline{K}'_{B,R}$ being the wavevector of the plasmon before and after its propagation respectively. Notice that the "-" sign refers to incident laser light (ω_0, $\underline{K}_o$) and the "+" sign to backpropagating light (ω_0, $-\underline{K}_o$). As it is evident from the equation (5) the plasmon propagation results in an additional shift of $3\omega/2$ radiation, the ordinary shift being an effect of the plasma temperature. Because a range of different α's is allowed the plasmon propagation is a source of $3\omega/2$ line broadening as well.

It is important to note that the measurements of $3\omega/2$ spectrum can give an estimation of electron temperature in the $n_c/4$ region, but the propagation of plasmons makes this diagnostic less accurate due to the variability of α.

Second Harmonic Generation from Underdense Plasmas

Close to the critical density different mechanisms developing during the laser-plasma interaction can produce 2ω radiation. In fact in this region ω_o plasmons can be generated either by resonant absorption or Electron Ion Decay. Such plasmons interacting in couples or with another pump photon can easily produce 2ω radiation at different angles[21]. There is also another way to produce 2ω radiation that consists in the interaction of the impinging electromagnetic wave with density gradients inside the plasma[22].

The last mechanism is the only one that can still work in underdense plasmas; but, due to the radial symmetry typical of laser produced plasmas, it provides a forward emission of 2ω radiation in a small angle from the laser beam axis [18,23]. 2ω radiation from underdense plasmas was detected at 90^0 only in few experiments. In one of them 2ω radiation at 90^0 was used to image filaments created in an underdense plasma. The authors[20] proposed a theory to explain such an emission as a result of the interaction of the impinging laser radiation with backreflected radiation at about the same frequency, in presence of density gradients in the plasma. According to the theory, the 2ω emission originates from the current density

$$\mathbf{j}_{2\omega} = (e^3/2\,im^2\omega^3)\,[\underline{E}_o\underline{E}_o e^{2iK_o z} + \underline{E}_B\underline{E}_B e^{-2iK_o z} + (\underline{E}_o\underline{E}_B+\underline{E}_B\underline{E}_o)]\cdot\nabla n \qquad (6)$$

where $\underline{E}_o$ and $\underline{E}_B$ are the amplitudes of electric fields of impinging and backreflected laser radiation and ∇n represents the density gradients in the plasma. The equation (6) shows that a radial component for ∇n is needed to allow 2ω emission.

This component is usually present in laser produced plasmas, in case enhanced by self-focusing and filamentation of the laser beam. As one can verify the first term contributes only for small angles ($\gamma \sim 0$) in an angular width $\Delta\gamma \sim (2\lambda/l_F)^{1/2}$; γ being the angle between the laser beam axis (z) and the direction of observation, and l_F the length of the emitting region.

The second term is responsible for the 2ω backscattered radiation, contributing for $\gamma \sim \pi$, while the third contributes only for $\gamma \sim \pi/2$ in an angular width $\Delta\gamma \sim \lambda/l_F$.

III. EXPERIMENTAL SETUP

The experimental setup is shown in Fig.1. A Nd laser (λ_o =1.064 µm, spectral width $\Delta\lambda \approx 0.7$ Å), delivering up to 3 J in 3 nsec (FWHM), was focused on target by an f/8 lens into a vacuum chamber. In a 60 µm diameter focal spot an intensity up to 5×10^{13} W/cm^2 was reached. Time modulations were present into the pulse, due to beating of longitudinal modes. Prepulse level was kept below 10^{-4} the main pulse and it was experimentally checked that this level could not produce any early plasma formation. This was not the case for a small number of shots for which the interaction showed completely different character.

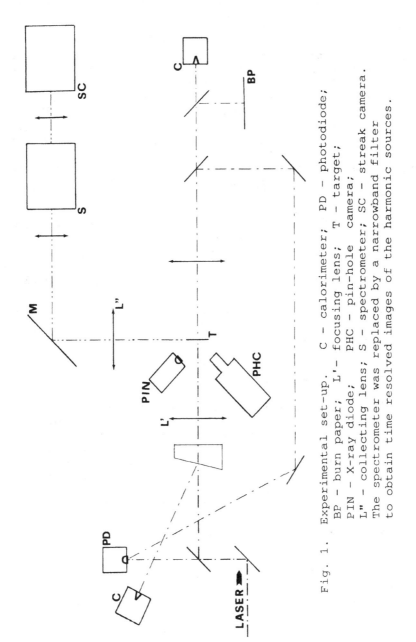

Fig. 1. Experimental set-up. C – calorimeter; PD – photodiode; BP – burn paper; L' – focusing lens; T – target; PIN – X-ray diode; PHC – pin-hole camera; L" – collecting lens; S – spectrometer; SC – streak camera. The spectrometer was replaced by a narrowband filter to obtain time resolved images of the harmonic sources.

The targets used consisted of thin formvar foils (polyvinyl formal [$C_5 H_{11} O_2$] ; density = 1.1 g/cm^3 ; $Z_a \approx 3$) , whose thickness ranged from 0.1 µm to 1.6 µm and was measured with an interferometric method [24]. The target was irradiated at normal incidence and the laser ligth was linearly polarized at 45° with respect to the irradiation/detection plane.

The radiation was collected at 90° from the beam axis, using an f/7 optics, and imaged into the entrance slit of a spectrometer coupled with a Hadland streak camera. The overall spectral resolution resulted 3 Å and the time resolution was 20 ps. The collecting optics was designed in order to have no spatial selection of the $3\omega/2$ and 2ω sources. Spectral calibration was performed with the 7032 Å, 7174 Å, 7245 Å lines of Ne for the $3\omega/2$ spectra and 5330 Å, 5341 Å, 5400 Å lines of Ne for the 2ω spectra. A wavelength fiducial was obtained putting a thin wire on the spectrometer output plane, perpendicularly to the dispersion axis. The position of the wire was in turn calibrated with the same spectral lines.

Alternatively, to obtain the time resolved images of the emitting regions, the spectrometer was replaced with a narrow band filter (100 Å FWHM centered at 7090 Å for the $3\omega/2$ and 30 Å FWHM at 5320 Å for the 2ω spectra) and the interaction region imaged directly into the entrance slit of the streak-camera,with a magnification G$\approx$10. The streak-camera slit was superimposed to the laser beam axis in the image of the plasma. The slit aperture was set in order to match the focal spot diameter and the spot image was carefully positioned. This set-up allowed to localize the position of the $3\omega/2$ and 2ω sources along the interaction region.

IV. EXPERIMENTAL RESULTS

In Fig.2 a spectrum, obtained at low dispersion, shows both $3\omega/2$ and 2ω lines. Together with the two harmonics the bremsstrahlung emission is apparent.

In Fig.3 and Fig.4 the spectra of the two harmonics, obtained at higher dispersion, are reported; Fig.5 and Fig.6 show the images of the sources of the two harmonics. A time modulation of the emission, resulting in spikes of tens of picoseconds is apparent. It is due to the intensity modulation already present in the laser pulse. The wavelength fiducial looks like an artificial intensity minimum in the two spectra. The dashed line parallel to the time-axis shows the position of the exact $3\omega_0/2$ and $2\omega_0$ lines respectively. It is interesting to observe that, while the $3\omega/2$ radiation was characterized by well defined regions of emission and a broad spectrum, the 2ω radiation was emitted from wide plasma regions with a narrow spectrum. This is an indication of completely different mechanisms of emission for the two cases. In fact $3\omega/2$ radiation can be emitted only in a density region close to $n_c/4$ via the interaction of e.m. waves with plasmons of broad energy spectrum. On the other hand 2ω radiation is generated at 90° by the interaction of two counterpropagating e.m. waves of about the same bandwidth as the laser light; this process can take place virtually in the whole laser plasma interaction volume.

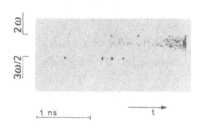

Fig.2. Low dispersion time resolved spectrum of light
collected perpendicularly to the beam axis. Shot
#061204; formvar foil thickness = 1.34 μm; laser
irradiance I = 1.3 x 10^{13} W/cm^2. Second harmonic
and 3/2 harmonic of the laser light are visible as
isolated spikes. Continuum plasma emission is
also shown as background more intense in the
yellow-green region of the spectrum because of
cathode sensitivity of the streak camera. 3ω/2
emission is much more intense than 2ω one. Only
few spikes show coincidence of two emissions, due
to the completely different mechanism of their
generation.

The laser intensity threshold to generate 3ω/2 light
resulted I_{th} ≈ 5 x 10^{12} W/cm^2 , but this harmonic completely
disappeared in case of prepulse level higher than 10^{-4} the main
pulse. This is due to the fact that, in such a condition, the
plasma is produced earlier and the electron density is lower than
$n_c/4$ at the time of the main pulse. 3ω/2 spectra showed a
strong variability from one spike to another, either for what
concerns shift and width. However all the spectra were red
shifted even if some of them showed a faint blue component too.
        From the images like that reported in Fig.6 we estimated
the scalelength of the density gradients in the $n_c/4$ region. We
found values of a few tens of micrometers, which reasonably agree
with the London-Rosen expansion model [25].
        Spatial modulations along the laser beam axis can be
observed in most of spikes in the 2ω images.

V. DISCUSSION AND CONCLUSION

    The spectra of 3ω/2 light emitted at 90° we obtained are
characterized by considerable values of both red shift and
bandwidth. Such values are very difficult to explain if the
propagation of TPD plasmons is not considered. As matter of
fact in this case both temperature and Doppler effects give
shifts lower than the mean shift of ≈35 Å we observed on a
large number of shots.
    More pronounced is the discrepancy for the 3ω/2 bandwidth.

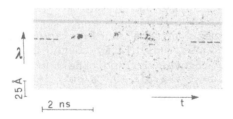

Fig. 3. Time resolved spectrum of 2ω emitted at $90°$.
Shot #231203; d = 1.50 μm; I = 1.1 x 10^{13} W/cm^2.
The dashed line matches the exact $\lambda_0/2$
wavelength.

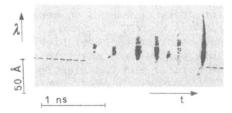

Fig. 4. Time resolved spectrum of $3\omega/2$ emitted at $90°$.
Shot #191202; d = 1.50 μm; I = 1.0 x 10^{13} W/cm^2.
The dashed line matches the exact $2\lambda_0/3$
wavelength.

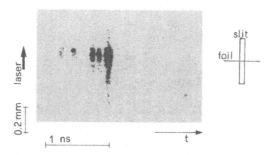

Fig. 5. Time resolved image of 2ω emitting sources taken
at 90° to the laser beam axis.   Shot #060204;
d = 1.39 μm; I = 1.5 x 10^{13} W/cm². The relative
position of foil target, laser beam and
streak   camera slit is shown.

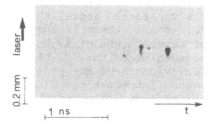

Fig. 6. Time resolved image of 3ω/2 emitting sources
taken at 90° to the laser beam axis.   Shot
#020204;   d = 0.8 μm; I = 1.1 x 10^{13} W/cm².

In fact in case of non-propagation we can take into account several causes of line broadening. The broadening due to Landau damping is negligible in our experimental conditions for $3\omega/2$ observed at 90°. Let us consider collisional broadening. The electron-ion collision frequency is related to the damping coefficient Γ_c of a plasma wave (of complex frequency ω) by the relation

$$\Gamma_c = \text{Im}(\omega) = \nu_{ei}/2 \quad .$$

In our conditions

$$\nu_{ei} \approx 6.7 \times 10^{-11} \; (Z \; n_e \; \ln\Lambda \; T_e^{-3/2}) \approx 2 \times 10^{12} \; \text{sec}^{-1}$$

and the collisional width is $\delta\omega = \nu_{ei}$, giving $\delta\lambda_c \approx 5$ Å.
    Finally the effect of the aperture of focusing ($\Theta_f \approx 8^0$) and collecting ($\Theta_c \approx 7^0$) optics results in an overall instrumental width $\Delta\lambda_{aper} \approx 3$ Å.
    Due to the effects considered above, the $3\omega/2$ line can be broadened of a few Å, a small extent if compared with the observed bandwidth of several tens of Å. We believe that the observed shift and broadening of the 3/2 harmonic in this experiment are a clear evidence of the propagation of the electron waves through an inhomogeneous plasma.
    From the analysis of Sect. II we know that the $3\omega/2$ radiation we detected at 90° is mostly a "direct" emission produced by coupling of the impinging laser radiation with red plasmons (ω_R, K_R).   However we will consider also the "indirect" emission which can be given by the coupling of $\omega/2$ plasmons reflected at the $n_c/4$ layer. In this case we intend with reflection a particular case of propagation, able to invert the component of the plasmon momentum along $\underline{K}_o$. Moreover, as in this experiment we evidenced a small amount of Brillouin scattering , we can not exclude the presence of a very weak "indirect" line from the coupling of SBS backscattered photons (ω_o, $-\underline{K}_o$) with blue plasmons (ω_B, $\underline{K}_B$) .   The blue component of the 3/2 harmonic is usually absent in our spectra, or very weak in a few spikes. This is not surprising because: i) the contribution of photons reflected at the critical layer is small (the plasma becoming underdense before the $n_c/4$ layer develops to suitable scalelength), ii) the intensity of Brillouin backscattered light is very weak, as already mentioned.   On the other hand the lack of a significant blue component also means that the reflection of $\omega/2$ plasmons propagating in the $n_c/4$ region is not an efficient process.

Finally some considerations about the threshold. We observed a threshold for 3/2 harmonic emission at nominal laser intensity of 5×10^{12} W/cm^2. This value is much lower than the expected threshold for TPD. Considering an $n_c/4$ layer scalelength of 30μm, which is a mean value of measurements taken from time resolved images, we find that the TPD threshold is dominated by plasma inhomogeneities and its value can be estimated of the order of 10^{14} W/cm^2. It is very reasonable that the discrepancy between the measured and calculated values of the threshold is due to the presence of hot spots and filaments into the interaction region which can lead to a local increase of the laser field. This point demands further experimental study to be clarified.

For what concerns the 2ω spectroscopic data, they are consistent with second-harmonic emission resulting from the sum of the laser frequency with the frequency of Brillouin backscattered light. The measured threshold of the 2ω emission agrees with the expected SBS threshold in our experimental condition. The main signature of this process comes from the red shift of the backscattered light as expected from SBS which in turn gives a red shift of the second-harmonic light. The frequency of 2ω field is exactly $\omega_o + \omega_B$ where ω_B is the frequency of the Brillouin backscattered wave, given by

$$\omega = \omega_o [1 - 2v_i/c \ [\ (1-n_e/n_c) \ / (1+4K_o^2 \lambda_D^2 (n_c/n_e - 1) \ )]^{1/2}] \qquad (7)$$

where $v_i = [(ZT_e + 3T_i)/m_i]^{1/2}$ is the speed of the ion acoustic wave, T_e and T_i are the electron and ion temperature, respectively, in energy units. Our data are consistent with this explanation provided one takes into account that

$$\Delta\omega/\omega \Big|_{2\omega_o} = 1/2 \ \Delta\omega/\omega \Big|_{\omega_o} \qquad (8)$$

The initial position of the 2ω line has a red shift $\Delta\lambda \approx 4$Å corresponding to $v_i = 2.5 \times 10^7$ cm sec^{-1}, i.e., $T_e \approx T_i \approx 600$ eV. This temperature, roughly costant in time after burnthough, i.e., during the peak of the laser pulse, is expected[25] in the given experimental condition when 2ω light is mostly emitted.

The increase in red shift observed during the emission is explainable in terms of Doppler shift due to the plasma motion in the direction of the beam axis (parallel to K_o). This motion is accelerated due to ablation pressure. From the red shift at 2ω the velocity v_p of the plasma motion can be evaluated if we put in equation (7) $v_i + v_p$ in place of v_i, taking $v_i = 2.5 \times 10^7$ cm sec^{-1}.

For all the spectra obtained with 1-1.5 μm targets we find $v_p \approx 2 \times 10^7$ cm sec^{-1} at delays of a few hundred picoseconds, up to $v_p \approx 6 \times 10^7$ cm sec^{-1} after ≈ 2 ns. These velocities are reasonable in our experimental condition and are comparable with direct measurements performed by other authors[26].

The 2ω line bandwidth showed a net increase during the emission, typical from 5 Å to about 20 Å. It is a consequence of the increasing bandwidth of the electromagnetic wave backscattered by the Brillouin process. Landau damping of ion sound waves could be in principle responsible for line broadening, but calculation leads to a narrower bandwidth than observed. We believe that the line bandwidth is essentially due to Doppler broadening related to plasma expansion. Consequently, the doublet structure apparent in most late-time spikes should indicate that two plasma regions with different velocities are more effective in SBS (and consequently in sum-frequency generation).

The frequency sum process, postulated by Stamper et al.[20] is now experimentally confirmed[27] as a nonlinear process occuring in laser-produced plasmas as well as already observed in crystals[22].

Finally let us consider the implications of our study of harmonic emission at 90° from the point of view of laser produced plasma diagnostics. Spectroscopy of 2ω light generated by sum frequency has in principle the same diagnostic power than the spectroscopy of Brillouin backscattered light. This latter is commonly used to measure electron temperature, estimate plasma turbolence and related parameters. Sum frequency light can give the same sort of information with the net experimental advantages of a larger spectral separation and a confortable angle of detection (perpendicular to the laser beam).

The diagnostic use of $3\omega/2$ light is first of all in localizing the $n_c/4$ layer and measure its scalelength. The shift of this harmonic from the $3\omega_0/2$ exact value can be used to measure the plasma temperature. The linear dependence of the shift on temperature makes these measurements very efficient. But from our experimental observations we realized that the propagation of the plasmons can modify the shift and strongly increase the brodening of the $3\omega/2$ line, resulting in a poor accuracy of temperature measurements.

Several features of 2ω and $3\omega/2$ spectra obtained in this experiment can be related to the influence of filamentation on the generation of these harmonics. Further studies could allow to use 2ω and $3\omega/2$ spectroscopy as a diagnostic of filamentation itself.

ACKNOWLEDGEMENTS

Thanks are due to I. Deha for valuable contributions and to S.Bartalini fo his creative technical support. We are also grateful to the target preparation group of the Laser Division of Rutherford Appleton Laboratory (UK) for the information on thin foil target fabrication.

The experiment was fully supported by Consiglio Nazionale delle Ricerche, Italy.

REFERENCES

1) R.P. Drake, E.A. Williams, P.E. Young, K. Estabrook, W.L.Kruer, H.A. Baldis and T.W. Johnston, Phys. Rev. Lett. **60**, 1018, (1988).
2) O. Willi, D. Basset, A. Giulietti and S.J. Karttunen, Opt. Comm., **70**, 487, (1989).
3) V. Yu. Bychenkov, V.P. Silin and V. T. Tikhonchuk, Sov. J. Plasma Phys. **3**, 730 (1977).

4) W. Seka, B.B. Afeyan, R. Boni, L.M. Goldman, R.W. Short, K.Tanaka and T. W. Johnston, Phys. Fluids **28**, 2570 (1985).

5) N.G. Basov, Yu. A. Zadharenkov, N.N. Zorev, A.A. Rupasov, G.V.Sklizkov, A.S. Shikanov, in Heating and compression of thermonuclear targets by laser beam, edited by N.G. Basov (Cambridge University Press,1986), and references therein.

6) S. Jackel, B. Perry and L. Lubin, Phys. Rev. Lett. **37**, 95 (1976).

7) Lin Zunqi, Tan Weihan, Gu Min, Mei Guang, Pan Chengming, Yu Wenyan and Deng Ximing, Las. Part. Beams **4**, 223 (1985).

8) T.A. Leonard and R. A. Cover, J. Appl. Phys. **50**, 3241 (1979).

9) P. D. Carter, S. M. L. Sim, H. C. Barr and R. G. Evans, Phys. Rev. Lett. **44**, 1407 (1980).

10) E. McGoldrick and S. M. L. Sim, Opt. Comm. **39**, 172 (1981).

11) L. V. Powers and R. J. Schroeder, Phys. Rev. **A 29**, 2298 (1984).

12) V. Aboites, T. P. Hughes, E. McGoldrick, S. M. L. Sim, S.J. Karttunen and R. G. Evans, Phys. Fluids **28**, 2555(1985).

13) F. Amiranoff, F. Briand and C. Labaune, Phys. Fluids **30**, 2221 (1987).

14) S.J. Karttunen, Laser and particle beams **3** , 157 (1985).

15) H.J. Herbst, J.A. Stamper, R.R. Whitlock, R.H. Lehmberg, and B.H. Ripin , Phys. Rev. Lett., **46**, 328, (1981).

16) N.G. Basov, V.Yu. Bychenkov, O.N. Krokhin, M.V. Osipov, A.A. Rupasov, V.P. Silin, G.V. Sklizkov, A.N. Starodub, V.T. Tikhonchuk, and A.S.Shikanov,Sov.Phys.Jept,**49**(6), (1979).

17) I.V. Alexandrova et al., Pis'ma Zh. Eksp. Teor. Fiz., **38**, 478, (1983);[JEPT Lett., **38**, 577, (1983).

18) D. Giulietti, G.P. Banfi, I. Deha, A. Giulietti, M. Lucchesi, L. Nocera, and Chen Zezun, Laser and Particle Beams, **6**, 141, (1988).

19) D. Batani, F. Bianconi, A. Giulietti, D. Giulietti, and L.Nocera, Opt. Comm, **70**, 38, (1989).

20) J.A. Stamper, R.H. Lehmberg, A. Schmitt, M.J. Herbst, .F.C.Young, J. H. Gardner, and S. P. Obenschain, Phys. Fluids, **28**, 2563, (985).

21) Gu Min and Tan Weihan, Laser and Particle Beams, **6**, 569, (1988).

22) Y.R. Shen, The Principles of Non linear Optics, (Wiley, New York, 1984).

23) A. Giulietti, D. Giulietti, L. Nocera, I. Deha, Chen Zezun, and G.P. Banfi, Laser Interaction,**8**,365,(1988).

24) Z. Z. Chen and E. Schifano, IFAM Internal Report, Jan. 1988.

25) R.A. London and M.D. Rosen, Phys. Fluids **29** , 3813 (1986).

26) K. Eidman, F. Amiranoff, R. Fedosejevs, A.G.M. Maaswinkel, R.Petsch, R. Sigel, G. Spindler, Yung-lu Teng, G. Tsakiris, and S. Witkowski, Phys. Rev., **A 30**, 2568, (1984).

27) A. Giulietti, D. Giulietti, D. Batani, V. Biancalana, L. Gizzi, L. Nocera and E. Schifano, Phys. Rev. Lett.,**63**, 524 (1989).

NONSTATIONARY STIMULATED BRILLOUIN BACKSCATTERING

IN INHOMOGENEOUS DENSITY PROFILES

S. Hüller, P. Mulser, and H. Schnabl

Technische Hochschule Darmstadt
Federal Republic of Germany

ABSTRACT

In inhomogeneous density profiles electromagnetic light is partially absorbed and reflected at the critical surface. In those profiles where the reflection is non-negligible the counterpropagating electromagnetic waves cause standing ion density fluctuations which act as a source for backscattering. To study this effect a simplified model based on the usual three wave interaction as well as an extended model using the nonlinear hydrodynamic description of the ion fluid have been investigated. In certain parameter regions at which absorption and reflection at the critical surface are of comparable magnitude no steady state in the evolution of the backscattered light can establish over long time intervals. The spectral composition of the backscattering signal shows a strong competition between the frequency shifted stimulated scattering (SBS) and the unshifted reflection process.

INTRODUCTION

In laser plasma theory it is standard to investigate stimulated Brillouin scattering (SBS) with the help of three coupled equations for the slowly varying complex envelopes (SVE) of two electromagnetic waves $\hat{E}_0, \hat{E}_1$ interacting with a low-frequency ion acoustic wave $\hat{n}_1$,

$$\frac{\partial \hat{E}_0}{\partial t} + (\frac{c^2 \vec{k}_0}{\omega}\nabla)\hat{E}_0 + \beta_+\hat{E}_0 = -i\frac{\omega_0}{4}\frac{\hat{n}_1}{n_c}\hat{E}_1\, e^{-i\int \vec{\sigma}_k d\vec{x}},$$

$$\frac{\partial \hat{E}_1}{\partial t} + (\frac{c^2 \vec{k}_1}{\omega}\nabla)\hat{E}_1 + \beta_-\hat{E}_1 = -i\frac{\omega_1}{4}\frac{\hat{n}_1^*}{n_c}\hat{E}_0\, e^{i\int \vec{\sigma}_k d\vec{x}},$$

(1)

$$\frac{\partial}{\partial t}(\frac{\partial \hat{n}_1}{\partial t} + 2\nu_1\hat{n}_1) - i2sk_a(\frac{\partial \hat{n}_1}{\partial t} + \frac{s^2\vec{k}\nabla}{\omega_a}\hat{n}_1 + \nu_1\hat{n}_1) \simeq -n_0\frac{kT_e}{m_i}k^2\hat{\varphi}_p\, e^{\,i\int \vec{\sigma}_k d\vec{x}}. \quad (2)$$

Herein, $\beta_\pm, \nu_1$ represent the damping constants, s the speed of sound, n_0 and n_c the equilibrium and critical densities of the plasma, and φ_p the normalized ponderomotive potential due to the high-frequency oscillations of the electrons,

$$\hat{\varphi}_p = -\frac{e^2}{m_e\omega_0\omega_1 kT_e}\hat{E}_0\hat{E}_1^* = -\frac{\hat{I}}{|\hat{E}_0|_{x=0}^2}\hat{E}_0\hat{E}_1^*. \tag{3}$$

$\vec{\sigma}_k = \vec{k}_0 - \vec{k}_1 - \vec{k}_a$ is a non-negligible function of the spatial coordinate in inhomogeneous plasmas. Because of the dispersion of electromagnetic waves, $\omega^2 = \omega_p^2(x)+c^2\vec{k}^2$, the validity of the matching condition $\vec{\sigma}_k = 0$ for a resonant interaction process is limited to a finite region in the profile. The mismatch factor $\exp\{i\int \vec{\sigma}_k d\vec{x}\}$ introduced in all three equations invalidates the model when the WKB approximation is no longer applicable.

In order to guarantee a correct treatment in the entire density profile it is necessary to dispense with the decomposition of the total electromagnetic field into an incident and a backscattered contribution,

$$\mathbf{E} = \hat{E}_0 e^{i\vec{k}_0\vec{x}} + \hat{E}_1 e^{i\vec{k}_1\vec{x}+i(\omega_0-\omega_1)t}. \tag{4}$$

The system of equations is contestable in a second crucial point which plays an important role even in homogeneous plasmas : Ion acoustic waves do not evolve like linear waves as was assumed for the derivation of eq.(2). An increment of the amplitude of a sound wave causes a deformation of the initial sinusoidal shape during the propagation. The description of ion acoustic waves can be improved by using either an equation for simple waves [8] or by considering the nonlinear hydrodynamic equations of the ion fluid [1] from which eq.(5) originates.

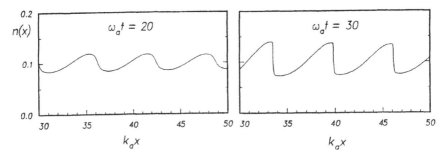

Figure 1. Sound wave in an early and a later stage showing the steepening of the profile in time due to the amplitude dependent propagation velocity [5].

288

EXTENDED HYDRODYNAMIC MODEL

With the following system of equations backscattering from inhomogeneous density profiles in x-direction can be treated in an adequate way [5] :

$$\frac{dv_\alpha}{dt} = -\frac{1}{n_0 m_\alpha}\frac{\partial p_i}{\partial x}\Big|_{x_\alpha} - \frac{kT_e}{m_\alpha}\frac{\partial \varphi}{\partial x}\Big|_{x_\alpha} - \nu v_\alpha + \mu\frac{\partial^2 v}{\partial x^2}\Big|_{x_\alpha}$$

(5)

$$\frac{dx_\alpha}{dt} = v_\alpha, \quad n_\alpha = \frac{n_0}{1 + \int v_\alpha dt},$$

$$\frac{\partial^2 \varphi}{\partial x^2} = \frac{1}{\lambda_D^2}(e^{\varphi - \varphi_p} - \frac{n_i}{n_0}),$$

(6)

$$-i2\omega_0\frac{\partial E}{\partial t} = c^2\frac{\partial^2 E}{\partial x^2} + \omega_0^2[1 - \frac{n_i}{n_c}(1 + i\frac{\nu_{ei}}{\omega_0})^{-1}]E,$$

(7)

and

$$\varphi_p = \frac{e^2}{m_e\omega_0\omega_1 kT_e} < E^2 > .$$

(8)

The full dynamics of the ion fluid is included. The momentum equations are written in Lagrangian coordinates (x_α, v_α) for each fluid element α [2]. It is assumed that the electrons form an isothermal background, leading to a Boltzmann factor in the resulting nonlinear Poisson's equation. The system is completed by an adiabatic law for the ions. When strong variations of the ion temperature occur, this law has to be replaced by an energy- or entropy equation. Nevertheless, dissipation in the fluid dynamics due to viscosity or heat conduction is included in both frictional terms.

Solely the fast time dependence according to the electron quiver motion is eliminated in the electromagnetic wave equation. The system can be applied also to inhomogeneous density profiles extending into the region where the electromagnetic wave becomes evanescent. In the case of normal incidence, which will be considered in the context of backscattering, the incident laser flux is partially absorbed and reflected at the critical surface where $\omega_p = \omega_0$ holds. For linear profiles without density fluctuations the portion of the reflected light can be determined analytically from Ginzburg's formula [4],

$$r = \Big|\frac{\hat{E}(-k_x)}{\hat{E}(k_x)}\Big| = \exp\{-\frac{4}{3}\frac{\nu_{ei}}{\omega}\frac{\omega}{c}L\}.$$

(9)

It is obvious that the specular reflection r decreases exponentially when the length of the linear profile or the collisional damping ν_{ei} is increased. When only a small fraction of the incoming light flux is absorbed, stimulated Brillouin scattering appears together with specular reflection; but in long scale plasmas with considerable absorption a tiny amount of the reversed light flux may seed the SBS process in a resonant region. Therefore, r is a further crucial parameter for SBS in laser plasmas.

As a consequence, one has to distinguish between strongly and weakly absorbing plasmas. In order to study the dynamic evolution of SBS in inhomogeneous plasmas eqs.(5-8) have to be solved numerically. The difference scheme used to solve eqs.(5,6) is well known from the publication of Sack and Schamel [7]. The electromagnetic wave equation (7) can be reduced to an ordinary differential equation of second order when the time derivative $\partial_t E$ is negligible compared to the second spatial derivative. A comparison with results from a Crank-Nicholson scheme should justify the validity of the simplification.

RESULTS

The classification into weakly and strongly absorbing plasmas with the help of Ginzburg's relation (9) is justified by the asymptotic behaviour of the solutions for both cases. As illustrated in figs.2-3, asymptotically stationary solutions of the backscattered signal (fig.5)

$$R^{1/2} = |\frac{E(t) - E_0}{E_0}|_{x=0},$$

($E_0(x = 0)$ is the incident electromagnetic field strength) can only evolve when the fraction of specular reflected light is below 2% in relation to the incident flux. The space-time evolution of the ion density shown in a 3D plot in fig.6 exhibits a continuous propagation of ion acoustic wave perturbations along their characteristics $x - st$ into the direction of the incident electromagnetic wave.

As long as the critical surface does not expand towards the low-density region the specularly reflected wave is just a fraction of the incident one, only reversed in its direction of propagation, but unshifted in the frequency. The superposition with the incident light causes a standing wave; this also acts on the ion fluid by its ponderomotive force which is also at rest. On a long time scale it causes density profile modifications. The ponderomotive force due to the SBS process moves in the frame of the ion acoustic wave. When specular reflection and SBS become of comparable magnitude, a strong competition between both processes (illustrated in fig.6) leads to a nonstationary behaviour of the backscattered light signal (fig.3) and to a broadening of its spectrum. In very short scale length plasmas specular reflection will dominate and will deform the profile. But, SBS becomes a dominating role when the inhomogeneity length is increased. This is obvious from fig.7; The data plotted as marks with error bars are obtained by the asymptotically observed mean values of numerical integrations of the system of eqs.(5-8) when applied to a linear ramp. Only the length of this ramp was changed in order to determine $R(L)$ whereas all other parameters were kept fixed. For comparison, the dependence $r^2(L)$ according to Ginzburg's formula eq.(9) is included and shows agreement with the numerical results for short L. The spectral shift of the main peak in the backscattered light proves also to be dependent on the portion r of light reversed at the critical surface. Fig.8 shows that the frequency shift $\omega_0 - \omega_1$ caused by simulated backscattering approaches its maximum value $\Delta\omega = 2s\omega_0/c$ ($k_0 < \omega_0/c$) when SBS dominates the specular reflection. In contrast to homogeneous plasmas the wavelength and consequently the frequency of the ion acoustic wave in inhomogeneous profiles is not uniquely determined for given temperatures. In the initial stage of the instability the light will be scattered on the largest existing coherent sound wave; but the asymptotic stage in time is dominated by an ion acoustic mode with the largest possible gain parameter G anywhere in the profile. G depends on physical quantities, such as (1) the incident flux $\hat{I}/\sqrt{1 - n_0/n_c}$, (2) the local density $n_0(x)$, and

290

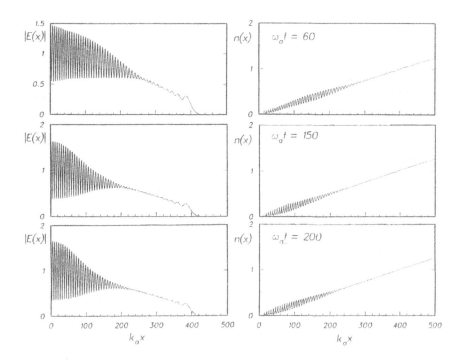

Figure 2. Spatial distribution of the total electromagnetic field E and the ion density at different time instants in case of a strongly absorbing plasma. The ion density is normalized to the critical density n_c. Parameters: $\hat{I} = 0.2$, $\nu_{ei}/\omega_0 = 0.02$, $2\omega_0 L/c = 400$ [5].

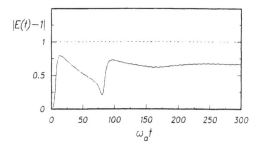

Figure 3. Signal of the backscattered wave at $x = 0$ in time normalized to $\omega_a = 2s\omega_0/c$ corresponding to fig.2 [5].

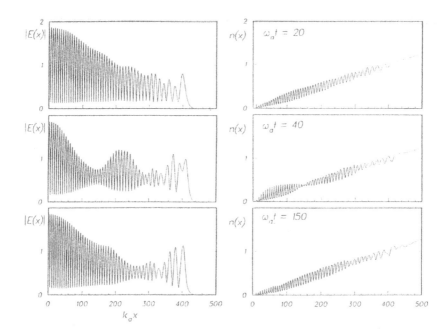

Figure 4. Electromagnetic field and ion density as in fig.2, but in a weakly absorbing case. Parameters: $\hat{I} = 0.2$, $\nu_{ei}/\omega_0 = 0.007$, $2\omega_0 L/c = 400$ [5].

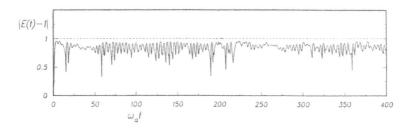

Figure 5. Signal of the backscattered wave at $x = 0$ in time normalized to $\omega_a = 2s\omega_0/c$ corresponding to fig.4 [5].

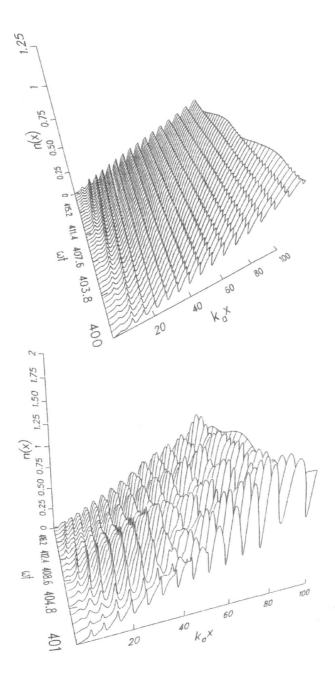

Figure 6. Space-time evolution of the ion density in a linear density ramp. Top: $\nu_{ei}/\omega_0 = 0.02$, $2\omega_0 L/c = 300$; thus $r^2 \simeq 0.5\%$. Bottom: $\nu_{ei}/\omega_0 = 0.02$, $2\omega_0 L/c = 200$; thus $r^2 \simeq 7\%$ [5].

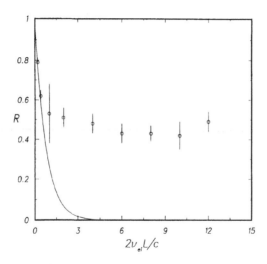

Figure 7. Mean value of the asymptotically attained reflectivity as a function of the length of a linear ramp. Single points with error marks result from numerical calculations. The solid curve shows the behaviour due to eq.(9) Parameters: $\nu_{ei}/\omega_0 = 0.02$, $\hat{I} = 0.2$. [5].

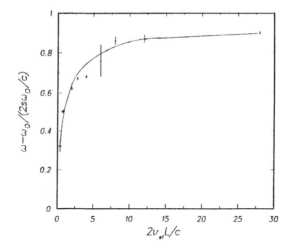

Figure 8. Frequency shift of the main peak normalized to its maximum value $2s\omega_0/c$ (in absence of expansion) as a function of $\nu_{ei}L/\omega_0$ [5].

(3) the effective interaction length L_{eff}. It is well known from homogeneous plasmas that G is proportional to $n_0 L_{eff}$; but there, both parameters are independent of the position. In the presence of an inhomogeneity and sufficient absorption the location where the most effective resonant backscattering can take place is balanced by the dependence of the interaction length L on n_0. The mismatch of the interaction process, $\exp\{i \int \sigma_k dx\}$, and the absorption $\exp\{-\int \beta_\pm dx\}$ cause L_{eff} to decrease more rapidly than n_0 grows as a function of x for a certain $x > x_0$; thus $G_{max} \sim n_0(x_0) L_{eff}(x_0)$. As a consequence, in weakly absorbing plasmas the most effective backscattering takes place at higher densities $(n_0 \to n_c)$ than in a strongly absorbing case. The frequency shift is regulated by the local refractive index, $\Delta\omega = (s\omega_0/c)\sqrt{1 - n_0/n_c}$. Since the absorption rate obeys the relation $\beta_\pm \sim \nu_{ei}$ the correlation between specular reflection r and the frequency shift is obvious.

REFERENCES

[1] Braginskij, S.I., "Transport Processes in Plasmas in Review of Plasma Physics", Vol. I, Eds.: A.M. Leontovich, Consultants Bureau, New York (1965)

[2] Davidson, R.C., "Methods in Nonlinear Plasma Theory", Academic Press, New York (1972), chapter 3

[3] Forslund, D.W., Kindel, J.M., Lindman, E.L.,Phys. Fluids 18(8), p.1017 (1975)

[4] Ginzburg, V.L., "The Propagation of Electromagnetic waves in Plasmas", Pergamon Press, Reading (1964), chapter IV

[5] Hüller, S., "Stimulierte Brillouin Rückstreuung an nichtlinearen Ionenschallwellen in Laserplasmen", Thesis (in German), Darmstadt (1990)

[6] Kruer, W.L., "The Physics of Laser Plasma Interactions", Addison-Wesley Publishing Comp. (1988), p.88

[7] Sack, Ch., Schamel, H., Physics Reports 156(6), pp.311-395 (1987)

[8] Whitham, G.B., "Linear and Nonlinear Waves", John Wiley & Sons (1974)

TURBULENCE IN VERY-HIGH MACH NUMBER, LASER-ACCELERATED MATERIAL

J. Grun, J. Stamper, J. Crawford[*], C. Manka, and B.H. Ripin

Space Plasma Branch

Plasma Physics Division

Naval Research Laboratory

Washington DC 20375

[*] Physics Department

SW Texas State University

San Marcos, Texas 78666

INTRODUCTION

Most flows in nature are turbulent. Often, turbulence critically influences the nature of an important event. For example, satellite communications can be interfered with by natural or man-made turbulence in the earth's atmosphere[1]; x and gamma rays in the 1987A supernova appeared prematurely because of turbulence[2]; the rate of star formation depends on the nature of interstellar turbulence[3]; turbulent mixing influences the efficiency of inertial-confinement-fusion pellets[4]. There are many other examples in aeronautics, chemistry, and combustion.

Turbulence has been studied for more than a century but it is still incompletely understood[5]. This is not surprising: Turbulence is by definition highly nonlinear, experiments are difficult to interpret quantitatively, and it is hard to relate experimental observations to theoretical predictions. Many experiments, such as those in shock tubes, suffer from extraneous effects caused by walls and membranes. However, with rapid advances in computer architecture, advances in computer aided visualization and analysis, and advances in experimental methods, rapid progress in understanding turbulence is now possible.

High-power lasers, such as those used for inertial-confinement-fusion research, are a new and very good tool for studying turbulence and other hydrodynamic phenomena[6,7,8]. Laser hydrodynamics experiments are not hampered by walls and membranes that make clean shock tube experiments difficult. Areas of parameter space not accessible by conventional methods can be reached easily with high-power lasers. For example, multi-megabar pressures, Mach numbers of a few hundred, temperatures of many electron volts can be achieved. Intense x-ray radiation can be turned on when needed. Also, it is relatively easy to vary parameters, such as pressure, flow speed, Mach number, temperature, fluid composition, etc., over a broad range. Laser experiments can duplicate many astrophysical parameters[9]. Also, many of the sophisticated diagnostics which have been developed over the past twenty five years for fusion research can be utilized in hydrodynamics experiments.

We use the N.R.L. Pharos III laser and target facility to study linear and nonlinear hydrodynamic flows. This paper describes our experimental work in very high Mach number turbulence.

DESCRIPTION OF EXPERIMENT

The Pharos III laser facility 10 has a three-beam glass laser with an output of 500 joules per beam at a wavelength of 1.054μm in a 2-6ns pulse. The laser wavelength can also be halved. Irradiances of 300 terawatts per square centimeter can be produced by focusing the laser a few hundred micron diameter spot. The spatial irradiance profile can be shaped and smoothed to a few percent level with the induced-spatial-incoherence (ISI) method[11]. Experiments are performed in a large, 80 cm radius chamber which can be evacuated or filled with gas. A magnetic field of up to 10 kgauss over more than 100 cm^3 surrounding the point of laser focus can be turned on. Figure 1 shows a photograph of the area around the Pharos III target chamber.

FIGURE 1.   Photograph of the area around the Pharos III target chamber. Laser focusing optics and parts of laser beam diagnostics are seen in the background.

In this work one beam of the laser is focused onto the surface of thin, 5-20μm, polystyrene (CH) or aluminium foils to produce an irradiance of 1-300 terawatts/cm^2 in a 100-1000μm diameter spot. The chamber is filled with gas but an external magnetic field is not used. This is what happens when the laser irradiates the foil:[12] The irradiated part of the foil (target) is heated on its surface to a temperature of about 800 eV[13], causing the surface to ablate. The ablated material expands toward the laser at about 700 kilometers per second in a narrow shell[14]. Half of the ablation mass is contained within a 40 degree half-cone angle[15,16]. Meanwhile, the background gas is photoionized by radiation from the target or by impact with ablation material. A Taylor-Sedov blastwave forms in front of the target when the ablated material sweeps up a gas mass equal to a few times its own[17]. The non-ablated part of the target is accelerated away from the laser by the rocket-like effect of the ablation material to speeds of 10-100 km/sec. During acceleration the target heats to 1-2 eV and becomes Rayleigh-Taylor unstable[6,7,8]. Later, the unstable target becomes turbulent and entrains background gas so that it too becomes turbulent. When the acceleration ceases the target decompresses thermally. An example of a target irradiated in a background gas, showing both the blast wave and the turbulence, is shown in Fig.2.

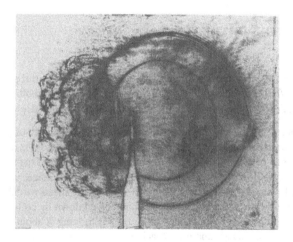

FIGURE 2.   A dual-time Schlieren shadowgraph of an aluminium foil irradiated with 36 joules in 5 torr of nitrogen. The two exposures were at 55ns and 160ns. Note the blastwave on the laser side of the foil and the turbulent region behind the target.

Our most important diagnostic tool is a laser probe (0.53μm, 350psec FWHM) which is transmitted through the turbulent region as shown in figure 3. The phase and amplitude of the probe may both, in theory, be affected by the turbulence. The disturbed probe light is recorded using various shadowgraphy methods, such as bright-field, Schlieren, or phase-contrast photography, and with holographic interferometry. In addition, visible light emission from the turbulent region is photographed with a gated optical imager (shuttered at 600psec) which is filtered so that it images emission from only the target material, only the background gas, or both. Emission spectra measured with a temporally and spatially resolved optical multichannel analyzer identify ionic species and help us estimate temperature.

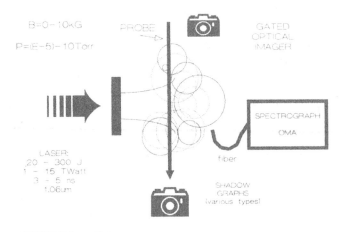

FIGURE 3.   Schematic of experiment and main diagnostics.

CHARACTERISTICS OF THE TURBULENT REGION

In this section we address the following questions about the turbulent region: How does it move ?; What is turbulent, the target, the background gas or both ?; What is its spatial wavelength spectrum?; and, Can the spectrum be manipulated?

How does it move?   Figure 4 shows a time sequence of phase-contrast photographs of the turbulent region. The speed of the turbulent front is determined from plots of the distance between the original foil position and the tip of the turbulent front versus the time of the observation (Fig. 5). From such time sequences we see that the front moves at a speed which is somewhat larger(< 20%) than the speed we calculate for the target material itself. Furthermore, the front is not slowed down by collisional interaction with the background gas. The front speed in figure 5 is moving at about Mach 70 with respect to the unheated background gas. Mach numbers of up to 200 were observed in other cases. The speed of the turbulent region does not depend strongly on the background gas pressure.

What is turbulent?   The fact that turbulence exists - and yet its motion is not perceptively slowed down by the background gas - may indicate that a gas-target interchange instability is not a dominant factor in its development. Rather, the target itself, which we know to be linearly unstable[6,7,8], may be the primary cause of the turbulence. This would explain why the turbulent front speed is somewhat higher than the target speed and why it does not depend significantly on the background gas pressure. The front is probably a shock driven slightly ahead of the target material - much like a shock moving ahead of a supersonic piston. An alternative picture, that a target-gas interchange instability causes the turbulence, is less likely. If this was the case, we would intuitively expect the background gas to slow the target. Since this does not happen we believe the first picture above to be more probable: But a definitive answer awaits more experiments and calculation.

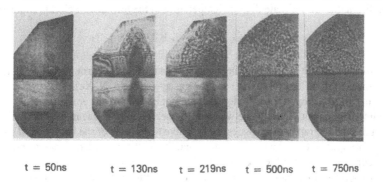

SHOT 88-546   SHOT 88-559   SHOT 88-554   SHOT 88-638   SHOT 88-639

t = 50ns   t = 130ns   t = 219ns   t = 500ns   t = 750ns

FIGURE 4.   Sample bright-field and phase-contrast time sequence of the
             turbulent region. The target is 20μm CH and the background gas is
             5 torr of nitrogen. Laser is incident from the bottom.

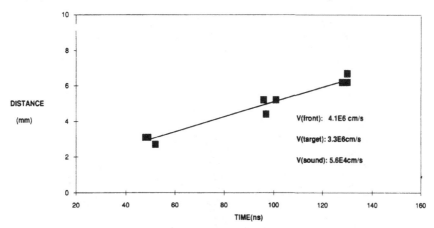

FIGURE 5.   Trajectory of turbulent region for the case of a 9 μm thick CH target
             irradiated with 5 terawatts/cm^2 in 5 torr of nitrogen. The turbulent
             region moves at Mach 70.

    Nevertheless, the background gas does mix with the target "piston", and is,
therefore, also turbulent. This can be seen by examining photographs of visible light
emission from the turbulent region and comparing them to shadowgraphy pictures taken at
the same time on the same shot. In front of the camera is a narrow band-pass (50A
FWHM) interference filter which passes the 5001A N^{1+} line from the background nitrogen
gas. In general, the overall shape of the emission region matches closely with the turbulent
region seen by shadowgraphy. But, small scale 10-100μm structures seen in the
shadowgraphs are washed out in the emission pictures, probably by integration of the
emission along the line-of-sight through the turbulent volume: The smallest emission
structures have a size of about 200μm. (Camera resolution is 50μm in the target plane.)
Limb brightening causes the outer edges of the emitting region to be brighter than the inner
regions. Figure 6a shows a sample emission picture and figure 6b a time integrated

spectrum from the turbulent region. No spectral lines from highly ionized states of are observed. From this we estimate that the temperature in the turbulent region is about 10 eV or less.

To further test the relationship between turbulence and background gas we varied the gas pressure (0.05 to 10,000 mTorr) and the gas species (He, N, Xe). Interestingly, below 200 mTorr of nitrogen the turbulence could not be seen in our emission and shadowgraph diagnostics. Turbulence was never observed in high vacuum. This may mean that the shadowgraphy diagnostics are sensitive to features of the background gas - perhaps density gradients in the swept up gas - and that collision with the background gas raises the temperature so that the region emits enough to be seen. This would have to be the case if the hypothesis that the target is turbulent without the presence of gas is correct. When the turbulence is seen it does not seem to be affected by the type or the pressure of gas. However, detailed spectral information has not yet been quantified.

a
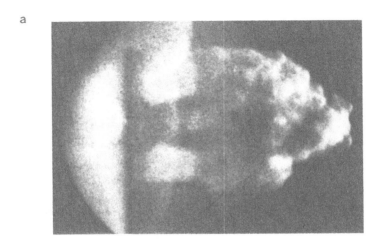

b
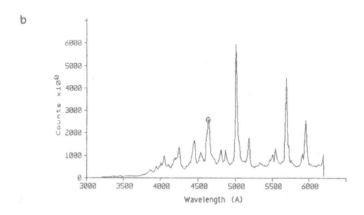

FIGURE 6.   (a) Sample emission photograph and (b) sample emission spectrum of the turbulent region.

Power Spectral Characteristics Nonuniformities in the turbulent region can be quantified by measuring their spatial frequency spectrum. Although such a spectrum cannot uniquely identify a turbulence mechanism, any theory that does not reproduce the spectrum can be eliminated. In addition, such spectra are used by designers of systems that must operate in turbulent surroundings. The challenge for experiments is to relate the observed structures to some underlying physical quantity such as density, energy, index of refraction, etc. In our work this is being done using phase-contrast shadowgraphy and holographic interferometry[18].

Phase-contrast shadowgraphy is an application of the Zernike phase-contrast microscope[19] to plasma research. The method is described with modern Fourier-transform mathematics by Cutrona et al.[20] Presby and Finkelstein[21], we recently discovered, were the first to use it in plasma research. The essential aspects of the method may be easily understood by considering the following simplified argument: Imagine the laser probe (Fig. 3) to be a planar electric wave of amplitude E. Assume that turbulence alters the phase, but not the amplitude, of the probe so that after the probe traverses the turbulence its electric field is given by

(1) $E = E_0 \exp\{ i(\varphi + \theta(x,y)) \}$

        $= E_0 \exp(i\varphi)\{ 1 + \Sigma(i\theta)^n/n! \}.$

Here φ is an averaged spatially independent part of the phase shift and $\theta(x,y)$ is a spatially varying part (which need not be less than 1). The probe is imaged onto a camera with dual-lens astronomical telescope as shown in Fig 7a. The first lens of this telescope Fourier transforms the probe's electric field at its focal point. The second lens picks up the Fourier transformed electric field and Fourier transforms it again at the location of the camera. The camera photographs the square of the twice Fourier transformed electric field. Mathematically, the first Fourier transform is given by

(2) $F[E] = E_0 \exp(i\varphi)\{ F[1] + F[\Sigma (\iota\theta)^n/n!] \}$;

and the second Fourier transform by

(3) $F[F[E]] = E_0 \exp(i\varphi)\{ F[F[1]] + F[F[\Sigma (\iota\theta)^n/n!]] \}$;

and the image on the film by

(4) $Image(x,y) = \{ F[F[E]] \}^2$.

(Magnification, time, and axial direction in the phase terms are, for simplicity, not included here.)

Now, the $F[1]$ term in equation 2 is the transform of the undisturbed part of the probe. Physically, it is a spot of light on the lens axis. This spot can be manipulated, thereby modifying the photograph given by equation 4.

For example, without any manipulation the photograph is given by

(5) $Image(x,y) = E_0 E_0^*$.

This is the so called bright-field shadowgraph. In the absence of absorption it is simply the image of the original probe. With absorption, it forms the shadow of the absorbing region.

If the central spot is eliminated the photograph is given by

(6) $\text{Image}(x,y) = E_0 E_0^* \{ \theta^2(-x,-y) + ... \}$,

which is a Schlieren shadowgraph sensitive to the square of phase fluctuations.

If the central spot is shifted by a quarter wavelength the photograph is given by

(7) $\text{Image}(x,y) = E_0 E_0^* \{ 1 + 2\theta(-x,-y) + ... \}$,

which is a phase-contrast shadowgraph sensitive to the first power of phase fluctuations.

From phase fluctuations we determine fluctuations in the index of refraction $N(x,y,z)$ integrated along the axial direction using

(8) $\int N(x,y,z) \, dz = \lambda_{probe} \theta(x,y)/2\pi$

where λ_{probe} is the probe wavelength. Assuming free electrons of density $n_e(x,y,z)$ dominate the index of refraction we get

(9) $\int n_e(x,y,z) \, dz = -\theta(x,y) \, (\lambda_{probe}^2 e^2 /m_e c^2)^{-1}$.

The phase-contrast method is subject to the following caveats:

·   Significant absorption or probe nonuniformity will invalidate the commutation in equation 2.

·   As a practical matter, one cannot eliminate $F[1]$ by itself. One also eliminates the DC and low frequency components in the terms containing θ. Therefore, measurements of low frequency fluctuations are not accurate[22].

·   If phase fluctuations larger than π occur then equation 7 cannot be uniquely inverted. This causes aliasing and spectrum distortions. Therefore, any phase excursion greater than π should not occur often and/or be at wavelengths not of interest to the analysis.

·   Optical distortions are neglected.

One nice feature of the method is that the probe is imaged right after leaving the turbulent region. Consequently, Fresnel diffraction effects, which would occur if the probe were to propagate a long distance, do not trouble us here. Fresnel oscillations, which distort phase measurements in atmospheric experiments[23] are not a problem.

Fig. 7b shows how the method is implemented in practice. The probe beam is split into two parts by a Wollaston prism after leaving the turbulent region. One part is imaged as a bright-field shadowgraph, while the other part is passed through a phase filter producing the phase-contrast shadowgraph of equation 7. The phase contrast filter is a flat ($\lambda/10$) fused silica window with a 1mm diameter by 2859 angstrom thick (1/4 λ) mesa on its surface. The film is calibrated by pre-exposing a part of it to the probe light through a Kodak step wedge.

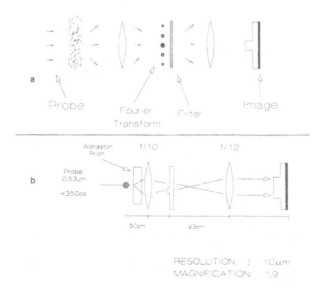

FIGURE 7.   (a) Schematic of the phase-contrast concept. (b) Implementation of the phase-contrast method.

Simultaneous bright-field and phase-contrast shadowgraphs verify that absorption or probe nonuniformity effects are not important in the region of interest. Also, photographs taken without turbulence or plasma being present let us balance the two split beams of the probe, and verify that extraneous effects from the mask or the optics are minimal. The diameter of the mesa implies that spatial wavelengths longer than $500\mu m$ (ie. $\lambda_{probe}f_{lens}/r_{mesa}$) are not measured accurately.

Figure 8 shows typical bright-field and phase-contrast shadowgraphs. The bright-field image shows a shadow of the unirradiated part of the foil, and a shadow of slow material ejected by shocks and heat leakage from the laser spot. Such slow material plays no role in turbulence formation. Also visible is a segment of the ablation-side blast wave which wrapped itself around the rear of the foil. This blastwave scatters light out of the optical path so that, for practical purposes, it behaves as an absorbing object. The turbulent region is barely visible in a few tiny spots in the bright-field image but is very visible in the phase-contrast shadow, proving that the turbulence is primarily a phase object. The phase-contrast shadow is also rich in other phase disturbing features such as a weak shock reflected from the foil holder, and sound waves outside the boundaries of the turbulence.

The phase-contrast photographs are digitized and transformed into equivalent phase images on a computer using the mathematics outlined above. Then the spatial power spectral density is determined from the expression

(10) $Power(k_x) = \Sigma_y[Abs(\int T(x)\ \theta(x,y)\ exp(-ik_xx)\ dx)]^2$,

where $T(x)$ is an 80% Tukey window[24].

305

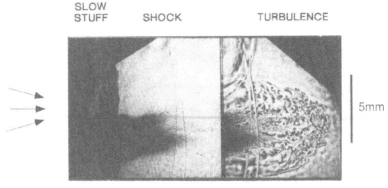

SLOW STUFF · SHOCK · TURBULENCE · 5mm · BRIGHT FIELD · PHASE CONTRAST

FIGURE 8.   A bright-field and a phase-contrast shadowgraph of a 10μm Al foil irradiated with 4.5 twatts/cm^2 in 5 torr of nitrogen. The pictures were taken 219ns after the peak of the laser pulse. Laser is incident from the left.

A typical one-dimensional, spatial power spectrum is shown in Fig. 9. Most turbulent power spectra can be characterized by three quantities: an 'outer scalelength' at which energy is pumped into the system, the 'inertial range' of wavelengths at which energy cascades from larger to smaller scales, and an 'inner scalelength' at which energy dissipates out of the system. In our case the outer scalelength is about 1mm which is similar to the diameter of the focal spot. (However, this number is not very reliable as explained above.) The inertial range has a $k_x^{-2.1}$ scaling. An inner scalelength is not observed at 20μm or larger wavelengths. Inner scalelengths smaller than 20μm are not unreasonable for the estimated Reynold's number in this flow. The scaling of the inertial range is independent of scan direction, reflecting symmetry in the fine turbulent structure.

A two-dimensional, k-space power spectrum, $P(k_x,k_y)$, can also be calculated. We find that $P(k_x,k_y=0) = k_x^{-3.2}$ and $P(k_x=0,k_y) = k_y^{-3.2}$. $P(k) = \int(P(k_x,k_y)k \, d\phi$, where ϕ is the azimuthal angle in k_x,k_y space, varies as $k^{-2.1}$. The various scalings are all consistent for spatially symmetric turbulence.

We have yet to show that unacceptable aliasing is not a problem with the phase-contrast measurements. This can be tested with holographic interferometry which measures absolute phase disturbances caused by phase objects, and can also provide an independent determination of the power spectrum. Such interferograms have been taken with a variety of background fringe spacings. Interferograms with fringe spacings of 1mm and 360μm show the outlines of the turbulent region clearly. However, the fringes cannot be followed into the turbulent region itself since they have a coarser scale than the turbulence fluctuations. But fringes in interferograms with 70μm spacing (Fig.10) can be followed through most of the turbulence. A visual analysis shows that aliasing is probably not significant, but a quantitative estimate has yet to be made. The mathematics for extracting two-dimensional phase information from holographic interferograms of laminar

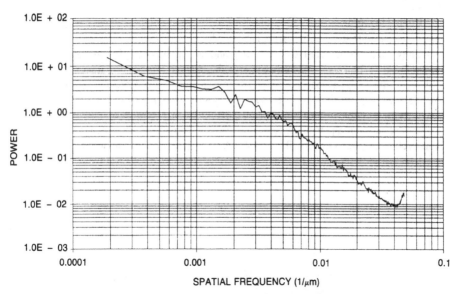

FIGURE 9.   One dimensional power spectrum of the turbulent region in a typical shot.

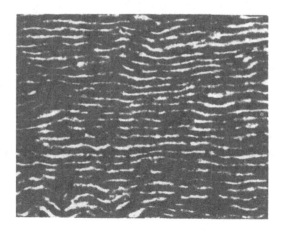

FIGURE 10.   Enlarged section of a holographic interferogram of the turbulent region. Reference fringes are spaced 70μm apart.

phase objects have been discussed by Takeda[25] and Nugent[26]. We are currently evaluating such methods for accuracy in a turbulent case.

Turbulence control   Power spectra alone are insufficient to unambiguously identify the responsible turbulence mechanism. Indeed, we have not proven that the nonuniformities are turbulent in the sense that energy is cascading from larger to smaller scales, or whether they are a frozen eddy structure with a spectrum determined by a linear instability. By manipulating the turbulence we may get further clues into its nature. For example, we can try to turn the turbulence on and off, or to control its power spectrum. If the Rayleigh-Taylor instability in the target is responsible for the turbulence, perhaps we can accelerate a stable target and see if the turbulence disappears. We have accomplished some of these goals in a few different ways. For example:

Burn Through: By focusing the laser beam to a tight spot we can raise its irradiance so that the entire target is heated by x rays, fast electrons and explodes. There is no ablative acceleration and the Rayleigh-Taylor instability is stabilized. When we irradiated a target under such conditions turbulence did not occur. Instead, we observed an expanding, hemispherical front similar to the blast wave normally associated with the target's laser-side (Fig. 11). Long wavelength nonuniformities, similar to weak regions on the surface of an expanding balloon, replaced the fine-scale turbulent structure associated with ablatively accelerated targets. This observation supports the turbulent target model.

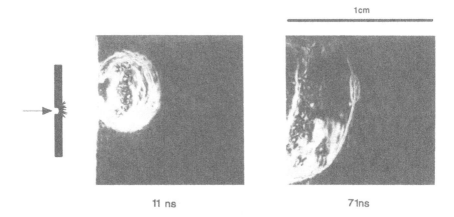

FIGURE 11.   Schlieren shadowgraphs of a 9μm CH foil in 5 torr of nitrogen irradiated by a tightly focused laser beam to 320 terawatts/cm^2. Note lack of turbulence.   Laser is incident from the left.

Structured targets: The Rayleigh-Taylor instability in flat targets develops from natural material imperfections or from nonuniformities in the laser beam. The wavelengths which grow can be controlled by purposefully imposing a large mass perturbation on the target - for example, by grooving its surface[6,7,8]. Now, if the turbulent spectrum is determined by the linear Rayleigh-Taylor instability then we should be able to alter the turbulent power spectrum by changing the wavelengths at which Rayleigh-Taylor grows. To test this we accelerated targets perturbed with a 100μm wavelength groove and compared the resulting turbulent power spectra to those of non-perturbed targets. We found that at early (<10ns) times a 100μm structure was visible in the shadowgraphs. But fully developed turbulence at 200ns had the same spectrum as that for flat targets. This supports the argument that we are observing turbulence and not a frozen eddy structure.

SUMMARY

This paper describes our studies of turbulence produced in a background gas by objects accelerated to very high velocities. We have developed methods to quantify features of the instability, in particular the power spectrum of the electron density. Also, we have measured the behavior of the turbulence under many different circumstances and determined its power spectrum. With these methods we are studying the nonlinear hydrodynamics of a system which duplicates in the laboratory many astrophysical properties. We intend to extend our methods to the study of hydrodynamics at less exotic conditions but without the complications of more classical experimental methods.

ACKNOWLEDGMENTS

We are grateful for the expert technical assistance of Nicholas Nocerino, Levi Daniels, Jim Ford, and Ray Burris who helped us conduct the experiments; Loretta Schirey who fabricated the targets; and Joshua Resnick who helped with some of the data analysis. Fruitful discussions with Drs. Cliff Prettie, Joe Huba, Wolf Seka, and T. Kessler are acknowledged. This work is supported by the Defence Nuclear Agency and the Office of Naval Research.

REFERENCES

[1]   L.A. Wittwer, The Propagation of Sattelite Signals in a Turbulent Environment, D.N.A. report AFWL-TR-77-183 (1978).

[2]   V. Trimble, Rev. of Modern Phys. 60, 859 (1988).

[3]   J. M. Scalo, in "Interstellar Processes", pg. 349:392, D. J. Hollenbach and H. A. Thronson Jr. editors, D. Reidel Publishing Company (1987).

[4]   D. L. Youngs, Numerical Simulation of Turbulent Mixing by Rayleigh-Taylor Instability, Physica 12D 32:44 (1984).

[5]   H. Tennekes and J.L. Lumley, "A First Course in Turbulence", The MIT Press, Cambridge (1987).

[6]   A. J. Cole, J.D. Kilkenny, P.T. Rumsby, R.G. Evans, C.J. Hooker, M.H. Key, Nature 299, 329 (1982).

7   J. Grun, M.H. Emery, S. Kacenjar, C.B. Opal, E.A. McLean, S.P. Obenschain, Phys. Rev. Lett. 53, 1352 (1984); J. Grun and S. Kacenjar, Appl. Phys. Lett. 44, 497 (1984).

8   J. Grun, M.H. Emery, C.K. Manka, T.N. Lee, E.A. McLean, A. Mostovych, J.Stamper, S. Bodner, S.P. Obenschain, B.H. Ripin, Phys. Rev. Lett. 58, 2672 (1987).

9   B.H. Ripin et al, in Laser and Particle Beams 8, (1990) [in press].

10 J. M. McMahon et al., IEEE J. Quant. Elec. QE17, 1629 (1981).

11 R. H. Lehmberg and S.P. Obenschain, Opt. Comm. 46, 27 (1983).

12 This description of the laser-foil interaction and the resulting turbulence is placed here to provide a frame of reference for the reader. Some of the assertions are based on extensive studies in the references, others are the subject of this article.

13 P.G. Burkhalter, M.J. Herbst, D. Duston, J. Gardner, M. Emery, R.R. hitlock, J.Grun, J.P. Apruzese, J. Davis, Phys. Fluids 26, 3650 (1983).

14 J. Grun, R. Stellingwerf, B.H. Ripin, Phys. Fluids 29, 3390 (1986).

15 J. Grun, R. Decoste, B.H. Ripin, J. Gardner, Appl. Phys. Lett. 39, 545 (1981).

16 J. Grun, S.P. Obenschain, B.H. Ripin, R.R. Whitlock, E.A. McLean, J. Gardner, M.J. Herbst, J.A. Stamper, Phys. Fluids,26, 588 (1983).

17 B.H. Ripin, A.W. Ali, H.R. Griem, J. Grun, S.T. Kacenjar, C.K. Manka, E.A. McLean, A.N. Mostovych, S.P. Obenschain, and J.A. Stamper, in Laser Interaction and Related Plasma Phenomena, Vol. 7, Edited by H. Hora and G.H. Miley (Plenum Publishing, 1986).

18 J. Crawford, J. Grun, J.A. Stamper, B.H. Ripin, and C.K. Manka, Bull. Am. Phys. Soc. 34, 1921 (1989); also to be published.

19 F. Zernike, Physica 1, 689 (1934); Physica 9, 686 (1942).

20 L. J. Cutrona, E.N. Leith, C.J. Palermo, L.J. Porcello, I.R.E. Transactions on Information Theory, June, 386:400 (1960).

[21]   H. M. Presby and D. Finkelstein, <u>The Review of Scientific Instruments 38</u>, 1563 (1967).

[22]   In extreme cases one can envision all of the probe light being deflected so that there is no DC term at all: The term 1 is cancelled by the DC components of the phase terms.(Cliff Prettie, private communication) This is minimized by overfilling the turbulent region so there is always a nondeflected part of the probe.

[23]   C. L. Rufenach, <u>J. of Geophysical Research 77</u>, 4761 (1972).

[24]   F. J. Harris, " On the Use of Windows for Harmonic Analysis with the Discrete Fourier Transform", <u>Proc. of the IEEE 66</u>, 51 (1978).

[25]   M. Takeda, H. Ina, S.Kobyashi, <u>J. Opt. Soc. Am. 72</u>, 156 (1982).

[26]   K. A. Nugent, <u>Applied Optics 24</u>, 3101 (1985).

ELECTRON MOTION IN PLASMA IRRADIATED BY STRONG LASER LIGHT

Keishiro Niu
Department of Energy Sciences, Graduate School at Nagatsuta
Tokyo Institute of Technology
Midori-ku, Yokohama 227, Japan

Peter Mulser
Institute für Angewandte Physik, Technische Hochschuhle Darmstadt
D6100, Darmstadt, Bundesrepublik Deutschland

Ladislav Drška
Faculty of Nuclear Science and Physical Engineering
Technical University of Prague
Přehova 7, 115 19 Praha 1, Czechoslovakia

ABSTRACT

Electron beam oscillates in a dense plasma irradiated by a strong laser light. When the frequency of the laser light is high and its intensity is large, the acceleration of oscillating electron becomes large and the electrons radiate electromagnetic waves. As the reaction, the electrons feel a damping force, whose effect on oscillating electron motions is investigated first. Second, the collective motion of electron beam induces the strong electromagnetic field by its own electric current when the electron number density is high. The induced electric field reduces the oscillation motion and deforms the beam.

INTRODUCTION

Recently, the pulse width of the Kripton-flouride (KrF) laser can be shorten to 10^{-15}s, hence the strong intensity of $4.42 \times 10^{23} \mathrm{W/m^2}$ (corresponding electric field is $1.29 \times 10^{13} \mathrm{V/m}$) accelerates an electron very close to the light speed during a quarter of pulse width.(R. Kristal et al. 1987). Owing to this strong acceleration, it radiates a retarded electromagnetic wave. The reacting force due to the radiation of this retarded electromagnetic wave acts back onto the electron motion.

When the laser light irradiates the target of the dense plasma, the accelerated electrons move to collectively forming a beam with high electric current density. The current induces the time-dependent azimuthal magnetic field which causes the electric field in the axial direction. This electric field suppresses the beam oscillation. The induced electric fields are divided into three. The first one comes from the time-dependent electron current density whose electron velocity changes with the temporally sinesoidal change in the laser intensity. This induced electric field reduces the oscillation velocity of the electron beam as a whole. The second electric field in the axial direction appears at the leading and trailing edges of the beam where the magnetic field in the azimuthal direction changes rapidly. This electric field decelerates the leading edge and accelerates the trailing edge. A third electric field comes from the charges. Space charges are induced by overlapping of the electrons with the background electrons or by being separated from the background ions. This electric field causes the beam to diverge.

MAXWELL EQUATIONS FOR RETARDED POTENTIALS

The Maxwell equations for the electric field $E(r,t)$ and the magnetic field $B(r,t)$ at the space coordinate r and at the time t are given by

$$\nabla \times \mathbf{E}(\mathbf{r},t) + \frac{\partial \mathbf{B}(\mathbf{r},t)}{\partial t} = 0, \tag{1}$$

$$\nabla \cdot \mathbf{B}(\mathbf{r},t) = 0, \tag{2}$$

$$\nabla \times \frac{\mathbf{B}(\mathbf{r},t)}{\mu} - \frac{\varepsilon \partial \mathbf{E}(\mathbf{r},t)}{\partial t} = \mathbf{j}(\mathbf{r},t), \tag{3}$$

$$\nabla \cdot \varepsilon \mathbf{E}(\mathbf{r},t) = \rho_e(\mathbf{r},t), \tag{4}$$

where $\mathbf{j}$ is the current density, ρ_e is the charge density, μ is the magnetic permeability in vacuum and ε is the dielectric constant in vacuum. If the electric potential ϕ and the magnetic potential $\mathbf{A}$ are used, $\mathbf{E}$ and $\mathbf{B}$ are given by

$$\mathbf{E}(\mathbf{r},t) = -\frac{\partial \mathbf{A}(\mathbf{r},t)}{\partial t} - \nabla \phi(\mathbf{r},t), \tag{5}$$

$$\mathbf{B} = \nabla \times \mathbf{A}(\mathbf{r},t), \tag{6}$$

and eqs.(1) − (4) reduce to

$$(\Delta - \frac{\partial^2}{c^2 \partial t^2})\mathbf{A}(\mathbf{r},t) = -\mu \mathbf{j}(\mathbf{r},t), \tag{7}$$

$$(\Delta - \frac{\partial^2}{c^2 \partial t^2})\phi(\mathbf{r},t) = -\rho_e(\mathbf{r},t)/\varepsilon, \tag{8}$$

provided that $\mathbf{A}$ and ϕ are expressed by the Lorentz gauge, that is, they satisfy

$$\nabla \cdot \mathbf{A}(\mathbf{r},t) + \frac{\partial \phi(\mathbf{r},t)}{c^2 \partial t} = 0. \tag{9}$$

The solutions of eqs.(7) and (8) are respectively given by

$$\mathbf{A}(\mathbf{r},t) = \int \frac{\mu \mathbf{j}(\mathbf{r}',t')}{4\pi|\mathbf{r} - \mathbf{r}'|} d\mathbf{r}, \tag{10}$$

$$\phi(\mathbf{r},t) = \int \frac{\rho_e(\mathbf{r}',t')}{4\pi\varepsilon|\mathbf{r} - \mathbf{r}'|} d\mathbf{r}, \tag{11}$$

where

$$t' = t - |\mathbf{r} - \mathbf{r}'|/c \tag{12}$$

EQUATION OF MOTION FOR ELECTRON

An electron is assumed to be a sphere whose radius is a_0. The charge e is distributed uniformly inside a_0. In the Maxwell equations (3) and (4), the current density $\mathbf{j}$ and charge density ρ_e are divided into two parts as follows,

$$\nabla \times \frac{\mathbf{B}(\mathbf{r},t)}{\mu} + \varepsilon \frac{\mathbf{E}(\mathbf{r},t)}{\partial t} = \mathbf{j}_0(\mathbf{r},t) + \sum_{i \neq 0} \mathbf{j}_i(\mathbf{r},t), \tag{13}$$

$$\nabla \cdot \varepsilon \mathbf{E}(\mathbf{r},t) = \rho_0 + \sum_{i \neq 0} \rho_i(\mathbf{r},t), \tag{14}$$

where suffix 0 refers to the electron itself and i to other one. The equation of motion of the electron is described as

$$m_0 \frac{d^2 \mathbf{z}(t)}{dt} = \int d\mathbf{r} [\rho_0(\mathbf{r},t)\mathbf{E}(\mathbf{r},t) + \mathbf{j}_0(\mathbf{r},t) \times \mathbf{B}(\mathbf{r},t)], \tag{15}$$

where m is the mass and $\mathbf{z}$ is the position of the particle centre. The electric field $\mathbf{E}$ and the magnetic field $\mathbf{B}$ are divided into two as

$$\mathbf{E}(\mathbf{r},t) = \mathbf{E}_0(\mathbf{r},t) + \mathbf{E}_1(\mathbf{r},t), \tag{16}$$

$$\mathbf{B}(\mathbf{r},t) = \mathbf{B}_0(\mathbf{r},t) + \mathbf{B}_1(\mathbf{r},t). \tag{17}$$

The electric field $\mathbf{E}$ with the suffix 0 or 1 and the magnetic field $\mathbf{B}$ with the suffix 0 or 1 respectively satisfies eqs.(1),(2),(13) and (14), for example,

$$\nabla \times \frac{\mathbf{B}_1(\mathbf{r},t)}{\mu} - \varepsilon \frac{\partial \mathbf{E}_1(\mathbf{r},t)}{\partial t} = \sum_{i \neq 0} \mathbf{j}_i(\mathbf{r},t), \tag{18}$$

$$\nabla \cdot \varepsilon \mathbf{E}_1(\mathbf{r},t) = \sum_{i \neq 0} \rho_i(\mathbf{r},t), \tag{19}$$

The force $\mathbf{F}$ is expressed as $\mathbf{F} = \mathbf{F}_0 + \mathbf{F}_1$ where

$$\mathbf{F}_0 = \int d^3[\rho_0(\mathbf{r},t)\mathbf{E}_0(\mathbf{r},t) + \mathbf{j}_0(\mathbf{r},t) \times \mathbf{B}_0(\mathbf{r},t)], \tag{20}$$

$$\mathbf{F}_1 = \int d^3[\rho_0(\mathbf{r},t)\mathbf{E}_1(\mathbf{r},t) + \mathbf{j}_0(\mathbf{r},t) \times \mathbf{B}_1(\mathbf{r},t)], \tag{21}$$

Now $\mathbf{E}_0$ and $\mathbf{B}_0$ in eq.(20) are expressed by $\mathbf{j}_0$ and ρ_0 as follows,

$$\mathbf{A}_0(\mathbf{r},t) = \frac{\mu}{4\pi} \int d\mathbf{r} \frac{\mathbf{j}_0(\mathbf{r}',t')}{|\mathbf{r} - \mathbf{r}'|}, \tag{22}$$

$$\phi_0(\mathbf{r},t) = \frac{1}{4\pi\varepsilon} \int d\mathbf{r} \frac{\rho_0(\mathbf{r}',t')}{|\mathbf{r} - \mathbf{r}'|}. \tag{23}$$

Since $|\mathbf{r} - \mathbf{r}'|$ in eq.(12) is small inside the electron, the terms $|\mathbf{r} - \mathbf{r}'|$ appearing in $\mathbf{j}_0$ and ρ_0 are Taylor expanded to delete t', then eq.(15) reduces to

$$m_0 \frac{d^2 \mathbf{z}(t)}{dt^2} = \mathbf{F}_1 - \frac{1}{4\pi\varepsilon} \int d\mathbf{r}\, \rho_0(\mathbf{r},t) \left[\nabla \int d\mathbf{r}' \frac{\rho_0\left(\mathbf{r}', t - \frac{|\mathbf{r}-\mathbf{r}'|}{c}\right)}{|\mathbf{r}-\mathbf{r}'|} \right.$$

$$\left. + \frac{1}{c^2} \frac{\partial}{\partial t} \int d\mathbf{r}' \frac{\mathbf{j}_0\left(\mathbf{r}', t - \frac{|\mathbf{r}-\mathbf{r}'|}{c}\right)}{|\mathbf{r}-\mathbf{r}'|} \right]$$

$$= \mathbf{F}_1 + \frac{1}{4\pi\varepsilon} \sum_{n=0}^{\infty} \frac{(-1)^n}{c^{n+1}} \frac{2}{3} \int d\mathbf{r}' \int d\mathbf{r}\, \rho_0(\mathbf{r}', t) \frac{d^{n+2}\mathbf{z}_0(t)}{dt^{n+1}}. \tag{24}$$

315

For n=0, we have

$$\mathbf{F}_0^{(0)} = -\frac{1}{4\pi\varepsilon}\frac{2}{3c^2}\int d\mathbf{r}\,\frac{\rho_0(\mathbf{r},t)\rho_0(\mathbf{r}',t)}{|\mathbf{r}-\mathbf{r}'|}\frac{d^3\mathbf{z}_0(t)}{dt^3}. \tag{25}$$

In this equation, the factor in the left hand side

$$\frac{1}{4\pi\varepsilon}\frac{2}{3c^2}\int d\mathbf{r}'\int d\mathbf{r}\,\frac{\rho_0(\mathbf{r},t)\rho_0(\mathbf{r}',t)}{|\mathbf{r}-\mathbf{r}'|} = \frac{e^2}{4\pi\varepsilon a_0} = W, \tag{26}$$

is the self-energy of the electron charge. Thus eq.(25) becomes

$$\mathbf{F}_0^{(0)} = -\frac{4}{3c^2}W\frac{d^2\mathbf{z}_0(t)}{dt^2}. \tag{27}$$

For n-1, we have

$$\mathbf{F}_0^{(1)} = -\frac{e^2}{6\pi\varepsilon c^3}\frac{d^4\mathbf{z}_0(t)}{dt^4}, \tag{28}$$

and for n>2, we obtain

$$\mathbf{F}_0^{(n)} = -\frac{1}{4\pi\varepsilon}\frac{(-1)^n}{n!c^{n+2}}\frac{2}{3}<a_0^{n-1}>\frac{d^{(n+3)}\mathbf{z}_0(t)}{dt^{n+3}}, \tag{29}$$

where

$$<a_0^{n-1}> = \int d\mathbf{r}'\int d\mathbf{r}\,\rho_0(\mathbf{r},t)|\mathbf{r}-\mathbf{r}'|^{n-1}\rho_0(\mathbf{r}',t). \tag{30}$$

Thus eq.(24) can be rewritten as

$$m_0\frac{d^2\mathbf{z}(t)}{dt^2} = \mathbf{F} - \frac{4}{32}W\frac{d^2\mathbf{z}(t)}{dt^2} + \frac{e^2}{6\pi\varepsilon c^3}\frac{d^3\mathbf{z}(t)}{dt^3} + \cdots, \tag{31'}$$

or

$$(m_0 + m_e)\frac{d^2\mathbf{z}(t)}{dt^2} = \mathbf{F}_1 + \frac{e^2}{6\pi\varepsilon c^3}\frac{d^3\mathbf{z}(t)}{dt^3} + \cdots, \tag{31}$$

where

$$m_e = \frac{4}{3c^2}W = \frac{e^2}{3\pi\varepsilon a_0 c^2}, \tag{32}$$

is the dress mass of the electron. Equation (31) has an infinite series of differential terms. The nature of a differential equation depends on the highest order of differentiation.

The equation of motion must not be covariant for the Lorentz invariant if it includeds the finite length a_0. In order that eq.(31) is the covariant, a_0 should be 0. In the case that $a_0 \to 0$, $<a_0^{n-1}>$ disappears for n>2, and eq.(31) reduces to the third order differential equation, although the dress mass tends to the infinity. This fact is well known as the divergence of the self energy of the charged particle. Here we consider that $a_0=0$ while $m_0+m_e=m$ is the finite to be 9.109×10^{-31}kg.

ELECTROMAGNETIC WAVE EMITTED BY ACCELERATED CHARGE

Now an accelerated motion (31) of a particle with a point charge e is considered. Then the charge and the current of the point charge are expressed by

$$\rho_e(\mathbf{r},t) = e\delta^3(\mathbf{r}-\mathbf{z}(t)), \tag{33}$$

$$j(r,t) = e\frac{dz}{dt}\delta^3(r - z(t)), \tag{34}$$

respectively. When eqs.(33) and (34) are substituted into eqs.(10) and (11), calculations lead to the power dw/dt of the electromagnetic wave emitted from the accelerated point charge,

$$\frac{dW}{dt} = \frac{e^2}{16\pi^2\varepsilon c^3}\frac{d^3z}{dd^3t}\int_0^\pi d\Omega\frac{\sin^2\theta}{(1 - \frac{dz}{cdt}\cos\theta)^5}, \tag{35}$$

where Ω is the solid angle and θ is the angle between the wave vector of emitted electromagnetic wave and the acceleration of the charged particle. If the velocity v=dz/dt of the charged particle is much slower than the light speed, i.e. v<<c, eq.(35) reduces to the Larmor formula,

$$\frac{dW}{dt} = \frac{e^2}{16\pi\varepsilon c^3}\left(\frac{d^3z}{dt^3}\right)^2. \tag{36}$$

The force on the electron as the reaction of radiating electromagnetic wave is calculated by using eq.(36) as

$$\int_{t_2}^{t_1}K(t)\frac{dz(t)}{dt}dt = -\int_{t_2}^{t_1}\frac{dW}{dt}dt = -\frac{e^2}{6\pi\varepsilon c^3}\int_{t_2}^{t_1}\left(\frac{d^2z}{dt^2}\right)^2dt$$

$$= -\frac{e^2}{6\pi\varepsilon c^3}\left[\frac{d^2z}{dt^2}\frac{dz}{dt}\Big|_{t_2}^{t_1} - \int_{t_2}^{t_1}\frac{d^3z}{dt^3}\frac{dz}{dt}ddt\right].$$

It is considered that dz/dt = 0 at t = t_1 and d^2z/dt^2 = 0 at t = t_2. Then K(t) can be written as

$$K(t) = \frac{e^2}{6\pi\varepsilon c^2}\frac{d^3z}{dt^3} = 1.88\times10^{-23}m\frac{d^3z}{dt^3} = T_0m\frac{d^3z}{dt^3}, \tag{37}$$

which is just the second term in the right handside of eq.(31). Similarly the reaction force H(t) based on eq.(35) is given by

$$H(t) = \frac{e^2}{8\pi\varepsilon c^3}\int_0^\pi d\theta\frac{d}{dt}\frac{\sin^3\theta\frac{d^2z}{dt^2}}{(1 - \frac{df\,z}{cdt}\cos\theta)^6}. \tag{38}$$

DAMPING OF ELECTRON MOTION DUE TO SELF-RADIATION

The equation of motion of an electron irradiated by laser **E** is given by

$$\frac{d^2z_0(t)}{dt^2} = -\frac{eE}{m} + T_0\frac{d^3z_0(t)}{dt^3}. \tag{39}$$

Here the Lorentz force due to the magnetic field of laser light is neglected and electric field is given by

$$E = E_*e^{i\omega t}, \tag{40}$$

where ω is the frequency and is chosen as $\omega=7.57\times10^{15}$rad/s for KrF laser. Since T_0 is the very short time period, the zeroth order solution z_0 is given by

$$\frac{d^2z_0}{dt^2} = -\frac{eE}{m}. \tag{41}$$

Equation (21) leads to

$$z_0 = z_{*0}e^{i\omega t} \qquad \left\{ (z_{*0} = -\frac{eE_*}{8m}(\pi/\omega)^2 \right\}. \tag{42}$$

As the approximate solution of eq.(39), we set

$$z_1 = (z_{*0} + \delta)e^{i(\omega t - \pi)+\gamma}, \tag{43}$$

where δ and γ are small quantities. When eq.(43) is substituted into eq.(39) the terms in the first order regarding T_0, δ and γ are derived as

$$-\omega^2 + 2i\omega\gamma z_{*0} - \omega^2 z_{*0}\gamma t = -iT_0\omega^3 z_{*0}, \tag{44}$$

from which we have for $\omega t = \pi/2$,

$$\gamma = -\frac{1}{2}\omega^2 T_0, \tag{45}$$

$$\delta = \frac{e\pi T_0 E_*}{4\omega m}. \tag{46}$$

The damping factor γ depends on the frequency ω only. Figure 1 shows γ versus ω. In the rage of γ-rays ($\omega > 7.57\times10^{22}$ rad/s), the damping factor becomes significant.

The solution of a differential equation has a different nature depending on its highest order of differentiation. Thus the method of obtaining an approximate solution of eq.(39), which is a differential equation of the third order, by starting from eq.(41), which is a differential equation of the second order, involves some risks. Fortunately we can derive the exact solution of eq.(39) as follows,

$$z = A + Bt + Ce^{t/T_0} + De^{i\omega t}, \tag{47}$$

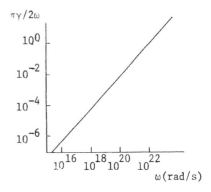

Fig. 1. Damping Factor γ due to the self-field emission of electron beam versus the frequency ω of the electron oscillation.

where A, B, C and D are constant. The first three terms of the right hand side of eq.(47) are the general solution of the homogeneous equation and the last term is a special solution of the inhomogeneous equation. The condition that z=finite at t=∞ leads to B=C=0. The initial condition that $z=z_{*0}$ at t=0 leads to $A+D=z_{*0}$. When solution (47) with these values of constants is substituted into eq.(36), we have

$$D = \frac{eE_*(1 + i\omega T_0)}{m\omega^2(1 + \omega^2 T_0^2)} = D_1 + iD_2 \tag{48}$$

The constant D is complex whose imaginary part is small because it includes T_0. The electron velocity is given by

$$\frac{dz}{dt} = -\omega D_1 \sin\omega t + \omega D_2 \cos\omega t. \tag{49}$$

When the electron has no retarded radiation, the velocity is given by the first term only. But the reaction on the electron due to the retarded radiation of electromagnetic wave reduces the absolute value of the velocity by the second term.

318

This reduction in this (absolute value of) velocity corresponds to γ in the approximation described above.

This damping phenomenon becomes different when the particle velocity approaches the light speed. For such a situation eq.(38) must be used for the second term in the right handside of eq.(39). For $\omega=7.57\times10^{15}\text{rad/s}$, $E=1.29\times10^{13}\text{V/m}$ and $\omega t=\pi/2$, the damping factor γ is given in Table 1 as a function of v/c.

Table 1. Damping factor of the electron when the particle velocity is close to the light velocity.

v/c	$\pi\gamma/2\omega$
1.00	-4.41×10^3
0.99	-2.43×10^0
0.98	-9.56×10^{-2}
0.97	-1.22×10^{-2}
0.96	-2.85×10^{-3}

As is clear from Table 1, the damping is significant when $v/c=1$. However the damping factor decrease rapidly with decreasing v/c. Soon as the particle velocity reduces to 0.98c, the damping becomes negligible.

Equation (19) expresses the equation of motion of a charged particle. The equation takes the retarded radiation from the electron itself into consideration. The effect of reaction due to this retarded radiation appears in the oscillatory motion of the electron when irradiated by the laser with a frequency higher than 10^{28}Hz. However, even in this case, the damping disappears when the electron velocity reduces to 0.98 times the light speed.

COLLECTIVE MOTION OF ELECTRON BEAM

The laser light of frequency ω and the intensity E_* induces the oscillation of the electron beam whose radius is r_6 and length is 2L in the target plasma. Here r_b is taken to be to the wave length $r_b=2.49\times10^{-7}\text{m}$ for KrF laser and L is the length of part of the target surface $L=1.0\times10^{-5}\text{m}$. Here E_* is chosen as $E_* = 1.29\times10^{13}\text{V/m}$. Because the oscillation amplitude of the electron beam is as small as $z_{*0} = 3.95\times10^{-8}$ m and is much less than L, the shift of the beam in the background plasma creates the charges at the both ends of the beam only. The charge ρ_e is given by

$$\rho_e = \begin{cases} en_{e'} & \text{for } -L + \int vd\,t < z < -L, \\ -en_{e'} & \text{for } L + \int vd\,t > z > L, \\ 0 & \text{for } z < -L \ \text{ or } \ z > L. \end{cases} \tag{50}$$

The current density of the beam depends on the electron number density n_e and is given by $j=en_ev=7.2\times10^{18}\text{A/m}^2$, for $n_e=4.5\text{x}10^{27}/\text{m}^3$ and $v = dz/dt \sim eE_z/m \ dt$. Thus the current density in eq.(10) is given by

$$\mathbf{j} = \mathbf{k}\,j(r',z',t)\,S(z';-L,L)\,S(r';0,r_b), \tag{51}$$

where k is the unit vector in the z direction (axial direction) and $S(x;A,B)$ is the step function defined by $S(x;A,B)=1$ for $A<x<B$ and $S(x;A,B)=0$ for $x<A$ and $x>B$. After eqs.(50) and (51) being substituted into eqs.(10) and (11), respectively, eqs.(5) and (6) together with $t=t'$ gives the distributions of the electric and magnetic fields. At every instant the coordinate system is chosen such that the centre of the beam

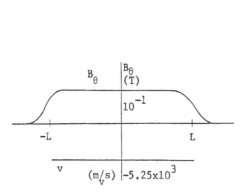

Fig. 2. The velocity V of the beam at $r=0$ and the magnetic field B_θ in the azimuthal direction at $r=r_b$ versus z at $\omega t=\pi/5$.

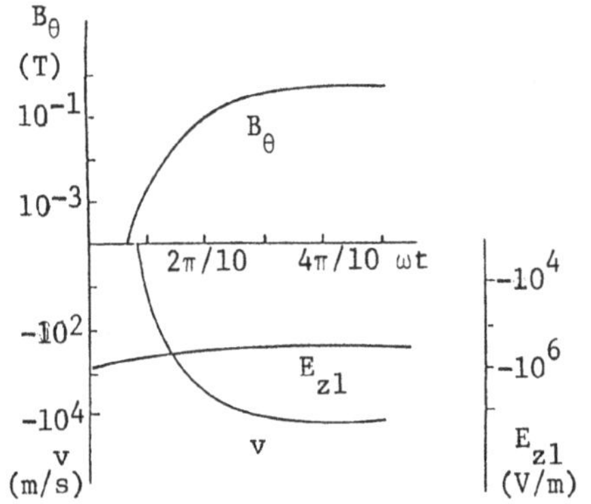

Fig. 3. The beam velocity V zt $z=0$ and $r=0$ the magnetic field B_θ in the azimuthal direction at $z=0$ and $r=r_b$ and the electric field E_{z1} induced by beam acceleration at $z=0$ and $r=0$ versus the time ωt.

is located at the origin. In Fig.2, the velocity v at $r=0$, and B_θ in the azimuthal direction at $r=r_b$ at $\omega t=\pi/2$ are plotted versus z. The beam velocity v increases with time between $\omega t=0$ and $\omega t=\pi/2$, hence the magnetic field B_θ increases with time, too. Figure 3 shows v at $z=0$ and $r=0$ and B_θ at $z=0$ and $r=r_b$ versus time t. At the instance $\omega t=\pi/2$ the total kinetic energy of beam is 2.89×10^4J, while the energy of the magnetic field induced in space by the beam current is 3.63×10^{-11}J. Thus the main part of the converted energy from laser is represented by the kinetic energy of the beam.

The temporal change in the magnetic field in the azimuthal direction induces the electric field in the axial direction. The electric field E_z in the axial direction consists of three parts,

$$E_z(r,z,t) = E_{z0} + E_{z1}(r,z,t) + E_{z2}(r,z,t) + E_{z3}(r,z,t), \tag{52}$$

where E_{z0} is the irradiating laser field. The field E_{z1} comes from the change in the velocity of the beam and takes the maximum value at the beam centre $r=0$ and $z=0$. In Fig.3, E_{z1} at $z=0$ and $r=0$ is plotted versus t, too. This field causes the deceleration of oscillating motion of electron beam. In our problem, the laser light accelerates the local electron beam. In our problem, the laser light accelerates the local electron cylinder. In connection with this collective motion of electrons the electric field E_{z1} as well as the magnetic field B_θ are induced. Here the intensity of magnetic field is not so high, but E_{z1} induced by the acceleration of the electron beam becomes rather high and E_{z1} cancels the electric field E_{z0} of the laser light. At $\omega t=\pi/2$, the beam reaches only the velocity of $=1.89\times10^4$m/s although without the induced electric field it should obtain the (negative) light speed. The third term in eq.(52) comes from the finite length of the electron beam. Near the leading edge the electrons are decelerated while near the trailing edge the electrons are accelerated. Thus the beam is bunched during the

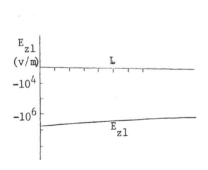 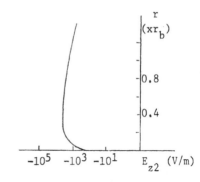

Fig. 4. The electric field E_{z1} induced by beam acceleration at r=0 and $\omega t=\pi/5$ versus z.

Fig. 5. The electric field E_{z2} induced by the motion of beam edge at z=L and $\omega t=\pi/2$ versus r.

motion. In Figs. 4 and 5 E_{z1} (r=0,z,$\omega t=\pi/2$) and E_{z2}(r,z=L,$\omega t=2$) are plotted, respectively. The electric field E_{z3} which comes from the charges at the beam edges is negligible in our case because the laser light has no high frequency that the translation of the beam during a quarter of period is very short in comparison with the beam length.

During the oscillation of the electron beam the beam velocity changes rapidly. When the velocity is small collisions among beam electrons and ions in the target plasma are important (Mulser, Niu and Arnold, 1989). However, the excursion distance z_{*0} of the electron during a quarter period of laser light is so small as $z_{*0}=3.95\times10^{-8}$m that this distance is smaller than the mean free path and hence the collision is not possibly dominant. The paper indicates that the absorption of laser light by the electrons in the material becomes inefficient due to the self-induced electric fields by the electron collective motion.

REFERENCES

Kristal R. et al. 1989 Laser Interaction and Related Plasma Phenomena 8, 9.
Mulser P., Niu K. & Arnold R. C. 1989 Proc. of Int. Symp. on Heavy Ion Inertial Fusion 89.

EARLY-TIME "SHINE-THROUGH" IN LASER IRRADIATED TARGETS

D. K. Bradley, T. Boehly, D. L. Brown, J. Delettrez,
W. Seka, and D. Smith

Laboratory for Laser Energetics
University of Rochester
250 East River Road, Rochester, NY 14623-1299

I.   INTRODUCTION

In direct-drive laser-fusion experiments, a deuterium or deuterium-tritium filled microballoon is imploded by direct illumination with a number of high-power laser beams. The implosion is driven by the ablation of material from the outer surface of the target. The ablation and compression occurs only after a plasma has been established at the surface of the target. This plasma, once established, dominates the interaction since the laser can no longer propagate past the critical density surface where the plasma frequency equals the laser frequency. Energy is absorbed at densities below critical and transported through the overdense region to the ablation front. Another important function of the plasma is to smooth, by thermal conduction, some of the the spatial non-uniformities in the laser energy deposition on the way to the ablation layer.

Laser fusion targets are primarily microballoons made of glass or CH. These targets are sometimes overcoated with a layer of low Z polymer which can also include additional "signature" layers. These layers are buried inside the target to provide information on energy transport or ablation dynamics.[1] Numerical modeling of these experiments usually assumes that a preformed plasma already exists at the target surface and that the laser light will penetrate only as far as the critical surface, where it is reflected. However, at early times, before plasma formation, the target and these coatings may be transparent to the laser light. At these times light can penetrate the target and deposit energy either throughout the bulk of the target, or preferentially at local absorption zones (i.e., buried layers and impurity sites). This so called "shine-through" can greatly affect the behavior of the target, despite containing energy which is nearly negligible as compared to that eventually absorbed by the target. Since shine-through occurs when no plasma is present, no smoothing will take place. Indeed it is possible that the nonuniformities in the beam could be made worse by self-focusing[2] or filamentation[3-5] of the incident beam before the plasma is formed. In Ref. 4, it was found that a 500 Å Al overcoating of the target eliminates all visible filamentation damage tracks.

It has been observed in previous experiments[6] that anomalously high laser burn-through rates occur in glass microballoons overcoated with several microns of plastic. Laser burn-through is defined as the time at which x-ray line emission from a buried signature layer can be detected by a spectrally dispersing streak camera. It has also been noted that the addition of a thin opaque "barrier" layer of Al or Au, to the outside of the target, reduces the observed burn-through rate to values much closer to those predicted by one-dimensional code simulations. The enhanced burn-through rates seen in the "bare" plastic coated targets have been attributed to the Rayleigh-Taylor instability, which can develop at both the ablation surface and the plastic-glass interface. It is also believed that the reduced burn-through rates seen when barrier layers are present may be caused by a reduction in the early-time shine-through, which will change the initial Rayleigh-Taylor seeds.

We present the results of planar-target experiments that demonstrate that shine-through does exist in many types of targets. First, we show that simple glass targets do indeed transmit the early portion of the beam before a plasma is formed. We then show the effect of various overcoatings on total transmitted energy and its spatial distribution. We see distinct differences in the behaviors of metallic and dielectric coatings. We also present time resolved transmitted light measurements for many of these coatings. Finally, we demonstrate methods to reduce shine-through and show the effect of one method on the performance of spherical laser-fusion targets. This is a presentation of work in progress and represents our understanding of the problem to date.

II.   EXPERIMENT

The experimental configuration used throughout these experiments is shown in Fig. 1. Targets were irradiated by 650 ps pulses of 351 nm light at peak intensities of 10^{11}–10^{14} W/cm^2 using the single beam GDL glass laser system[7] at the University of Rochester's Laboratory for Laser Energetics. The beam was focused using a 0.6 m f/3 lens together with a distributed phase plate (DPP)[8] to give a highly reproducible focal spot. The range of incident irradiances were obtained by varying the laser output energy. Light transmitted through the target was collected by a second f/3 lens and imaged, with a 7.5 magnification, onto both a film pack containing Kodak T-Max 400 film (two-dimensional time integrated image) and the input slit of a UV-sensitive streak camera (one-dimensional

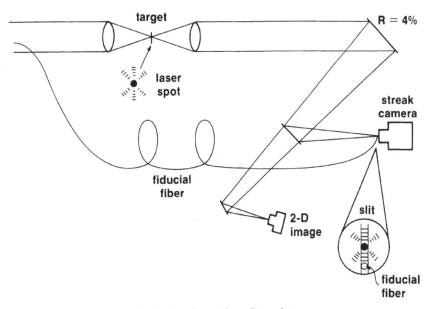

Fig. 1 Experimental configuration.

image with temporal resolution). Both cameras were filtered by a narrow bandpass filter centered at 351 nm. A small fraction of the incident beam was collected in an optical fiber and transported to the streak camera slit as a timing fiducial. The T-Max film was calibrated at 351 nm by comparing the pixel histograms of a series of identical images recorded at known different exposure levels. Each target was mounted behind a 900-μm diameter hole in a ~100-μm thick Ti foil. Since the DPP spot contains significant energy beyond the central region, the aperture is needed to reduce the possibility of detecting a ghost image from one of our wedged glass pick-offs.

The intensity distribution at the best focus of the lens/DPP system is shown in Fig. 2. This was recorded on the time-integrating camera on a low energy shot with no target present. A vertical lineout through the center of the image (i.e., along the axis of the streak camera slit) is shown alongside. Most of the laser energy (~83%) is contained in the central spot, with the first minimum occurring at a radius of 194 μm. The secondary peaks, which occur at increasingly lower intensities, provide a useful means of observing the effects of several different irradiances on a single shot. Figure 3 shows a streak camera record of a shot similar to that shown in Fig. 2, i.e., without a target. The fiducial signal is not timed to be coincident with the main pulse (note offset) and the pincushion distortion seen is a consequence of the image intensifier used. The spatial lineout, taken through the peak of the pulse, shows a similar shape to that of Fig. 2 although only the first of the subsidiary maxima is visible.

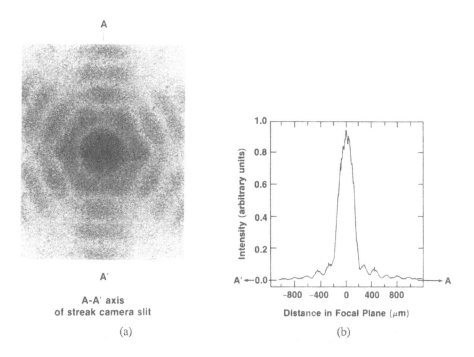

Fig. 2 Intensity distribution at best focus using a DPP. (a) image from time-integrating camera; (b) lineout through center of image, along A A[1].

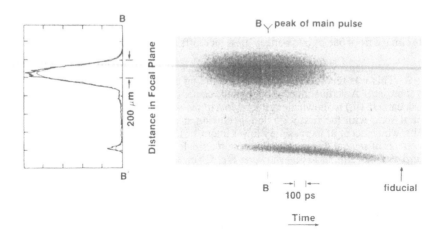

Fig. 3 Streak camera image of focal spot with no target present.

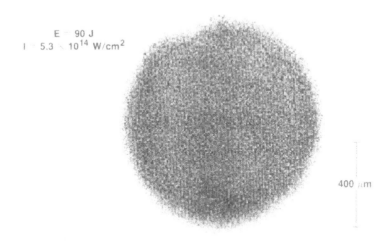

Fig. 4 Time-integrated intensity distribution recorded through a glass cover slip.

Figure 4 shows a time integrated image of the 351-nm light transmitted through a 150-µm glass microscope cover-slip. The incident energy was 90 J, corresponding to a peak irradiance of 5.3 x 10^{14} W/cm^2 in the central spot. There is little evidence of the DPP focal distribution of Fig. 2, the only structure apparent being laser speckle. This complete lack of features in the transmitted light signal, despite the large intensity variations in the incident beam profile (> a factor of 20) suggests that the total transmitted energy is insensitive to the incident intensity over a wide range of values. This assertion is confirmed by the data in Table 1, which lists the measured transmitted energy (E_{trans}) in the central DPP spot region for a range of laser energies, on different shots, corresponding to (central spot) intensities between 3.0 x 10^{12} and 5.3 x 10^{14} w/cm^2. The measured transmitted energies range from 1.8 mJ to 5.0 mJ, and are essentially independent of the incident energy or intensity.

Table 1
Measured transmitted energies in the central laser spot through glass targets at different irradiances

E_{laser} (J)	I_{laser} (W/cm^2)	E_{trans} (mJ)
0.5	3.0 x 10^{12}	5.0
4.4	2.6 x 10^{13}	2.0
53	3.1 x 10^{14}	1.8
76	4.5 x 10^{14}	2.8
90	5.3 x 10^{14}	2.5

If we assume that all the incident light is transmitted through the target until a certain threshold is reached, and then calculate the instantaneous intensity at which the integrated laser energy equals E_{trans}, we obtain "cut-off" intensities in the range 1.0–2.3 x 10^{11} W/cm^2. Figure 5 shows the streak camera record from a shot with 5 x 10^{14} W/cm^2 incident on a glass target. Our current streak camera calibration is insufficient to give accurate timing values. However, it can be seen from the figure that throughout the DPP distribution, the "turn-on" time (the detection threshold of the streak camera system) and the "turn-off" time (due to the formation of a critical density surface) of the transmitted signal both follow a locus of constant intensity. Also, in contrast to Fig. 3, the strong spatial modulation of the DPP pattern is not as evident.

A selection of cover-slips, similar to the one used in Fig. 4 were coated with a thin (500 Å) layer of aluminum. Their UV transmission properties were then measured using a Perkin Elmer Lambda-9 spectrophotometer, showing that the aluminum layer causes an attenuation of $10^{2.2}$ at 351 nm compared to the bare glass slide. Figure 6 shows images of the transmitted light for two shots where the coated targets were irradiated on the Al-coated side. The first image, corresponding to an incident irradiance of 1.5 x 10^{12} W/cm^2 (0.25 J), shows the central spot region to be attenuated down to the level of the film background noise. However, the subsidiary lobes (which correspond to significantly lower intensities) have been transmitted much more strongly. The second image, which was recorded from an incident intensity of 2.9 x 10^{14} W/cm^2 (48 J), seems almost a complete reversal of the low energy shot. In this case, there is little or no apparent transmission in the outer lobes, but a region of anomalously high transmission has occurred at the center of the main spot.

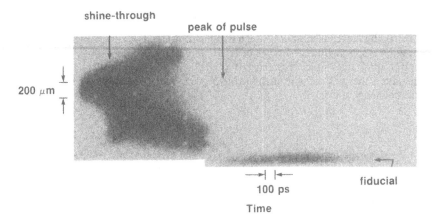

Fig. 5 Time-resolved image recorded through glass target.

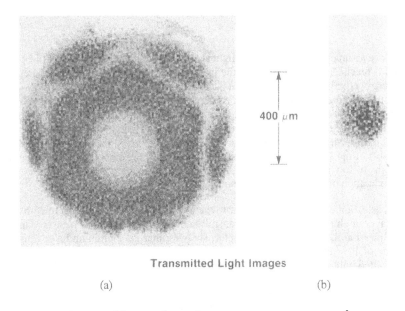

Transmitted Light Images

(a)   (b)

Fig. 6 Time-integrated images from glass targets coated with 500 Å Al. (a) 0.25 J
incident energy; (b) 48 J incident energy.

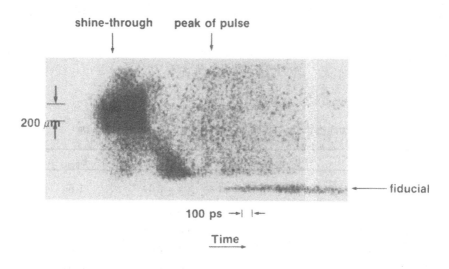

Fig. 7 Time-resolved image of intensity recorded through Al-coated target.

A time resolved image for a target similar to those shown in Fig. 6 is shown in Fig. 7. In this case the incident intensity was 5×10^{14} W/cm^2. The streaked image is quite different to that obtained for the bare glass target at the same intensity (see Fig. 5). In this case the transmission at the center of the distribution dominates with little transmission observed elsewhere which is consistent with the time-integrated images. Also, the time at which the transmitted light is detected, as a function of spatial position, does not seem to follow the same incident intensity distribution as Fig. 5. The turn-off time of the central spot appears to be consistent with that seen for the uncoated glass.

Other coating materials were investigated, including carbon and Parylene, and the data from these shots are summarized in Table 2. The parylene layer, which was measured to have a cw transmission at 351 nm of 80%, gives a measured transmission of ~0.6 mJ, compared to 1.6 mJ for the uncoated glass. No significant difference was noted between front and rear surface coatings. We believe that the reduction in E_{trans} observed with the parylene coatings is due to the lower breakdown threshold of parylene compared to glass. The similarity between the behavior of front and rear side coatings indicates that the initial breakdown occurs in the parylene in both cases. In contrast, the carbon layer, which had a cw transmission of 4% compared to bare glass, allows 0.05 mJ transmission when coated on the front surface (facing the laser) of the target and 0.23 mJ when coated on the rear surface. The reason for the difference between a front and back surface carbon coating is not yet understood.

In some direct-drive target applications, it is important to be able to view through the target before laser irradiation, e.g. to monitor the formation of cryogenic fuel layers.[9] In such a case, the use of an opaque material such as aluminum as a shine-through barrier could be a significant problem. Ideally, one would use a material which was non-transmissive at the drive laser wavelength, but transparent enough to allow optical viewing or probing at a different wavelength prior to the shot. To investigate this we had a number of glass targets coated with a dielectric mirror, similar to the coatings on the high-damage-threshold UV mirrors routinely used on GDL. The coatings consisted of 10 layers of TiO$_2$/SiO$_2$, each of one quarter-wave optical thickness, chosen to give a 1% transmission at 351 nm, with high transmission elsewhere. A streak camera record of the light transmitted through one of these targets irradiated at 3×10^{14} W/cm^2 is shown in Fig. 8. The image displays similar features to the uncoated target of Fig. 5, but in this case the signal is much

Table 2

Light transmitted as a function of coating material and location

Target	E_{laser} (J)	I_{laser}	E_{trans} (mJ)
Glass	90	5.3×10^{14}	1.6
6 μm CH/Glass	84	5.0×10^{14}	0.54
Glass/6 μm CH	91	5.4×10^{14}	0.67
950 Å C/Glass	73	4.3×10^{14}	0.053
Glass/950 Å C	100	5.9×10^{14}	0.23

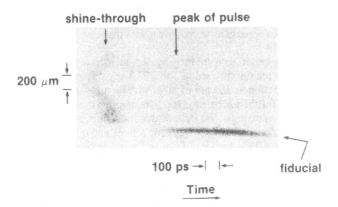

Fig. 8 Streak image of light transmitted through a target with a dielectric mirror coating.

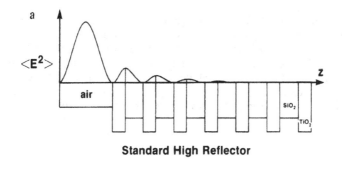

Standard High Reflector

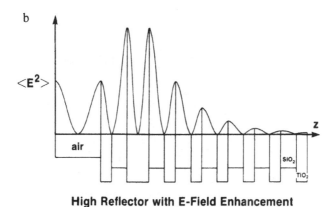

High Reflector with E-Field Enhancement

Fig. 9 Electric field inside dielectric mirror coatings. (a) Highly damage-resistant coating; (b) Coating with E field deliberately enhanced.

lower and the time between initial transmitted light detection and "turn-off" is much less. Again, since our streak speed is not accurately calibrated, it is difficult to give accurate comparisons, but the "turn-off" time of this target appears to be similar to that of the bare glass target. The time-integrated image recorded on this shot was barely above the background noise level.

Dielectric mirror coatings normally reduce the electric field to zero inside the mirror, as shown in Fig. 9(a), in order to increase the coating's damage threshold at high laser powers. However, as a means of reducing early-time shine-through, this is, in fact, the opposite of what we would wish to do. Ideally the coating will act as a high reflector but will also break down and create a plasma early in the pulse, preferably as close to the surface as possible. A coating designed to do that is shown in Fig. 9(b). In this case the TiO_2 and SiO_2 layer thicknesses have been chosen to enhance the electric field strengths at the layer boundaries. These high fields should cause dielectric breakdown to occur at much lower intensities, and hence, earlier in the pulse, than the coating of Fig. 9(a). A streak image recorded through such a coating is shown in Fig. 10. In this case there is no transmission recorded in the early part of the pulse at all, the only emission being a low level scattered light signal detected around the peak of the pulse.

Finally, to test whether shine-through barrier layers can actually have an effect on direct-drive laser driven implosions, a number of deuterium filled glass microballoons, both with and without a 500 Å Al barrier layer, were imploded using the 24-beam OMEGA laser system. The normalized neutron yields obtained from this series are shown in Fig. 11 as a function of initial glass shell thickness. The targets with the Al coating show a clear yield improvement over the uncoated targets. Al was used because of the ease with which it can be coated. We are currently investigating methods of coating spherical targets with more complex layers.

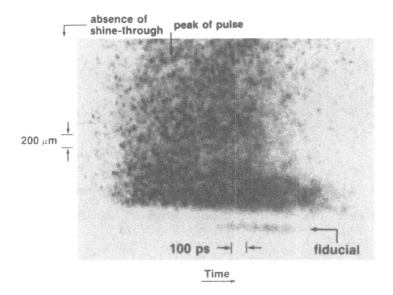

Fig. 10   Time-resolved image of light transmitted through target with E field enhancing coating. Note absence of early shine-through signal. The low level light signal observed around the peak of the pulse is due to scattered light.

III.   SUMMARY

We have observed that shine-through can occur in a variety of different targets. The total amount of light transmitted through glass targets does not appear to be a function of either incident laser energy or intensity and is typically of the order of a few mJ. This may be simply an effect of the electric field (intensity) required to initiate breakdown in the bulk of the material, and may also be related to the amount of energy required to ionize enough material to form a critical surface. Although the energy involved in the shine-through is small, it can promote the growth of the Rayleigh-Taylor instability in two ways. First, breakdown, which occurs randomly in the bulk material or at interfaces, can add to the initial perturbations that seed the instability. Second, if breakdown occurs at the interface between plastic and signature layers, a cavity is created in the plastic, leading to

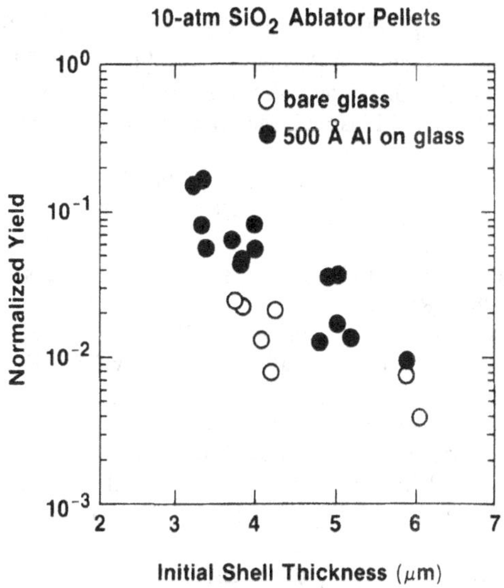

10-atm SiO$_2$ Ablator Pellets

Fig. 11   Comparison of normalized neutron yields obtained from D$_2$ filled targets with and without barrier layers.

an increase in the growth rate of the instability at that interface.[10] Self focusing or other mechanisms which can result in uneven breakdown will only further enhance the Rayleigh-Taylor instability. Finally, we have shown that shine-through is significantly reduced when highly reflective dielectric layers are coated onto the substrate, and that this reduction is even greater when the layers are chosen so as to reduce the threshold for dielectric breakdown near the target surface.

ACKNOWLEDGMENT

This work was supported by the U.S. Department of Energy Division of Inertial Fusion under agreement No. DE-FC03-85DP40200 and by the Laser Fusion Feasibility Project at the Laboratory for Laser Energetics which has the following sponsors: Empire State Electric Energy Research Corporation, New York State Energy Research and Development Authority, Ontario Hydro, and the University of Rochester.

REFERENCES

1.   J. Delettrez, R. Epstein, M. C. Richardson, P. A. Jaanimagi, and B. L. Henke, "Effect of Laser Illumination Nonuniformity on the Analysis of Time-Resolved X-Ray Measurements in UV Spherical Transport Experiments," Phys. Rev. A 36:3926 (1987).

2.   A. J. Palmer, "Stimulated Scattering and Self-Focusing in Laser-Produced Plasmas," Phys. Fluids 14:2714 (1971).

3.   P. K. Kaw, G. Schmidt, and T. Wilcox, "Filamentation and Trapping of Electromagnetic Radiation in Plasmas," Phys. Fluids 16:1522 (1973).

4.   F. E. Balmer, T. P. Donaldson, W. Seka, and J. A. Zimmermann, "Self-Focusing and the Initial Stages of Plasma Generation by Short Laser Pulses," Opt. Commun. 24, 109 (1979).

5.   D. L. Brown, W. Seka, and T. Boehly, "Self Focusing and Dielectric Laser Breakdown," Bull. Am. Phys. Soc. 34:1938 (1989).

6.   J. Delettrez, D. K. Bradley, P. A. Jaanimagi, and C. P. Verdon, "Effect of Barrier Layers in Burn-Through Experiments with 351-nm Laser Illumination," to be published in Phys. Rev. A.

7.   W. Seka, J. M. Soures, S. D. Jacobs, L. D. Lund, and R. S. Craxton, "GDL: A High-Power 0.35 μm Laser Irradiation Facility," IEEE J. Quantum Electron. QE-17:1689 (1981).

8.   Laboratory for Laser Energetics LLE Review 33, Quarterly Report No. DOE/DP40200-65, 1987 (unpublished), pp. 1–10.

9.   H.-J. Kong, M. D. Wittman, and H. Kim, "New Shearing Interferometer for Real-Time Characterization of Cryogenic Laser-Fusion Targets," Appl. Phys. Lett. 55:2274 (1989).

10. Laboratory for Laser Energetics LLE Review 40, Quarterly Report No. DOE/DP40200-102, 1989 (unpublished), pp. 173–184.

LASER INDUCED BREAKDOWN AND HIGH VOLTAGE INDUCED

VACUUM BREAKDOWN ON METAL SURFACES

Fred Schwirzke

Department of Physics
Naval Postgraduate School
Monterey, CA 93943

ABSTRACT

Breakdown and plasma formation on surfaces are fundamental processes in laser target interaction experiments as well as in other areas of pulsed power technology. The initial plasma formation on the surface of a laser irradiated metal target is very non-uniform. Micron-sized plasma spots form within nanoseconds. Quite similar, the initial plasma formation on the surface of a cathode of a vacuum arc, vacuum diode, and many other discharges is highly non-uniform. Micron-sized cathode spots form within nanoseconds. The concept of explosive electron emission from a cathode spot is well established in the literature. However, the details of the breakdown process were not well understood. Unipolar arcing represents a discharge form which easily leads to explosive plasma formation. Power dissipation for an unipolar arc is considerably higher than for field emitted or space charge limited current flow. Using a laser produced plasma it has been demonstrated that unipolar arcs ignite and burn on a nanosecond time scale without any external electric field being applied. Similar unipolar arc craters have now been observed on the cathode surface of a pulsed vacuum diode with an externally applied field of $E=1MV/2.5$ cm. The experimental results show that cathodes spots are formed by unipolar arcing. The localized build-up of plasma above an electron emitting spot naturally leads to a pressure gradient and electric field distribution which drives the unipolar arc. The high current density of an unipolar arc provides explosive plasma formation.

INTRODUCTION

Laser induced vacuum breakdown and high voltage induced vacuum breakdown do not seem to have much more in common than the end result: a plasma layer is formed near a surface. The breakdown process in a vacuum diode is most commonly considered to be initiated by the field emission of electrons from a cathode whisker or other emitting micropoints on the surface due to the externally applied electric field. The field emission current density, given by the Fowler-Nordheim equation is assumed to be large enough to cause the whisker to explode due to joule heating effects, forming a dense plasma. The plasma surface then acts as a virtual cathode and the diode current density becomes space charge limited. The externally applied electric field vanishes near the cathode surface. Hence, E changes rapidly during the breakdown process from a maximum value on the tip of the whisker at the onset of the field emission current with density j_{FE} to $E \approx 0$ once a sufficiently dense plasma is formed and the current density is space charge limited j_{CL}, as given by Child-Langmuir's law.

The variation of the diode current with time is schematically shown in Fig. 1. Opinions differ concerning the detailed processes which lead to the explosive formation of a dense cathode spot plasma and its structure. The current density j and the electric field E influence field emission of electrons, joule heating, and plasma formation. Estimates of j for a cathode spot vary by orders of magnitude between 10^5 A/cm^2,[1], and a 10^9 A/cm^2,[2]. This indicates that the details of the breakdown process were not yet well defined as the following quotations from the literature show: "Processes in a cathode spot are extremely complicated and in many respects are still not understood".[2] "Explosion of a microtip is accompanied by transition of the cathode matter into a dense plasma. The mechanism of this phase transition is not completely understood up to now".[3] "At present, the mechanism of transport in the near-cathode region of a vacuum discharge is far from being revealed. There is a large discrepancy (up to three orders of magnitude) in experimental data on current density... However (as far as we are aware), there is no paper that proposed a physical model that, at least qualitatively, describes with due regard for the dominant role of the explosion mechanism, and that explains the numerous experimental data from a single viewpoint."[4]. "Cathode spots are complicated objects of both solid state physics and plasma physics. But on the plasma side, the experimental knowledge is rather scarce. More data is needed on the plasma dynamics...."[5]

The problem then is to understand the rapid formation of a dense cathode spot plasma which facilitates the transition $j_{FE} \to j_{CL}$ while $E_{ext} \to 0$ within the plasma. This paper describes how the transition can occur by the mechanism of unipolar arcing.

Unipolar arcing[6] represents a discharge form which easily leads to explosive plasma formation. Power dissipation for an arc is considerably higher than for field emitted or space charge limited current flow. Using a laser produced plasma it has been demonstrated that

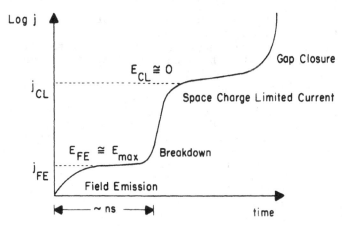

Fig. 1. Increase of current density in a high voltage diode. j_{FE} is the prebreakdown field emitted current density. j_{CL} is the Child-Langmuir space charge limited current density. The electric fields are at or very near the cathode surface corresponding to the curve plateaus. Gap closure defines the the point at which the breakdown arc extends across the diode gap.

unipolar arcs ignite and burn on a nanosecond time scale without any external electric field being present. This paper presents a model and experimental results which show that unipolar arcing is the basic breakdown process which leads to the formation of a laser produced plasma as well as to the formation of cathode spots in a vacuum diode (and presumingly also to the formation of cathode spots in other discharges).

LASER INDUCED BREAKDOWN

Laser-induced breakdown of various materials has been studied extensively as function of irradiance, wavelength, pulse duration and other parameters.[7] Crater formation, called laser pitting, was often observed but not well understood. An unresolved issue was the mechanism by which localized heating can lead to plasma formation. Unipolar arcing, being primarily an electric interaction with thermo-mechanical aspects, was identified as the basic laser damage mechanism which causes pits and craters.[8]

Laser beams interact with surfaces by a variety of thermal, impulse and electrical effects. Energy coupling is considerably enhanced once surface electrical breakdown occurs. Laser induced breakdown on surfaces was studied with a Q-switched neodymium laser of 25 ns FWHM.[9] The laser pulse was directed onto targets placed in a vacuum chamber, providing pressures of the order of 10^{-6} Torr. Incident laser energy was varied by inserting neutral density filters of varying transmittance in the beam path. For low energy shots the laser amplifier was not fired. Laser energies on target between 0.0075 and 10 joules were obtained by these techniques. The beam was focused to various spot sizes on the target to provide further variation in power density. Low power, defocused laser pulses were used to study the onset of surface breakdown. A polaroid camera was positioned above the target to note plasma formation by recording the attendant light. Fig. 2 shows the damage on a stainless steel surface once breakdown occurs which is characterized by light emission from the plasma. The craters result from electrical microarcs burning between the plasma and the surface. The hole at the center is the electron emitting cathode spot of the arc, the surrounding ring with crater rim is the electron receiving anode area. Since both "electrodes" are located on the same metal surface the arc is called "unipolar".

A review of the target surface reveals that the damage is not evenly distributed. Besides the severe deformation caused by arcing, there is no other direct laser damage (like uniform surface melting) observable. All damage is in the form of arc craters. Obviously, the plasma manages to concentrate the laser-plasma energy towards the small cathode spots. Small cathode holes have a diameter of

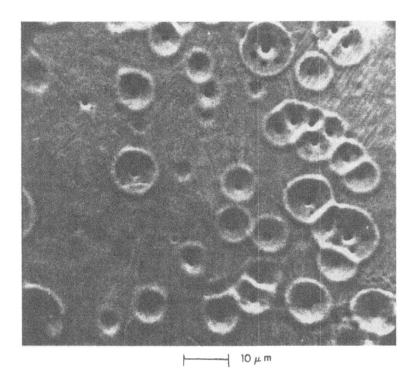

$\longmapsto$ 10 μm

Fig. 2. Laser produced unipolar arc craters on stainless steel.

about $\gtrsim 0.7$ μm which is smaller than the wavelength of the Nd laser, $\lambda = 1.06$ μm. Using the 10.6 μm radiation of a CO_2 laser the same effect was observed, unipolar arc craters with cathode holes of ~ 1 μm $\ll 10.6$ μm, the laser wavelengths. The data confirm that unipolar arcing is a plasma-surface interaction process, and not a self-focusing effect. The laser heats the plasma, and the plasma causes the arcing.

The initial breakdown phase may be characterized by localized desorption of contaminants from laser heated spots on the surface of a metal. Microprojections or whiskers, dust particles, absorbing inclusions, metal grain boundaries etc, can serve as such spots. While most of the laser light will be reflected from a metal surface, some absorption occurs within the skin depth. Threshold intensities for laser induced breakdown fall in the range of 10^7 - 10^8 W/cm^2 which corresponds to wave electric fields of 6x10^4 to 2x10^5 V/cm. Broad area electrodes and rf cavities show enhanced electron emission[10] with enhancement factors $\beta \approx 10^2$ to 10^3 at relatively low dc and rf fields of E $\approx 10^5$ V/cm. Whiskers, adsorbates and dust can further enhance the field emission of electrons. As a consequence "vacuum" breakdown can occur already at electric fields as low as E $\gtrsim 10^4$ V/cm. Localized heating, enhanced electric fields, and the emission of electrons[11] lead to desorption of a few monolayers within a few ns forming an expanding high density neutral gas cloud as indicated in Fig. 3. Electron heating and ionization of neutral particles will preferentially occur at a distance $\lambda/4$ from the surface where for normal incidence the first maximum in the electric field occurs due to the superposition of the incidence wave and the wave reflected on a metal surface. Once the electrons gain sufficient energy to ionize, the locally increased electron density further increases the absorption of laser radiation. To assure quasi-neutrality, the plasma in contact with the surface immediately forms a sheath. The width of the sheath is of the order of the Debye length $\lambda_D = (\epsilon_0 kT_e /e^2 n_e)^{1/2}$. Independent of n_e the plasma assumes a positive potential with respect to the wall as given by the floating potential

$$V_f = -\frac{kT_e}{2e} \ln \left[\frac{M_i}{2\pi m_e} \right] . \tag{1}$$

The average electric field in the sheath is approximately

$$E_s = \frac{V_f}{\lambda_D} = (n_e kT_e)^{1/2} (1/4\epsilon_0)^{1/2} \ln (M_i /2\pi m_e) \tag{2}$$

E_s approaches zero near the plasma boundary and assumes its maximal value at the surface.

In accordance with Eq. 2 the sheath electric field increases with the local build up of the plasma pressure as $E \propto P_e^{1/2}$. The dielectric strength of air at atmospheric pressure is 30 kV/cm. If we take this as a typical value for onset of sparking we find from Eq. 2 that a plasma layer of $n_e \geq 2$ x 10^{12} cm^{-3} and $kT_e = 10$ eV can provide already a sheath electric field of $E \geq 30$ kV/cm to start spark formation. Enhanced field emission and finally, thermionic electron emission from a cathode spot will lead to laser induced unipolar arcing[6 and 9], even though there is no external electric field being applied. Micron sized arcs burn between the plasma and a metal surface driven by pressure gradients and local variations of the sheath potential. Similar unipolar arcs have been observed on the cathode surface of a pulsed vacuum diode.

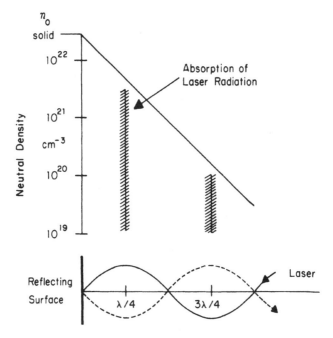

Fig. 3. Absorption of laser radiation near a reflecting
         surface. The neutral density gradient in front
         of the surface is caused by blow- off of surface
         material.

HIGH VOLTAGE INDUCED VACUUM BREAKDOWN

Highly non- uniform cathode spots form and control many kinds of vacuum discharges. The concept of electron emission from an exploding "whisker" is well established in the literature[1-5]. However, since the estimates of current density j from a cathode spot vary by several orders of magnitude[12], the details of the actual mechanism of explosive emission and formation of cathode spots are not yet fully understood.

The expression "whisker" is meant to represent any electron emitting spot on the surface of a cathode, such as a microprojection, adsorbed contaminant, metal grain boundary, etc., where the electric field and the field emission current are enhanced by one to several orders of magnitude.

The Whisker Explosive Emission Model

This standard model shows the importance of the current density j for the formation of the cathode spot. A high voltage pulse applied to a vacuum diode leads to enhanced field emission of electrons, i_{FE} from the tip of a whisker located on the cathode surface. The field emission current density is given by an equation similar to the Fowler-Nordheim equation

$$j_{FE} = c_1 \beta^2 E^2 \, \exp \, [- c_2 / \beta E] . \tag{3}$$

β is the field enhancement factor, c_1 and c_2 are constants.

After a delay of the order of a few ns the whisker "explodes" into a dense expanding plasma and a crater is formed at the cathode surface. The current jumps to the space charge limited value i_{CL}, Fig. 1. Joule heating, $\Delta Q/\Delta t$, of the whisker is assumed to lead to its explosion. For a whisker of cylindrical shape (radius r, length L) and neglecting any heat losses

$$\frac{\Delta Q}{\Delta t} = iV = j \pi r^2 L E_w = j^2 \rho v$$

where $E_w = j\rho$ is the electric field in the whisker, and $v = \pi r^2 L$ its volume. For stainless steel $\rho = 7 \times 10^{-6}$ ohm-cm, and the particle density is n = 8.2×10^{22} cm^{-3}. The energy/atom provided by ohmic heating is

$$\frac{\Delta Q}{atom} = \frac{j^2 \rho v}{n v} \Delta t \text{ or in } \frac{eV}{atom} = 5.3 \times 10^{-10} \, j^2 \, \Delta t$$

where j is measured in A/cm^2. To estimate rough values for j and Δt it may be assumed that the explosive transition from a whisker to a dense plasma requires about $\gtrsim$ 5.3 eV/atom. Thus $j^2 \, \Delta t \gtrsim 10^{10}$ A^2 sec/cm^4. The enhanced field emission of electrons depends strongly on E and the surface temperature of the emitting spot. In fact, the prebreakdown current density for a field of about 10^7 V/cm from a cold spot should be quite small. Even, if the prebreakdown enhanced field emission currents should, by whatever mechanism, increase to $j_{FE} \sim 10^7$-10^8 A/cm^2, the time required to provide bulk ohmic heating of the whisker would be $\Delta t = 10^{-6}$ to 10^{-4} sec. Experimentally, the plasma formation in a fast pulsed diode occurs on a nanosecond time scale. Consequently, higher current densities would be required $j \sim 10^9$ A/cm^2.

The whisker explosive emission model requires that the field emitted current heats the whisker within ns until an explosive formation of plasma occurs. From there on the current to the anode is space charge limited with the current density j_{CL}. Since the initial plasma layer forms very near the cathode surface and is very thin, about 10^{-4} cm, it might be justified to consider a cross section near the surface where the transition $j_{FE} \rightarrow j_{CL}$ is supposed to occur. In general one would expect $j_{FE} \ll j_{CL}$. The Child-Langmuir current to the anode represents an upper limit. For example, for a diode voltage of 1 MV and a gap of d = 1cm, assuming a uniform E = 10^6 V/cm, j_{CL} = 2.3 x 10^3 A/cm^2, which is much smaller than the $j_{FE} \approx 10^8$-10^9 A/cm^2 required to explode a whisker by joule heating within a few nanoseconds. The field emitted electron current will become space charge limited at a value which is insufficient for the explosive like transition of the whisker into a dense plasma. Furthermore, the externally applied E vanishes near the cathode surface when the current becomes space charge limited. This should turn-off the field emission current. The question then is, how can the relatively small field emission current lead to the "explosive" formation of a cathode spot within nanoseconds?

Onset of Breakdown

Electrode surfaces are usually far from ultra high vacuum clean. Adsorbates are weakly bound to the surface by van der Waals interactions, ~ 0.25 eV/molecule. Whiskers, dust particles, oxide spots and similar microscopic nonuniformities provide enhanced field emission of electrons if a sufficiently high electric field, $E \gtrsim 10^5$ V/cm is applied. The electric field, the emission of electrons, and the impact of ions stimulate desorption[11] and [13]. Moving with sound velocity $v \approx 3.3 \times 10^4$ cm/s a suddenly released monolayer of 2×10^{15} molecules/cm^2 forms a dense neutral gas layer. After 3 ns the average particle density of the neutrals is $n_0 = (2 \times 10^{15}/\text{cm}^2)/vt = 2 \times 10^{19}$ cm^{-3} with vt = 10^{-4} cm. For comparison, the atmospheric particle density at 0° C is 2.7×10^{19} cm^{-3}. The electron mean free path length for ionizing a neutral $\lambda = 1/(n_0 \sigma_0)$ depends on the ionization cross section σ_0 which is a function of the electron energy. For O_2, H_2 (and other gases) the ionization cross section has a maximum value of about $\sigma_0 \approx 10^{-16}$ cm^2 for electrons with an energy of about 100 eV. In an electric field of 10^6 V/cm a field emitted electron has gained 100 eV at the distance of 10^{-4} cm (Fig. 4). Thus $\lambda \approx 5 \times 10^{-4}$ cm and about 20% of the emitted electrons have a chance for an ionizing collision within d $\approx 10^{-4}$ cm. For a diode the onset of breakdown is typically delayed by 1 to 10 ns which corresponds to the expansion time of the neutrals for reaching the maximum of the ionizing zone, i.e., where the electrons have about 100 eV.

If the emitted electrons have negligible initial velocity then they will be accelerated until they collide at t_{coll} having gained a velocity of $v_e = (e/m)Et_{coll}$. Electrons in a field of $E = 10^6$ V/cm need $t = 3.45 \times 10^{-13}$ sec to gain 100 eV over a distance of $d = 10^{-4}$ cm. Ions produced at $d = 10^{-4}$ cm are accelerated towards the cathode. For an oxygen ion the time of flight to the cathode surface would be 5.8×10^{-11} s while the two electrons would need a time of flight of 3.7×10^{-11} s to reach the anode 1 cm away. The positive ions produced at $d = 10^{-4}$ cm spend more time in front of the cathode.

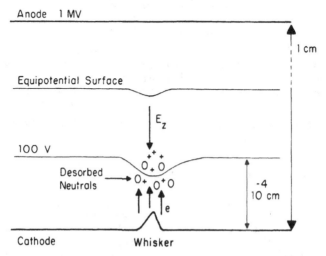

Fig. 4. Ionization of desorbed neutral particles near the 100 V equipotential surface where the ionization cross section has a maximum value.

The build-up of positive space charge near the tip of the electron emitting spot enhances E and thus strongly j_{FE} which will further increase the ionization rate and so on. The ions hit the cathode surface with about 100 eV and recombine. The increasing ion bombardment and release of recombination energy lead to surface heating and further release of adsorbed gases[13], i.e., "discharge cleaning". While initially a linear potential drop was assumed between anode and cathode, the increasing positive space charge in front of the electron emitting spot forms a cathode fall potential which implies a higher E. The maximum ionization zone, where the electrons have about 100 eV, moves more closely to the cathode surface and into a region of higher neutral density. As the ionization rate further increases, a double layer forms between the ions moving from the ionization zone to the cathode and the electrons moving to the anode. The electric field of the double layer reduces the externally applied E_{ext} between the ionization zone and the anode until the electron flow becomes space charge limited j_{CL}. A plasma has formed which now screens E_{ext} from reaching the cathode. However, the plasma in "contact" with the cathode still forms a positive space charge sheath of width given by

$$\lambda_d = (\epsilon_0 V_P / n_e e^2)^{1/2} \qquad (4)$$

where V_P is the plasma potential. The sheath electric field $E_s \approx V_P / \lambda_d$ controls now the electron emission. Essentially the same situation exists for a laser produced plasma in contact with a metal surface. It forms a positive space charge sheath with a width given by the Debye length and $E_s = V_f / \lambda_D$ where V_f is the floating potential, Eq. 2. E_s ignites unipolar arcs. As the plasma density increases, so does the surface heating of the cathode spot by ion bombardment. However, for the considered one dimensional geometry the electron emission is limited by j_{CL}. In reality, the electrons can flow sideways and even back to the cathode surface as explained by the unipolar arc model.

FORMATION OF CATHODE SPOTS BY UNIPOLAR ARCING

Electron emission from a spot on the surface and desorption and ionization of adsorbates lead to the formation of a small dense plasma blob above the electron emitting spot (Fig. 5). The increased plasma pressure P_e above the cathode spot leads to a pressure gradient and an electric field E_r in radial direction, tangential to the surface.

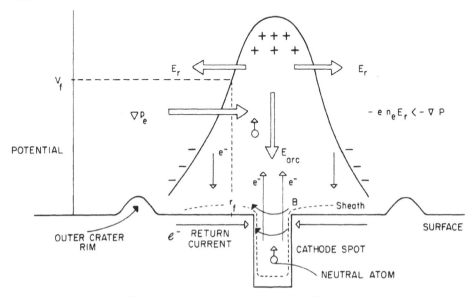

Fig. 5. Unipolar Arc Model.

Without any current flowing this field would be the ambipolar electric field

$$\vec{E}_{amb} = -\nabla P_e / en_e .\tag{5}$$

Associated with this field the plasma potential decreases in radial direction. Consequently, the plasma sheath potential is also reduced in a ring-like area A around the cathode spot. The sheath forms as the radially expanding plasma sweeps over the metal surface. The sheath potential distribution will be such that the quasineutrality of the plasma cloud is assured. At some radial distance r_f from the cathode spot the sheath potential will be equal to the floating potential, Eq. 1, providing equal ion and electron flow rates to the surface at this location, i.e., the net current through the sheath is zero,

$$i_s = +en_i \ A \left[\frac{kT_e}{M_i}\right]^{1/2} - \frac{1}{4} \ |e| \, n_e \bar{v} \ A \left\{\exp\left[-\frac{|e| V_f}{kT_e}\right]\right\} = 0 \tag{6}$$

$\bar{v}$ is the average magnitude of the speed of the electrons. The first term corresponds to the ion saturation current which remains essentially constant, independent of the sheath potential for $V_s > 0$ (Fig. 6). The second term gives the electron flow through the sheath. At distances $r \lessgtr r_f$ the plasma potential, and thus the sheath potential $V_s \gtrless V_f$, and the electron flow for a Maxwellian velocity distribution will change exponentially. The net wall current density is then $j_s > 0$ for $V_s > V_f$, and $j_s < 0$ for $V_s < V_f$, while $j_s = 0$ for $V_s = V_f$.

$$j_s = -\frac{1}{4} \ |e| \, n_e \bar{v} \left\{\exp\left[-\frac{|e| V_s}{kT_e}\right] - \exp\left[-\frac{|e| V_f}{kT_e}\right]\right\} \tag{7}$$

Electron   Ion
Current Density

The ion saturation current has been expressed by Eq. 6. $j_s > 0$ implies that more ions than electrons leave the plasma and the current is in direction of the sheath electric field $\vec{E}_s$, from the positive plasma potential to the wall at zero potential. $j_s < 0$ means that more electrons than ions reach the wall. In the conventional sense, $\vec{j}_s$ is in the opposite direction to $\vec{E}_s$. The conventional current flows "uphill", from zero potential at the wall to the reduced but still positive $V_s < V_f$. At the front of the radially expanding plasma V_s will approach zero, i.e., the radial potential drop ΔV from r_f to the outer edge of the plasma will be $\Delta V(r) \simeq V_f$. From the radial pressure gradient, Eq. 5, the amount of sheath reduction is found by integrating the radial electric field distribution resulting in $\Delta V(r) = [kT_e/e]\ln[n_e(r_f)/n_{eo}]$. $n_e(r_f)$ is the plasma electron density at r_f, where $V_s = V_f$. $n_{eo}(r)$ is the plasma electron density at the outer edge. Equating ΔV with the sheaths floating potential, Eq. 1,

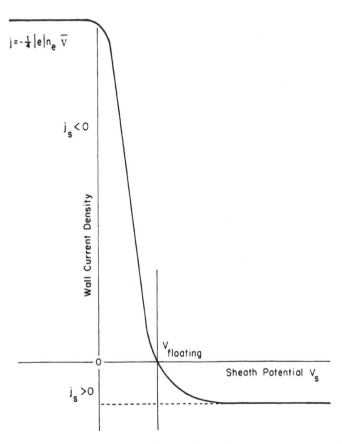

$j = -\frac{1}{4}|e|n_e \overline{v}$

$j_s < 0$

Wall Current Density

0

$V_{floating}$

Sheath Potential V_s

$j_s > 0$

Fig. 6.   Current density from the plasma to the wall as
function of sheath potential V_s.

$$\Delta V = V_f$$

$$\frac{kT_e}{e} \ln \frac{n_e(r_f)}{n_{e\,0}} = \frac{kT_e}{2e} \ln \frac{Mi}{2\pi m_e}$$

we find that independent of the electron temperature the sheath potential can approach zero when

$$\frac{n_e(r_f)}{n_{e\,0}} = \left[\frac{Mi}{2\pi m_e} \right]^{1/2}$$

In this case, the electron return current to the surface is determined by the electron saturation current $i_s = -(1/4)|e|\bar{n}\bar{v}A$ (Fig. 6). Actually, the increased flow of electrons to the surface at $r > r_f$ due to $V_s < V_f$ implies a reduction of the negative space charge. This results in a reduced radial electric field $E_r(r) = -\Delta V/\Delta r < E_{amb}$. Consequently, the outward directed electron pressure gradient force becomes larger than the electric field force which holds the electrons back $|-\nabla P_e| > |-en_e \vec{E}_r|$. A net force, $\vec{F}_{net}$, acting on the electron fluid is pointing outward in radial direction

$$\vec{F}_{net} = -\nabla P_e - |e|n_e \vec{E}_r \tag{8}$$

This is the driving force of the unipolar arc. The current density j of the arc follows from the equation of motion for the electron fluid $-|e|n\vec{E} - \nabla P_e + |e|n\vec{j}/\sigma = 0$. σ is the electrical conductivity. The inertial term is assumed insignificant over the time scale of the arc. Solving for $\vec{j}$ and assuming a constant kT_e, independent of r, we find

$$\vec{j} = \sigma \left[\frac{\nabla P_e}{en_e} - \frac{dV}{dr} \right] = \sigma \left[\frac{kT_e}{|e|n_e} \frac{dn_e}{dr} - \frac{dV}{dr} \right] \tag{9}$$

Again, $j = 0$ if the expression in the parentheses is zero, i.e., $-dV/dr = E_{amb}$. It is essentially the mass difference between electrons and ions which causes a difference between the two terms on the right hand side. Due to the condition of quasineutrality, the plasma density profile is determined by the dynamics of the slow ions, $dn_e/dr = dn_i/dr$, while dV/dr is essentially determined by the more mobile electrons. From Poisson's law follows that small changes of the electron density lead to a change of the curvature of the potential curve,

$$\frac{d}{dr} \left[\frac{dV}{dr} \right] = \frac{|e|}{\epsilon_0} (n_e - n_i). \tag{10}$$

For a strictly one dimensional density profile the curvature of the associated potential profile is proportional to $(n_e - n_i)$. It is concave downwards if $n_i > n_e$, Fig. 7. At the location r_f where $n_i = n_e$, $(d^2V/dr^2) = 0$. The curvature is concave upwards for $r > r_f$, i.e., $n_e > n_i$. In the two-dimensional geometry of the unipolar arc, the loss of electrons to the wall due to the reduced sheath potential $V_s < V_f$ in the region $r > r_f$ leads to a slight

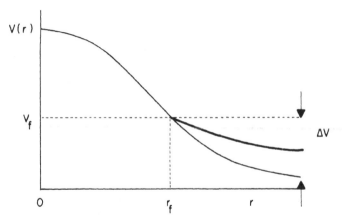

Fig. 7. Plot of plasma sheath potential as a function of
radial plasma dimension. Light curve is without
electron losses to the surface. Heavy curve is
with electron losses. The electron emitting
cathode spot is at r = 0. r_f gives the location
of the floating potential V_f.

decrease of n_e by Δn_e. Consequently the curvature will be
reduced in this region.

$$\frac{d^2 V}{dr^2} = \frac{e}{\epsilon_0} \left[(n_e - \Delta n_e) - n_i \right] > 0$$

still $|n_e - \Delta n_e| > n_i$. The more pronounced curve in Fig. 7
schematically shows this change of the potential profile.
As more electrons flow from the plasma to the surface,
$\Delta V(r)$ is decreased and the second term on the right hand
side of Eq. 9 $|dV/dr| < |(kT_e/en_e)(dn_e/dr)|$. Since
$(dn/dr) < 0$ a $\vec{j} < 0$ results, i.e., $\vec{j}$ has the same sign as
the current through the sheath $j_s < 0$ in the region $r > r_f$.
The current driving term in Eq. 9 is equivalent to the emf
of a battery. The "internal" resistance of the plasma
battery implies that $|dV/dr|$ decreases with increasing
current. This positive feedback mechanism leads to a rapid
increase of $\vec{j}$. $\Delta V(r)$ and the sheath potential $V_s(r)$ will
adjust in the region $r > r_f$ to values which provide a
continuous, increasing current flow as more neutrals become
ionized thus increasing the driving ∇P_e term. The
increasing temperature of the surface of the electron
emitting spot by ion bombardment and ohmic heating leads
then to thermionic electron emission and a large arc
current begins to circulate. This is the mechanism by
which a cathode spot forms. The high current density of
the unipolar arc leads to "explosive" plasma formation.
The sequence of events leading to the explosive formation
of a cathode spot is summarized in Fig. 8.

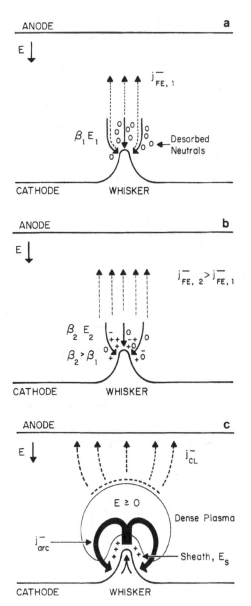

Figs. 8 a, b, c. Transition from field emitted electron
current density j_{FE}^-, Figs. 8a and 8b,
to space charge limited electron flow
j_{CL}^- and electron arc current j_{arc}^- in
Fig. 8c. $j_{FE}^- < j_{CL}^- \ll j_{arc}^-$. β is the
electric field enhancement factor, and
E_s the sheath electric field.

EXPERIMENTS

In order to determine whether the vacuum diode breakdown mechanism is cathode spot formation via unipolar arcing, a Model 112A Pulserad generator was used with diode voltages between 0.84 - 1.1 MV, gap spacing 2 - 2.5 cm, E = 2.9 - 5.8 x 10^5 V/cm, diode currents 14-21 kA and a pulse length of 20 ns FWHM[14] and [15]. The cathode was a stainless steel cylindrical rod with a diameter of 3.18 cm. The cathode was repolished metallographically using a fine slurry of 0.05 μm Al_2O_3 after each shot. Characteristic unipolar arcing craters were observed on all exposed cathodes. However, as expected, the higher voltage shots displayed a greater density of craters and a higher degree of cathode surface melting than did the lower voltage shots. Fig. 9 shows the cathode surface after one low voltage shot. The dark area is the undamaged polished metal. The light areas are the sections of surface that underwent arcing. Overlapping unipolar arc craters are visible in the heavily damaged area. Individual craters (not visible) appear in the fringes of the light areas at the limit of the plasma formation. Fig. 10 shows a close-up of unipolar craters on a cathode surface after one 20 ns FWHM duration shot. Grain boundaries are visible on the undamaged background area. This should be compared with Fig. 2 which is a close-up of laser induced unipolar arcs on the surface of a stainless steel target taken after one shot from a Neodymium laser of 25 ns FWHM with an average irradiance of 5.4 MW/cm^2. Fig. 11 shows a cathode surface after a similar shot at a higher voltage. Fig.s 12a and 12b show respectively the areas near the edges of heavily damaged section on the cathode of Fig. 9 and on a target exposed to unfocused laser radiation. Figs. 13 and 13b show individual laser and diode produced unipolar arc craters. The structure and size of the crater rim and cathode hole are essentially identical.

Cathode surfaces exposed to the greatest diode potential difference (1.1 MV) exhibited the most extensive cratering. The exposed areas were 100% cratered with many craters overlapping and severe surface melting in evidence (Fig. 11). Cathode surfaces which were exposed to a lower potential difference (0.7 MV) exhibited the least cratering. As can be seen in Fig. 9, approximately 15% of these cathode surfaces were cratered with surface melting evident only in the most heavily damaged area. In the lightly damaged area of Fig. 10 individual craters are visible on an otherwise undamaged surface.

This experimental result implies that unipolar arcing is the primary breakdown process. The similarity of the crater patterns on all of these surfaces leads to the conclusion that the cathode spots in each case were formed by the unipolar arcing process, and that plasma formation by this process leads to vacuum breakdown.

CONCLUSIONS

Unipolar arcing can supply the necessary current density

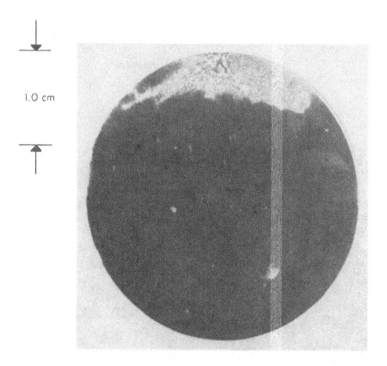

i.0 cm

Fig. 9. Photograph of diode cathode surface after one
discharge of 0.7 MV at 1.91 cm gap spacing,
E=3.7x10^5 V/cm, and a diode current of 14 kA.

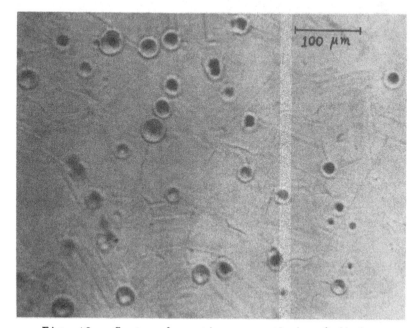

100 μm

Fig. 10. Crater formation on cathode of diode.

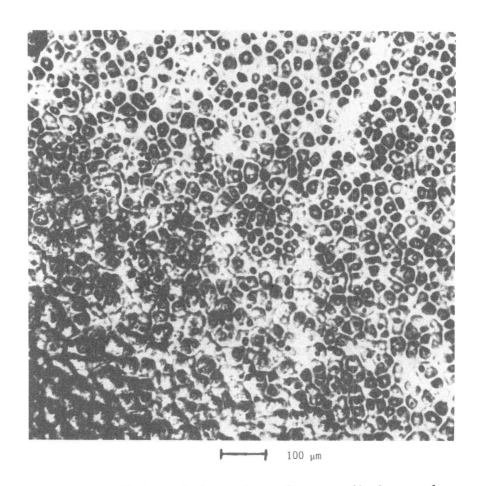

├────────────┤ 100 μm

Fig. 11. Diode cathode surface after one discharge of
1.1 MV at 2.54 cm gap spacing, E = 4.3 x 10⁵
V/cm, and a diode current of 21 kA.

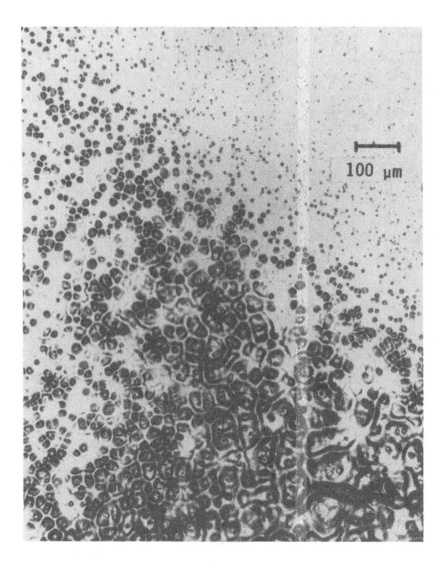

Fig. 12a. Unipolar arc craters near the edge of the
heavily damaged area of the cathode of Fig. 9
single small craters can be seen in the upper
right hand corner.

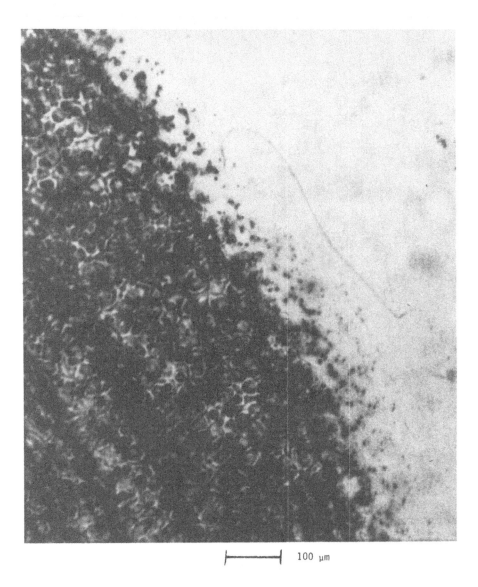

100 μm

Fig. 12b. Laser produced unipolar arcing.

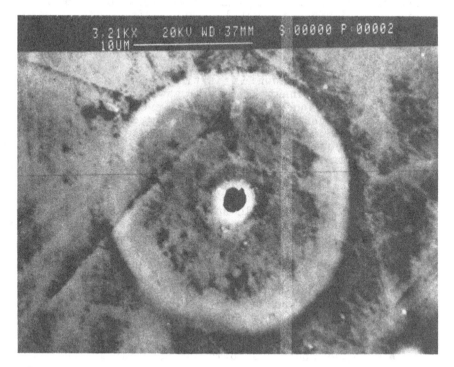

Fig. 13a. Unipolar arc crater on cathode surface.

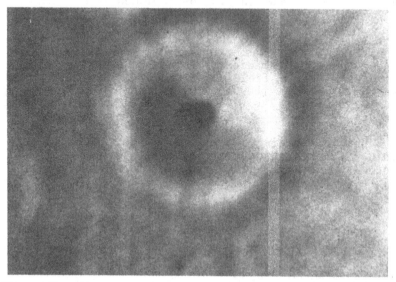

Fig. 13b. Laser produced arc crater. Rim to rim
diameter 5.5 μm, cathode spot size 0.7 μm.

for explosive plasma formation and electron emission from a cathode surface spot in a nanosecond time frame. Unipolar arcing does occur on the surface of a vacuum diode cathode with an externally applied electric field, and it is remarkably similar to the interaction between a laser produced plasma and a metal target surface with no external field. The following conclusions are also made:

- Unipolar arcing is the primary breakdown process.

- Surface breakdown is initiated by the ionization of desorbed contaminants, mostly gases, by field emitted electrons. Since this requires energy deposition only within a few contaminant monolayers at a time instead of an entire whisker volume, and since the neutral contaminants are only loosely bound to the surface by relatively weak Van der Waal forces, the onset of surface breakdown by this mechanism requires much less current and energy than vaporization and ionization of the entire cathode whisker.

- The localized build-up of plasma above an electron emitting spot naturally leads to pressure and electric field distributions which drive the unipolar arc.

- Since the external field is screened from the cathode surface, the whisker current is no longer determined by the space charge limited current. Hence the unipolar arc current density can be many orders of magnitude greater than the Child-Langmuir space charge limited diode current density.

- Cathodes spots are formed by unipolar arcing.

- The large variation in published values for cathode spot current densities can be explained by the fact that the unipolar arc current is not included in measurements of the overall diode current density.

- The closure of the diode gap results from the expansion of the plasma ions under the influence of the plasma pressure gradient. This can occur since the external electric field is effectively screened from the dense plasma when the current to the anode becomes space charge limited. The ions thus see no retarding external field.

- Unipolar arcing represents a basic form of a small scale discharge which contributes to phenomena like laser induced breakdown and formation of cathode spots in a unique way.

ACKNOWLEDGEMENTS

This work was sponsored by the Naval Research Laboratory and the Naval Postgraduate School.

REFERENCES

1. V. I. Rakhovsky, IEEE Transactions on Plasma Science, PS-12, 199 (1984).
2. E. A. Litvinov, G. A. Mesyats, and D. I. Proskurovski, Sov. Phys. Usp. 26, 138 (1983).
3. G. N. Fursey, IEEE Transactions on Electrical Insulation, EI-20, 659 (1985).
4. V. I. Rakhovsky, IEEE Transactions on Plasma Science, PS-15, 481 (1987).
5. B. Jüttner, IEEE Transactions on Plasma Science, PS-15, 474 (1987).
6. F. Schwirzke, J. Nucl. Mat. 128 & 129, 609 (1984).
7. "Laser-Induced Plasmas and Applications", L. J. Radziemski and D. A. Cremens, eds., Marcel Dekker, Inc., New York and Basel (1989).
8. F. Schwirzke, Unipolar Arcing, A Basic Laser Damage Mechanism, in: "Laser Induced Damage in Optical Materials", H. E. Bennett, A. H. Guenther, D. Milam and B. E. Newnam, eds., U.S. Department of Commerce, National Bureau of Standards, Special Publication 669, pp. 458-478 (January 1984).
9. F. Schwirzke, Laser Induced Unipolar Arcing, in: "Laser Interaction and Related Plasma Phenomena", Vol. 6, pp. 335-352, H. Hora and G. H. Miley, eds., Plenum Publishing Corporation, New York (1984).
10. J. Halbritter, IEEE Transactions on Electrical Insulation, EI-18, 253 (1983).
11. J. Halbritter, IEEE Transactions on Electrical Insulation, EI-20, 671 (1985).
12. E. Hantzsche and B. Jüttner, IEEE Transactions on Plasma Science, PS-13, 230 (1985).
13. F. Schwirzke, H. Brinkschulte and M. Hashmi, J. Appl. Phys. 46, 4891 (1975).
14. S. A. Minnick, "Unipolar Arcing on the Cathode Surface of a High Voltage Diode", M. S. Thesis, Naval Postgraduate School, December 1989.
15. F. Schwirzke, X. K. Maruyama and S. A. Minnick, Bull. Am. Phys. Soc. 34, 2103 (1989).

NEW BASIC THEORY FROM LASER-PLASMA DOUBLE LAYERS:

GENERALIZATION TO DEGENERATE ELECTRONS AND NUCLEI[*]

H. Hora[a,b], S. Eliezer[a,c], R.S. Pease[a,d], A. Scharmann[b], and D. Schwabe[b]

a) Department of Theoretical Physics, Univ. NSW, Sydney, Australia
b) 1st Institute of Physics, Justus-Liebig-University, 6300 Giessen, Germany
c) SOREQ NRC, Yavne 70600, Israel
d) Culham Laboratories, Abingdon, Oxon, England

ABSTRACT

The study of laser-plasma interaction resulted in several extensions and new developments of plasma theory including the general formulation of the nonlinear force of laser-plasma interaction, the importance of collisions, quantum collisions and to the discovery of dynamic internal electric fields and double layers in inhomogeneous plasmas. The resulting surface tension in cavitons and at plasma boundaries (due to the faster emitted electrons) results in stabilization against Raleigh-Taylor instability. The same occurs with the degenerate electron gas within the ion lattice of a metal: the electrons try to leave the ion lattice with the Fermi energy until a double layer is being built up. The resulting surface tension immediately agrees with measured values from metals. This can be applied to explain the driving surface force of the Marangoni flow by the flow the electrons in the electron layer swimming at the metal surface. We further derive Hofstadter's charge decay in nuclei from a Debye length and calculate the measured surface energy of nuclei.

1. INTRODUCTION

Plasma theory generally is considered to be well established [1] based on - admittedly - less fascinating classical physics only, and apart from some minor peripheric modifications, one may usually be satisfied about the existing theories of plasma dynamics, instabilities, radiation processes and thermal properties including the equation of state [2]. Contrary to this rosy picture, only few of the plasma physicists are disturbed by the fact that not all observations of plasmas can be covered by theory. This discrepancy is taken to be not dramatic since one is aware that plasma physics is very complicated and one is not

[*] Supported by the WE-Heraeus Foundation

worried similarly to the situation in meteorology when the observations of the weather can rather not be followed up by theoretical models.

This complicated situation should not prevent us to ask how the basic laws of plasma theory may need changes or even fundamental corrections. The study of laser produced plasmas is a typical case to learn how new theory had to be developed. One example is the fact that intense laser radiation irradiating a plasma surface is reflected in a stuttering way, changing the reflectivity between few and nearly 100% irregularly with periods of 10 to 30 picoseconds. As predicted numerically (see Fig. 7.7a and b of Ref. 3) and measured by M. Lubin [4,5] and re-discovered as a consequence of the observed periodicity of scattered spectra [6] in many details by Luther-Davies [7].

This fact was the long years trouble in laser fusion for which the ingenious way out for smooth transfer of optical energy to plasmas was given by Nuckolls [8] to convert the laser radiation into x-rays and let them drive the laser fusion (indirect drive). The stuttering interaction is now overcome by the use of random phase plates (RPP), induced spatial incoherence (ISI), or broad band or fly eye optics [9] such that the very efficient smooth direct drive compression of laser irradiated pellets is possible in future. The suppression of the pulsation by the standing wave and density ripple mechanisms [Fig. 7.7a & b of Ref. 3] and its hydrodynamic relaxation [10,11] confirms the nature of the pulsations and the reason for its suppression. It was clarified that the appearance of instabilities are not primarily of influence for this process.

A less dramatic and more conservative development of plasma theory was accompanied with the study of laser plasma interaction in order to gain our basic knowledge in plasma physics from laser-plasma interaction. There were not only relativistic effects to be included, e.g. when the oscillation energy of the electrons in a very intensive laser field arrived or exceeded the rest mass energy mc^2 of the electrons [12] also a basic quantum effect of plasma resistivity appeared about which more will be dealt below.

Another basic property was related to the internal electric fields and double layers in plasmas at their surfaces and in any place of inhomogeneity as seen from the studies of laser produced plasmas [13]. These fields in the plasma interior seemed to be like a heresy against the plasma fundamentalists since it is assumed from the times of Tonks and Langmuir that plasmas are - apart from microscopic fluctuations - space charge neutral and practically all preceding theory was based on this theorem. How difficult this was to be taken by the established plasma theorist can be seen from the statement of Kulsrud [14] when refereeing a book of Hannes Alfven [14] where evidence was given about the internal electric fields in the inhomogeneous space plasmas, that the sentence was printed "Alfven's electric fields whose origin is not intuitively clear".

This paper reports on the recent developments of the double layers embedded into the general development of plasma theory due to laser-plasma interaction where the recognitions of the double layers came from. This development not only advanced and improved plasma theory it was also possible recently to use these results of plasma models to explain properties of solids, the surface tension of metals, and even more, the surface tension of nuclei in a straightforward way in agreement with experimental values. The last application to nuclei however opened another problem to nuclear theory, as far as the role of the electron charge or the fine structure constant appears while these should not be involved with nuclear forces.

2. NEW PLASMA THEORY DERIVED FROM LASER INTERACTION

The more than 20 years conferences LASER INTERACTION AND RELATED PLASMA PHENOMENA of this series of proceedings contains nearly all steps continuously developing the basic equations of plasma description. A short summary will be presented first before the double layer physics originating from these developments will be described. For some details discussion, reference is given to a preceding paper [16].

a. Nonlinear Force

The theoretical methods in treating fully ionized plasmas are given by three different ways. The first one uses the computational description of the motion of a large number N of particles, e.g. 1 million electrons and ions, where however the use of Coulomb collisions is highly restricted if not at all neglected. Electric fields are well included and all double layer effects result automatically though not much attention has been given to this property. Only very recently with the biggest computers, the collisions were included in some rudimentary form. The second concept is to summarize the N-particle problem into a distribution function for which the Boltzmann or Vlasov equation determines its change by collisions. While this model, called kinetic theory, is able to treat very specific problems very successfully, its general formulation with nonlinear terms and a general collisional term is very complex, and mostly very limited approximations can be treated in a sufficient transparent way.

The momentum integrals of these kinetic equations arrive at the macroscopic hydrodynamic equations, the equation of continuity, of energy conservation, and of the conservation of momentum or the equation of motion. It is remarkable that it was not the integration of the Boltzmann equation but an algebraical evaluation and appropriate intepretation of the Euler equations for electrons and ions which lead Schlüter [17] to an equation of motion of a plasma given by the force density.

$$f = m_i n_i \left(\frac{\partial}{\partial t} \underline{v} + \underline{v} \cdot \underline{\nabla} \underline{v} \right) = - \nabla P + \frac{1}{c} \, \underline{j} \times \underline{H} + \frac{\tilde{n}^2 - 1}{4\pi} \, \underline{E} \cdot \underline{\nabla} \underline{E} \qquad (1)$$

where $\underline{v}$ is the plasma velocity, m_i is the ion mass n_i is the ion density, P is the thermic pressure, $\underline{j}$ is the electric current density, $\underline{H}$ the magnetic field and $\underline{E}$ the electric field in plasma. The complex refractive index $\tilde{n}$ is given by

$$\tilde{n}^2 = 1 - \omega_p^2/\omega^2(1 - iv/\omega) \qquad (2)$$

where ω_p is the plasma frequency, v is the collision frequency and ω is the frequency of the incident laser radiation. The last nonlinear term in Eq. (1) was expressed by current densities only and the conversion to electric fields is possible only in the case of a periodic electric field [12,18]. This nonlinear term did not appear in the derivation from the Boltzmann equation [19] and is called "the Schlüter term".

Equation (1) was derived strictly under the conditions of space charge neutrality. When applying Eq. (1) to derive the forces in laser produced plasmas an insufficiently appeared at oblique incidence of the laser radiation: for collisionless plasma, a net force appeared along the plasma surface what could not be possible because of momentum conservation. It was then possible to derive the missing nonlinear terms for the force density [12,18] from the criterion of momentum conservation. The part of the force density $\underline{f}$ which was not due to the gasdynamic pressure (the thermokinetic term in Eq. (1)) but exclusively given by the electrodynamic quantities $\underline{j}, \underline{H}$ and $\underline{E}$, which terms are all of quadratic form, is called the nonlinear force [12,18,21]

$$f_{NL} = \frac{1}{c} \underline{j} \times \underline{H} + \frac{1}{4\pi} \, \underline{E} \, \underline{\nabla} \cdot \underline{E} + \frac{1}{4\pi} \left(1 + \frac{1}{\omega} \frac{\partial}{\partial t} \right) \underline{\nabla} \cdot \underline{E} \, \underline{E} \, (\tilde{n}^2 - 1) \qquad (3)$$

The proof of completeness of the stationary solution (3) - with no temporal derivative in Eq. (3) - was established on the mentioned momentum conservation for oblique incidence of laser radiation on plasmas [18], while the correct transient formulation with the temporal derivative was subject to a long years controversy [20]. Its final form in Eq. (3) first derived in 1985 [21] was checked to be of general validity from its Lorentz and gauge invariance [22]. The last term in Eq. (3) splits into three by differentiation of the triple product; one term is the Schlüter term, two others were new.

The nonlinear force contains terms which are the former known ponderomotive force and further nonponderomotive terms not known before 1969 [12,18]. What was basically new in the derivation of Eq. (3) was that the electric field term appeared after the Lorentz term which intentionally had been neglected by Schlüter because of space charge neutrality. This was a first indication that at least for high frequency fields, space charge neutrality was no longer possible, otherwise the neglection of this term violated the described momentum conservation.

b. Refractive Index and Need Not to Neglect Collisions

The refractive index of Eq. (2) led to some confusion about the presumption of its derivation. The model of Schlüter's derivation [17] was based on the hydrodynamic equations for electrons and ions and the resulting electrical current density (Schlüter's second equation apart from Eq. (1) which is a generalized Ohm's law) is simply a result of hydrodynamics. Its combination with the Lorentz form of the Maxwellian equations (i.e. using pure vacuum and only positive and negative charges or dipoles in this vacuum such that all properties of dielectric or magnetic response are suppressed) however led to the appearance of a dielectric constant as given by the refractive index (2) [23]. The discrepancy to the conceptual assumptions was easily suppressed by the result, that this was just representing the measured values of the dielectric constants as measured by microwave or optically from plasmas.

A question was further what collision frequency (v_c) has to be used. The value of v_c in Schlüter's initial assumptions was given from the viscosity type of collisions between the electron and ion fluid. It was no surprize that this collision frequency determined the direct current electric resistivity in plasmas in agreement with electrical measurements in plasmas. The question was whether this could be generalized to the collisions in high frequency fields. It was then remarkable that the use of this dc (or viscosity) collision frequency taken for evaluating the optical constants of plasmas for laser irradiation in 1964 [14] based on Spitzer's theory [19] including the limitation of invalidity for low Coulomb logarithms and under exclusion of degenerate plasmas, led to optical absorption constants for low density plasmas without collective effects as exactly given by the quantum electodynamical QED theory. The higher density case with collective effects and with the appearance of the metallic optical constants of plasmas was possible only on the basis of plasma theory [12].

The QED theory tracing back to Gaunt 1932 derived [12] the inverse bremsstrahlung of colliding electrons with fully ionized ions in a plasma by evaluating the free-free transitions as needed in astrophysics. The experimentally correct result is the following dependence on the plasma temperature T dependence of the collision frequency of the kind of

$$v \approx T^{-3/2} \qquad (4)$$

appeared only in the QED treatment if stimulated emission was included. Neglecting stimulated emission, the collision frequency would have been

$$v \approx T^{-1/2} \qquad (5)$$

This correlation of the results with Spitzer's theory was a remarkable connection of the ordinary v_c collision frequency with the general QED theory. The numerical agreement - apart from unimportant logarithmic factors [24] between the differently derived theories was unimportant. The values agreed with the optical measurements.

While the neglection of collisions was nonsurroundable in most cases of the N-particle simulation or in kinetic theory it led to dangerous situations in the theory of laser produced plasmas. Since the most important energy transfer in plasmas by collisions appears near the critical density, where the poles of the functions for the refractive index are located, strong gradients in the functions and in the derivatives are significant. The change of the density by few percents near the critical density in a plasma of 10 keV temperature results in a change of the absorption constant by orders of three magnitudes (see Fig. 6.3 of Ref. 12). In this case collisionless theory is worthless.

Another drastic example [12,13,16] is the effective refractive index at the Försterling-Denisov resonance absorption [25] which has a pole at minus infinity. Adding a tiny amount of absorption, e.g. by Landau damping, causes a jump of the function from minus infinity to nearly plus infinity (see Fig. 11.6 of Ref. 12).

While this resonance absorption was a long years topic to try to explain the complex mechanisms of laser-plasma interaction it could not explain resonance mechanisms at perpendicular incidence. The Försterling-Denisov resonance absorption appeared only - if at all - for oblique incidence of p-polarized electromagnetic waves. There were experimental indications of resonance mechanisms at perpendicular incidence and all kinds of models were vaguely introduced to assume that the interaction interface had a lateral ripple just to get some oblique incidence.

As we have derived later on the basis of the double layers, there is indeed a very strong new resonance at perpendicular incidence [26] in the superdense plasma which can be reached by laser fields if a very steep density gradient is produced at nonlinear force dominated laser-plasma interaction as confirmed by numerous experiments [27]. This resonance again essentially depends on the collision frequency. Its neglection, however, runs into poles of the involved functions such that only a careful inclusion of damping arrives at realistic results.

c. Quantum Collisions

It was remarkable to see from the QED theory that there was a basically different dependence of the optical absorption on the plasma temperature when stimulated emission was included or not. See Eqs. (4) & (5). This is a remarkable phenomena and would not have appeared in such a drastic form if not a hefty discussion about the correct absorption constants would have risen by Nicholson-Florence [28] where some misunderstanding was so helpful to drastically highlight the situation.

What is clear [29] is that the absorption of photons of much higher energy than the plasma temperature should follow the absorption without stimulated emission, Eq. (5).

Such cases, however, cannot be easily measured. It would mean that e.g. x-ray laser beams with keV photon energy are transmitting plasmas of an electron density of more than 10^{25} cm^3 with a plasma temperature of 50 eV. The normal case is the other way round. Laser photons of few eV penetrate plasma of appropriately high densities of temperatures in the 100 eV or more range.

What is remarkable is that the beam has a stronger penetration than in the case without stimulated emission. There does exist stimulated emission which prevents a strong absorption, though there is no population inversion (sic!) but only equilibrium Maxwellian thermal electron distributions. This is a rather remarkable case of stimulated emission without inversion, in which connection the result by Fearn & Scully [30] should be mentoned where laser emission is discussed without population inversion. What was searched for before the laser was verified, to produce stimulated emission, is being done automatically in plasmas at thermal equilibrium in the same way as a stimulated emission is being produced by equilibrium blackbody radiation resulting in Planck's radiation law.

Being aware of these relations and profound connections, one has to understand the result that the classical Coulomb collision of electrons with fully ionized ions follows the relation [4] ignoring at this stage the modification by Spitzer's Coulomb logarithm

$$v = \frac{Zn_e\pi e^4}{3^{3/2}m^{1/2}(kT)^{3/2}} \quad = v_{class} \tag{6}$$

where Z is the charge of the fully ionized ions, n_e is the electron density, e the electron charge, m the electron mass and k Boltzmann's constant in agreement with Eq. (4). This derivation, [24] is very primitively given from the 90 degree hyperbola for the motion of an electron arriving at a central distance r, the impact parameter, towards the z-times charged nucleus.

This is evident only if the electron's velocity (corresponding to the temperature T of the plasma) is small enough. It has been pointed out in words by Bethe [31] and it was evaluated in numbers by Marshak 1941 [32] that this classical Coulomb collision has to change if the plasma temperature is higher when the de Broglie wavelength of the electron becomes the same size as twice the impact parameter. From then on for higher energies, the wave mechanical interaction is stronger resulting in a 90 degree wave diffraction. The change happens at the following conditions resulting in a general formula [12,33]

$$v = \sqrt{\frac{3kT}{m}} \frac{\pi n_e r^2_{Bohr}}{4Z^2\sqrt{1+4T/T^*-1}^2} \tag{7a}$$

which splits into the branches

$$v = \begin{cases} v_{class} & ; \text{ if } T < T* \\ \\ v_{class} \, T/T*; & \text{ if } T > T* \end{cases} \qquad (7b)$$

where the discriminating temperature is

$$T* = 4Z^2 \, mc^2 \, \alpha^2/(3k) = Z^2 \, 38eV \qquad (7c)$$

using the Bohr radius $r_{Bohr} = h^2/me^2$ and the fine structure constant $\alpha = e^2/hc$.

It is remarkable that - as a consequence of Eqs. (7) and (7a) - even for the free-free-collisions of electrons with fully ionized nuclei, the Bohr radius has a significance and that plasmas a high temperature are basically quantum mechanical and no longer classical with respect to their Coulomb collisions. This result however, is not generally accepted. While Spitzer [19] was using the change of classical and quantum mechanical impact parameter correctly as in Eq. (7a) in his Coulomb logarithm, he did not use it for the total Coulomb collision cross section.

At this stage it is a philosophical question whether the classical or the quantum cross section is valid. One may argue that the classical cross section is always valid even if the wave mechanical diffraction would require a larger cross section. The wave mechanical scattering would then be a Mie scattering as known from optics and the shorter classical cross section would survive. Philosophically the situation, however, may be different to the optical Mie scattering analogy. One may argue that always that cross section is valid which is larger. In this case, Eqs. (7) and (7a) have to be taken with the different branches.

Since thermal equilibrium plasmas at temperatures above 100 eV are not long available yet it may be understandable that the quantum branch has not been a point of experiments in the past. In the following we are presenting experimental examples which confirm the quantum properties of plasmas at high temperatures.

Just at the time when the quantum collisions became evident [33], the experiments with the stellarator plasmas were achieved at high temperatures. These experiments of magnetic confinement fusion with zero current plasmas in producing fusion energy were rather transparent and understandable. However the temperatures achieved were very low, about 50 to 80 eV only for a long time. Most of these experiments were closed down, e.g., in Princeton, where L. Spitzer et al had initiated this type of magnetic confinement fusion, after the high current toroidal magnetic field experiments of the Kurchatov Institute in Moscow initiated by Yawlenski and pushed forward finally by Artsimovich turned out to be successful. The excellent heating of such tokamak plasmas to rather high temperatures due to ohmic resistivity should be mentioned later again.

The stellarator experiments were nevertheless continued - despite broad criticisms - by G. Grieger and his team at the Max-Planck-Institute for plasma physics in Garching, Germany based on the high class experimental technics and craftsmanship, and motivated by the attempts to understand the Pfirsch-Schlüter diffusion of the plasma across the magnetic field contrary to the disappointing Bohm diffusion which was considered to be so disadvantageous in Princeton and the other stellarator experiments. Grieger et al succeeded [34] - just before they would have been forced to dismantle their experiments under the pressure of the success of tokamaks and of other constellations - that their deuterium stellarator was producing long confinement times and high temperatures up to keV in a rather transparent conditions of ideally fully ionized deuterium plasma. Nuclear fusion gains were respectable and confirmed the achievements of the high temperatures.

The diffusion across the magnetic field was nevertheless much faster than that of classical diffusion. This diffusion is due to the collisions in plasma otherwise the ideal confinement of plasma in magnetic fields would be rather close to be realized in collisionless plasmas trapping the electrons and ions by the magnetic fields. The collisions disturb this trapping and a diffusion across the magnetic field appears. What was measured in a rather transparent and precise way was that the diffusion across the magnetic field was 20 times faster than the classical value for plasma temperatures of 800 eV. Taking into account Eqs. (7) and (7a), the transition temperature between the low temperature classical case and the quantum collision is just 40 eV. The diffusion is proportional to the collision frequency [33] and the factor T/T^* for the plasma temperature of $T = 800$ eV is then just 20 exactly as measured. This indicates immediately the mechanism of the quantum collisions.

Another example is taken from the measurements of the ohmic conductivity of plasmas. The measurements are highly complex and very rare cases supply the sufficiently accurate information. Taking the case of a hydrogen tokamak at the center line [35], the ohmic conductivity at 2.21 keV electron temperature was measured to be 3.96×10^4 Ohm^{-1} cm^{-1}. Using the Ohmic conductivity according to [33].

$$\eta = \begin{cases} \eta_d & \text{, if } T < T^* \\ \eta_d \dfrac{T^*}{T}, & \text{if } T > T^* \end{cases} \tag{8a}$$

with the use of Spitzer's Ohmic conductivity [Eq. 5.35 of Ref. 19]

$$\eta_d = T^{3/2}/(3.80 \times 10^3 Z \ln\Lambda) \tag{8b}$$

one finds from Eq. (8a) the value of 2.805×10^4 Ohm^{-1} cm^{-1} which is about 30% below the experimental value. However the classical Spitzer value (8b) is 1.27082×10^6 Ohm^{-1} cm^{-1} and therefore more than 30 times higher than the measured value. This is a convincing result in favour of the quantum collisions. In order to save the classical collisions for explaining the resistivity in plasmas, one assumes that there are low concentrations of highly charged heavy ions as contamination in the plasma to reproduce the measured conductivities. Though the whole complicated analysis of the experiments is still not transparent easily and any differentiation is to be discussed, it may have been shown that plasmas at high temperature have an anomalous resistivity given by the quantum collisions, and that this quantum property of the plasma is the reason for the achievement of the high temperature by Ohmic heating in tokamaks.

Another example of tokamak plasmas is given to explain the quantum collision at high plasma temperatures from the measurements of the thermal conductivity parallel to the magnetic field. This direction avoids the complication of the magnetic field and one can then relate simply to conditions of homogeneous plasmas free from the difficulties of double layer effects well known from the reduction of the thermal conductivity by double layers [13]. In these conditions of the experiments the plasmas were homogeneous and no double layers possible [36].

Taking the measurements for thermal conduction parallel to the magnetic field of the mentioned rare and difficult case with a high beta value from Fig. 5 of Ref. [36] (of the inner part of the tokamak only for $r < a/2$), Fig. 1 contains these experimental values of the thermal electron conductivity depending on the electron temperature. This is compared (drawn plots) with the absolute values of the anomalous branch of the generalized thermal conductivity κ of the following equations and the absolute values of the classical κ_c.

$$\kappa = \kappa_c \ \{(T^*/2T) \ [(1 + 4T/T^*)^{1/2} - 1]\}^2 \tag{9}$$

which can be approximated by

$$\kappa = \begin{cases} \kappa_c & \text{if } T < T^* \\ \\ \kappa_c.T^*/T, & \text{if } T > T^* \end{cases} \tag{10}$$

with the classical thermal conductivity

$$\kappa_c = 4.67 \times 10^{-12} \ \varepsilon\delta_T \ \frac{T^{5/2}}{Z \ln \Lambda} \ \text{cal/sec deg cm} \tag{11}$$

where Spitzer's Coulomb logarithm in $\ln\Lambda$ was used. Other measurements for low beta or for the turbulent range $r > a/2$ are basically different with a decreasing dependence on the temperature and other much different behaviour.

The comparison in Fig. 1 shows that the high-beta tokamak is fitting the exact theory within 40%, Eq. (9), quite well in absolute numbers and it well represents the slope of a $T^{3/2}$ law of equation (10). The classical branch (intentionally extended as a plot in Fig. 1 to temperatures above T^* is higher than the measurements by a factor 71 at 1 keV and generally by one or two orders of magnitude and has a different slope of $T^{5/2}$. Though there is no total fit of the experimental values, the difference in the slope and in the absolute values clearly confirms the quantum collisions, if one takes into account the difficulty of the experiments.

It is remarkable that the scaling law of the energy confinement time τ_E for tokamaks derived from the T-11 group [36] apart from a purely geometric factor with the dimensions of the tokamak and a weak factor $q\,(a)$ was reported (using the electron density n_e)

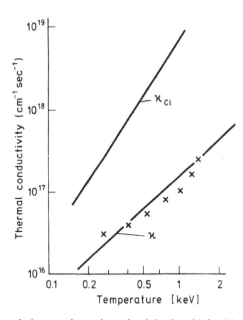

Fig. 1. Measured electron thermal conductivity in a high q(a) = 1.3 tokamak parallel to the magnetic field in the nonturbulent central range (r < a/2) depending on the electron temperautre [36] compared with the quantum generalized thermal conduction κ, formula (9b), in its anomalous conductivity branch. This is in contrast to the strongly deviating classical conductivity κ_c with a clearly different slope and more than 10 times higher values.

$$\tau_E \approx n_e/T^{1/2} \approx \nu \tag{12}$$

which corresponds to the collision frequency ν taking into account its quantum modified anomalous branch, Eq. (7a), contrary to the classical branch given by $n_e/T^{3/2}$. The quantum branch Eq. (11) is valid only for electron temperatures $T > T^*$. The result of Eq. (12) would very directly represent the expectation that the confinement time is proportional to the collision frequency ν responsible for the drift of the plasma across the magnetic field, but given at high temperatures by the anomalous values of equations (7) and (7a).

These examples indicate how experiments - definitely uninfluenced by any theoretical prejudice - may prove the prediction of a change of the classical Coulomb collision of electrons with fully ionized ions into the quantum branch at higher electron energies. The electron energy of this change is given by the electron rest mass and the square of the fine structure constant, Eq. (7b).

d. Double Layers

With reference to reviews about the whole historical development of double layers in plasmas [13] it should be noted that this is a question since the beginning of the physical description of plasmas by Tonks and Langmuir 1929. It was indeed a marginal question related to the surface of plasmas only or to interfaces towards walls of vessels or probes or to injected solid thermonuclear fuel pellets injected into magnetically confined plasmas (see Section 5 of Ref. 12). For the plasma interior, space charge neutrality was assumed and the theoretical models were based on this assumption.

Numerous plasma pioneers were discussing the plasma surface properties and the double layers as e.g. Bohm and Bernstein, and a first transparent picture of a double layer in one dimensional spatial and velocity coordinates was given by Knorr and Goerz [37] as a result that they tried to argue against the results of Bernstein, but arriving then at an excellent agreement and a very much more advanced presentation and a transparent picture. The fully capable experimental verification of the double layer was then established in the double or triple plasma devices of Hershkowitz and collaborators and by colleagues in other centres [38] where low density low temperature plasmas were separated at different voltages by grids and where the potential differences between the parts necessarily produced electric double layers. A realization of double layers and electric fields inside of plasmas was then discovered from numerical investigations of turbulent plasmas [39]. And since then a rather respectable group of plasma physicists represented in special meetings [40] are gaining more and more general interest.

All this was not at all new to the space physicists of the school of Hannes Alfven [41] who even after Langmuir during the forties could not easily accept the space charge neutrality of plasma according to Langmuir since they knew the contrary from the ionospheric plasmas and the solar activities at the earth's poles since the pioneering work of Birkeland. Plasma do have fields in the interior, as long as there are inhomogeneities in density and/or temperature. All this should have been known not only as a marginal property since the existence of the ambipolar fields in plasmas were

$$E = - \frac{1}{en_e} \nabla \frac{3}{2} nkT \qquad (13)$$

were well known e.g. and shown in the classical monograph of Spitzer [19] or as a special term in the generalized Ohm's law, the second Schlüter equation of his space charge neutral two fluid plasma model where the first equation, the equation of motion, was mentioned before, Eq. (1) [17].

Laser produced plasmas touched the questions of surfaces and double layers from the very beginning in 1964 as reported in the monographs [3] and [12]. While the irradiation of solid targets in vacuum with the early spiking laser pulses of up to 100 KW power, produced plasma with temperatures of 30,000 degrees and above, such that ions were emitted with the then thermal energies of few eV, Linlor's first use of the 10 MW Q-switch laser pulses resulted in 5 keV and more ion energies [12]. For the analysis it was assumed that the plasma had well a temperature of few eV only and that the faster expansion of the light electrons in the plasma surface may electrostatically speed up the ions to the keV energies.

The discrepancy came only from the small number of keV ions measured. We had evaluated that the number of ions can only be of the amount of ions in the Debye length thick double layer at the surface of the plasma expanding against the vacuum. This number was about 10^{10} ions or less while Linlor's and the numerous repeated last measurements resulted in 10^{15} and many more ions. Also the emission of electrons was of about 100 mA/cm^2 in agreement with the space charge limitation of the Child-Langmuir law when the laser pulses were 100 kW or less, while the 10 MW laser pulses arrive at electron currents of kA/cm^2 completely beyond any classical limit.

While the acceleration of the ions in the double layer due to the faster escaping electrons was evident theoretically [42] in all details and while later experiments clearly showed the appearance of this group of electrostatically speeded up ions [see Wagli and Donaldson in Ref. 27], this observed number just corresponded to the rather small number we had concluded from the geometry of the Debye length determining the double layer. The fact of the very much higher ion number was the reason to search for another explanation.

This other explanation led to the derivation of the nonlinear force [12,21], where however for the conditions of the low laser powers of 10 MW, the mechanism of the ponderomotive self-focussing had to be derived first. The quantitative result [43] agreed with the just then microscopically measured filaments of self-focussing in gases [44]. This same mechanism could then be concluded in all the laser irradiation experiments using solid targets where the nonlinear forces not only explained the ion energies of up to 10 keV and more immediately, but also the fact that this linear force acceleration mechanisms in the filaments resulted in a separation of the ions by their space charge with a linear dependence of the ion energy on the space charge as measured first by Peacock et al [45] and repeated again and again.

This separation by the space charge resulted from the knowledge [46] of the initial model of the nonlinear force that the laser light in the irradiated inhomogeneous plasma with its decreasing refractive index produced a quiver motion with increasing amplitude and increasing oscillation energy. Due to the phase shift between the E- and the H-vector of the light due to the inhomogeneity as given from the WKB approximation for the laser wave field, a quiver drift of the electron occurs towards the lower plasma density converting the excess of the increased oscillation energy $\varepsilon_{osc,max}$ into translative energy of the electrons. However, since the very dense electron cloud cannot move alone, the ions will be attached electrostatically to follow the electrons, however with a separation of higher and less high ionized ions resulting in ion energies ε_i

$$\varepsilon_i = Z\varepsilon_{osc,max}/2 \qquad (14)$$

where Z is the ion charge. This result has been cofirmed always in a large number of experiments of various groups [12].

While the derivation of the nonlinear force on the basis of the space charge neutral two fluid theory of Schlüter [17], resulting in the then generalized Eq. (3), could not explain the mentioned mechanism of the electron cloud and of the ion cloud by definition, we were waiting long until a direct evidence of the space charge separation mechanism and the expected high electric fields between the electron and ion could be demonstrated directly. This was possible by using one of the first available CD7600 computers for this purpose including several ingenious numerical steps and a basically new solution of the boundary-initial value problem in numerical mathematics by P. Lalousis [47].

It turned out that all plasmas contain electric field in their interior and that starting from any inhomogeneous configuration of an electron and ion cloud hydrodynamically expanding a strong oscillation of the electrons appeared even if no laser field was incident. The computation of this genuine two fluid motion where instead of the degrading space charge limitation the correct Gauss-Poisson equation was used, resulted in a very realistic description of a laser produced plasma, well including the nonlinear intensity dependent optical constants, thermal equipartition between the electron and ion fluid, thermal conductivity, and viscosity for fully realistic real time conditions. The computations were very extensive because of the time steps which had to be less than one tenth of the shortest plasma oscillation - in our case 0.1 fsec or less - and the integration had to go up to 20 psec and more. The results are well known [13,47,48] and led to the appearance of strong dynamically changing double layers with longitudinal electric field amplitude of more than 10^8 V/cm for neodymium glass laser irradiation of 10^{16} W/cm^2.

The caviton generation due to the nonlinear force action was reproduced as known from the numerous preceding numerical calculations beginning with the famous work of Shearer, Kidder and Zink [49]. In the case of the caviton generation, however, the normal and also the inverted double layers appeared [48] in agreement with the just then published measurements [50].

The semi-analytical evaluation of the generated longitudinal electric fields E_s at laser irradiation with electric field amplitudes $\underline{E_L}$ and $\underline{H_L}$ arrived at the following relation [26].

$$
\begin{aligned}
|\underline{E}| = \frac{4\pi e}{\omega^2_p} \Big[& \frac{\partial}{\partial x} \left(\frac{3n_e kT_e}{m_e} + n_e v_e{}^2 \right) \\
& + \frac{\partial}{\partial x} \left(\frac{3n_i kT_i}{m_i} + Zn_i v_i{}^2 \right) \\
& + \frac{1}{m_e} \frac{\partial}{\partial x'} (E_L{}^2 + H^2_L/8\pi)] \left(1 - \exp - \frac{v}{2} t \, \cos \omega_p t \right) \\
& + \frac{\omega_p{}^2 - 4\omega^2}{(\omega_p{}^2 - 4\omega^2)^2 + v^2\omega^2} \frac{4\pi e}{m_e} \frac{\partial}{\partial x} \frac{\partial}{\partial x} (E^2_L + H^2_L) \cos 2\omega t \\
& + \frac{2v\omega}{(\omega_p{}^2 - 4\omega^2)^2 + v^2\omega^2} \frac{4\pi e}{m_e} \frac{\partial}{\partial x} (E^2_L + H^2_L) \sin 2\omega t \qquad (15)
\end{aligned}
$$

where the very first term in the first bracket corresponds to the well known ambipolar field, Eq. (12), modified however that it can oscillate with the local plasma frequency damped by the collision frequency around the former known constant value. All other terms had to be derived newly.

The terms next to the ambipolar field term are well recognizable pressure terms, the static electron pressure, the kinetic pressure (first realized by Dragila [51]) and a kind of "ponderomotive pressure". The last expression is used very frequently but it has to be realized that this expressions is not a conservative term in general therefore the word "potential" is wrong. The last term is the new resonance for perpendicular incidence [26] and the second last term was the discovery [13,52] of the widespread second harmonics emission from the laser irradiated plasma corona which is nearly of uniform intensity despite a strong decay of the plasma density. This was just measured by Aleksandova, Brunner et al

[53] with a periodicity of 0.05 mm and by Gu Min [53] with a periodicity of 4 µm. Both periodicities were immediately reproduced from the theory [52].

Consequences of the convincing and clearly evident double layer effects were the derivation of surface tension in plasmas of considerable amount which stabilize the surface of expanding laser produced plasmas or stabilize against Rayleigh-Taylor instabilities [13]. We derived the fact that inhomogeneous plasma cannot have Langmuir waves, only Langmuir oscillations with locally changing frequency and nevertheless reasonably energy transport in the way of **pseudo-Langmuir waves** [13]. The surface tension also led to a stabilization of surface waves [13]. All these double layer effects were reported in these proceedings or reviewed [13] before. New results are presented in the following section.

3. EXTENSION OF DOUBLE LAYER AND SURFACE TENSION EFFECTS TO DEGENERATE ELECTRONS

In the preceding summary of reviews of double layer effects it was noted that the appearance of surface tension in plasmas was a surprise since surface tension in condensed materials is due to unsaturated cohesion of dipole energies of the atoms and molecules. In a fully ionized plasma, only Coulomb forces are acting between the particles which alone should not cause the mentioned effects of solids. Nevertheless, the electric field energy in a double layer or in any interface of inhomogeneous plasma per area produces surface tension which was discussed before with respect to stabilizing against the Rayleigh-Taylor instability and stabilizing surface waves of a length below a given threshold. We are now applying this plasma surface model to the processes of the degenerate electrons in a metal and shall see that there is an immediate agreement with the measured value.

It should be mentioned that the following considerations of a plasma double layer at metal surfaces has a similarity to the jellium model. The difference, however can be seen immediately when the jellium model is applied to calculate the surface tension of metals where it was shown that the model "fails completely" giving the wrong sign for higher density metals, such as aluminium [54]. Only the introduction of first order pseudo-potential calculations and by additional appropriate electrostatic lattice energies, agreement with measurements was possible. The difference to the following model can be simply seen from the fact that the plasma double layer model immediately arrives at a rather convincing agreement of the theory with the measurements.

Surface tension on condensed matter is produced by the dipoles of the molecules of the matter. For metals, the surface tension is described by special models using the methods of solid state theory. In plasmas, there are only Coulomb forces between the particles and it seems, at first glance, that no surface tension is possible. However, when electrostatic fields are present inside an inhomogeneous plasma, especially at the surface of a plasma expanding against vacuum [3,12,13], one can simply define an electrostatic energy in the volume of the surface range of the plasma divided by the surface area to arrive at a surface tension. This is yet another application of the double layer process pioneered by Birkeland and Alfven [55] in the study of laboratory, space, and astrophysical plasmas.

a. Electric Fields and Electron Pressure Gradients

In the absence of electric currents and of magnetic fields, electron pressure gradients in a plasma are sustained by electric fields (the well known ambipolar fields)[13] where only the electron pressure, P_e, electron density and electron temperature is considered. The electrostatic energy associated with the pressure gradient scale length $L_p = P_e/\nabla P_e$ is $\approx L_p$ $E^2/8\pi$.

The ratio of this electrostatic energy density to the electron thermal energy per unit volume nkT_e is $\frac{2}{3}(\lambda_D/L_p)^2$, where λ_D is the Debye length $(kT_e/4\pi n_e e^2)^{1/2}$. Consequently in circumstances when the scale length of the inhomogeneities approach the Debye length, the electrostatic energy may become significant. A generalization of Eq.(13) for the temporally varying, longitudinal electric field in a plasma along the pressure gradients due to irradiation of electromagnetic radiation of frequency ω and transversal field amplitudes $\underline{E}_L$ and $\underline{H}_L$ is given by Eq.[15] [26]

The electric energy due to the dynamic longitudinal field E of Eqs. (13) or (15) multiplied by L_p has the dimension of energy per unit area. For a uniform mass of plasma surrounded by a narrow boundary region, the surface tension may be defined by

$$\frac{E^2}{8\pi} L_P \cong \sigma_E \tag{16}$$

where σ_E is surface tension. The determination of L_p is a matter of rather detailed calculation, based on models and on computer solutions [47,48,56].

b. Electrical Double Layer Model

A Double-layer model (Nernst's double layer theory) of the edge of a plasma was applied to liquid electrolytes as early as 1903 and 1913 by Gouy [57] and Chapman [58]. In these analyses, local thermal equilibrium was assumed and the ion motion was taken to be determined by ion mobility[59].

The expression for the surface energy in an electrolyte of temperature T and a Debye length λ_d is given [14] by

$$\sigma_E = \frac{1}{8\pi} (kT/e)^2 \lambda_d^{-1} \left[\frac{8}{\pi}\right] [\text{Cosh} (\phi_0 e/2kT) - 1] \qquad (17)$$

where ϕ_0 is the total potential drop across the double layer in the electrolyte.

For the edge of a gaseous plasma (Tonks and Langmuir [60]), ion motion in the electric field is important. If a fixed ion distribution is assumed, ie.,

$$n_i = n_{io} (f(x)), \qquad (18)$$

then the fields can be obtained from the electric potential ϕ. Assuming a plane slab geometry; we have

$$\frac{\partial^2 \phi}{\partial x^2} = 4\pi(n_i Z - n_e)e; \qquad (19)$$

$$n_e = n_{eo} \text{Exp}(-e\phi/kT_e); \qquad (20)$$

$$n_i = n_{io}.f(x) \qquad (21)$$

These equations are then used to evaluate the quantity

$$\sigma_E = \int_{-\infty}^{+\infty} \frac{1}{8\pi} \left(\frac{\partial \phi}{\partial x}\right)^2 \qquad (22)$$

The additional subscript 0 indicates values deep inside the plasma where the potential ϕ and its derivatives are zero. The quantity Ze is the ionic charge. There is no net charge on the plasma, so that the electric fields are zero at $+\infty$ and the electrostatic energy calculated is just that associated with the double layer.

The model used for the double layer calculation is shown in Figure 2. In the first case (Fig. 2a) n_i is a step function and n_e decreases linearly. In the second case (Fig. 2b), the electric field is taken to be a Gaussian function of x with the same maximum value as that found for the first case. These two models yield, respectively,

$$\sigma_E = 0.36 \text{ x } \frac{1}{8\pi} \left(\frac{kT_e}{e}\right)^2 . g^{3/2}/\lambda_d ; \qquad (23)$$

$$\sigma_E = 0.27 \text{ x } \frac{1}{8\pi} \left(\frac{kT_e}{e}\right)^2 g^{3/2}/\lambda_d \qquad (24)$$

Here the total potential drops across the double layer $\phi_{+\infty} - \phi_{-\infty}$ is taken as $g.\left(\frac{kT_e}{e}\right)$, where $(\phi_0 e/kT = g;$ and $[\cosh(\phi e/2kT)-1] \approx \gamma^2/8$ for small g.

Because both these models assume very sharp boundaries for the ion number density, the upper limit of the electrostatic energy associated with the double layer is

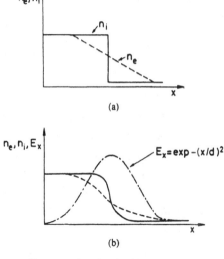

Fig. 2. Density profiles used for the double-layer calculations in equations (10) and (11).

$$\sigma_E \approx 0.3 \times \frac{1}{8\pi} \left(\frac{kT_e}{e} \right)^2 g^{3/2}/\lambda_d \tag{25}$$

c. Metallic Surfaces

At the surface of a metal, the ion density does indeed have a sharp edge, although the ion lattice spacing may be somewhat elongated towards the edge. The electron energy has a maximum of the Fermi energy E_F', and the electrostatic screening length is given by

$$\lambda_{dF}^2 = \frac{2}{3} \left[E_F/4\pi n_o e^2 \right] \tag{26}$$

There must be a potential difference of the order E_F/e at the surface to contain the electron gas. Unlike the Maxwellian distribution, the Fermi distribution cuts off rather sharply at room temperature as there are no high energy electrons to be contained (at least before the surface is irradiated with laser light). Consequently an estimate of the electrostatic energy from equation 25 is

$$\sigma_E = \frac{1}{8\pi} \left[E_F/e \right]^2 . 0.27/\lambda_{dF} \tag{27}$$

where the numerical factor g has to be taken to be close to unity.

Table 1 compares this calculated quantity σ_E with the observed surface tension of both liquid and solid metals.

The purpose of Table 1 is to show that the right orders of magnitude are obtained. Detailed wave mechanical calculation of surface tensions are reported in the literature, [61, 62] which include terms for the distortion of the ion distribution and the exchange and correlation energies of the electrons. The surprisingly close agreement between observed surface tensions and that calculated from Eq. (27) shown in Table 1 is presumably a reflection of the rather simple physics, namely that there must be a potential difference of (E_F/e) to restrain the electrons, and it must act over a distance of about the atomic diameter. The screening length λ_{dF} is about 0.5 A in the above cases. If the ion lattice is strained so that the electric field is reduced, then the strain energy will offset the loss of electrostatic energy.

d. Numerical Calculation of Surface Tension in Laser Produced Plasma

When laser light is applied to the surface of a metal, the first effect, before the ions start expanding, is to raise the electron temperatures and hence the quantity s_E. Thus, the

Table 1. Surface tensions of liquid and solid metals.

Metal	Observed Surface[22]* Tension (ergs/cm^2)	^{S}E from Eq. 14 ergs/cm^2
Li	390	480
Nd	210	190
K	111	90
Al	.948	3700/711+
Cu	1325	1100
Ag	1140	730

---

* Where the surface tension is given in the form s = A - BT where T is temperature, and the quantity A has been taken since BT << A at room temperature.

+ The two numbers assume, respectively, 3 and 1 free electrons per atom.

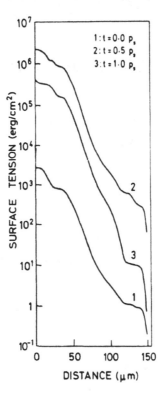

Fig. 3. Surface tension versus distance at times t = 0, 500, and 1000 fs due to the laser produced longitudinal electric fields in a 150 μm deep perpendicularly plasma of parabolic initial density and keV initial temperatures at 10^{15} W/cm^2 perpendicular CO_2 laser irradiation.

maximum values of the electrostatic energy at 10^5 ergs/cm^2 in a gaseous plasma (Eq. 25, g=3, n_e-10^{21}cm^{-3},T_e=10^3eV) exceed the room temperature metallic values, shown in Table 1. Once the ions start moving under the influence of the electric fields, the double layer is broadened and the electrostatic energy is reduced. This effect lasts only a short time, of the order of the reciprocal of the ion plasma frequency.

The generation of double layers (DLs) and their associate electric fields in laser produced plasma have been studied by the model consisting of two macroscopic electron and ion fluids coupled by Poisson's equation and interacting with an external electromagnetic field [13, 49, 56]. Recently, calculations for Nd and CO_2 laser irradiance of $I\lambda^2 = 10^{16} \approx 10^{18}$ Wμm^2/cm^2 have been performed [56], which show an electrostatic field as high as 10^9 and 10^{10} V/cm and an energy gain of 80 eV for accelerated electrons in this field. It is because there exists an electrostatic field associated with a double layer of the order of a

Debye length on the surface of a laser produced plasma that it is possible to produce an effect of surface tension [63].

The surface tension σ_E defined by the infinitesimal work done by a small increase of the area of separation can be expresed as [63]

$$\sigma_E = \int \frac{E^2}{8\pi} dR$$

where $\dfrac{E^2}{8\pi}$ is an energy density of the electrostatic field in DL's and dR is an infinitesimal distance in plasma. Usually the "electrostatic" field E is space and time dependent [13,26,48,56]. Therefore the value of σ_E at a section of the plasma may be calculated by the following formula.

$$\sigma_E(x) = \int_{x_0}^{x} \frac{E^2}{8\pi} dR \qquad 0 \leq x \leq x_0$$

here x_0 denotes the whole plasma depth.

To obtain the absolute value $\sigma_E(x)$, we utilize a numerical integration procedure [48,56] for the case of CO_2 laser irradiance of $10^{15} W/cm^2$ on a 0.15 mm thick hydrogen plasma slab of initial $10^7 {}^{\circ}K$ temperature. Initially, the expansion velocity for both electrons and ions is assumed to be zero, and ion and electron number density profiles are taken to be parabolic shape with a maximum cut-off density of $10^{19} cm^{-3}$ (center of the parabola). An overdense plasma region is assumed at the side opposite the laser light. In our simulation, pretreated plasma is allowed to relax for 0.5 psec without the application of the external laser irradiance source [18]. After the relaxation process, a CO_2 laser field with a characteristic wavelength of 10.6 µm is applied at the plasma-vacuum boundary (x=0). Time steps of 0.5 fsec and space steps of 1 µm are used, respectively.

Based on these results, electrostatic fields E with amplitudes as high as 10^8 V/cm. Space dependences of surface tension σ_E at three different times of these calculations are plotted in Fig. 3. A number of interesting phenomena are observable in this Figure.

When t=0, at the moment of laser application, the surface tension is of order of 10^3 ergs/cm^2 at the DL surface (x=0), dropping monotonically to zero as the depth increases. After 0.5 psec of laser application, the value of σ_E reaches at a maxima of 10^6 ergs/cm^2 at the plasma surface. This is a consequence of the high electrostatic field near the DL surface. Although the surface tension decreases at time t = 1.0 psec, its value remains above 10^5 ergs/cm^2.

The surface tension σ_E at different times shows three distinct spatial variation regions (see Fig. 3). The first region extends from x=0 to x = 40 μm, where σ_E has its highest value. This phenomenon may be caused by a nonlinear force [2] which produces a known density minima (cavitons) near x = 40 μm, producing a strong electrostatic field (see Fig. 2). The second region, having a monotonically decreasing $\sigma_E(x)$ extends to x = 100 μm. In the third region where x > 100 μm, $\sigma_E(x)$ flattens out. at a small value. This is due to the fact that the plasma in this area is overdense near (x = 150μm) and the electrostatic field approaches zero. At x = 0, the curves in Fig. 3 are nearly horizontal due to the fact of low density merging into vacuum for x<0.

The realization of the swimming electron layer of about 1 Angstrom thickness on top of the metal crystal is the reason for an increase of the deuterium concentration on a palladium or titanium lattice and for a two dimensional strong surface screening effect [63] in order to explain the cold fusion reactions at the surface [64] of these metals. This swimming electron layer is also an immediate reason for understanding the Marangoni flow in a liquid metal with local heat source, e.g., from an incident laser or electron beam [65]. The phenomenologically given force by the lateral gradient of the surface tension in the metal can then be immediately recognised by the locally different surface tensions resulting in an electrostatic driving of the swimming electron layer to produce the Marangoni flow in this case.

e. Application to the Dynamics of Gaseous Plasmas

Two well-known examples of surface tension in fluid dynamics were mentioned in ref [13], and will be briefly discussed here.

At a sharp boundary between semi-infinite fluids of density ρ_1 and ρ_2, in a gravitational field g, the surface tension σ stabilizes the boundary for all wavelengths λ of disturbance less than a critical value λ_c where

$$\lambda_c^2 = 4\pi^2\sigma/g(\rho_1 - \rho_2). \tag{28}$$

Taking a = 10^{16} cm/s^2, (as a typical acceleration in laser produced plasmas) r = 10^{22} m_p g/cm^{-3}where m_p is the proton mass, σ = 10^5 erg/cm^2 yields λ_c = 2 μm. This distance is comparable with the minimum scale length of the density gradient in a laser produced plasma, which can approach the collision-free skin depth $(m_e c^2/4\pi n e^2)^{1/2}$ ie., close to the wavelength of the incident light [66, Fig. 163]. In these circumstances, the sharp boundary approximation is poor. The growth-rate of Rayleigh-Taylor instability is reduced, and some stabilizing might be expected. This is similar to the electric field produced stabilization for finite Larmor radius what we may interpret as an effect of surface tension. The Rayleigh criterion for the stability of a spherical droplet of radius r_0 charged to a potential of V e.su. is [67]

$$V < 4\pi r_0 \sigma \qquad (29)$$

Without surface tension σ the droplets are unstable to all V. The potential V is quite independent of the potential across the double layer, although it may be of the same order of magnitude (as is the contact potential difference in the case of metals). With $r_0 = 10^{-2}$cm, $\sigma = 10^5$ erg/cm^2, a droplet is stable for V < 200 e.s.u., ie., about 60 kV. The Rayleigh calculation is for a static equilibrium state. This is far from the case for the spherical targets ablatively imploded by laser interaction. A stabilizing effect in ablatively accelerated plasma has been found using a single-fluid plasma model [68]. The critical wavelength found there is larger than that calculated above from the electrostatic surface tension, but the two become comparable when the scale length in the ablative layer becomes close to the Debye length.

f. Application to the Surface Tension of Nuclei

The successful explanation of surface tension in metals by plasma double layers described in the preceding section [69] is extended to nuclei in order to explain the earlier measured [70] decay length of the charge distribution in the surface of nuclei as a kind of Debye length and to derive then the well-know measured [71] nuclear surface energy.

The Debye length λ_{dF} for the degenerate electron gas is given in Eq. (26). The surface tension is then [69] given by Eq. (27) and the resulting values for the surface tension of metals are within less than 20% in agreement with the measurements. It should be underlined that our earlier described [13] derivation of the surface tension in plasmas is based on the immediate calculation of the electric-field energy (per surface area) as it is produced by the faster electrons leaving the plasma, or similarly by the degenerate electron gas leaving the ion lattice of the metal [13,69]. The analogy to the conditions of an electrolyte is based on the modification of the Debye-Hückel theory which was used as the Thomas-Fermi model and further extended by inclusion of exchange interaction to the Thomas-Fermi-Dirac model [2]. The description of the surface tension of electrolytes is based on thermodynamic functions, dating back to 1903 [57] and treated extensively [73], and results in the relation between the Debye length λ_d and the surface tension σ in analogy to Eq. (2) as in [13]

382

$$\sigma_E = \frac{(kT/e)^2[\text{Cosh}(\phi_0 e/2kT) - 1]}{8^{1/2}\pi^{3/2}\lambda_d}$$

where T is the temperature and ϕ_0 is the potential drop across the double layer. One important assumption to our derivation of Eq. (27) is that the ion density drops as a step function describing the rigid ion lattice and the electron distribution causes an electric field to be taken as a Gaussian function. The less realistic case of a linear decay of the electron density results in a numerical factor 0.36 instead of 0.27 in Eq. (27) [69].

In nuclei, there is no space charge neutral state but it has been shown experimentally [70] that the positive charge density in heavy nuclei is constant in the interior, as seen, e.g., from plots of the charge density which is first constant and then decays to zero within 4.5 to 5 fm, defining a half-width which allows us to determine the radius of the bismuth nucleus of $r_n = 6.5$ fm (Fig. 4).

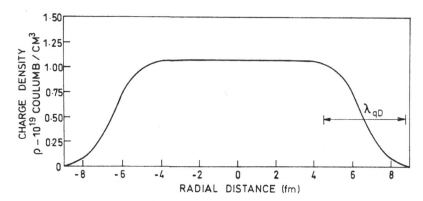

Fig. 4. Measurement of the positive charge density in a bismuth nucleus [70] depending on the radius. The peripheric decay may be expressed by the given by λ_{qD}, Eq. (30) which is determined by a kind of degenerate Debye length, however where the compensation of the charge is not given by charges of the other kind but by nuclear forces. This quasi-Debye length is therefore reduced by a factor too in the charges.

The nucleus cannot immediately be described as a plasma - not considering here the gluon plasmas mode [74] - but it seems that the nuclear forces produce a state for the positive charges that is similar to plasma properties of a positron gas. Using the charge number 83 for bismuth, the density n_0 of the positive charges is 7.92×10^{37} cm^{-3} in agreement with Hofstadter's [71] measured charge density of 1.26×10^{19} C cm^{-3}.

It is remarkable that the length (l) for such charge density for the empirical Fermi energy $E_F = 40$ MeV [72] obtains

$$\lambda_{qD} = 4.32 \text{ fm} \qquad (30)$$

in good agreement with the measured 4.5 fm length of the charge decay at the surface of the bismuth nucleus. This can be interpreted as a quasi-Debye length in a very general sense only as given by these considerations. Again, without having the full analogy to the plasma in metals [since only the gas of the positive charges is involved and no opposite charges, which then reduces Eq (27) by a factor of 0.5], the value of the surface tension of bismuth nuclei is obtained in the form [using Eqs. (27) and (30)].

$$\sigma_n = 4.42 \times 10^{20} \text{ erg cm}^{-2} \qquad (31)$$

This results in a surface energy

$$E_{surf} = 4\pi r_n^2 \sigma_n = 6.6 \times 10^8 \text{ eV} \qquad (32)$$

while the empirical value [72] is 6.5×10^8 eV.

It is remarkable that the nucleon saturation theory [72] resulted in a surface energy proportional to the Fermi energy E_F while our plasma model results in a proportionality with $E_F^{3/2}$. Though more interpretation and understanding is necessary, the agreement with the measurements and the immediate derivation from the principles may be a point of interest.

4. CONCLUSIONS

Our conclusions are as follows:

1. Where the electron pressure of a plasma is restrained by electric fields, the associated electrostatic energy density can be comparable to the thermal energy density of the electrons when the pressure gradient scale length is comparable to the Debye length. The electric field has been detected directly by the deflection of ions moving through the double layer at grazing incidence [75].

2. By integrating this electrostatic energy across a double layer (or plasma sheath) at the surface of a plasma, an electrostatic energy per unit area is obtained which could play a role in plasma dynamics like that of surface tension in fluid dynamics.

3. The double layer model has long been applied to the surface of electrolytes; a direct application to the surface of metals indicates the correct order of magnitude for the surface energy (Table 1), if the double layer of the degenerate electron gas is calculated due to the pressure of the electron gas as it was verified from the compressibility (see Fig. 1.2 of Ref. [2]) as an expression of the electrostatic quantum model of the atoms [12, p.29]. Using the relativistic Fermi energy of nuclei, Hofstadter's measurement of the charge distribution can be explained with a decay given by a Debye length and resulting in a surface energy of the nucleus as measured..

4. In gaseous plasma, the electrostatic energy itself arises in circumstances when the ions are being accelerated thermally.

5. Some stabilizing effects at short wavelengths of surface waves might be observed in laser-produced plasmas, and especially at short times.

6. The application to other cases of sharp plasma boundaries include space and cosmical plasmas, where regions of different plasma densities, temperatures, chemical composition, and magnetization are often sharply delineated by thin surfaces of charge separation [55].

REFERENCES

[1] N.A. Krall and A.W. Trivelpiece, Principles of Plasma Physics, McGraw-Hill, New York, 1973).
[2] S. Eliezer, A.K. Ghatak, H. Hora and E. Teller, Equations of State, (Cambridge University Press, Cambridge, 1986).
[3] H. Hora, Laser Plasmas and Nuclear Energy (Plenum, New York, 1975).
[4] M. Lubin, ECLIM 74th Conference, Garching, April 1974.
[5] S. Jaeckel, B. Perry and M. Lubin, Phys. Rev. Lett. 37, 95 (1976).
[6] R. Dragila, R.A.M. Maddever, and B. Luther-Davies, Phys. Rev. A 36, 5292 (1987).
[7] A. Maddever, B. Luther-Davies, and R. Dragila, Phys. Rev. A (1990), A. Maddever, PhD Thesis Australian National University 1988.
[8] J.H. Nuckolls, Physics Today 35, (No. 9) 24 (1982).
[9] Y. Kato, K. Mima et al, Phys. Rev. Lett. 53, 1057 (1984); Obenschain et al, Phys. Rev. Lett. 56, 2807 (1986); Deng Ximing and Yu Wenyan, Advances in Inertial Confinement Fusion Res. (IAEA, Kobe, Nov. 1983), C. Yamanaka ed. (ILE Osaka 1984) p. 66. (1982); Tan Weihan et al, Laser and Part. Beams 3, 237 (1985).
[10] Gu Min and H. Hora, Chinese Journal of Lasers 16, 656 (1989).
[11] H. Hora, L. Cicchitelli, Gu Min, G.H. Miley, G. Kasotakis, and R.J. Stening, Laser Interaction and Related Plasma Phenomena, H. Hora and G.H. Miley eds. (Plenum, New York 1990). Vol. 9, p.
[12] H. Hora, Physics of Laser Driven Plasmas (Wiley, New York 1981).
[13] S. Eliezer and H. Hora, Physics Reports 172 339 (1989); Fusion Technol. 16, 419 (1989); H. Hora, P. Lalousis, and S. Eliezer, Phys. Rev. Lett. 53, 1650 (1984).
[14] R.M. Kulsrud, Physics Today, 34 (No. 4) 56 (1983) 3rd col., 7th line.
[15] H. Alfven, Cosmic Plasmas (Reidel, Dordrecht, 1981).

[16] H. Hora, S. Eliezer, M.P. Goldsworthy, R.J. Stening, and H. Szichman, Laser Interaction and Related Plasma Phenomena, H. Hora and G.H. Miley eds. (Plenum, New York, 1988) Vol. 8, p.293.

[17] A. Schlüter, Z. Naturforsch, 5A, 72 (1950).

[18] H. Hora, Phys. Fluids 12, 182 (1969).

[19] L. Spitzer, Physics of Fully Ionized Plasmas (Wiley, New York, 1959).

[20] A. Zeidler, H. Schnabe, and P. Mulser, Phys. Fluids 28, 372 (1985).

[21] H. Hora, Phys. Fluids 28, 3705 (1985).

[22] T. Rowlands, Plasma Physics (1990).

[23] R. Lüst, Z. Astrophhys. 37, 67 (1955).

[24] H. Hora, Inst. Plasmaphysik, Garching Reports 6-23 and 6-27 (1964); H. Hora, and H. Wilhelm, Nuclear Fusion 10, 111 (1970).

[25] R.B. White and F.F. Chen, Plasma Physics, 16, 567 (1974).

[26] H. Hora, and A.K. Ghatak, Phys. Rev. A31, 3473 (1985); M.P. Goldsworthy, F. Green, P. Lalousis, R.J. Stening, S. Eliezer and H. Hora, IEEE Trans. Plasma Sc. PS-14, 823 (1986).

[27] R. Fedosejevs et al, Phys. Rev. Lett. 39, 932 (1977); H. Azechi et al, Phys. Rev. Lett. 932 (1977), P. Wagli, and T.P. Donaldson, Phys. Rev. Lett. 40, 875 (1978); F.J. Mayer et al, Phys. Rev. Lett. 40, 30 (1978); B. Luther-Davies and J.L. Hughes, Opt. Comm. 18, 605 (1976); A. Montes, and O. Willi, Plasma Phys. 24, 67 (1982).

[28] T.P. Hughes and M.B. Nicholson-Florence, J. Phys. A1, 588 (1968).

[29] H. Hora, Opt. Comm. 41, 268 (1982).

[30] Heidi Fearn, and M.O. Scully, Laser Interaction and Related Plasma Phenomena, H. Hora and G.H. Miley eds., (Plenum, New York 1990) Vol. 9, p.

[31] H.A. Bethe, Handbuch der Physik, H. Geiger, and L. Scheel eds., (Springer, Berlin 1933) Vol. 24, p.497.

[32] R.E. Marshak, Ann. N.Y. Acad. Sci. 41, 49 (1941).

[33] H. Hora, Nuovo Cimento, 64B, 1 (1981)

[34] G. Grieger and Wendelstein VII - Team, Plasma Physics and Controlled Nuclear Fusion Research 1980 (IAEA Vienna 1981) Vol. I, p.173 and p. 185.

[35] J.Hugill, Nuclear Fusion 23, 331 (1983), Analyzing results given there by Dimock et al, Fig. 5.

[36] K.A. Razumova Plasma Physics 26, 37 (1984).

[37] G. Knorr and Ch. Goerz, Astrophys. Space Sci. 31, 209 (1974).

[38] N. Hershkowitz, Space Science Rev. 41, 351 (1985); R. Schrittwieser, and G. Eder eds. Double Layers (Inst.) Theor. Phys. Innsbruck 1984).

[39] T. Sato and H. Okuda, Phys. Rev. Lett. 44, 740 (1980); T.K. Yabe, K. Mima, K. Yoshikawa, H. Takabe, and M. Hawano, Nucl. Fusion 21, 803 (1981).

[40] A.C. Williams, ed., Laser and Particle Beams 5, No. 2 (1987).

[41] A.L. Peratt, Laser and Particle Beams, 6, 474 (1988), IEEE Trans. Plasma Sc. PS-17, 65 (1989).

[42] S.I. Anisimov, and Yu. V. Medvedev, Sov. Phys. JETP 49, 62 (1979).

[43] H. Hora, Z. Phys. 226 (1969).

[44] V.V. Krobkin, and A.J. Alcock, Phys. Rev. Lett. 21, 1433 (1968); M.C. Richardson and A.J. Alcock, Appl. Phys. Lett. 18, 357 (1971).

[45] F.E. Irons, R.W. McWhirter, and N.J. Peacock, J. Phys. B5, 1975 (1972).

[46] H. Hora, Laser Interaction and Related Plasma Phenomena, H. Schwarz et al eds. (Plenum, New York 1971), Vol. 1, p.387.

[47] P. Lalousis, PhD Thesis, Univ. New South Wales 1983).

[48] P. Lalousis and H. Hora, Laser and Particle Beams, 11, (1983), H. Hora, P. Lalousis and S. Eliezer, Phys. Rev. Lett. 53, 1650 (1984).

[49] J.W. Shearer, R.E. Kidder and J.W. Zin, Bull. Amer. Phys. Soc. 15, 1483 (1970).

[50] S. Eliezer and A. Ludmirski, Laser and Particle Beams 1, 251 (1983).

[51] R. Dragila, Laser and Particle Beams 3, 79 (1985).

[52] M. Goldsworthy, H. Hora and R.J. Stening, Plasma Physics and Controlled Nuclear Fusion Research 1988 (IAEA Vienna 1989) Vol. III, p.181.

[53] Aleksandrova, Brunner et al, Laser and Particle Beams 3, 197, (1985); Tan Weihan, Lin Zunqi, Gu Min et al, Phys. Fl. 30, 1510 (1987).

[54] N.G. Land, and W.I. Kohn, Phys. Rev. B1, 4555 (1970).

[55] H. Alfven, Laser and Particle Beams 6, 389 (1988), A.L. Peratt ed, Laser and Particle Beams 6 No. 3 (1988).

[56] H. Szichman, Phys. Fluids, <u>31</u>, 17002 (1988).

[57] G. Gouy, J. Chim. Phys., <u>29</u>, 145 (1903).

[58] D.L. Chapman, Phil. Mag., <u>25</u>, 475 (1913).

[59] M.P. Tosi, Liquid Surfaces and Solid-Liquid Interfaces in Amorphous Solids and the Liquid State, E. March et al., Eds. (Plenum, New York, 1985) p.125.

[60] L. Tonks and I. Langmuir, Phys. Rev., <u>33</u>, 195 (1929).

[61] S. Lundquist, "Electrical response of surfaces," Surface Sci., vol. 6, p,.331, 1974.

[62] G. Rickayzen, Liquid Surfaces and Solid-Liquid Interfaces in Amorphous Solids and the Liquid State, E. March et al., Eds. New York: Plenum, 1985, p.523.

[63] Z. Henis, S. Eliezer and A. Zigler, J. Phys. <u>G15</u>, L219 (1989).

[64] H. Hora, L. Cicchitelli, G.H. Miley, M. Ragheb, A. Scharmann and W. Scheid, Nuovo Cimento <u>12D</u>, 393 (1990).

[65] G. Tsotridi, H. Rother, and E.D. Hondros, Naturwissenschenften, 76, 216 (1989); D. Schwabe, Phys. Chem., Hydrodyn. <u>2</u>, 263 (1981).

[66] H.G. Ahlstrom, Physics of laser fusion", Lawrence Livermore Nat. Lab., Livermore, CA, Tech. Rep., 1982, Montes and Willi, see Ref. [27].

[67] G.I. Taylor, "Disintegration of water drops in an electric field", Proc. Roy. Soc., Ser. A, vol. A280, p.383, 1964.

[68] H. Takabe, K. Mima, L. Montierth, and R.L. Morse, Phys. Fluids, <u>28</u>, 3676 (1985).

[69] H. Hora, Gu Min, S. Eliezer, P. Lalousis, R.S. Pease, and H. Szichman, IEEE Trans. Plasma Sci. PS-17, (1989).

[70] H. Hora, Naturwissenschaften <u>76</u>, 214 (1989).

[71] R. Hofstadter: Australian Phys. <u>24</u>, 236 (1987).

[72] S.A. Moszkovski in: Handbuch der Physik, Vol. 34, p.432. (S. Flügge, ed.). Berlin: Springer 1957.

[73] S. Ono, S. Kondo, in: Handbuch der Physik, Vol. X, p.134. (S. Flügge, ed.). Berlin: Springer 1960.

[74] R. Stock, CERN Courier (Dez. 1987); J. Rafelski, Phys. Lett. <u>B207</u>, 371 (1988).

[75] C.W. Mendel, J.N. Olsen: Phys. Rev. Lett. 34, 859 (1975); see also Ref. [50].

CAVITON GENERATION WITH TWO-TEMPERATURE

EQUATION-OF-STATE EFFECTS IN LASER PRODUCED PLASMAS

H. Szichman, S. Eliezer and A. Zigler

Plasma Physics Department
Soreq Nuclear Research Center
Yavneh 70600, Israel

ABSTRACT

We developed a two-temperature equation-of-state (EOS) for
a plasma medium. The cold electron temperature is taken from
SESAME tables, while the thermal contribution of the electrons
are calculated from the Thomas-Fermi-Dirac (TFD) model. The ion
EOS is obtained by using the Gruneisen-Debye solid-gas interpolation method. These EOS are well behaved and are smoothed over
the whole temperature and density regions, so that also the
thermodynamic derivatives are well behaved. Moreover, these EOS
were used to calculate the average ionization $\langle Z \rangle$ and $\langle Z^2 \rangle$ of
the plasma medium. Furthermore, the thermal conductivities have
been calculated using an extrapolation between the conductivities in the solid and plasma (Spitzer) states. The two-tempera-
ture EOS, the $\langle Z \rangle$ and $\langle Z^2 \rangle$ and the thermal conductivities for
electrons and ions were introduced into a two fluid hydrodynamic
code to calculate the laser plasma interaction in aluminum and
gold slab targets. It was found that the two EOS is important
mainly from the ablation surface outwards (toward the laser).
In particular the creation of cavitons in the distribution of
electrons is here predicted, specially for light materials as
aluminum.

INTRODUCTION

The physics of the simulation of the processes inherent to
the interaction of laser light with matter is based on the hydrodynamic equations of one or more fluids. In order to solve
these equations a further knowledge on the transport coefficients (such as thermal and electrical conductivities, radiation
opacities, etc.) besides the EOS is necessary, when these thermodynamical parameters are to be described over many orders of
magnitude of temperature and density (or pressure)[1]. Thus, by
inspecting the requirements and goals of the research for inertial confinement fusion, it may be concluded that one needs the
data of EOS for many materials in the domain of

$$0 < kT < 100 \text{ keV}; \quad 10^{-4} < \rho/\rho_0 < 10^4; \quad 0 < P < 10 \text{Mbar} \tag{1}$$

where ρ_0 is the initial liquid or solid density of the material under consideration, ρ/ρ_0 is the compression and P and T are the pressure and temperature respectively.

The requirements on the selection of the EOS become very severe, if one takes into account the fact that for a correct functioning of the simulation code, not only the EOS should be accurate but also conserve thermodynamical consistency over the wide range dictated by (1), a condition that no known theory is able to fulfill.

In this paper, we consider the interaction between a laser driven shock wave and planar target. The central point in our studies is the use of a "realistic" two temperature EOS, which have high accuracy, particularly in the 0.01-1 Mbar region. We built-up also improved tables for the average degree of ionization and thermal conductivities having higher accuracy, particularly in the compressed portion of the target, than other algorithms used in similar codes. Using our 1D Langrangian hydrocode, we present results obtained for planar targets of aluminum and gold irradiated by low laser intensities (of about 10^{12} W/cm^2), so that the problems of hot electrons and energy radiation may be neglected.

THE MODEL

The basic physical processes of the laser-plasma interaction are computed by our 1-D hydrodynamic Lagrangian two-fluid code[2]. This code simulates the plasma time evolution by solving at each Langrangian cell the three basic conservation laws: (i) Mass conservation (which is automatically satisfied in the Lagrangian mesh system); (ii) Momentum conservation given by the linearized Navier-Stokes equation[2] including artificial viscosity in order to mantain numerical stability; (iii) Energy conservation, when there are two energy equations, one for the electron subsystem and one for the ion subsystem.

In our simulations, the laser light is absorbed in the plasma by an inverse bremsstrahlung process. The radiation which is not absorbed by inverse bremsstrahlung, is assumed to be resonantly absorbed at the critical surface. The code includes also procedures describing collisional heat exchange between the ions and the electrons, as well as classical heat transport (with optional flux limiter) by the thermal electrons and ions.

In order to calculate the above mentioned processes, we need separate EOS for the electrons and ions, the average values of the ionization <Z> and <Z^2>, as well as thermal conductivity coefficients for the whole range of temperatures and densities relevant to the simulations. As the overall accuracy of the computation greatly depends on the accuracy of these parameters, special effort was applied to obtain reliable values for them. The description of the methods for the evaluation of the tables of the EOS, the average degree of ionization <Z> and the thermal conductivities will now be discussed.

Two temperature EOS and the TFD model

One of the reasons of the more general use of one-tempera-

ture codes is based in the use of the SESAME tables[3] for the
implementation of EOS in the hydrocode. The main difficulty of
these tables is the fact that they are accurate mainly in one-
temperature tables which cannot be implemented easily into two-
temperature codes, meanwhile most of the separate ion and electron tables describe the low and intermediate isotherms with numerical noise which makes them inadequate for use in practical
computations. A second way of calculating separate EOS is by
using the Thomas-Fermi-Dirac (TFD) model[1,4]. While this model
is very successful in describing the electronic part of the EOS
at high values of temperature and density, its accuracy is nevertheless defficient at low temperatures where the ionic components of the EOS are important. Moreover, the TFD model does
not describe properly electronic excitations in the corona region. However in the present work we suggest a model which suc-
cesfully combines the above methods and overcomes their difficulties.

Our tables for EOS of matter are constructed as a superposition of three distinct parts[3]: (i) a cold curve due to electronic forces at T = 0K, based on the SESAME tables[3] and described by means of semi-empirical formulas[5]; (ii) the contribution of thermal electronic excitations (deduced from the TFD
theory) and (iii) that of ionic excitations. The total pressure
can be written as

$$P(\rho,T_e,T_i)=P_c+[P_{TFD}(\rho,T_e)-P_{TFD}(\rho,0)]+P_{ion}(\rho,T_i) \qquad (2)$$

where P_{TFD} is the pressure calculated in the TFD model.

In order to preserve the thermodynamic consistency, a
similar equation holds for the internal energy.

In the region of the shock front, the main components in
Eq. (5) are the cold curve and the ion thermal terms. The electron thermal contribution, which determines the momentum transfer from the corona into the target, has more influence in the
outer portion of the plasma. It turns out, that all the three
terms in Eq. (5) should be known to fairly good accuracy, particularly at low laser intensities when the behavior of (5) is
far from the linear dependence in T and as predicted by the
ideal gas theory.

The cold pressure P_c, may be obtained from the semiempir-
ical fitting of phenomenological formulas to experimental shock-
wave data. The cold curves given by the SESAME tables are
reasonably accurate for use in hydrodynamics simulations, as was
indeed done in our work.

For the description of thermal ion contribution we used the
solid-gas interpolation method[6,7], which was shown to give excellent agreement with shock wave experiments[5], meanwhile the
TFD model was used to obtain the thermal electronic contribution
to the EOS, using the method described in Ref. 8. The TFD model
was preferred over the more conventional TF model, because it
includes the Dirac exchange forces between electrons, thereby
improving the overall accuracy. In the same Ref. 8 it was shown
that the TFD atom may be solved in a straightforward and compact
way which greatly simplifies the computation of the EOS. The
high accuracy of this method is manifested by the fact that the
virial theorem is satisfied by these computations to better than

10^{-4} for temperatures around 100 eV. The solution of the TFD atom is carried-out by a three step procedure that solves simultaneously the following equations:

(1) The electronic contribution to the potential energy at a distance r from the nucleus is computed by means of the spherically symmetric solution of the integral Maxwell equation,

$$-eV(r) = \mu + \frac{4\pi e^2}{r} \int_r^{r_0} n(r')(r'-r)r'dr' \tag{3}$$

(2) The total potential energy ξ (=potential+exchange energies) is implicitly given by the solution of

$$\xi = -eV(r) + \frac{e^2}{\pi} \left[\frac{2mkT}{\hbar^2}\right]^{1/2} I'_{1/2}(\xi/kT) \tag{4}$$

where the second term on the right hand side is the contribution of the exchange potential.

(3) The local electron number density n(r) is computed by assuming a Fermi-Dirac statistical distribution of the electrons,

$$n(r) = \frac{1}{2\pi^2} \left[\frac{2mkT}{\hbar^2}\right]^{3/2} I_{1/2}(\xi/kT) \tag{5}$$

Here,

$$I_j(x) = \int_0^\infty \frac{y^j dy}{\exp(y-x)+1} \tag{6}$$

is the Fermi-Dirac integral of order j. Finally, the chemical potential μ is determined from the neutrality condition,

$$Z = \int_v n(r,\mu) \cdot d^3r \tag{7}$$

This formalism of the TFD model permits the description of the EOS by means of closed formulae. Thus the pressure is obtained simply by the differentiation of the Helmholtz free energy[8] at the atom boundaries[9] and the final expression is

$$P = \frac{kT}{3\pi^2} \left[\frac{2mkT}{\hbar^2}\right]^{3/2} I_{3/2}(\xi_0/kT) -$$

$$\frac{e^2}{2\pi^3} \left[\frac{2mkT}{\hbar^2}\right]^2 \left[I_{1/2}(\xi_0/kT) \cdot I'_{1/2}(\xi_0/kT) - X(\xi_0/kT)\right] \tag{8}$$

where ξ_0 is the energy parameter of Eq. (4) at the atom surface and $X(\xi)$ is the exchange integral defined by

$$X(\xi) = \int_{-\infty}^{\xi} \left[I'_{1/2}(y) \right] \, dy \qquad (9)$$

The energy is given according to Cowan and Ashkin[4], by the addition of three terms,

$$E = E_{kin} + E_{pot} + E_{xch} \qquad (10)$$

where,

$$E_{kin} = kT \int_{v} \frac{I_{3/2}(\eta)}{I_{1/2}(\eta)} \, n(r) \, d^3r \qquad (11)$$

$$E_{pot} = -\frac{e}{2} \int_{v} \left(-V[r] - \frac{\mu}{e} + \frac{Z}{r} \right) n(r) \, d^3r \qquad (12)$$

$$E_{xch} = -\frac{e^2}{\pi} \left[\frac{2mkT}{\hbar^2} \right]^{1/2} \int_{v} \frac{X(\eta)}{I_{1/2}(\eta)} \, n(r) \, d^3r \qquad (13)$$

here, $\eta = \xi(r)/kT$.

Finally, the virial theorem is also verified in the TFD model[4] and the energy is then additionally being given by

$$E = (3PV + Epot + Exch)/2 \qquad (14)$$

In summary, Eq. (2) is a two-temperature EOS, which may be implemented in a laser-plasma interaction code. It is expected that the above described method yields more accurate results (particularly at low temperatures) than other previously used simpler models.

As a representative test of our EOS tables, we compare in Fig. 1 results of our calculations of the electron thermal contributions to the pressure in gold with similar data obtained[3] from the two-temperature EOS SESAME tables. An excellent agreement was observed for isotherms of temperature 10 eV and higher in Au as well as in Al. At the same time, one can observe the irregular behavior up to 10 eV of the isotherms taken from the SESAME tables, while our calculations are "smooth" functions ready to use for the hydrocode simulations.

Average degree of ionization

A second set of parameters, whose values strongly affect the accuracy of the two-fluid simulations, are $\langle Z \rangle$ and $\langle Z^2 \rangle$, the first and second moments of the ionization state distribution in the plasma. Their values appear explicitly in the computation of inverse bremsstrahlung, thermal conductivity, radiative cooling and others. These parameters, as function of density and temperature, were read into the program from separate tables, which were computed to rather high precision.

In these tables, $\langle Z \rangle$ and $\langle Z^2 \rangle$ were calculated by using the TFD model[9], thus obtaining overall consistency between these ones and the EOS tables (see above Paragraph). In the TFD statistical model, the atomic degree of ionization is given by

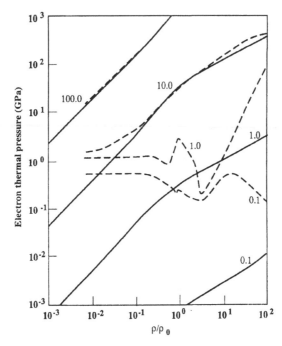

Fig. 1. Electron thermal pressure for gold according
to present TFD model(solid line) compared with
SESAME tables data[3](dashed line).

the number of electrons in the atom wich are in a positive
energy state ($\epsilon - \mu \gtrsim 0$, see Eq. (3)). This fact is expressed
numerically by,

$$\langle Z \rangle = \frac{1}{2\pi^2} \left[\frac{2mkT}{\hbar^2} \right]^{3/2} \int_V F_{1/2} \left[\frac{\epsilon}{kT} , \left| \frac{\epsilon - \mu}{kT} \right| \right] d^3r \qquad (15)$$

where,

$$F_j(x, \beta) = \int_\beta^\infty \frac{y^j dy}{\exp(y-x)+1} \qquad (16)$$

is the incomplete Fermi-Dirac integral. As was shown[7] a plausible generalization of Eq. (15) leads to

$$\langle Z \rangle = Z \langle F_{1/2}(x,\beta)/I_{1/2}(x) \rangle_{Av} \qquad (17)$$

$$\langle Z^2 \rangle = Z \cdot \langle Z \rangle \langle F_1(x,\beta)/I_1(x) \rangle_{Av} \qquad (18)$$

Fig. 2 shows the results of $\langle Z \rangle$ obtained for Al, as a
function of the temperature and compression of the atom. Comparison of these results of this model show agreement with the
estimations of the Saha model in their common validity
domain[7]. It is worth to be mentioned that Eqs. (17) and (18)
predict correctly the appearence of pressure ionization in the
region of highly compressed plasma (See Fig. 2).

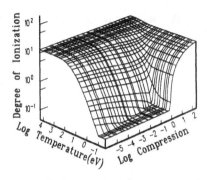

Fig. 2. Degree of ionization of Al as a function
of temperature and compression

Thermal conductivities

In our simulations the thermal conductivities are calculated for the plasma phase using the Spitzer[10] expression for a
neutral Lorentz gas. Near and below the melting point, where
the Spitzer formula is inadequate, we used a procedure to ex-
trapolate the experimental value at the normal conditions of
temperature and compression. This extrapolation was carried-out
as follows. From the Wiedemann-Franz law relating the thermal
and electrical conductivities in solids one gets,

$$\kappa_e = \frac{\pi^2}{3} \langle Z \rangle n_i k (kT) \left[\frac{\tau}{m} \right] \tag{19}$$

where κ_e is the electron thermal conductivity, n_i is the ion
number density, τ is the electron relaxation time and m is the
electron mass. One can approximately establish that,

$$\upsilon = \Lambda / \tau \tag{20}$$

where Λ and υ are the electron mean free path and velocity
within the atom borders. By analysing the multiscattering of
the electron in the atom, Ziman[11] arrived to the conclusion
that for temperatures below the melting point, the mean free
path scales like

$$\Lambda = 50 R_o (T_m / T) \tag{21}$$

where R_o is the interatomic distance and T_m is the melting
temperature.

Combining Eq. (21) with the Lindemann melting point law expressed as

$$\frac{T_m}{T_{mo}} = \frac{1}{\eta^{2/3}} \left[\frac{\theta}{\theta_o}\right]^z \tag{22}$$

where, η is the compression and T_{mo} is the melting temperature for $\eta = 1$, one finally arrives to the following phenomenological formula for the thermal conductivity

$$\kappa_e = \langle Z/Z_o \rangle \left[\frac{\upsilon_0}{\upsilon}\right] \left[\frac{\theta}{\theta_o}\right]^z \kappa_{eo} \tag{23}$$

where $\kappa_{eo}, Z_o, \upsilon_0$ are respectively the thermal conductivity, average Z and the electron mean velocity at room temperature and normal density.

The Fermi energy is used in order to evaluate the mean electron velocities and the ratio of the Debye temperatures θ/θ_o is determined from the ionic EOS as follows

$$\theta(\eta) = \theta_o \exp\left[\int_1^\eta \Gamma(x) \frac{dx}{x}\right] \tag{24}$$

where Γ is the same Gruneisen coefficient used in calculating the ion EOS by means of the solid-gas interpolation method[6,7].

The final expression for κ_e is given by an extrapolation between (23) and the Spitzer value for the plasma state. In Fig. 3 we present the results obtained for gold at normal density showing three regions of our conductivity model. In this same Figure we compare values obtained for the thermal conductivity according to three different models proposed by Lee and More[12], Spitzer[10] and the present paper. The calculations of Lee and More coincide practically with those obtained from the Spitzer's formula. Whereas at high temperatures all three models predict similar results, at the solid region the thermal conductivity, as computed by our method, is by as much as two orders of magnitude larger than that predicted by the Spitzer's formula[10].

RESULTS AND DISCUSSION

In the present simulations, a trapezoidal pulse with rise and fall times of 1 ns and a plateau of 3 ns was used, similar to the pulse shape used in our laboratory. This type of trapezoidal pulse was used in our laboratory for the generation of shock waves[13]. The range of laser intensities investigated was between 10^{12} and 10^{13} W/cm^2. Our main aim was the analysis of various physical effects introduced by the simulations of a two-temperature code in gold and aluminum, particularly the phenomena ocurring between the resonance and ablation surfaces. In Fig. 4, the distributions of temperatures, pressure and thermal conductivity are described as a function of the local plasma electron density for an aluminum slab target irra-

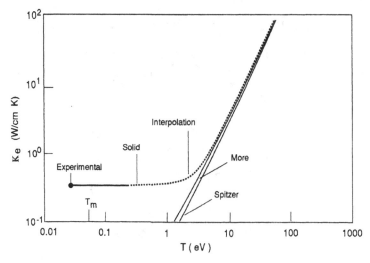

Fig. 3. Comparison of the conductivity κ_e for gold at
normal density by means of various models.
(A) Experimental value; (a) Lee and More[12];
(b)Spitzer[10]; (c)Present model.

diated by a 10^{12} W/cm^2 Nd laser beam. These distributions are
given at a time of 1.0 ns (the stage where the laser pulse
reaches its maximum). Similary Fig. 5 is describing the results
obtained for a gold slab irradiated by an intensity of $3\cdot10^{12}$
W/cm^2.

From Figs. 4 and 5 one can estimate the importance of two
temperature fluid codes by analyzing the difference between the
electron and ion temperatures and correspondingly, the gap between partial pressures and thermal conductivities. In the
corona, where n $\leq 10^{21}$ cm^{-3}, the difference in the thermodynamic quantities (as calculated by one or two fluid models)
may be larger by a factor of two or more. Therefore, most of
the phenomena in the corona should be studied by means of at
least two fluid model. In the domain between the corona and the
ablation surface the discrepancy between the thermodynamics
quantities of the two fluids differ by less than 30%. However,
from our calculations it is evident that one fluid code is appropriate for densities larger than those existing at the ablation surface, provided that one knows the time dependent evolution of the boundary conditions at this surface. Therefore, the
two temperature EOS developed in this paper are of major importance in analysing the plasma effects in the coronal region and
less predominantly in the ablation one, though for the comprehensive understanding of the phenomena occuring in the ablation
surface, it is required a detailed computation of the thermal
conductivity as it has been done here.

The simulation results obtained in the gold targets do not
significantly differ from those in aluminum. Thus it is concluded that two temperature EOS are important for most of the
materials in the study of the coronal region phenomena.

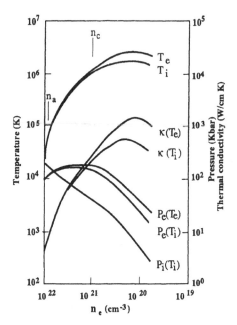

Fig. 4. T, P and profiles in a laser-produced aluminum
plasma at a Nd laser irradiance of 10^{12} W/cm².
Distributions are given for a time of 1.0 ns
from the beggining of the laser pulse.

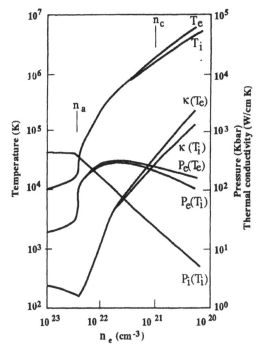

Fig. 5. T, P and profiles in a laser-produced gold plasma
at a Nd laser irradiance of $3 \cdot 10^{12}$ W/cm².
Distributions are given for a time of 1.0 ns
from the beggining of the laser pulse.

In Fig. 6 we show for the same conditions as described above, the distribution of the electron number density in an aluminum target as function of the plasma depth for different time intervals of 0.4, 1.0 and 1.6 ns. As can be clearly seen in Fig. 6 a caviton in the electron density is created in the vicinity of the ablation surface. The existence of a caviton is possible in a two fluid code while it can not be achieved in a one fluid simulation. The caviton is created in our model due to the behavior of the average degree of ionization $\langle Z \rangle$ (see Fig. 2): $n_e = \langle Z \rangle n_i$, where the ion density number n_i is a smooth function according to our simulations and the caviton in the n_e distribution is created by the effect of pressure ionization in the liquid phase. This phenomenon is described in our TFD model, while it cannot be achieved by for example by using an ideal gas EOS or the Saha rate equations for the calculation of $\langle Z \rangle$, as these last models do not include ionization pressure nor the continuum lowering of the ionization potential[9,14].

This computational predictions are qualitatively, in agreement with measurements carried-out by Briand et al.[15]. In that paper, nevertheless, the cavitons are reported to be originated in the corona. We do not see them in the corona because the pressure ionization effect is too small in this region in order to be statistically counted by the TFD model. In order to simulate the creation of cavitons in the corona, one has to introduce special components in the hydrocode, like the ponderomotive forces as done by Hora et al.[16], or perhaps the consideration of Friedel phase shifts[17] may conduct to the same effects.

In conclusion, the use in laser-matter interaction simulations of a detailed EOS quantum-mechanical treatment, where shell effects and other material properties are taken into account, needs huge computers and is very time consuming. In order to investigate various phenomena in this field of research it is important nevertheless, to have an accessible method of calculation by means of medium computers. We developed a com-

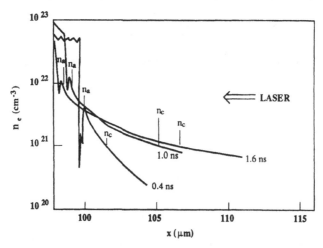

Fig. 6. The distribution of the electron number density as a function of the plasma depth calculated for a laser-produced plasma at a Nd laser irradiance of 10^{12} W/cm^2.

pact scheme in order to achieve this goal and for that purposes we developed a two-temperature EOS for a plasma medium. In doing so it was found that the two EOS is important mainly from the ablation surface outwards (toward the laser). In particular the creation of cavitons in the distribution of electrons was predicted, specially for light materials as aluminum.

REFERENCES

1. S. Eliezer, A. Ghatak and H. Hora, Introduction to Equations of State: Theory and Applications, (Cambridge Univ. Press, Eng., 1986).
2. A.D. Krumbein, H. Szichman, D. Salzmann and S. Eliezer, Israel AEC Report IA-1396, 1985 (Unpublished).
3. N.G. Cooper, Los Alamos National Laboratory Report LASL-79-62 (Unpublished).
4. R.D. Cowan, and J. Ashkin, Phys. Rev. 105, 144 (1957).
5. H. Szichman, and D. Krumbein, Phys. Rev. A 33, 706 (1986).
6. S.B. Kormer, A.I. Funtikov, V.D. Urlin,and A.N. Kolesnikova, Zh. Eksp. Teor. Fiz. 42, 686 (1962)[Sov. Phys.-JETP 15, 77 (1962)].
7. S.L. Thompson, and H.S. Lawson, Sandia Corporation Report SC-RR-710714 (Unpublished).
8. H. Szichman, S. Eliezer, and D. Salzmann, JQSRT 38, 281 (1987).
9. H. Szichman, and H. Hora, J. Phys. C 34, 5847 (1989).
10. L. Spitzer, Physics of Fully Ionized Gases, (Interscience, N.Y., 1962).
11. J.M. Ziman, Principle of Theory of Solids, (Cambridge,1969).
12. Y.T. Lee, and R.M. More, Phys. Fluids 27, 1273 (1984).
13. S. Jackel, R. Laluz, Y. Paiss, H. Szichman, B. Arad, S. Eliezer, Y. Gazit, H.M. Loebenstein, and A. Zigler, J. of Phys. E 15, 255 (1982).
14. R.M. More, Atomic and molecular physics of controlled thermonuclear fusion, edited by C.J. Joachim and D.E. Post, (Plenum, New York, 1983), p. 359-440.
15. J. Briand, M. El Tamer, V. Adrian, A. Goenes, Y. Quemener, J. C. Kiepper and J. P. Dinguirard, Phys. Fluids 27, 2588 (1984).
16. H. Hora, P. Lalousis, and S. Eliezer, Phys. Rev. Lett. 53, 1650 (1984).
17. C. Kittel, Quantum Theory of Solids, (Wiley, N.Y., 1963).

RADIATION DAMAGE IN SINGLE CRYSTAL CsI(Tl) AND POLYCRYSTAL CsI

O. Barnouin, A. Procoli, H. Chung and G.H. Miley

Department of Nuclear Engineering
University of Illinois at Urbana-Champaign
103 South Goodwin Avenue
Urbana, IL 61801

1. Introduction

CsI is a material that can be used for infrared detector windows and for making scintillators for high energy physics experiments (it is then doped with Thallium). This crystal may have to work in a harsh radiation environment which can cause an undesired signal to be sent to the detector or modify its calibration. It is therefore important to measure the response of this crystal to a flux of neutrons and gamma rays and to study the effects of dose accumulation. Such a study was conducted at the TRIGA nuclear reactor of the University of Illinois.

Radiation damage is most often studied through absorption measurements over the visible spectrum,[1] rarely through radioluminescence measurements. Radioluminescence occurs[2] when an electron excited to the conduction band deexcites radiatively, either directly to the valence band, or to a defect state (indirect, radiative deexcitation). Indirect deexcitation, however, can also occur without emission of a photon and the energy of the electron is lost through heating of the crystal lattice. The radioluminescent intensity then results from a competition between these processes. High energy radiations, by modifying the number of electrons at defect sites, modify the relative contribution of these deexcitation processes to the global deexcitation and leave the material in a state different from that before irradiation. One then speaks of radiation damage.

The investigation of radiation damage on CsI(Tl) has already been performed,[3] although not extensively, with 1.25 MeV photons from a ^{60}Co source as gamma-ray source. The radioluminescence was due to irradiation by a ^{137}Cs source. Although the total dose and dose rate were very much lower than in the present study (~500 rads, 100 rads per minute), a sizable reduction of the radioluminescence was observed (35% decrease of the emission at 500 rads). The damage was also found to be permanent.

It is the purpose of this paper to present results on radioluminescence measurements in both the visible and near infrared regions during a high-dose, pulsed irradiation of single crystal CsI(Tl) and polycrystal CsI by neutrons and gamma rays. It was found that, in both crystals, the radioluminescence decreases at first with increasing dose, but stabilizes after a few pulses. Also, heating the sample makes the radioluminescence decrease more rapidly in the single crystal CsI(Tl) than in the polycrystal. Finally, a cycle of heating and cooling is shown to repair the damage.

2. Experimental Setup

The samples underwent a series of pulses from the TRIGA pulsed nuclear reactor of the University of Illinois. The main characteristics of the pulse are shown in table 1 and the experimental setup is shown in figure 1.

Table 1. Pulse characteristics of the
University of Illinois' TRIGA nuclear reactor

Pulse width	12.5 ms
Total power in a pulse	5.8 kWh
Total dose per pulse (SiO_2)	100 krads
Dose components: neutrons	5 krads
gammas	95 krads
Peak dose rate	6.7 Mrad/s
Pulse rate	4 pulses/hour

The CsI(Tl) sample was manufactured by Harshaw Company. It was grown from a solution and is a disk 25.4 mm diam., 2 mm thick. The polycrystal was made at the Oak Ridge National Laboratory and is made of a powder compressed uniaxially at 100 MPa at a temperature of 150°C. It is 25.4 mm in diameter and 10 mm in thickness.

The sample under test is placed in the middle of a nine-inch long aluminum cylinder mounted on a carriage. This carriage is positioned in the middle of a thruport adjacent to the reactor core (fig. 1). The cylinder is wrapped in a resistive heating tape used for raising the temperature of the sample and is long enough to ensure a uniform temperature profile on the sample surface during the temperature rise. A thermocouple is attached to the carriage and is in direct contact with the sample to monitor its temperature carefully.

The light emitted by the sample impinges on a four-inch diameter, focussing aluminum mirror located at the exit of the thruport. The light is focussed onto an infrared, PbSe detector (detection range 0.7-4.5 μm) and onto a visible range, silicon detector (detection range 0.2-1 μm). A light chopper was used to account for the effect of the radiation on the detectors.

A typical reactor pulse provides a 0.1 Mrad dose to a SiO_2 sample. 0.5% of that dose (Table 1) comes from neutrons and 95% from gamma rays. The absorbed dose was not measured again in CsI, but a quick look at the respective gamma

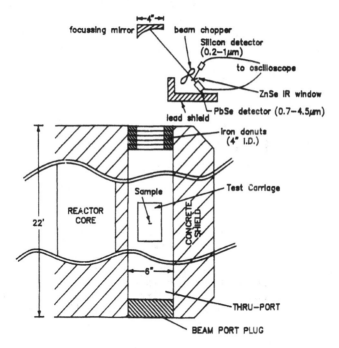

Fig.1. Experimental setup (Note: not to scale.)

absorption coefficients of SiO_2, Cs and I shows that the dose in CsI should not be larger than that in SiO_2 by more than a factor of two. Due to the higher atomic numbers of Cs and I with respect to SiO_2, the role of gamma rays will still be accentuated in CsI, making the neutron contribution to the radioluminescence and to the damage negligible.

3. Experiments Performed and Results

Repetitive pulsing was performed for two consecutive days on the single crystal CsI(Tl) and for one day on the polycrystal. The pulses were spaced 17 minutes apart (this minimum time is imposed by NRC regulations).

3.1 Single crystal CsI(Tl)

The sample of single crystal CsI that had been delivered was found to contain Thallium impurities (20 to 30 ppm[4]). An emission spectrum was obtained with an OMA and revealed an emission peak at around 0.54 μm which is characteristic of the presence of Thallium[4,5] (whereas pure CsI only emits around 0.305 μm[4]). Thallium increases the intensity of emission at 0.54 μm to the detriment of that at 0.305 μm. It also "protects" against radiation damage "by scavenging free carriers before they can find a more damaging site".[6] Further experiments will be needed with pure CsI to assess the role played by Thallium in the present results.

The change in radioluminescence for increasing dose in shown in figure 2 for the visible and infrared ranges (data taken at the peak of the pulse). In both ranges, the radioluminescence from CsI(Tl) at the peak of the pulse decreases and stabilizes after around nine pulses. The

403

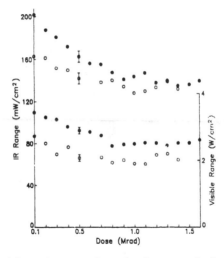

Fig.2. Radiation damage in single crystal CsI(Tl) at
room temperature. Solid circles are data taken
the first day, open circles to data taken the
next day. Infrared data are at the top.

decreases amount to ~ 30% of the initial signal in the
infrared and ~ 26% in the visible range for data taken the
first day. Due to the scatter of the data, a total of 16
pulses were taken to firmly establish the presence of this
saturation level before the temperature of the sample was
raised.

After that first series of pulses, the temperature of the
sample was raised and pulses were taken all along the
temperature rise. The results as a function of temperature
are shown in figure 3 for both wavelength ranges. The
emission decreases by factors of 30-40 for a temperature rise
of 100-125°C above room temperature.

Data taken during the second day (fig. 2) show what may
look like a recovery of the radioluminescence after the first
day. The radioluminescence during the first pulse is higher
than the stabilization level reached the day before. It then
decreases for increasing dose until a constant level is
reached that is slightly lower than observed the day before.
This might be due, however, to condensation of moisture on the
sample during the night.

3.2 Polycrystal CsI

This sample was first irradiated with a series of nine
pulses. The resulting radioluminescence change at the peak of
the pulse is shown in figure 4.a for the visible range
emission. In contrast with the polycrystal, no measurable
infrared emission was observed. The drop in figure 4.a is
around 60% of the original emission and takes places within
six pulses (compared with the 25% emission drop within nine
pulses for CsI(Tl)). After stabilization of the
radioluminescence (the scatter of the data is obviously less
in this case than for the single crystal), the temperature was
slowly increased to 170°C and pulses were taken as the

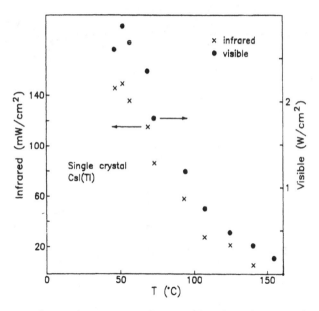

Fig.3. Radioluminescence intensity for increasing
temperature. Data taken right after the first
series of experiments at room temperature.

temperature was increased (fig. 4.b). The decrease of the
emission is then due to thermal quenching. In contrast with
the measurements on the single crystal, raising the
temperature decreases the emission (relatively to the
stabilization level obtained after 6 pulses) by only 60%.
This represents a decrease by a factor of 2.5, whereas a
diminution by 30 or 40 was observed for the single crystal.
The emission from the polycrystal also remains constant
between 150°C and 170°C.

The temperature was then brought down and some pulses
were taken around room temperature. Obviously, the
radioluminescence has recovered from the first series of
pulses. It however decreases again when the irradiation goes
on at room temperature (fig. 4.c) and it stabilizes at the
same level as that observed in figure 4.a.

As a final experiment, the temperature was increased
quickly (in about 15 minutes) to 135°C and then brought down
again after a pulse was taken at that temperature. The result
is indicated by the two crosses in figure 4.b. Provided that
there is no large temperature gradient at the surface of the
sample (as might arise when heating the sample so fast), this
experiment shows that the radioluminescence at higher
temperature depends on the heating rate (fig. 4.b, lower
cross). The annealing of the defect sites, which should
contribute to an increase of the radioluminescence, is lagging
behind the temperature, so that the radioluminescence is lower
than if the temperature was raised slowly. However, the
cooling rate was the same as during the first cycle, and,
since the defects were given enough time to be annealed, the
radioluminescence then coincides with the previous data (fig.
4.b, upper cross).

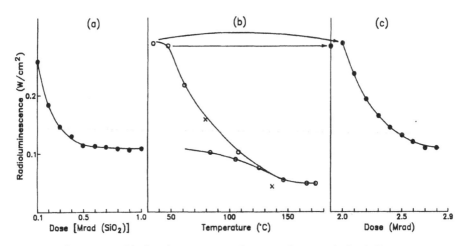

Fig.4. Radioluminescence from polycrystal CsI.
Chronological order is from left to right.

4. Analysis

These experiments show that CsI, either single or polycrystal, does suffer radiation damage when irradiated by a pulse of neutrons and gamma rays. This radiation damage is evidenced by a decrease of the radioluminescent intensity for increasing dose. However, two important conclusions have been reached: 1) the radioluminescent intensity stabilizes after a few pulses and, 2) the damage can be repaired by heating and cooling the sample, as is shown by the recovery of the radioluminescence after cooling.

The initial decrease of the radioluminescence might be due to several phenomena. First, an increase of the self-absorption during the irradiation. Absorption experiments performed at 0.633 μm show that the absorption at that wavelength increases during the whole pulse. However, the fact that the radioluminescence pulse is symmetric with respect to the peak of the pulse and is always proportional to the radiation pulse tells us that self-absorption does not play a more important role in the second half of the pulse than in the first half.

Second possibility, non-radiative recombinations can be favored by irradiation at high-energy. Available literature[7] does not give more precise explanations for the decrease of the radioluminescence for increasing dose. This decrease, together with the asymptotic behavior observed after 0.6-0.9 Mrad, can, however, be compared to the decrease and stabilization of the infrared transmission spectrum recorded after successive steady-state irradiations (0 Mrad (fresh sample), 1 Mrad,...) of a similar sample (fig. 5). The transmission decreases up to a dose of 1 Mrad (or less, since no data is available before 1 Mrad), and then remains constant for larger doses. Although the spectral range is different from that of the radioluminescence data, both kinds of results nevertheless evidence a saturation effect for the same total dose.

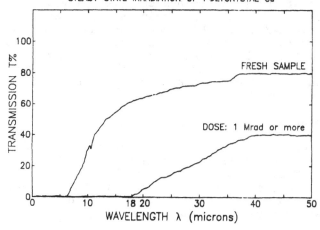

Fig.5. Transmission spectrum of polycrystal CsI before
and after several irradiation from the TRIGA
reactor at steady-state power.

It is impossible for us to evidence the exact saturation
mechanism from the data presented above. Considering the
small contribution of neutrons to the total dose and the small
kerma of CsI with respect to SiO_2, neutron creation of defects
is not a determining factor in our results. However, traps
can be filled by recombination of electrons located in the
conduction band after ionization by gamma-rays. They can also
be emptied by excitation. The number of electrons in traps
influences the relative rates for radiative and non-radiative
recombinations, hence the total radioluminescent emission.
Whatever the real cause, figures 4 and 5 tell us that, after ~
6 pulses, a kind of balance is reached in which the color
centers would lose as many electrons by excitation as they
would gain through electronic capture. An increase in
temperature alters that balance by exciting electrons out of
the traps (quenching).

The remark about the recovery of the radioluminescence
intensity after heating comes from the results on the
polycrystal (fig. 4.c). The temperature rise, apart from
quenching the emission, also modifies the number of electrons
in traps, decreasing the possibilities for charge carriers to
recombine non-radiatively, thereby increasing the
radioluminescence at room temperature. This can also explain
the recovery of the radioluminescence in the single crystal at
the beginning of the second run, which took place after a
temperature rise was performed at the end of the first day.
The fact that the stabilization level reached during the
second day is lower than that observed during the first day
might be a real feature of radiation damage or be due to
increased self absorption in the sample (due in particular to
moisture absorbed during the night).

Thermal quenching of the emission is not as effective in
the polycrystal as in the single crystal (fig. 3, 4.b). A

calculation of the radioluminescent intensity at 150°C, however, shows that it is approximately the same for both crystals (~40 mW/cm^2). This indicates that the crystal structure no longer matters at that high a temperature. This might be explained by assuming that all defects responsible for the visible emission have been annealed and that emission occurs by direct electron transition from the conduction band to the valence band.

5. Conclusions

This work shows that radiation damage in polycrystal CsI is stronger that in single crystal CsI(Tl). However, the presence of Thallium as a dopant in the single crystal clouds the issue. Further measurements on pure CsI are needed to determine whether Thallium effectively protects the crystal from radiation damage (or the opposite).

Several features of the radioluminescence exhibited by the samples are noteworthy: 1) the radioluminescence decreases in both the single and polycrystal CsI as a result of radiation damage; 2) the decrease is larger for the polycrystal than for the single crystal, but stabilizes in either case after a dose of 0.6-0.9 Mrads has been provided at room temperature; 3) a heating-cooling cycle repairs the damage by annealing some of the defects and, 4) heating the sample 120°C above room temperature decreases the emission in the polycrystal by only 60%, versus 98% in the case of the single crystal. The repair of the damage by heating is important since it means that the damage cannot be considered permanent, as long as one can afford to warm up the crystal.

6. Acknowledgements

This work was performed under contract with the Oak Ridge National Laboratory and the U.S.A. Strategic Defense Command.

We thank the University of Illinois' TRIGA reactor crew for their assistance in the experiment.

7. References

1.   For example, Kubo, K. "Radiation Effects in LiF Crystals," Journal of the Physical Society of Japan 16; 2294; 1961.

2.   See, for example, McKeever, "Thermoluminescence of Solids", Cambridge University Press 1985.

3.   Bobbink, G.J.; Engler, A.; Kraemer, R.W. "Study of Radiation Damage to long BGO Crystals," Carnegie Mellon University Report CMU-HEP 83-13; 1983. Cited by Ref.5.

4.   Brad Utts, Harshaw Company, Solon (OH), private communication, September 21, 1989.

5.   Grassmann, H.; Lorenz, E.; Moser, H.G. "Properties of CsI(Tl) Renaissance of an Old Scintillation Material," Nuclear Instruments and Methods in Physics Research 228; 323; 1985.

6.   Williams, R.T. "Nature and Defects Generation in Optical
      Crystals," SPIE Radiation Effects in Optical Materials
      <u>541</u>; 25; 1985.

7.   Levy, P. "Overview of Nuclear Radiation Damage Processes:
      Phenomenological Features of Radiation Damage in Crystals
      and Glasses," SPIE Radiation Effects in Optical Materials
      <u>541</u>; 2; 1985.

ACTIVE BEAM-CONTROL AND ACTIVE LASER-DIAGNOSTIC OF

INTENSE PULSED ION SOURCES*

K. Kasuya**, K. Horioka, Y. Saito, K. Matsuura, N. Tazima,
N. Matsuura, S. Kato*** and Y. Goino***

Tokyo Institute of Technology, Department of Energy Sciences
The Graduate School at Nagatsuta
Nagatsuta 4259, Midori-ku, Yokohama, Kanagawa 227, Japan

ABSTRACT

Recent results of the experimental research works in Tokyo
Institute of Technology concerning the intense pulsed ion beam are
described in this paper. The contents are divided into two major
subjects, one is a diode operation with an actively produced anode
plasmas, and the other is an active laser light diagnostics of the
diode. These are performed with the aim of making better diode
operations, which can be applied to obtain higher intensity diodes for
the future target irradiation experiments or so.

INTRODUCTION

To utilize the full potential of intense pulsed ion beams for
inertial confinement fusion, it is necessary to transport and bunch the
ion beams before the target irradiation. To do so, it is also
inevitable to make an active control of the beam production and to
investigate the fundamental particle behaviors in the diode gaps with an
active laser diagnostics beforehand. So that, the both subjects as are
shown in the title are described in this paper.

In our recent presentation (Kasuya et al., 1989), we proposed a new
method to pre-trigger and to make initial anode plasmas of pulsed ion
sources. With an additional pre-discharge along the anode surface
before the arrival of the diode main pulse at the anode, we could change
the turn-on delay and improve the diode and beam characteristics. After
this presentation, we continued the same kinds of experiments to gather
the detailed results, while a revised version of pre-discharge method
with another external energy source was tried also to improve the impedance characteristics of the diodes.

In our recent another presentation (Horioka et al., 1989), we
showed our preliminary results concerning the second subject, an active
laser diagnostics of the diode gaps. The time and spatial distributions
of the neutral particles and electrons were observed with a resonant
laser interferometry or a resonant laser scattering. After this presentation, we continued the same kinds of experiments with higher voltage

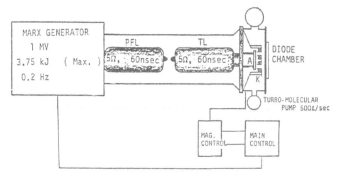

Fig. 1. Schematic diagram of pulsed power apparatus "PICA-3".

of the diodes after the revision of the pulsed power and the diode sections.

For the both items, we show our recent results in more detail in the following paragraphs.

PULSED ION DIODES WITH ACTIVELY PRODUCED ANODE PLASMAS

It is necessary to make the anode plasmas before the arrival of the main high voltage pulse at the intense ion diode, because there is turn on delay time between the high voltage pulse applied to the diode and the beam extraction from the diode without the sufficient anode plasmas beforehand. In the conventional diodes, the anode plasmas are produced passively with the application of the diode main high voltage, so that the controls of the diode and beam characteristics are not easy with this method.

In this paper, we adopt two kinds of methods to obtain high voltage pulses which are used to produce pre-plasmas as the anode source plasmas of the diodes. In the first method, we extract a small amount of energy from the voltage divider within the pulse forming line section of the main pulsed power apparatus for the diode. This voltage is applied directly to the flash board which itself works also as the anode plate.

In the second method, we use an additional and separate capacitors to get the high voltage to produce the anode pre-plasmas. The flash board is placed at a certain distance behind the anode plate, and the

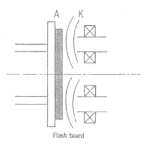

Fig. 2. Diode with flash board for pre-discharge on anode surface.

plasma is injected from the flash board to the anode. To exclude the disturbance by the neutral particles in the first method, the second is preferrable to the first, while the more dense plasma is obtained with the first method rather than the second method under a moderate consumption of energy for the pre-discharge.

Pulsed Power Apparatus for Diode

A schematic diagram of experimental apparatus is shown in Fig.1, which we call "PICA-3". The nominal specification of the Marx generator is 1MV, 3.75KJ and 0.2Hz (repetition frequency). The line impedance of the PFL and TL is 5 ohm, while the pulse width of them is 60ns. A water gap switch is used between the PFL and TL, and a magnetically insulated diode of beam extraction type is installed in the vacuum chamber. The anode is fed by this pulsed power apparatus and the beam is extracted in a ring shape. The electron current in the diode is suppressed by a pair of magnetic coils fed by an external capacitor bank. So that, both the Marx generator and the bank are triggered with a proper time delay. The vacuum chamber is evacuated by a turbo-molecular pump or an oil diffusion pump, and the anode is operated at the room temperature in our experiment.

Diode with Plasmas Produced by Flash-Discharges along Anode Surface

Diode and Flash Board of Active Anode. We use a beam extraction type, magnetically insulated diode which schematic diagram is shown in Fig. 2. A pulsed radial magnetic field is applied within the anode-cathode gap region of 8mm width. The diameter and the thickness of the anode base plate made of brass are 210mm and 12mm. On this metal plate, a dielectric flash board of 200mm diameter and 1.6mm thickness is attached. On the surface of this flash board, a train of metal thin film electrodes is located along the azimuthal direction. The total number of flash gaps on the anode surface is from 80 to 100. The whole azimutal angle is divided into 4 segments, respective of which is connected with one of the feed cables. The pre-formed plasma with this pre-discharge is supposed to act as the source plasma, which grows up with the main diode pulse afterward. The cathode is made with a pair of thin coaxial SUS pipes, between which the pulsed ion beam is extracted. A pair of magnetic field coils are located at the inner and outer places of the cathode pipes. The inner and outer radii of the both cathode pipes are 90 and 130mm. The number of turns of the inner and outer coils are 24 and 4. A capacitor bank of 500µF and 5KV is used for these coils.

High Voltage Circuit for Pre-Discharge on Anode Surface. Instead of the conventional anode of the pulsed ion diode, we use a flash discharge type anode, which is fed by a high voltage circuit excited by a voltage divider installed in the PFL at a place near to the water gap switch.

An equivalent circuit diagram of the high voltage circuit to feed the anode flash board is shown in Fig. 3. Here C_p, R_1, R_2, G, C and L are the capacitance of the PFL, the resistances of the both divider resistors, the circuit air gap switch, the capacitance of the intermediate capacitor and the inductance of the load inductor to bridge the anode feeding cables. In this case, the anode flash board is segmented into 4 quarter parts, and we use 4 cables to make the anode flash discharges.

The PFL is charged up to about 600KV in about 500ns by the Marx generator, and a part of the energy is extracted by the sodium thiosul-

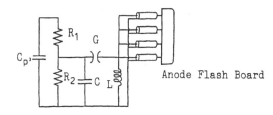

Anode Flash Board

Fig. 3. Equivalent circuit of power source for pre-discharge on anode
surface.

fate ($Na_2S_2O_3 \cdot 5H_2O$) voltage divider shown in Fig. 3. The maximum charging voltage of C (a ceramic peaking capacitor in Fig. 3, which is placed
at outside of the diode chamber) is about 40KV. By the adjustment of the
air gap length, the charging voltage of the capacitor and the cable
length, the voltage and the timing of feeding pulse is controlled. The
inductance L in Fig. 3 is necessary to fix the DC potential of the right
hand electrode of the air gap at ground potential.

The length of the 50 ohm cables is 7m, and the capacitance and inductance of the respective cable are 0.47nF and 2.2µH. As the propagation time per unit cable length is 5ns/m, the time for the pulse to make
a round trip of the whole cable length is 70ns. As we need about 300ns
to get the peak voltage of C, the cable and anode flash board behave
like a distributed constant circuit.

Diagnostic Methods. We use a sodium thiosulfate voltage divider to
measure the output voltage of the TL. The diode voltage and current are
measured with shunt inductors and current pickups, while the extracted
ion beam current is measured with a biased ion collector (BIC). An
electric probe or high speed camera are also used to measure the pre-
produced plasmas or the uniformity of the plasma brightness.

Experimental Results. An open-shutter photograph of the anode
flash over (end-on view) is taken, and a relatively uniform brightness

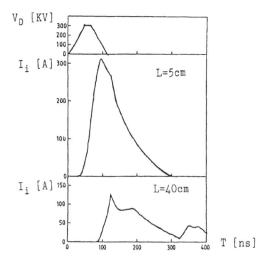

Fig. 4. Traces of diode voltage and beam ion current measured at 5 and
40cm from anode, with pre-discharge and with delay time of
60ns.

414

is assured along the azimuthal direction, while the radial uniformity is not enough so good. From the voltage trace at the both ends of the single (quarter angle) flash board segment, the pre-discharged plasma is produced within several 10ns in this case.

To make a rough estimate of the plasma density, the plasma current is observed with a biased ion collector. The results, for example, the peak current vs. capacitor voltage, the plasma expansion speed vs. the voltage and the amount of produced plasma vs. the voltage are shown in another paper, Kasuya et al. (1989). We can say that the estimated amount of plasma are nearly proportional to the capacitor voltage.

The beam characteristics with and without pre-discharges are also investigated in another paper, mentioned above. The standard conditions of the experiments are as follows. The output voltage of the Marx generator is 600 KV, the anode cathode gap length is 8mm, and the capacitor voltage for the pre-discharge is 40KV. The delay time between the pre-discharge and the main diode pulse is changed (20-200ns), and the corresponding change of the extracted beam characteristics are examined. Examples of the diode voltage (the more precisely, the output voltage of the TL) and the extracted ion current vs. time are shown in Fig. 4, where the biased ion collector is located at 5 and 40cm from the anode surface, and the delay time is 60ns.

With many other these kinds of traces as are shown in Fig. 4, we can explain the diode operation as follows. Without the pre-discharge, the ion current rises late, so that the ions are accelerated at the decreasing part of the diode voltage. This corresponds with a long turn-on time of the anode plasmas. On the other hand, with the delay time between 20 and 60ns, the numbers of ions accelerated at the rising part of the diode voltage increase with the delay time. This means that the anode source plasmas are supplied in early time and the turn-on of the ion beam is improved. With the delay time of more than 90ns, the pre-discharged plasmas expand widely in the anode-cathode gap, and the diode voltage does not become high as before under the nearly shorted condi-

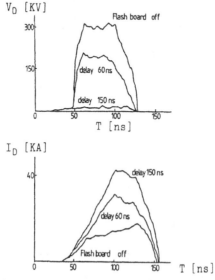

Fig. 5. Traces of diode voltage and diode current with and without pre-discharge, and with delay time as a parameter.

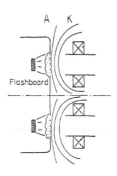

Fig. 6. Diode with plasma injection to supply anode pre-plasmas.

tion of the gap. Only the low energy beam is extracted, and we can not get an enough amount of the high energy beam.

Another kind of traces are gathered in Fig. 5, where the diode voltage and diode current vs. time are measured with the same diode. The parameter is the delay time (60 or 150ns), and the case without the pre-discharge is also shown. The optimum delay time in our experiments is said to be about 60ns with these figures.

Diode with Injected Plasmas from behind Anode Surface

During our investigations of the particle distributions within the anode-cathode gap region and the diode characteristics described above, it becomes clear that the separation of charged particles from the neutral particles within the neighborhood of the anode surface is preferrable from the point of view of the better diode and beam characteristics. So that, we modified the above diode as follows.

Diode and Flash Board of Active Anode. We use the same kind of magnetically insulated beam-extraction type diode, which is shown above. The cross-sectional view of the diode is shown in Fig. 6, where A and K are anode and cathode. Here the anode face is made of a mesh-like structure, and the anode source plasma is injected from behind the mesh. As is shown in the figure, the flash board is placed at the left side of the anode face with some distance. When the flash discharge occurs and the plasmas move toward the right direction, the neutral particles are retarded to arrive at the mesh surface because of the difference of the expansion speed. If the timing of the diode main

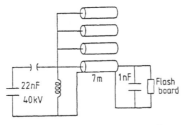

Fig. 7. Equivalent circuit of power source for plasma injection type flash board.

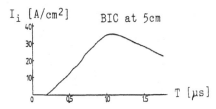

Fig. 8. Plasma current measured by biased ion collector.

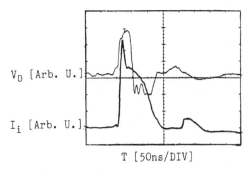

T [50ns/DIV]

Fig. 9. Traces of diode voltage and beam ion current measured at 3cm
from anode, with plasma injection and with delay time of 60ns.

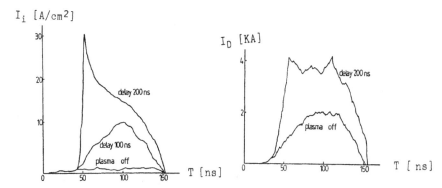

Fig. 10. Traces of beam ion current density and diode current with and
without plasma injection and with delay time as a parameter.

pulse is adequate, the neutrals do not disturbe the clean conditions of
the anode region. Only the charged particles pile up just at the left
side of the mesh, which act as the ion source.

The experimental conditions for this diode is as follows. The anode
-cathode gap length is 8mm, the radial spread of the mesh is 10mm, the
area for beam extraction is $20cm^2$, and the nominal time for the plasma
and the neutral particle to arrive at the mesh position are 200ns and
1,000ns. The ratio of the magnetic field strength applied to the anode-
cathode gap and the same critical value to insulate the electrons is 1.6
normally. Except for the methods to supply the anode pre-plasmas, the
both diodes in this section and the former section are same altogether.
And the diagnostic methods used for this diode is also same with the
former one.

<u>High Voltage Circuit for Pre-Discharge on Anode Surface.</u> As is
shown in Fig. 7, we use a modified version of the former circuit. The
left capacitor in the figure is a DC-charged external one, and the air

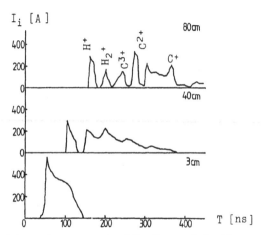

Fig. 11. Traces of beam ion current with plasma injection and with different location of ion collector (3, 40 and 80cm).

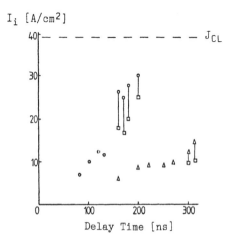

Fig. 12. Beam ion current density at 3cm from anode as a function of delay time.

gap is triggered by the output voltage of the voltage divider in Fig. 3. The inductor and the flash board here are the same one as the former diode, while the peaking capacitor is placed just at the nearest position of the flash board. This circuit is operated also without the peaking capacitor. In Fig. 7, only the quater part of the whole circuit is shown for the right side of the feeding cables.

Experimental Results. The ion flux of the injected plasma by the flash board is measured by a biased ion collector, which result is shown in Fig. 8. The collector is placed at 5cm distance from the anode mesh. When the magnetic field for the diode is applied, the signal is not measured at all, because the plasma is suppressed to expand into the diode region.

Examples of diode voltage and beam ion current are shown in Fig. 9, where the biased ion collector is placed at 3cm from the anode surface and the delay time between the pre-discharge and the main diode pulse is about 200ns. A very sharp rise of the ion current is observed in the same figure and there is no turn-on delay of the diode of course.

Another traces of the beam ion current density and the diode current are shown in Fig. 10, where the parameter is the delay time of the both high voltage pulses, and the traces without the pre-plasmas are also shown. With our pre-discharge circuit adopted here, the maximum delay time is 200ns. Within this limitation, the longer the delay time is, the larger beam ion current density is obtained.

Serialized traces of total beam ion current are shown in Fig. 11, where the distance of the biased ion collector from the anode is changed from 3 to 80cm. There is no clear separation of lighter and heavier particles at 3cm, while the separation by the time of flight becomes clear with the distance. Respective species are seemingly separated at 80cm. Anyhow, the rising part of the signal is very sharp, especially for the proton. The experiments with longer delay time are expected after the revison of the triggering circuit of our apparatus in the near future.

The representative beam ion current density from a serialized traces as are shown in the former figures at 3cm distance from the anode are plotted as a function of the delay time in Fig. 12. In the same figure, circles correspond to the proton peaks in the beam ion current density traces, and the squares connected with the circles correspond to the heavier particle shoulders in the traces. These are for the case of with the peaking capacitor, while the remainders are for the case of without the peaking capacitor. The triangles and the connected squares correspond to the same ones as are described above. The longest delay time for the case without the peaking capacitor is 300ns, while the one for the case of with the capacitor is 200ns. The horizontal dotted line in Fig. 12 corresponds to the Child-Langmuir ion current density in this case. As all of the data points stay below this line, the anode plasma density is not high enough to supply the diode needs.

LASER DIAGNOSTICS OF PARTICLES BETWEEN ANODE AND CATHODE

As one of the main reasons of the diode falling impedance with time is seemingly the unexpected fast expansion of the neutral particles within the diode anode-cathode gap, it is very urgent and also very important to investigate time and space resolved behaviors of the neutral and charged particles within the gap. Therefore we show our recent results of active laser diagnostics for this purpose in this section. We adopt two kinds of diagnostic methods, the high speed optical interferometry with tunable laser light and the laser induced resonant scattering, which are described in the following sections.

Optical Interferometry with Tunable Laser Light

Experimental Apparatus. A schematic diagram of the experimental apparatus is shown in Fig. 13. The same kind of diode described above is driven with the same kind of pulsed power apparatus of Blumlein line type called "PICA-2" by us. The anode in this case is made of paraffin wax $(CH_2)_n$, from which ion beam is extracted. The typical diode operation condition is also same with the former one.

A tunable dye laser pumped by a nitrogen laser with a LC-inversion type is used as the light source of the Michelson interferometer. The nitrogen laser of 5ns pulse width is triggered by a voltage pulse from a voltage divider at the end of the Blumlein line.

Experimental Results. The resonant interferometry is useful when we need high sensitivity. As the pulsed power apparatus and the diode

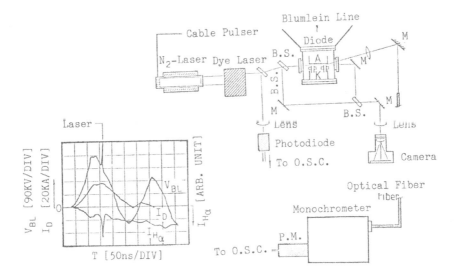

Fig. 13. Schematic diagram of active laser diagnostic for intense diode.

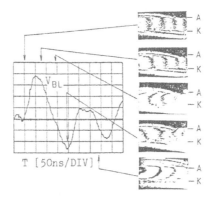

Fig. 14. Serialized interferograms at different diode time sequences accompanied by trace of Blumlein line voltage.

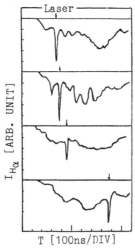

Fig. 15. Traces of serialized scattered light signals at different diode time sequences.

in our laboratory are not so strong as the ones in larger laboratories in the world, the absolute value of particle density is not enough high to use the normal type of interferometer.

As the relations between the number of fringe shift and the respective particle density are described in a paper, for example, Horioka et al. (1989), it is repeated here only that the directions of the fringe shift by electrons and neutral particles are opposite, which is used to say qualitatively which particles exist dominantly within the diode gap with the interferogram.

Examples of the interferograms are shown in Fig. 14, where the laser light is of 0.07 nm bandwidth and tuned to the wavelength of at 0.08 nm longer than the center of the H_α line. A typical trace of the Blumlein line voltage and the timing of the laser light are also shown in the same figure. As can be seen in these interferograms, the electron density buildup precedes the neutral hydrogen buildup, which result supports qualitatively the plasma expansion model for this kind of diode. Namely, the fast plasma expansion witin the diode gap is induced by the charge exchange of the high energy neutral particles.

Laser Induced Resonant Scattering by Particles in Diode

As the interferometry mentioned above has no space resolution in the direction of the optical path length, the resonant scattering of the laser light by the local particles is tested here to estimate the particle distribution in the diode more quantitatively.

Experimental Apparatus. A schematic diagram of the apparatus is shown in Fig. 13. The spectral width of the tuned dye laser light is 0.07nm centered at H_α line. The induced-scattered light is focused on the optical fiber which has a rectangular cross section and 7m length. Then the light is guided into a 50cm monochrometer (Jasco CT-50C) and observed with a photomultiplier (Hamamatsu R2257). The image of the diode region is formed on the front end of the fiber. The scattered light from the diode plasma of 1mm thichness is detected together with the background H_α line intensity. The spectral bandwidth of the monochrometer is 0.6nm, so as to detect the whole H_α spectrum. Typical osciloscope traces of the Blumlein line voltage, the diode current and the resonant scattered light in this case are shown in Fig. 13. These are explained in the following section.

Experimental Results. Typical osciloscope traces of the photomultiplier signals are shown in Fig. 15, where T is the time from the diode voltage rise. Compared with the background H_α level, the (two or three times) enhanced scattered light is observed in response to the incident laser light.

When the incident laser light is intense enough to saturate the the resonant excitation, we can estimate the initial ratio of upper and lower level populations before the laser excitation from the ratio of the peak scattered light intensity and the background light intensity, Horioka et al. (1989). This is not so easy when the saturation is not realized. For the moment, we are not sure that the signals in Fig. 15 are obtained under the saturation condition or not. So that, in the near futeure this must be assured at first to make the observation more quantitatively.

CONCLUDING REMARKS

    With two kinds of methods to make pre-discharged anode plasmas, it
is demonstrated that we can control the turn-on time of the pulsed ion
beam and the impedance characteristic of the diode. There seems to be a
most preferable timing to make a best pre-discharged plasma before the
main diode pulse.

    To obtain more suitable anode plasmas for this kind of pulsed ion
diodes, it is necessary to continue the same R and D. Exactly necessary
amount of anode plasmas to extract the full ion beams which become to be
starved at the decreasing part of the diode voltage pulse are expected
to be produced to make the diode impedance characteristic more preferable, in the near future.

    To make clear the behaviors of the various particles in the diode
gap, two kinds of active laser diagnostics are tried preliminary.   In
principle, the resonant interferometry and the resonant scattering make
us possible to estimate numerically the difference and the ratio between
the corresponding lower and upper level populations to the laser wavelength. So that, when we make the both diagnostics simultaneously with
the same diode operation, we can decide the particle number density in
the both levels. Although the results shown here are very preliminary
for the both diagnostics, we will continue our works and we hope that we
will be able to make the diode physics clearer in the near future.

ACKNOWLEDGEMENTS

    The supports by the Grant-in-aid for Scientific Research from the
Ministry of Education, Science and Culture in Japan are greatly
acknowledged by the authors. They also thank to the support by the
Japan Society for Promotion of Science, Musashi Institute of Technology,
Nissin Electric Company Limited and Tokyo Institute of Technology. One
of the pulsed power apparatus, "PICA-3" is installed in a Building for
Cooperations in Tokyo Institute of Technology at Nagatsuta Campus, which
members concerned the authors acknowledge also. The technical supports
by the students, A.Hagiwara, T.Aso, N.Hikida, H.Iida and M.Murazi in our
laboratory must be acknowledged lastly.

REFERENCES

Horioka, K., Tazima, N., and Kasuya, K., 1989, Prooeedings of 2nd Int.
        Conf. on Ion Sources, July 10-14, Berkeley, CA, U.S.A. (to be
        published).
Kasuya, K., Horioka, K., Saito, Y., Kato, S., Goino, Y. and the
        PICA-3 Group, 1989, Proceedings of 7th IEEE Pulsed Power Conf.,
        June 11-14, Monterey, CA, U.S.A. (to be published).

*   Supported by a Grant-in-aid for Scientific Research and a Grant-in-
     aid for Promotion of Cooperation between University and Private
     Company both by the Ministry of Education, Science and Culture in
     Japan and Tokyo Institute of Technology.
**   Presented at the 9th International Workshop on "Laser Interaction
     and Related Plasma Phenomena" held at Naval Postgraduate School,
     Monterey, California, U.S.A., November 6-10, 1989.
*** Permanent address : Beam Application Division, R & D Department,
     Nissin Electric Co., Ltd., 47, Umezu-Takase-cho, Ukyo-ku, Kyoto,
     Japan 615.

SATURABLE MAGNETICS FOR LASER AND PLASMA INTERACTIONS

H. Kislev and G. H. Miley

Fusion Studies Laboratory
Department of Nuclear Engineering
University of Illinois
Urbana, Illinois 61801

INTRODUCTION

Recently there has been a growing interest in sharpening electric and magnetic pulses in order to improve a wide variety of pulsed sources like compact particle accelerators[1], X-ray sources etc. An essential tool for pulse sharpening is the saturable magnetic switch. Some of the unique properties of this device includes negligible energy absorption, fast recovery (i.e., high repetition rate), low specific volume (liter/kW), and an improved design flexibility compared to conventional spark gap switches.

Here we present potential new applications for magnetic switches in the area of laser and plasma interaction. First a numerical method we used to estimate the response of circuits with magnetic switches is presented. Then, the response of a ferrite loaded single-turn coil connected to a laser-driven capacitor plate is analyzed in some detail. Finally, the design of a 50-Hz magnetic switched pulser for a repetitive Z-pinch device is presented along with energy balance issues.

BACKGROUND

The key property of a magnetic switch is the complete magnetization saturation of the core. The B-H curve and the magnetic permeability μ_r vs. H curve are shown in Fig. 1a for a typical ferrite. An increase in the magnetic field intensity H takes the core from the remnant field B_r along the solid curve to the saturation field B_s for a total flux swing of $\Delta B = B_r + B_s$. The core permeability first increases as the core comes out of saturation, reaches a peak corresponding to a high-impedance state, then is "switched off" to a low impedance as the current raises further. The initial state of the core can be regulated using a reset circuit. (A thorough discussion of saturable inductors and magnetic switches is contained in Ref. 2.)

The degree that a core below the Curie point can hold off a voltage V (volts) for a time t (seconds) depends on the flux swing B(Tesla) in the core, the cross sectional area of the core in m^2 and N, the number of turns around the core (Fig. 1b).

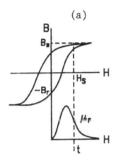

(a)

Valid when $T \leqslant T_C$

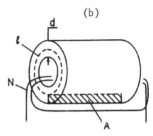

(b)

$$L = \frac{\mu_r \mu_o N^2 A}{\ell}$$

$$\int V dt = 2\Delta BAN$$

$$\tau \geqslant \frac{d}{\sqrt{\mu\varepsilon}}$$

Fig. 1. (a) Magnetic field (B) and magnetic permeability
(μ) as a function of magnetic field intensity (H).
(b) Core parameters.

$$V dt = 2\Delta BAN$$

The inductance of the core is given by

$$L = \mu_o \mu_r \, AN^2 \, /\ell$$

In this expression, μ_o is the permeability of free space, μ_r is the relative permeability of the core material, and ℓ is the mean magnetic field path length. In fast magnetic switching, the switch's saturation inductance and the number of turns must be minimized.

MAGNETIC SWITCHED CIRCUIT DESIGN USING PSPICE

A valuable tool for magnetic switch design is a circuit analysis code. Such a computer code allows the designer to enter the switching circuit parameters and subsequently monitor its response without actually building it. One such code, PSPICE[3] has been in use for some years. It is simple to use yet powerful and can be operated both in personal and mainframe computers.

The saturable inductor "device" in PSPICE is based on the Jiles-Atherton model[4] and uses the following parameters: M and M_s are the actual and saturation magnetization, a is the mean field parameter, α is the slope while C and K are the domain wall flexing and pinning constants, respectively. First one has to fit the saturable core model parameters to the B-H curve of the selected ferrite. The anhysteric B-H curve is obtained in PSPICE by loading the saturable core with a triangular current pulse while setting the initial permeability using the equation

$$\mu_r = 1/(3\alpha M_s - a)$$

while setting K to zero. Then, K is set to a non-zero value to create
hysteresis and finally C is fixed by desired initial permeability.

Second, the circuit is wired using the given (or estimated)
components values. Allowances for the potential drop in switches,
internal leaks and losses may be also needed. Unfortunately, magnetic
switched circuits are prone to numerical instabilities. To circumvent
this problem, some small inductors and resistors are added in series
with the saturable core. In addition, the current tolerance must be
kept well below 10^{-5} in order to avoid numerical instabilities and to
maintain energy conservation. With these improvements, the simulation
circuits described here are typically solved in a few minutes using an
IBM-AT equipped with a 80287 co-processor.

MODIFICATION OF CO_2 LASER DRIVEN MAGNETIC PULSER

Some laser-plasma interactions induce brief but very intense
electrical phenomena. Magnetic switches seem to be optimal for such an
environment. They can hold large voltages for a short time and yet are
extremely fast and compact. The switching period is roughly $d/\sqrt{\epsilon_0\mu_0\mu_r}$
where d is the core thickness. This period is typically a few hundreds
picoseconds for a d = 3 mm core.[6] Next we describe a possible
application for magnetic switches in select types of laser targets.

Recently, it was found that a capacitor plate "target" illuminated
by an intense focused 100-J CO_2 laser pulse is charged to several 100's
kV. This capacitor can in turn discharge intense currents into a tiny
coil coupled to the target (see Fig. 2a).[5] This type of target (LCR
target) produces magnetic microfields which can reach several hundred
Tesla.

We have constructed a simple circuit to simulate the LCR target[5]
using PSPICE (Fig. 2b). The twin plates capacitor is charged negatively
and then positively to simulate electron effects. Its capacitance was
set to be 1 nF following energy and resonance frequency considerations.
In other words, the plasma dielectric constant must be about three
orders of magnitude higher than that of vacuum during charging of the
capacitor. A variable resistor is installed in parallel with the
capacitor to simulate the gap shorting process.

We have tested this simulation circuit by comparing its response
to the results of a published experiment.[5] The measured and simulated
voltage and current traces are shown in Fig. 2c for a 2-mm diameter wire
coil (4.5 nHy, 0.18 ohm). Apparently, there is a good agreement between
the simulated and measured current traces during the first three
nanoseconds. The peak magnetic field is 55 T while the peak dB/dt
reaches 10^{11} T/sec at t = 1.1 nsec.

The energy transfer from the capacitor into the single turn coil
is described in Fig. 3a. Note that the capacitor energy is transferred
to the coil only after the gap is shorted. When the single-turn coil
is replaced with a strip coil (0.3 nHy, 0.005 ohm), 45% of the stored
energy in the capacitor is converted to magnetic energy (see Fig. 3b).
(Note that the traces are meaningless after the capacitor became shorted
at t =2 nsec.) The simulated peak magnetic field is increased to 200 T
(compared to the measured 400 T) and the peak dB/dt is $5*10^{11}$ T/sec.
This difference in B_{max} may be attributed to a delayed contraction of
the foil's inner surface in the actual experiment.

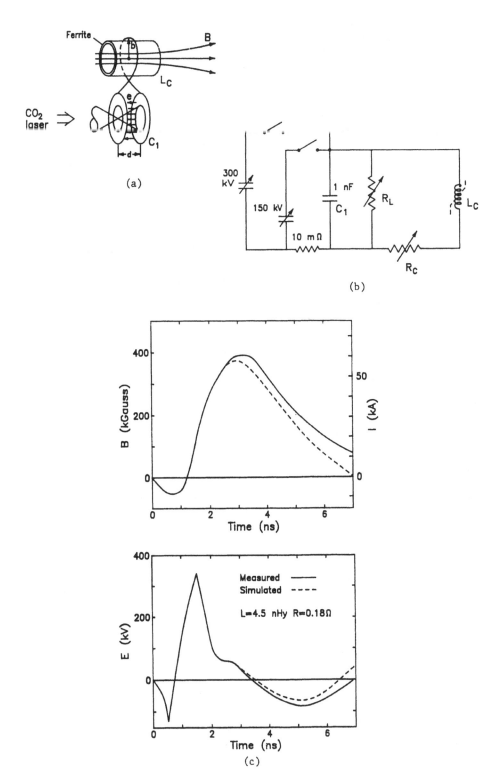

Fig. 2. LCR laser target analysis (a) Schematic target diagram (b) Equivalent circuit for several inductive loads (c) Simulated and measured voltage and current traces from a target containing a single turn wire coil.

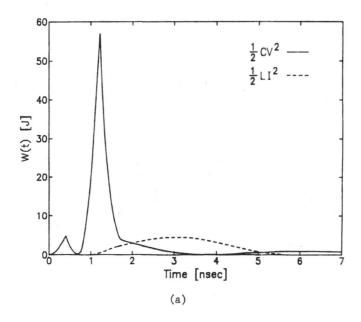

(a)

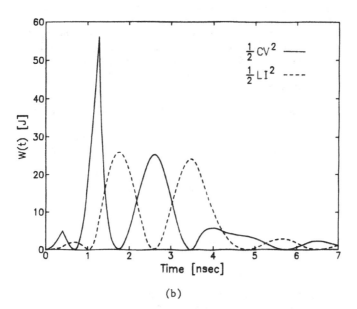

(b)

Fig. 3. Electrical energy transfer in a LCR laser target (a) with a single
turn wire coil and (b) with a low-inductance strip coil.

The magnetic switching effect was estimated by inserting a tiny core ($\Delta B = 3$ T, A = 0.001 cm^2 and l = 0.3 cm) into with the single-turn wire coil. The dB/dt histories, with and without switching, are shown in Fig. 4a. One narrow peak ($2*10^{12}$ T/sec) can be observed on the switched trace during the switch saturation. The dB/dt peak FWHM is roughly 7 psec which equals to the switching time (in PSPICE d = $\sqrt{A/\pi}$). The heat penetration into the ferrite layer (> T_c) following exposure to the ohmic heated coil (100 eV)[5] is found to be negligible (e.g. ~0.3 μ after 10 nsec).

The sudden change in dI/dt during on--off switching induces high frequency components in the current history. We have calculated the multi-GHz spectral content of the current trace in strip coil type LCR targets with the core ($\Delta B = 3$ T, A = 0.01 cm^2 l = 0.3 cm) and without switching. With switching, a 100-psec 5-kA period of 5 - 7 GHz (and additional higher harmonics) is added to the fundamental 0.5 - 1 GHz peak of the non-switched current history (Fig. 4b). Since it is known that cylindrical laser targets emit an RF spectrum which correlates well with the illuminating laser power.[7] The present finding is particularly interesting. By coupling such a coil with a proper resonator, we may be able to produce extremely short (100 -10 psec) RF pulses at a desired frequency range. Similar results could be expected in larger strip coils coupled to capacitor banks.

MAGNETIC SWITCHED SUPPLY FOR FAST Z-PINCH DEVICES

Z-pinches are useful as keV X-ray pulsed sources. Their applications include X-ray sub-micron lithography[8] and X-ray laser pump schemes.[9] In some Z-pinch devices, the soft X-rays are generated during a multi-kA discharge in a gas puff or in evaporated material from a capillary. Much research is currently invested in upgrading the average power of the Z-pinch by either increasing the energy per pulse or raising the device repetition rate. We will describe two potential schemes here that use magnetic switching to increase the average power of such Z-pinch devices.

For the present work we have simulated the circuit response during a capillary Z-pinch discharge experiment[9] using "optimal" component values (see Fig. 5a). The simulation circuit consists of a 20-kV (compared to 25 kV in Ref. 9) 1.8-μF capacitor bank coupled to an inductive resistive load (2.5-cm long NaF capillary). The simulated and measured plasma current traces are shown in Fig. 5b. Although about 90% of the total discharge ohmic energy is deposited during the first half cycle (1.5 μsec FWHM), the experiment showed that most X-ray photons are emitted during the second and third half cycles.

In this experiment, a sodium bearing discharge was produced from NaF plasma ablated from the capillary wall during the discharge. The sodium emits in the Na x $1s^2$ $1S_0$-$1s^2p$ $^1P_1$ line at 11.0027 A which can be used to pump the Ne 11.0003 A line. Analysis of Faraday cup signals (Fig. 5c) and X-ray pinhole images revealed non-uniformities in both the radial and axial directions. This is presumably caused by successive plasma fronts, associated with the periodic nature of the current. Since a stable and uniform pumping is an essential for lasing, the authors of Ref. 7 suggested driving the Z-pinch with a wider square current pulse.

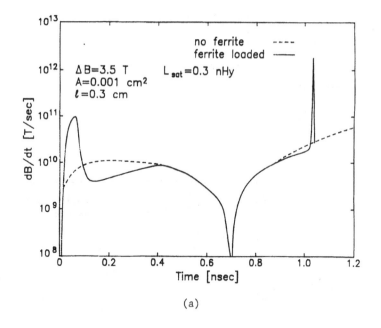

(a)

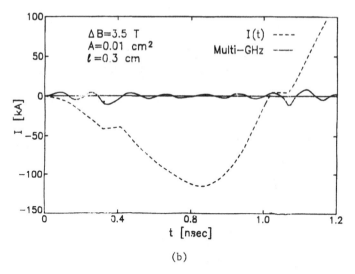

(b)

Fig. 4.   (a) dB/dt histories near LCR laser targets with both bare and
ferrite-loaded single turn coils.   (b) Total and multi-GHz
current trace for a LCR laser target with a ferrite-loaded strip
coil.

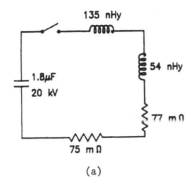

(a)

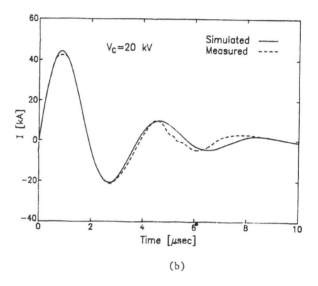

(b)

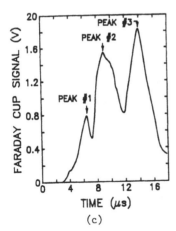

(c)

Fig. 5. Sodium bearing capillary Z-pinch discharge experiment (a) Simulated circuit. (b) Simulated and measured current traces from
this circuit. (c) Voltage signal from Faraday cup located 15 cm
from the Z-pinch.

To accomodate this requirement, we modified the original circuit by placing an additional 1.8 μF capacitor C_2 in parallel with the first capacitor (See Fig. 6a) while retaining the same load parameters. Only when C_1 is mostly discharged, the magnetic switch S_2 is opened and C_2 is discharged into the load. Both capacitors are fed from a single 3.6 μF intermediate capacitor which in turn can be charged from a slower driver. As seen in Fig. 6b, the current pulse shape expands to 3 μsec FWHM 40-kA peak (compared to 2-usec 60-kA peak using a single 3.6 μF capacitor). This wider pulse shape appears to offer better hydrodynamic stability and to provide a more uniform heating of the ablated plasma from the capillary.

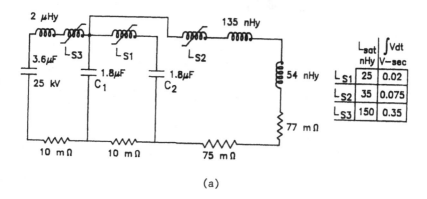

	L_{sat} nHy	$\int Vdt$ V–sec
L_{S1}	25	0.02
L_{S2}	35	0.075
L_{S3}	150	0.35

(a)

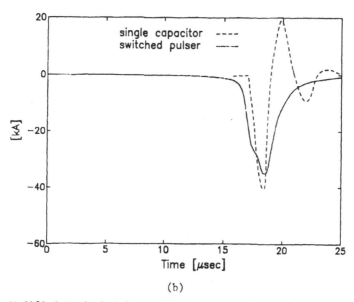

(b)

Fig. 6. Modified Z-pinch driver with an extended pulse width (a) Modified circuit with an intermediate capacitor (b) Current trace "obtained" from the modified circuit.

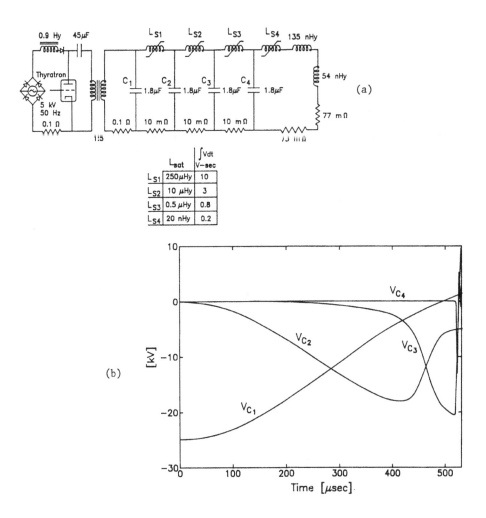

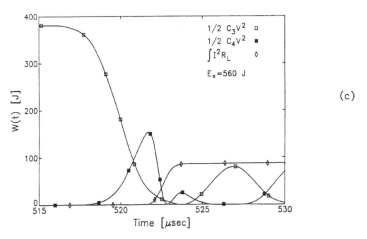

Fig. 7. Magnetic-switched 200-J 50-Hz pulser for Z-pinch devices (a)
Electrical diagram. (b) Charging voltage traces of the final
capacitors. (c) Energy transfer through the final driver
components.

In order to drive the Z-pinch at 50 Hz we "constructed" a Melville pulse compression line.[2] This circuit consists of a resonant charging unit followed by several magnetic compression stages which end up with the load (see Fig. 7a). As the capacitor C_1 charges up, the volt-second applied to the switch L_{s1} increases until saturation occurs. At saturation, the switch's inductance drops drastically, allowing C_1 to discharge resonantly through the switch to charge C_2. L_{s1} is designed to be much smaller than L_0 so the discharge's time constant $\pi \sqrt{L_{s1}C_1/2}$ is much shorter than the charging time of C_1, resulting in pulse compression. The process is repeated successively until the last capacitor is discharged through the Z-pinch load.

Fig. 7b depict[5] the charging voltage of the four capacitors closest to the load. The discharge time of the final capacitor (90% to 10% of peak voltage) is about 3 μsec compared to 450 μsec of the first one. Apparently, some isolation (rectifier-transformer) must be installed between the resonant charging section and the remainder of the circuit to prevent premature excitation[1] of the fast magnetic switches.

We have also estimated the magnetic switches cooling requirements and found them reasonable. Our estimates indicate that a naive scaling of this type of pulser to the MJ range (as in Ref. 10) may not be possible without careful consideration of the above difficulties.

The energy transfer through the final modulator components is depicted in Fig. 7c. Apparently, about 70 J are restored in C_3 after the discharge is completed. The modulator's single-pulse efficiency (energy deposited into Z-pinch divided by stored energy in C_1) is about 15% compared to 32% in the original circuit. However, this efficiency can easily be doubled by utilizing the restored electrical energy. The capillary structure described in Ref. 9 is not suitable for repetitive operation. However, a possible scheme for producing NaF plasma clouds at 50 Hz is by pulsed laser illumination of a NaF-coated Z-pinch anode.[11]

CONCLUSIONS

In this paper we have presented some new applications for magnetic switches in the laser and plasma interaction fields. We used the PSPICE circuit solver to calculate the electric response of laser driven LCR targets[12] with and without magnetic switches. It is shown that intense ($2*10^{12}$T/sec) dB/dt pulses can be produced by loading the tiny coil with a small ferrite core. Furthermore, the sudden dI/dt changes during the core switching adds a multi-GHz component to the current history. This effect could be used to convert a CO_2 laser pulse into extremely short multi-GHz RF emission.

We have also presented two schemes based on magnetic switches that might improve and increase the repetition rate of small Z-pinches. The capability to increase the current pulse FWHM from 1.5 to 3 μsec at the same current using magnetic switched circuit was demonstrated. A 50-Hz 200-J driver for this Z-pinch was "constructed" using a Melville compression line. The driver's single-pulse efficiency (last four stages) is about 15% compared to 32% in the original circuit. However, this efficiency could be doubled by electrical energy recovery.

ACKNOWLEDGEMENTS

One of the authors (HK) would like to acknowledge useful discussions with members of the Nuclear Research Centre - Negev. Typing and drafting support by the Fusion Studies Laboratory at the U. of Illinoins is also acknowledged.

REFERENCES

1. H. Rhinehart, R. Dougel, and W. C. Nunnally, Proc, 4th IEEE Int. Conf. on Pulsed Power (1985) p. 660
2. W. S. Mellville, Proc. IEE 98(3) (1951) p. 185
3. PSPICE User Manual, Microsim Corp. (1987).
4. D. C. Jiles and D. L. Atherton, J. Magnetism & Magnetic Materials, 61 p. 48 (1986).
5. S. Nakai, H. Daido, M. Fujita, K. Terai, F. Miki, K. Nishihara, M. Murakami, K. Mima, A. Hasagawa, and C. Yamanaka, Laser Interactions and Related Plasma Phenomena, Vol. 7 p. 213, G. H. Miley and H. Hora, ed. (Plenum Press, NY), 1986.
6. N. Seddon and E. Thornton, Rev. Sci. Instrum 59(11) (1988) p. 2497.
7. D. R. Bach, D. W. Forslund, M. A. Yates, S. Evans, H. Kohler, and R. Zacharias, Laser Interactions and Related Plasma Phenomena, Vol. 7 p. 213, G. H. Miley and H. Hora, ed. (Plenum Press, NY, 1986).
8. G. D. Lougheed, M. M. Kakez, J. H. W. Lau and R. P. Gupta, J. Appl. Phys., 65(3) (1989). p. 978
9. B. L. Welch, F. C. Young, R. J. Commisso, D. D. Hinshelwood, D. Mosher and B. V. Weber, J. Appl. Phys. 65(7) (1989) p. 2664
10. A. J. Power, H. R. Bolton, and W. Bessel, Proc. 5th IEEE Int. Conf. on Pulsed Power (1986) p. 696.
11. V. A. Gribkov, S. Denus et al., Sov. J. Plasma Physics 11(1) (1985) p. 70.

NUCLEAR DRIVEN UV FLUORESCENCE FOR

STIMULATION OF THE ATOMIC IODINE LASER

W. H. Williams, G. H. Miley, and H. J. Chung

Univ. of Illinois, Fusion Studies Laboratory
Urbana, IL 61801

ABSTRACT

Results to date of an experimental study of the nuclear-induced excitation of the excimer XeBr are presented. The excitation source used was the reaction $^{10}B(n,^4He)^7Li + 2.8$ MeV. Pressures up to 800 torr Xe/Br/Ar were studied, using Br_2, $CHBr_3$, and BBr_3 as bromine donors. Gas fill mix ratios were varied to maximize the excimer line output at 282 nm. Absolute fluorescence efficiencies (light output at 282 nm/energy deposited in cell) were measured, with an optimum of 25-38% found. The application of this fluorescence in the development of an atomic iodine photodissociation laser is discussed.

INTRODUCTION

The potential of using nuclear reaction energy to drive a laser has been studied for many years in a variety of laser systems. Ref. 1 gives some of the highlights of this research. Studies have shown that Nuclear Pumped Lasers (NPL's) have the potential of being scaled to very high powers[2,3,4], due to the ability to pump large laser volumes (good penetration of neutrons through materials) and the potential for high efficiencies (no Carnot cycle limitations). As a result, NPL's also have a significant advantage in high power space-based applications where the nuclear fuel represents a much smaller, lighter energy storage system than is required for large electrically pumped lasers.

The specific type of NPL of interest here utilizes nuclear reactions to cause fluorescence in a gas mixture. This fluorescence is then coupled into a separate laser cell, acting as the light source for photolytic pumping of the laser. The component of this system which uses nuclear reaction energy to initiate the fluorescence of a gas mixture is termed a Nuclear Driven Flashlamp (NDF).

The three principle reactions considered for NDF testing, as well as most NPL work, are:

$$n + {}^{10}B \longrightarrow {}^4He + {}^7Li + 2.8 \text{ MeV};$$
$$n + {}^3He \longrightarrow T + p + 0.8 \text{ MeV};$$
$$n + {}^{235}U \longrightarrow ff + 200 \text{ MeV}.$$

The neutron absorption cross section for the last reaction is smaller than the other two, but when its much larger energy release is factored in the ^{235}U reaction is able to provide the largest power density per unit neutron flux. Because of special use restrictions with enriched uranium, however, and contamination problems of the gas by fission fragments, the first two reactions are being used in the present work. ^{10}B is incorporated as a thin wall coating on the inside of the test cell, while ^3He serves as a fluorescent gas diluent.

This experimental effort is supported by an earlier kinetics study done by Wilson[5] showing that nuclear excitation of the excimer XeBr should provide an excellent UV light source for the atomic iodine photo-dissociation laser. The atomic iodine laser involves photodissociation of appropriate iodine containing gas molecules, e.g. C_3F_7I or CF_3I, giving I* lasing at 1.31 micron, and has been used in high power laser systems[6].

This work attempts to substitute the traditional electrically-driven UV source with an NDF. Use of an excimer in a UV NDF has been suggested by Miley and Prelas, et al[7,8]. Excimers, in general, are a good choice for UV fluorescence generation because of the high conversion efficiencies with which excimer mixtures convert deposited energy to ultra-violet light (up to 50%). The excimer XeBr* is chosen for the present application because it emits at 282 nm, which is near the peak of the absorption cross-sections for the two common atomic iodine lasants, C_3F_7I and CF_3I.

In these experiments the excimer is used as an NDF gas to power a photolytic laser. An attempt to stimulate lasing of the excimer itself is discouraged for two reasons: 1) test facilities for this work cannot induce the high power densities needed for a UV laser; and 2) the reactor radiation environment causes transient and permanent transmission losses in optical materials. This induced loss is higher at shorter wavelengths. Previous work by our group[8] has shown the radiation-induced loss in fused silica in the UV to be sufficiently small that a single light pass is not significantly attenuated through a quartz window. The multiple passes

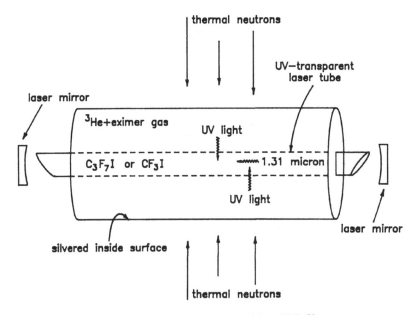

Fig. 1. Planned laser system to utilize NDF fluorescence.

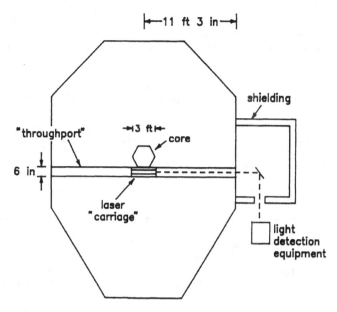

Fig. 2. Experimental set-up at the Univ. of Illinois TRIGA.

Since it is desirable to change gas fills remotely (i.e., with the cell in the throughport) due to neutron-induced activation of the cell structural materials, a series of pneumatic valves and vacuum/gas fill lines are attached to the cell. In addition, in order to measure the energy deposited in the cell gas, for fluorescence efficiency calculations, a piezoresistive pressure transducer (Kulite IPT1100-250A*) was attached to the cell, a technique used by other investigators[12]. This technique yields the average thermal energy deposited in the gas as this energy heats the gas and raises the pressure in the closed cell volume. This transducer proved to be sufficiently resistant to radiation to allow a large number of shots (ca. 75 shots = 7.5 MRad) before replacement.

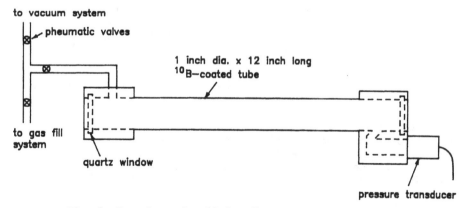

Fig. 3. Experimental cell for fluorescence measurements.

----

*Identification of equipment manufacturers is for information only and does not constitute an endorsement for that item.

through laser cavity end windows, however, would represent a large loss for the system. A very high gain medium would be required to overcome this loss. The losses are much smaller in the IR, however, so lasing at 1.31 mμ should be easier.

Figure 1 shows a schematic of the planned laser cell. The cell utilizes two concentric tubes, the inner containing the lasant gas, and the outer containing the fluorescing excimer with ^{3}He. The diameter of the outer cell is approximately 5 cm, and the length approximately 1 meter. The cell will be placed in an intense thermal neutron flux from the Univ. of Illinois research reactor. UV light generated by the XeBr is coupled into the lasant in the inner cell. To maximize this coupling efficiency, the outer cell is mirrored to allow UV photons to be reflected into the inner cell. This design necessitates the use of ^{3}He, which is transparent to UV photons, and serves as the reactant. (^{10}B or ^{235}U are useable only as poorly reflective wall coatings.) Analysis of the maximum coupling efficiency attainable is currently under investigation. Three reports studied for this type of application predict coupling efficiencies from 18 to 80%, depending on geometry and lasant pressure[8,10,11]. This spread indicates the large number of uncertainties involved. Careful design analysis should allow accurate determination of this number for our particular cell configuration.

Testing to date has been designed to determine XeBr* fluorescence efficiencies. Mix ratios of fluorescing gas constituents have been varied, using three different bromine donors under nuclear excitation.

EXPERIMENTAL SET-UP

All experiments for this work is being done in the Univ. of Illinois TRIGA reactor. The TRIGA is a light water reactor capable of running in steady state or pulsed mode. Pulsed operation used for this work provides a transient thermal neutron flux 1000 times that of steady state operation. The characteristics of a pulse are shown in Table 1.

Figure 2 shows a schematic of the reactor test set-up. This shows the special suitability of this facility for doing NDF testing in that a six-inch diameter "throughport" runs through the reactor shielding, adjacent to the core. Test cells can be placed in this port, and emitted light is directed into detection equipment with the use of mirrors. The core is large enough that a test cell 1 meter in length can receive a reasonably uniform irradiation, the flux ranging from 1 to 3 x 10^{15} n/cm^{2}sec.

Figure 3 shows the test cell constructed for the present phase of fluorescence tests. The experiment consists of a 1 inch diameter aluminum tube, coated on the interior with a thin ^{10}B coating. The cell ends are closed with fused silica windows. A ^{10}B coated cell was used for this phase because it allows flexibility to study a large number of gas mixes. Utilizing the optimum gas mixture as determined from these experiments, tests will be conducted using ^{3}He, which has a significant cost, to determine fluorescence efficiencies under ^{3}He excitation.

Table 1. TRIGA Pulsing Characteristics
(for a \$3 reactivity insertion)

- 12 msec fwhm
- 3 x 10^{15} n/cm^2-sec peak thermal flux
- 4 pulses/hour

438

Two types of fluorescence analyses are conducted on each test: a time-integrated spectrum of the light is obtained with an Optical Multi-channel Analyzer (OMA), and a time-dependant trace of one spectral line is obtained with a monochromator and photomultiplier tube. Two different OMA's were used, Princeton Applied Research models 1450/1452 and 1460/1456. The monochromator was a CGA-McPherson model EU-700 with RCA 031034 photomultiplier. The OMA is used for fluorescence efficiency determinations. The test set-up was designed and aligned to allow absolute determination of the light generated in the cell. The time-dependent photomultiplier measurement is used to investigate dose rate or total dose dependence in the fluorescence during the pulse.

Research grade Xe and Ar were used as received. A/R grade Br_2 (99.6%), $CHBr_3$ (99+%), and BBr_3 (99.99%) were used after treatment with several freeze/vacuum evacuate/thaw cycles to remove entrapped gases. The test cell was evacuated to approximately 10^{-3} torr prior to each fill.

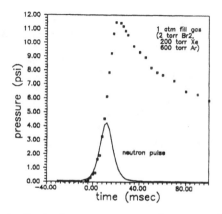

Fig. 4. Typical pressure transducer results.

EXPERIMENTAL RESULTS

A typical trace of the pressure rise during a pulse is shown in Figure 4. The peak pressure occurs near the end of the pulse, with pressure decreasing quickly thereafter as the heated gas (approx. 250 deg. C.) cools against the cell wall. This cooling rate produces a significant temperature change on the time scale of the pulse, so the peak pressure measured does not fully represent the total thermal energy deposited. A correction for this cooling is done by assuming the gas cools exponentially with a time constant of 0.009 sec^{-1}. If the pulse is approximated by a gaussian, the curve in Fig. 4 can be fit and the actual pressure rise that would have been seen without cooling back-calculated as 12.6 psi, or slightly higher than the measured 11.6 psi. The radiation-induced voltage spike in the pressure transducer, as measured on an empty cell, is also subtracted from the signal, yielding a measured energy deposition of 0.1

J/cm^3. Added to this, however, must be the fluorescence energy release. Based on measurements discussed below, this was a maximum of 0.10-0.17 J/cm^3, indicating a total energy deposition in the gas of 0.20-0.27 J/cm^3. By way of comparison, Chung & Prelas[13] predicted an energy deposition of 0.15 J/cm^3 for 760 torr He for this geometry and neutron fluence. Elsayed-Ali & Miley[14] predicted 0.24 J/cm^3 and measured 0.14 J/cm^3 for 400 torr O_2, using another experimental technique, also for the same geometry and neutron fluence.

Figure 5 shows a typical spectrum obtained with the OMA 1450/1452. The indicated peaks were also found in e-beam induced fluorescence data[15,16], though relative peak differ somewhat.

The fluorescence efficiency at 282 nm was determined by a knowledge of the absolute light intensity generated in the test cell, divided by the energy deposited in the cell, based on pressure transducer measurements. Determination of absolute light intensities were based on vendor-supplied sensitivities of the OMA detector elements, and spectral throughput of the spectrograph used. An independent calibration will also be done using a light source calibrated for spectral irradiance.

The variation of the 282 nm fluorescence efficiency with varying gas mixes is shown in Figure 6. This is for a fixed Br_2 pressure of 1 torr, and a total pressure of 800 torr. (The indicated fluorescence for no Xe actually occurs at about 290 nm, and is attributed to Br_2^*. This peak decreases with the addition of Xe.) It is seen that a broad maximum in the fluorescence occurs near 100 torr Xe. Two or three consecutive pulses were also performed on each fill, these indicated by the letters A, B, and C; and two separate fills done at 250 torr Xe. It is seen that the fluorescence intensity appears to decrease with continued pulsing. The cause for this has not been fully determined. It is speculated that impurities adsorbed on the cell wall may be driven off during a pulse and this contaminated subsequent pulses. An alternate possibility is depletion

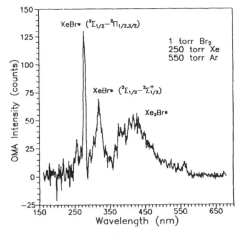

Fig. 5. Typical fluorescence spectrum.

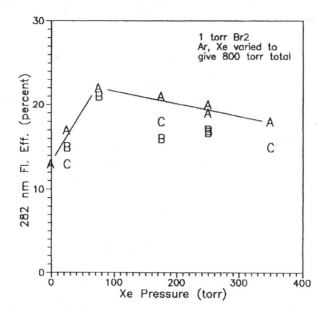

Fig. 6. Peak fluorescence spectrum at 282 nm. The letters A, B, C represent sequential pulses on one fill. Solid lines are to visually help organize data.

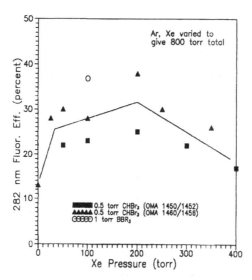

Fig. 7. Peak fluorescence efficiencies for CHBr$_3$ and BBr$_3$.

of the bromine concentration. Bromine radicals formed during the pulse may react with boron on the wall to form stable BBr$_3$. Future testing with ^{3}He fill and no wall coating should distinguish between these two possible causes for the decrease in intensity.

Figure 7 shows results with two other bromine donors, CHBr$_3$, and BBr$_3$. Results from two different OMA systems, having separate calibrations, are shown in the figure, indicating the uncertainty in the measurements. It is felt the true efficiency lies between the two sets of data. Calibrations with an intensity-calibrated lamp should clarify these measurements.

Comparing Figures 6 and 7 it is observed that CHBr$_3$ and BBr$_3$ are better fluorescers than Br$_2$, although only one pulse on BBr$_3$ was recorded. The CHBr$_3$ peaks at a lower concentration and higher Xe to bromine donor ratio than does Br$_2$. These dependencies on gas composition agree well with that found in e-beam pumping[16].

The peak efficiency (25-38%) for the best donor (CHBr$_3$) is somewhat higher than the 13-15% predicted for nuclear pumping in ref. 5. The 22% peak for Br$_2$ is also higher than the 11% measured in Ref. 15 from e-beam pumping. No firm explanations have been found for these discrepancies at present. They may be due to processes in a nuclear-pumped plasma which do not occur in an electron beam excitation, and which have not been theoretically accounted for. The quantum efficiencies for this fluorescence are 36% through the ionic channel, and 53% through the metastable channel[17], indicating the present results are quite significant in terms of efficient energy channeling into the excimer state.

The suitability of this fluorescence to the atomic iodine laser is shown in Figure 8 where it is seen that the major XeBr* line corresponds very well with the peak of the photodissociation cross-sections of two of the common lasants.

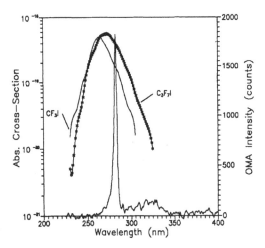

Fig. 8. Comparison of nuclear-induced XeBr* fluorescence with the photoabsorption cross sections of two atomic iodine lasants.

Given these measured fluorescence efficiencies an estimation can be made of the suitability of this system for laser testing in the TRIGA reactor. Calculations[18] predict average power deposition in ^{3}He to be approximately three times that measured for ^{10}B. Using the laser geometry shown in Fig. 1 with a laser tube diameter of 1.5 cm, a fluorescing tube diameter of 5 cm, and a coupling efficiency of 50%, the UV intensity on the laser cell would be approx. 50 W/cm^2. Reported experimental threshold values of 0.07 and 0.5 W/cm^2 (for somewhat different UV sources and geometries)[19,20] indicate such a system should achieve lasing in the TRIGA.

It was found that the fluorescence output measured on the monochromator was generally synchronous with the temporal shape of the reactor pulse. This indicates minimal dose rate or dose effects for the gas mixes and doses studied.

CONCLUSIONS

The fluorescence of the excimer XeBr under nuclear reaction-induced excitation is being studied. The ^{10}B(n,^{4}He)^{7}Li reaction has been used to excite the mixture to date. Parametric testing has been done with the Br$_2$/Xe/Ar, CHBr$_3$/Xe/Ar, and BBr$_3$/Xe/Ar systems These experiments have shown CHBr$_3$ and BBr$_3$ to be good bromine donors. Future testing will be done with ^{3}He as the reactant. Following this, a laser will be tested utilizing this fluorescence to pump the photolytic atomic iodine laser.

ACKNOWLEDGEMENT

This material was prepared with the support of the U.S. Dept. of Energy (DOE) , Agreement No. DOE DEFG07-88ER12821. However, any opinions, findings, conclusions, or recommendations expressed herein are those of the authors and do not necessarily reflect the views of DOE.

REFERENCES

1. G. H. Miley, et al., 50 Years with Nuclear Fission, 1 (American Nuclear Society, LaGrange Park, IL, 1989), pg. 333.

2. S. Balog, et al., Fusion Technology, 15, 383 (1989).

3. D. E. Wessol, et al., Laser Interaction and Related Plasma Phenomena, 8 (Plenum Publishing, 1988), pg. 27.

4. G. H. Miley, Atomkernenergie/Kerntechnik, 45, 14 (1984).

5. J. W. Wilson, Appl. Phys. Lett. 37(8), 695 (1980).

6. F. M. Abzaev, et al., Sov. Phys. JETP 55(2), 263 (1982).

7. G. H. Miley, et al., Bult. APS, 24, 117 (1978).

8. M. A. Prelas, et al., Laser and Particle Beams 6(1), 25 (1988).

9. G. H. Miley, et al., presented at the Symposium on Optical Materials for High Power Lasers, Boulder (1987).

10. J. W. Wilson, A. Shapirro, J. Appl. Phys. 51(5),2391 (1980).

11. M. A. Prelas, G. L. Jones, J. Appl, Phys, 53(1),165 (1982).

12. C. Duzy, J. Boness, IEEE J. Quant. Elec. QE-16(6), 640 (1980).

13. A. K. Chung, M. A. Prelas, Proc. of the IEEE Intl. Conf. on Plasma Science, San Diego, CA 1983 (IEEE, New York, 1983) p 135.

14. H. E. Elsayed-Ali, G. H. Miley, Rev. Sci. Instr. 55(8), 1353 (1984).

15. J. C. Swingle, et al., Appl. Phys. Lett. 28(7), 387 (1976).

16. W. L. Wilson, Jr., et al., J. Chem. Phys., 77(4), 1830 (1982).

17. J. T. Verdeyan, Laser Electronics, 2nd Ed. (Prentice Hall, NJ), 1989, pg. 346.

18. J. W. Wilson, R. J. De Young, J. Appl. Phys., 49(3), 980 (1978).

19. R. T. Witte, Optics Comm. 28(2), 202 (1979).

20. R. J. De Young, E. J. Conway, Proc. of Intl. Conf. on Lasers '85, 467 (1985).

A HIGHLY REFLECTING MICROWAVE PLASMA MIRROR

H. Figueroa

University of Southern California
Los Angeles, California 90089-0484

INTRODUCTION

Lately there has been a strong interest in developing efficient plasma mirrors to reflect microwave beams[1]. This interest has arisen mostly from the possibility to redirect radar and satellite signals by the interaction with a plasma grating created in the ionosphere. Since the index of refraction of a plasma medium can be considerably less than 1, high reflectivities should be expected. The excitation of these gratings, however, depends on the efficiency of coupling between the plasma and the interference pattern of the two pump waves that are writing the grating. An efficient coupling is in most cases difficult to achieve which results in a low efficiency reflecting grating.

In laboratory plasmas, gratings with very large density variations can be created by properly arranging the electrodes of a gaseous discharge. The reflective efficiency of these gratings will thus depend on the reliability of the source to generate dense plasmas and reproducible plasma geometries. The development of this technology could have an impact on the free electron laser technology, since a high reflectivity plasma mirror could withstand high power radiation and simultaneously would allow for easy injection of the electron beam into an optical cavity.

In this paper we analyze the reflectivity characteristics of a plasma mirror based on the back lighted thyratron (BLT)[2]. The BLT is capable of generating highly reproducible plasmas with peak densities on the order of 10^{15} cm^{-3}. This density corresponds to the cutoff density for a 1 mm wavelength electromagnetic wave. Such high densities thus, are ideal to reflect sub-millimeter wavelength radiation, since high reflectivities could be obtained while keeping the absorption at low levels.

We will consider two different mirror geometries. The first geometry consists of a stack of plasma layers separated from each other by vacuum interspacings. The plasma layers are assumed to be slabs with perfectly sharp boundaries. We then study the reflectivity characteristics of the mirror as a function of the number of layers in the stack and also as a function of the layer thickness and plasma density. We find that since the index of refraction of a plasma can be made considerably less than 1, very few plasma layers are needed to obtain high reflectivities even at relatively low densities, provided that the Bragg scattering condition for maximum reflectivity is nearly satisfied. We also find that absorption due to inverse bremsstrahlung does not appreciably reduce the levels of the reflected light except in the cases when the density of the plasma layers becomes very close to the critical density.

The second geometry we consider is that in which the density of the plasma layers does not present sharp boundaries but rather varies sinusoidally along axis of the stack. We find that for frequencies satisfying the relation

$$\omega^2 = \omega_p^2 + \left(\frac{\pi c}{\lambda_i}\right)^2$$

where λ_i is the period of the plasma density variation, the reflected intensity can be even higher than in the previous case. The reason for that being that now there will be reflected electromagnetic waves from every point along the stack rather than from a finite number of places which will interfere constructively.

I. REFLECTION FROM A STACK OF PLASMA SLABS WITH SHARP BOUNDARIES

Consider a single plasma slab of thickness h and index of refraction n_1 surrounded by a medium whose index of refraction is n_i and n_t to the left and right of the slab as shown in Fig.1. An electromagnetic wave traveling through the plasma slab will suffer multiple reflections, and after each round trip part of it will be reflected and part of it will be transmitted. In the Figure, E_i, E_r, and E_t represent the incident, total reflected, and total transmitted electric fields after an infinite number of round trips. The electric fields are assumed to be parallel to the boundary surfaces.

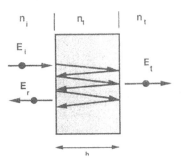

Figure 1. Electromagnetic wave propagating through a transparent slab.

It is well known from thin film theory[3] that the total electromagnetic fields on the left side of the slab (incident plus reflected) can be related with the transmited fields through the characteristic matrix of the slab as given by the relation

$$\begin{bmatrix} E_1 \\ B_1 \end{bmatrix} = \begin{bmatrix} m_{11} & m_{12} \\ m_{21} & m_{22} \end{bmatrix} \begin{bmatrix} E_2 \\ B_2 \end{bmatrix} = M \begin{bmatrix} E_2 \\ B_2 \end{bmatrix} \tag{1}$$

where

$$m_{11} = m_{22} = \cos(k_o n_1 h)$$

$$m_{12} = -\frac{i}{n_1} \sin(k_o n_1 h)$$

and

$$\hspace{12cm}(2)$$

$$m_{21} = -i n_1 \sin(k_o n_1 h)$$

Here, E_1 and B_1 represent the total fields on the left hand side (incident plus reflected) and E_2 and B_2 represent the transmitted fields; M is the characteristic matrix of the slab. The reflection coefficient is then given by

$$r = \frac{n_i m_{11} + n_i n_t m_{12} - m_{21} - m_{22}}{n_i m_{11} + n_i n_t m_{12} + m_{21} + n_t m_{22}} \hspace{3cm}(3)$$

For a stack composed of p plasma slabs such as the one shown in Fig. 2, the fields on the incident and transmitted sides are related through the characteristic matrix of the stack which is the product of the characteristic matrices of each slab by separate

$$\begin{bmatrix} E_1 \\ B_1 \end{bmatrix} = M_1 M_2 M_3 \cdot M_p \begin{bmatrix} E_{p+1} \\ B_{p+1} \end{bmatrix} = M \begin{bmatrix} E_{p+1} \\ B_{p+1} \end{bmatrix} \hspace{2cm}(4)$$

Here, E_1 and B_1 are the total fields (incident plus reflected) to the left of the stack and E_{p+1} and B_{p+1} are the transmited fields after the pth slab; M_i represents the characteristic matrix of the ith slab.

The reflection coefficient is again given by the relation (3) except that now the matrix elements m_{ij} represent the whole stack rather than a single slab.

In this section we will analyze the reflectivities of a stack composed of an odd number of plasma layers. The stack will contain only two types of slabs with indices of refraction n_1 and n_2 as shown in Fig. 3. We will assume the stack to be surrounded by vacuum.

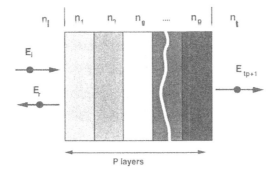

Figure 2. Stack of p plasma layers; n_j represents the index of refraction of the jth plasma slab

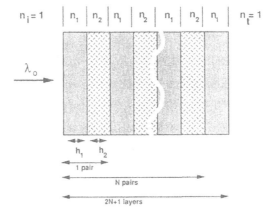

Figure 3. Stack composed of an odd number of plasma layers containing only two types of slabs with indices of refraction n_1 and n_2

The physical thickness of each slab type will be h_1 and h_2 respectively, and the incident electromagnetic wavelength will be denoted by λ_0. For optical thicknesses such that $n_i h_i = (2m+1)\lambda_0/4$, where m is an integer, the reflectivity of the stack is given by [3]

$$R = r^2 = \left[\frac{1 - n_1^2 \left[\dfrac{n_1}{n_2} \right]^{2N}}{1 + n_1^2 \left[\dfrac{n_1}{n_2} \right]^{2N}} \right]^2 \qquad (5)$$

Since the reflectivity increases as the ratio n_1/n_2 decreases, then maximum reflectivities can be achieved in the limit when n_1 tends to zero and n_2 tends to 1. In this limiting case the stack would look as composed of alternating regions of dense plasma and vacuum as shown in Fig. 4.

At low intensities, the reflectivity of the stack will be limited by inverse bremsstrahlung absorption which will heat the plasma and therefore, lower the reflectivity. At high intensities, on the other hand, radiation pressure and also the onset of instabilities such as stimulated

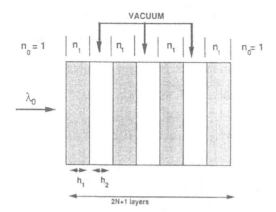

Figure 4. Stack composed by alternating regions of plasma and vacuum. The plasma density should be close to the critical density for maximum reflectivity. The shaded regions represent the plasma layers.

Brillouin scattering and filamentation, could significantly affect the mirror performance. We will consider the cases where the intensity of the e.m. wave is low enough, although not necessarily its power, so that the mirror geometry remains unchanged.

In Figure 5 we have plotted the reflectivity of such a stack as a function of the plasma density N. The number of plasma layers NS has been varied from 3 to 7 and absorption due to inverse bremsstrahlung has been neglected. The optical thickness of each of the vacuum and plasma layers has been kept at $0.25\lambda_0$. Notice that even with densities as low as $0.5N_c$, where N_c is the critical density, reflectivities higher than 50% can be achieved with only 3 plasma layers.

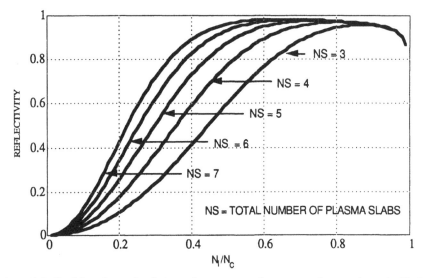

Figure 5. Reflectivity of a stack of plasma layers separated vacuum regions as shown in Fig.4 . Absorption due to inverse bremsstrahlung has been neglected.

It is common in gas discharge experiments to generate plasmas with temperatures of only a few eV. In those cases, the intensity levels of the reflected e.m. wave could be considerably reduced by collisional absorption. In order to calculate the effects of absorption on the intensity levels of the reflected wave, we included an absorption term as given by Johnston and Dawson[4] in the expression for the reflection coefficient. These results are shown in Figure 6. In this figure, we show the reflectivity of a stack of plasma layers as a function of the plasma density for a variable number of plasma layers. The optical thickness of both the plasma layers and the vacuum regions has been kept at 0.25 λ_0 where λ_0 is taken to be 1 cm. The plasma temperature is 1 eV. Notice that even at such low temperatures the effects of absorption on the reflected intensities is negligible except when the plasma density gets to be very close to the critical density. For densities lower than $0.7N_c$ absorption can be neglected.

As the optical thickness of the plasma layers and/or the vacuum regions depart from the condition for maximum reflectivity, the reflected fraction of the e.m. wave will decay very rapidly as shown in Figure 7. In this figure, we have plotted the reflectivity of a stack of 6 plasma slabs as a function of their optical thickness. The thickness of the vacuum regions is $0.25\lambda_0$, and the density of the plasma slabs is $0.6N_c$. The plasma temperature is 10 eV and the e.m. wavelength is 1 cm. Notice that the reflectivity is maximum when $n_1 h_1 = (2m+1)\lambda_0/4$, where m is an integer. Although the reflectivity decreases very rapidly as the plasma slab thickness moves away from these optimum values, reflectivities higher than 90% can still be obtained within a bandwidth of $0.10\lambda_0$. Notice also that the effects due to inverse bremsstrahlung absorption are negligible, since the reflectivity levels remain unchanged as the plasma slab thickness increases.

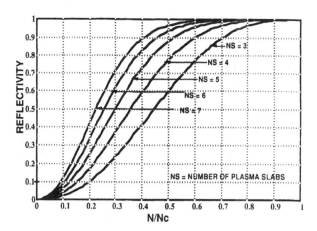

Figure 6. Reflectivity levels from a stack of plasma layers with temperature of 1 eV. The e.m. wavelength is 1 cm.

For colder plasmas, however, and shorter e.m. wavelengths, absorption can considerably reduce the levels of the reflected wave as the thickness of the plasma layers is increased. This is shown in Figure 8. Here, we show the reflectivity of a stack composed of 6 plasma layers of density $0.6N_c$ as a function of the plasma layer thickness. The plasma temperature is taken to be 1 eV and the electromagnetic wave is 1 mm in wavelength. The thickness of the vacuum regions is $0.25\lambda_0$. Notice that in this case the peak reflectivity decreases from 90% to 50% as the thickness of the plasma layers increases from $0.25\lambda_0$ to $1.75\lambda_0$.

Experimentally, one can attempt to design a plasma mirror by shaping the electrodes of a gaseous discharge such that the discharge will create plasma layers with the desired interspacing as shown in Fig. 9. One inconvenience with this approach is that the location of the breakdown discharge is not known precisely, especially in the case of wide electrodes.

A more reliable way to design a plasma mirror is by positioning a series of BLT's next to each other, separated by a given distance as shown in Figure 10. Since the discharge occurs along the longest path, which is through the center hole, one will be able to minimize the uncertainty in both the relative position and also the thickness of the discharges. The thickness of the discharge can be chosen by shaping the center hole into a slot of given width. Experiments are underway to test these ideas.

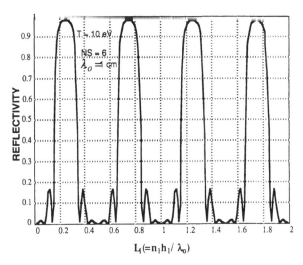

Figure 7. Reflectivity of a stack of 6 plasma layers as a function of the slab thickness. The plasma temperature is 10 eV and the e.m. wavelength is 1 cm.

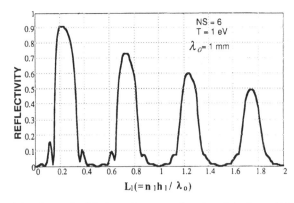

Figure 8. Reflectivity of a stack of 6 plasma layers as a function of the layer thickness. The plasma temperature is 1 eV and the e.m. wavelength is 1 mm.

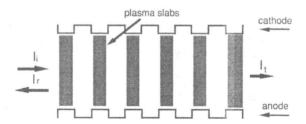

Figure 9. Generation of a stack of plasma layers by shaping the electrodes of a gaseous discharge.

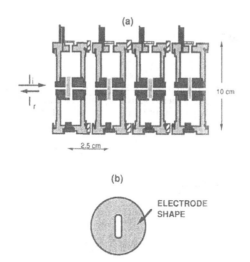

Figure 10. (a) Plasma mirror composed of a series of BLT's stacked one next to the other and separated from each other by a given distance. The discharges occur through the center holes. The figure also shows typical BLT dimensions which can be varied to suit the mirror requirements. (b) BLT electrode showing the center hole shaped into a slot.

II. REFLECTION FROM A SINUSOIDALLY VARYING PLASMA DENSITY

The high reflectivities obtained from a stack of plasma layers with sharp boundaries are the result of constructive interference of the waves reflected off those boundaries which are finite in number. When the boundaries are diffused, however, reflection will occur over an infinite number of places, everywhere where the plasma shows a density gradient. For a properly taylored plasma variation, the reflected waves could add constructively at every point, thereby increasing the intensity of the reflected wave even further.

As was shown by Figueroa and Joshi[5], a plasma with a periodic density variation introduces forbidden bands of frequencies on the propagation of e.m. waves. These bands are centered around the frequencies given by

$$\omega^2 = \omega_p^2 + c^2 \left(\frac{mk_i}{2}\right)^2 \tag{6}$$

where ω_p is the average plasma frequency, $k_i = 2\pi/\lambda_i$, and λ_i is the spatial period of the density variation. Here, m is an integer. The highest reflectivity occurs within the first forbidden band (m=1).

We will assume that the frequency of the e.m. wave lies within the 1st forbidden band and also that satisfies the relation

$$\omega^2 = \omega_p^2 + \frac{c^2 k_i^2}{4} \tag{7}$$

The intensity of the reflected e.m. wave will depend on the amplitude of the density ripple. We will consider the case where the average plasma density is equal to the amplitude of the density variation as shown in Figure 11. The incident, reflected, and transmited e.m. waves are indicated by I_i, I_r, and I_t respectively. Since

$$\beta^2 = \frac{8\omega_p^2}{c^2 k_i^2}$$

as defined by eqns. (19.a) and (15) of reference 5, we obtain that

$$\frac{\omega^2}{\omega_p^2} = 1 + \frac{2}{\beta^2} \tag{8}$$

We will consider the cases in which the peak plasma density is considerably less than the critical density of the incident e.m. wave, so that absorption can be neglected. Thus, from eqn. (8) we obtain that $\beta^2 < 2$. For that range of values of β^2 we can approximate the imaginary part of the e.m. wavevector by

454

$$\text{Im} (k) \approx 0.06 \, k_i \beta^2 \qquad\qquad (9)$$

as can be seen from Fig. 7 of reference 5. The term Im(k) represents the attenuation of the e.m. wave due to reflections as it propagates through the plasma density ripple.

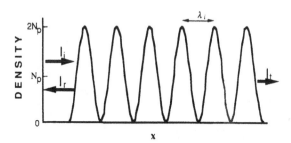

Figure 11. Plasma density variation with a peak density twice that the average density. The spatial ripple period is indicated by λ_i.

Thus, from relations (8) and (9) one can obtain the fraction of the e.m. wave that is reflected from a ripple with a given number of periods and a given amplitude. This is shown in Figure 12. In this figure, the number of ripple periods has been varied from 3 to 7, and absorption has been neglected. Notice that the reflectivity levels have significantly increased from the case of perfect slabs as can be seen by comparing Figures 12 and 5.

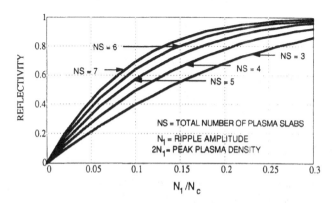

Figure 12. Reflected fraction of an incident e.m. wave with wavevector $k=\pi/\lambda_i$ on a sinusoidally varying plasma density with period λ_i. The number of ripples has been varied from 3 to 7.

At present, we are designing a plasma source which will provide us with a sinusoidally varying plasma density of controllable ripple period. The source is based on the multiple-gap BLT[6] as shown in Figure 13. The plasma source consists of a stack of alternating electrodes and insulating plates of variable thicknesses. A charging voltage of several keV's is applied between the cathode and the anode. All the other electrodes are floating. The electrodes have a center hole with a typical diameter of 3 mm. Since this device operates on the left hand side of the Paschen curve, the discharge will occur along the center holes, thus creating a long plasma channel whose density will vary periodically along its axis. By timing the arrival of the e.m. wave with the triggering of the source, one would be able to choose the desired peak plasma density.

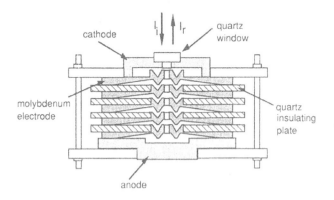

Figure 13. Multiple-gap BLT showing the insulating plates and the Mo floating electrodes. The typical thickness of the electrodes and insulating plates is 3 mm.

In conclusion, we have examined the reflectivity levels of a mirror composed of a periodically varying plasma density in the low intensity limit . Because of the densities available in the laboratory, our studies were restricted to the microwave and millimeter wave regions. We found that with the mirror geometries presented, reflectivities higher than 90% could be achieved with a small number of plasma periods. We also found that in most instances, absorption due to inverse bremsstrahlung could be neglected. The development of this new technology could have an impact on the fields of high power microwave sources, in which high power, low intensity beams are of common occurance.

ACKNOWLEDGEMENT

This work was sponsored by the Army Office Research and the Air Force ATRI Program at UCLA.

REFERENCES

1. Y.S. Zhang and S.P. Kuo, Abstract of the 1989 IEEE International Conference on Plasma Science, Buffalo, N.Y. 1989.

2. K. Frank, E. Boggasch, J. Christiansen, A. Goertler, W. Hartmann, C. Kozlik, G. Kirkman, C. Braun, V. Dominic, M.A. Gundersen, H. Riege, and G. Mechtersheimer, IEEE Trans. Plasma Sci., Vol. 16, No. 2, 317(1988).

3. See for example, M. Born and E. Wolf, "Principles of Optics," Pergamon Press, N.Y., 1965.

4. T. Johnston and J.M. Dawson, Phys. Fluids **16**, 722 (1973).

5. H. Figueroa and C. Joshi, "Laser Interaction and Related Plasma Phenomena" **7**, p. 241, Plenum Press, N.Y. (1986).

6. T-Y. Hsu, G. Kirkman, H. Figueroa, R. Liou, H. Bauer, and M.A. Gundersen,"Proceedings of the Pulsed Power Conference," Monterey, CA 1989. (to be published)

Particle acceleration with the axial electric field of a TEM10 mode laser beam

F. Caspers
E. Jensen

CERN
CH – 1211 Geneva 21
Switzerland

Abstract

Due to their finite spot size, optical beams have small axial field components. For a Gaussian TEM_{10} mode, the axial electric field has a maximum on the optical axis while the transverse field vanishes there. The possible use of this axial field for the acceleration of highly relativistic particles has been studied.

The ratio of the axial to the maximum transverse electric field is inverse proportional to the spot size radius; for a spot size radius of $w_0 \approx 100 \, \lambda$, it is in the order of $4 \, 10^{-3}$. But the transverse field components seen by the highly relativistic particles are significantly reduced by the Lorentz transformation, thus the axial field components become important.

Finite spot size and axial field are inevitably connected to a phase velocity along the axis which is by approximately a factor of $(1 - \pi/(kw_0)^2)^{-1}$ above c. Consequently synchronism between wave and particle beam cannot be sustained for more than half a RF period. But due to the relativistic Doppler shift, particles can travel approximately the confocal length, kw_0^2, before the field reverses sign. The particle energy increase is proportional to the square root of the laser power, and lies in the order of 10 MeV for laser powers of 10^{12} W.[1]

---

[1] Presented at the "9th International Workshop on Laser Interaction and Related Plasma Phenomena", 6. – 10. Nov. 1989, Monterey, CA

1. Introduction

The high fields which exist in the focus region of high power lasers cannot be used directly for acceleration of unbound particles in vacuum. For a single homogeneous plane wave, the particle will always lag in phase if it is not highly relativistic *and* moving in parallel with the wave. But in this latter case, the fields are transverse to its direction of propagation, and thus no energy exchange occurs. Extending these thoughts to the superposition of plane waves, we cite A. M. Sessler [1]:

> The field pattern produced by *any* array of optical elements, provided one is not near a surface (more than a wavelength away) and not in a medium, is simply a superposition of plane waves. It is not very difficult to generalize the considerations made for a single plane wave and conclude that for a relativistic particle, which moves with essentially constant speed and in a straight line, there is no net acceleration.

To obtain synchronism, the phase velocity of the wave has to be slowed down to the particle velocity, i.e. the component of the wave vector $\vec{k}$ in the direction of particle motion $k_\parallel$, has to be greater than the free space wave number, $k_\parallel > k$. Due to $k^2 = k_\parallel^2 + k_\perp^2$, this is possible with an imaginary transverse component of $\vec{k}$, $k_\perp = j\alpha$. These evanescent waves decay perpendicular to their direction of propagation. They exist in the immediate vicinity of dielectric or grating structures. For highly relativistic particles, this imaginary $k_\perp$ has to be only small, but still the necessary distance between these surfaces and the particle beam may become inconveniently narrow.

In this paper, we will consider a laser beam propagating in z-direction, acting on a single highly relativistic charged particle propagating in parallel to it. Even though it is not guided near a surface or dielectric, such a beam has some similarity to evanescent waves: it decays exponentially in the transverse direction, it has axial fields, and there exist even straight lines parallel to the axis where $v_\varphi < c$. What is the effect of the small axial electric field that exists in the laser beam on the particle?

2. The field of a Gaussian beam wave

The transverse electric field of a TEM_{lp} mode of a Gaussian beam wave with $\cos l\varphi$-dependence and linear polarization in x-direction is given in cylindrical coordinates (ρ, φ, and z) by [2]

$$\frac{E_x}{E_0} = \frac{w_0}{w} \left(\sqrt{2}\frac{\rho}{w}\right)^l L_p^{(l)}\left(2\left(\frac{\rho}{w}\right)^2\right) e^{-\left(\frac{\rho}{w}\right)^2} \cos l\varphi \; e^{-j\Phi}$$

$$\Phi = kz - (2p+l+1)\arctan\frac{2z}{L} + \frac{2z}{L}\left(\frac{\rho}{w}\right)^2,$$

where w_0 denotes the beam waist radius and

$$w(z) = w_0 \sqrt{1 + \left(\frac{2z}{L}\right)^2}$$

is the beam shape with the "confocal length" $L = kw_0^2$, the meaning of L being the length of a confocal optical resonator supporting this beam. $L_p^{(l)}$ denotes the generalized Laguerre polynomial [3], and

$j^2 = -1$. l and p are the azimuthal and radial order of the mode, respectively. The approximation is valid for infinite apertures and large spot sizes, $\lambda/w_0 \ll 1$.

In Fig. 1, the transverse electric field of the TEM_{10} mode, polarized in x-direction and with $\cos l\varphi$-dependence, is plotted. The transverse coordinates are normalized to w; the length of the arrows is proportional to the electric field.

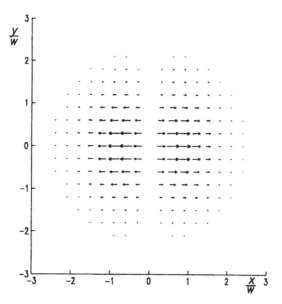

Fig. 1.   Transverse electric field of a TEM_{10} mode

Due to the transverse variation of the transverse fields, an axial field must exist, i.e. there has to be an axial field near the center of Fig. 1 for the electric field lines to be closed. This follows directly from Maxwell's equations

$$\nabla \times Z_0 \vec{B} = j k \vec{E},$$

$$-\nabla \times \vec{E} = j k Z_0 \vec{B},$$

where $k = \omega/c$ and $Z_0 = 377\,\Omega$. If we write $\nabla = \nabla_\perp + \vec{u}_z \dfrac{\partial}{\partial z}$ and assume again a linearily polarized wave, $E_x = Z_0 B_y$, we have

$$j k \vec{E} = -\vec{u}_x \frac{\partial}{\partial z} Z_0 B_y + \nabla_\perp \times \vec{u}_y Z_0 B_y,$$

$$j k Z_0 \vec{B} = -\vec{u}_y \frac{\partial}{\partial z} E_x - \nabla_\perp \times \vec{u}_x E_x.$$

The last term in each of these two equations gives rise to axial field components. The TEM_{lp} Gaussian modes are thus not exactly "transverse electromagnetic"; the name "quasi TEM mode" is more

461

appropriate. The axial fields can in good approximation be computed from the transverse variations of the transverse fields, as long as the ratio λ/w_0 is small. The axial fields are of first order in λ/w_0 while correction terms for the transverse fields in the above equations are of second order.

The normalized transverse and axial electric fields of the TEM_{10} mode are plotted in Fig. 2 for $y = 0$ (Note the different scales!). While the transverse field has a zero in the center, the axial field has a significant peak there. E_0 is calculated from the laser beam power P by the formula

$$E_0 = \sqrt{\frac{2 Z_0 P}{\pi w_0^2}} \ .$$

The ratio of the maximum axial electric field to the maximum transverse field is $\sqrt{2} \exp(0.5)/(kw)$.

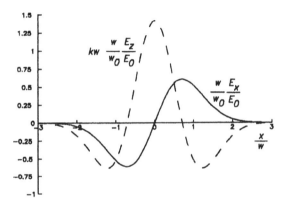

Fig. 2.   Normalized transverse (solid line) and axial electric fields (dashed line) of the TEM_{10} mode

3. Lorentz transformation of the fields

The difference in the scales of the two curves in Fig. 2, kw, is usually a large number. For a beam waist radius of some hundred λ, e.g., it is in the order of 10^3. The axial fields are by this factor smaller than the transverse fields. But they become important for highly relativistic particles due to the Lorentz transformation of the fields.

Given the fields in the laboratory frame (unprimed quantities) and the particle velocity βc and energy $\gamma = (1 - \beta_2)^{-1/2}$, we get in the rest frame of the particle (primed quantities) [4]:

$$E'_x = \gamma (E_x - \beta Z_0 B_y), \quad B'_x = \gamma (B_x + \beta E_y/Z_0),$$
$$E'_y = \gamma (E_y + \beta Z_0 B_x), \quad B'_y = \gamma (B_y - \beta E_x/Z_0),$$
$$E'_z = E_z, \qquad\qquad\qquad B'_z = B_z.$$

In our case of a linearly polarized wave travelling with the particle, this becomes

$$E'_x \approx \frac{1}{2\gamma} E_x$$
$$E'_z = E_z$$

for the electric field. For 1 GeV electrons, e.g., the transverse field will be decreased by a factor of about 4000. Thus the axial field plays in fact a dominant role.

4. Phase relations

Two effects lead to non-synchronism between particle and wave. First there is the deviation of the particle velocity v from c. Let us define

$$X \equiv 1 - \frac{v}{c}.$$

For a highly relativistic particle, we obtain

$$X \approx \frac{1}{2\gamma^2}.$$

For a 1 GeV electron, e.g., X would be in the order of 10^{-7}.

The second effect is the deviation of the phase velocity v_φ of the optical beam from c, for which we define

$$Y \equiv \frac{v_\varphi}{c} - 1.$$

v_φ is calculated from $\Phi - \omega t = const$ along the particle trajectory. Along the axis we obtain

$$Y = \frac{2(2p + l + 1)}{(kw)^2 - 2(2p + l + 1)} \approx \frac{2(2p + l + 1)}{(kw)^2}.$$

The approximation is valid for small λ/w_0. For a TEM_{10} mode with $w_0/\lambda = 100$, Y is approximately $5 \cdot 10^{-6}$.

Because both X and Y are generally positive quantities, the particle will lag in phase behind the wave, and there will be in fact no "net" acceleration, in full agreement with the statement by Sessler cited above. But on how long a distance could the particle be kept in phase? In the rest frame of the particle, the laser frequency would appear Doppler downshifted by the factor

$$\frac{\omega'}{\omega} = \frac{X+Y}{\sqrt{2X}} .$$

The nominator describes the non-relativistic Doppler shift, while the denominator accounts for the change of the proper time. After half a period, the particle would slip out of phase, this is in the particle frame

$$t' = \frac{1}{2}\frac{2\pi}{\omega'} .$$

Back in the laboratory frame, this time is

$$t = \frac{1}{\sqrt{2X}}t' = \frac{T}{2}\frac{1}{X+Y} ,$$

where T is the original RF period. For the TEM_{10} mode, it turns out that a highly relativistic particle ($X \ll Y$) can travel just the confocal length L during this time, independent of the waist radius.

5. Achievable energy gain

The maximum energy the particle could gain on its path is given by eV_{acc}, where e is the charge of the particle and the total accelerating voltage is given by the integral

$$V_{acc} = \int_{-z}^{z} E_z e^{j\frac{\omega}{v}\zeta} d\zeta .$$

V_{acc} is plotted versus z in Fig. 3 for the TEM_{10} mode and a highly relativistic particle moving along the axis. The maximum occurs always at $z = L/2$, as stated above. The maximum value is $\sqrt{Z_0 P/\pi} = E_0 w_0/\sqrt{2}$. For a given laser power, it is indepedent of the parameters of the optical beam. For 1 TW, e.g., and a spotsize of $3 \cdot 10^{-6}$ m^2 ($w_0 = 100\lambda$, $\lambda = 10 \mu$m), E_0 would be 15.5 GV/m. The maximum value of E_z would be 25 MV/m, and the maximum of V_{acc} would be 11 MV.

For other particle trajectories and other modes, one would have slightly different formulae, but the result never seems to exceed significantly the "single-cycle acceleration", $\sqrt{Z_0 P/\pi}$ which is small even for very high laser powers.

In [5], similar results were obtained for the single-cycle electron acceleration for electrons passing a plane wave in an oblique angle.

The proportionality between the acceleration and the square root of the laser power (to the laser amplitude) is a remarkable difference between the acceleration scheme considered here and the one used in "far field accelerators" [6] where the acceleration is directly proportional to the laser power. Below a certain threshold power (which might be very high), the effect of the axial fields is greater than of the transverse fields and must not be neglected.

464

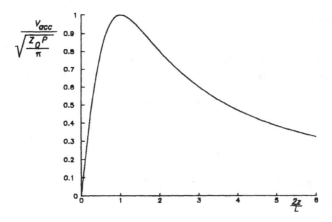

$$\frac{V_{acc}}{\sqrt{\frac{Z_0 P}{\pi}}}$$

Fig. 3.   The total accelerating voltage V_{acc} for a highly relativistic particle moving along the axis in a TEM_{10}-mode versus z.

6. Refocussing

The low energy gain achieved by this scheme could possibly be increased by using a periodic structure to refocus the optical beam. The scheme sketched in Fig. 4 is not meant to be a technical solution, but just to illustrate the idea of refocussing; it might as well be some mirror arrangement. The phase shift in the lenses would have to be such that the particle, moving along the axis, is in the accelerating phase in each cell. In its periodicity, this structure resembles again a conventional linac. Keeping in mind that there are two types of "near field", one in which evanescent waves exist (which normally extends only the order of λ from the structure), and a second one where the field might already be described by a superposition of homogeneous plane waves (characterized by a dependence of the radiation characteristic on the distance), the proposed scheme clearly belongs to the group of "near field accelerators".

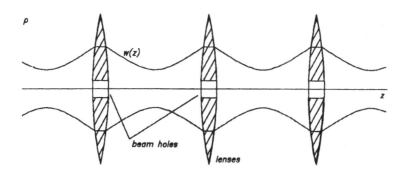

Fig. 4.   Refocussing of the beam by a set of lenses

7. Conclusions

- For a highly relativistic particle travelling along with a laser beam, the transverse fields Lorentz transform down by a factor $(2\gamma)^{-1}$, thus making the small axial fields important.

- Using these axial fields in a TEM_{10} mode, the maximum achievable energy gain of a highly relativistic particle travelling on the optical axis is $e\sqrt{Z_0 P/\pi}$. This value is obtained on the "confocal length" L and is independent of the spot size w_0. For $P = 1$ TW and highly relativistic electrons, this would be approximately 11 MeV.

- The effect is of first order in $\sqrt{P}$, and thus has to be considered when second order effects are investigated, for example in "far field accelerators".

- For the time being the axial field components of laser beams do not seem to lead to competitive particle acceleration for practically available laser powers.

Acknowledgment

The authors would like to thank H. Hora and E. Wilson for stimulating and enlighting discussions, and A. M. Sessler for his valuable comments.

References

1. A. M. Sessler *The quest for ultrahigh energies* Am. J. Phys., Vol. **54**, No. 6, 505 (1986)

2. H. Kogelnik, T. Li *Laser beams and resonators* Proceedings of the IEEE, Vol. **54**, No. 10, 1312(1966)

3. M. Abramowitz, I. A. Stegun *Handbook of Mathematical Functions* New York: Dover Publications (1970)

4. J. D. Jackson *Classical Electrodynamics* New York: Wiley (1975)

5. M. J. Feldman, R. Y. Chiao *Single-cycle electron acceleration in focused laser fields* Physical Review A, Vol. **4**, No. 1, 352 (1971)

6. J. D. Lawson *Laser Accelerators: Where do we stand?* CAS − ECFA − INFN Workshop, Frascati, (1984)

ACCELERATION OF ELECTRONS TO TeV ENERGY BY LASERS IN VACUUM[*]

L. Cicchitelli[1], H. Hora[1,2], A. Scharmann[2], and W. Scheid[2]

1) Dept. Theor. Physics, Univ. NSW, Kensington-Sydney, Australia
2) Faculty of Physics, Justus Liebig University, Giessen, Germany-W.

ABSTRACT

For the different acceleration schemes of electrons by lasers to very high energies, schemes without effects by surfaces, dielectrics, or plasmas are considered, only using vacuum interaction between lasers and electrons. One scheme follows the lateral nonlinear forces in laser beams, a second one the motion of electrons in accelerated intensity minima of laser fields, and a third one is based on the exact relativistic and Maxwellian equations of motions of electrons in plane propagating waves where no complete numbers of waves are acting. Conditions are elaborated under which electrons should achieve energies in the TeV range.

1. INTRODUCTION

For the acceleration of charged particles (electrons, nuclei etc.) to very high energies (TeV) with high luminosity [1,2], various schemes are being discussed where secondary media are involved (plasmas, boundary effects, image forces, Cerenkov effects etc.) [3,4]. This paper discusses the aims of the best possible exclusion of these media. One scheme in this direction is the inverse Free Electron Laser [5] where only the far-field of magnetic coils cause such an interaction between a laser beam and an electron beam in vacuum that increases the electron energy at the cost of laser energy. We shall not discuss this scheme.

One scheme of electron acceleration by lasers without the unusual external wiggler field of in the standard FELs, is the experiment by Boreham [6,7] where electrons are emitted radially from a laser beam leading to energies of keV. No plasma processes are involved since the Debye length [6] is much larger than the length along which the laser field accelerates the electrons. This scheme can be used for a wiggler-free FEL or as an acceleration scheme where a sweeping laser beam accelerates (kicks) electrons sidewise [8]. The theory for this scheme is quite complex since it turned out that only the Maxwellian exact description of the laser beam including the (sic!) longitudinal field components arrive at an experimentally agreeing result (see section 12.3, ref [9], [10]). This scheme is discussed in section 2.

The two further laser acceleration schemes in vacuum are:

---

[*] This work was supported by the Wilhelm Heinrich and Else Heraeus Foundation.

I. The acceleration of electrons in the dynamically moving minima of superimposed laser fields, and

II. Acceleration in propagating laser waves which are rectified or consist only in one half-wave pulse.

The first scheme is discussed in the following sections (3 to 6) and the second scheme in the sections (7 to 10). In Section 12, an estimation for using laser beams of finite diameter instead of plane waves is discussed.

2. LATERAL KICKING OF ELECTRONS BY SWEEPING LASER BEAMS

Contrary to the general belief that a laser beam in vacuum cannot produce a net acceleration to electrons apart from scattering mechanisms as e.g. Thompson scattering or stimulated Thompson scattering in the Kapitza-Dirac effect, a net acceleration was discovered first in an experiment [6] where electrons were emitted with drift energies of up to keV radially from the focus of a neodymium glass laser beams of about 10^{16} W/cm^2 maximum intensity I. It was immediately verified that the quiver energy of the electrons in these laser field was twice the value of drift energy as expected from the general relation of the electron energies gained by nonlinear forces in an inhomogeneous laser field [9].

Similar experiments with measuring of the electron energy of the electrons emitted from a laser pulse were performed where, however the laser intensity was so low that the quiver energy was in the range of few eV only [7] In this case the net energy transfer by the nonlinear force was of the same values as expected from the theory [8] but additionally, the maxima in the energy spectrum of the electrons were an immediated evidence of the multiphon ionization process of the atoms from where the electrons were originating in the laser focus. As known from private communications [7], higher laser intensities cause a diminishing or washing out of the multiphoton maxima which were not at all detected at the keV electron energies [6].

The explanation is very simple: at high intensities one can detect the Keldysh tunnel ionization [see Hawkes Ref. 7] which is semiclassical. Since the intensity threshold of the correspondence principle of electromagnetic interaction is just between the mentioned high and low intensities, it is understandable that the low intensity range relates to the quantum process of multiphoton ionization and the high intensity results in the classical ionization.

Another fundamental question was touched by the experiments [6]. While the energies of keV electrons emitted radially from the laser focus were a straightforward result of the energy integral description of the nonlinear force theory [8], the description by the quiver drift of the Maxwellian stress tensor did not agree [see page 231 of Ref. 8] with the experiment [6]. The reason was found from the common use of only transversal components of the laser field, in the beam. These fields are not the Maxwellian exact solution and one had to derive the longitudinal (sic!) components of the light field. This little addition changed the theory from "no acceleration" or 100% polarization dependence" into full action without polarization dependence as measured. The same strong polarization dependence of the theory dropped when adding the Maxwellian exact very small longitudinal components in Gaussian laser beams [11].

The value of the results [6] was not only to demonstrate how laser beams in vacuum without plasma mechanism or external fields are able to accelerate electrons. This was a nonlinear collisionless dissipation process where Laser energy was transferred to kinetic energy of electrons. The axial momentum transfer was necessary showing that the electrons were not exactly perpendicularly emitted but with a forward direction. This forward momentum was exactly the momentum of the optical energy taken out of the beam. The compensation of the radial momentum can be explained by the angular momentum of the laser beam energy. This result is an emission of the electrons from the beam in a non axial symmetry but with a spiral-like structure.

The inverse process has been discussed [8] where the injection of electrons into a laser beam can convert translative mechanical energy of electrons into optical energy [12]. It is necessary then to use laser pulses in order to gain the energy from the transition processes similar to the wake field mechanisms derived by Sprangle et al [see Ref. 5]. For cw-beams however, an electron injected into the axis of a laser beam either transmits it or is reflected. For reaching the axis one has to take into account the spiral-like radial momentum transfer. Instead of firing an electron into a resting cw-laser beam one can use a laterally sweeping laser beam interacting with a resting electron. If the relative sweep velocity expressed for an electron energy is just a little less than the half of the maximum oscillation energy of an electron at the beam axis, the electron will be kicked in radial beam direction and will gain twice the mentioned relative velocity. (Fig. 1a.)

For relativistic case of this electron acceleration of an electron by a laser beam through sidewise kicking, one simply has to take into account the relativistic quiver energy [8].

3. DYNAMIC INTENSITY MINIMA FOR ACCELERATION

Weibel and others [13] considered the standing wave field of microwaves and concluded a confinement of electrons in the minima (nodes) of the standing wave. The

Fig. 1a. Acceleration of a resting electron by a sidewise sweeping of a laser beam and kicking on the electron due to the nonlinear force mechanisms (6,8,12).

forces for driving the electrons were simply of the ∇E^2 –type which corresponds to a high frequency electrostriction or to ponderomotive forces and were called "Miller forces" [14]. We shall discuss these forces in the following especially under the paradox by F.F. Chen [15] that the theory of the Maxwellian stress tensor should not at all result in forces. After solving the paradox we present the first detailed computation of the motion of electrons in these fields.

The initial motivation was [8, 16, 17] that a motion and acceleration of such intensity minima (by a dynamisation of the otherwise "standing" waves) should provide an acceleration scheme for electrons in vacuum. It was necessary [18] that these minima should move with less than the speed of light c even with only a very tiny difference to c. It was evaluated how the superposition of two travelling laser waves with a temporally varying phase or frequency could reach this condition by an approximate solution of Maxwells equations. Its realisation - indeed - includes complex conditions of variations of dielectric materials or other surface interactions such that the scheme would be reduced to the earlier mentioned material interaction problems.

Though the following discussion of the acceleration of electrons in the intensity minima of a standing wave field [16,19] is using Maxwellian exact solutions and the just mentioned problems with approximations do not apply, we mention here another scheme

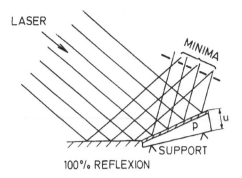

LASER

MINIMA

u

p

SUPPORT

100% REFLEXION

Fig. 1b. Generation of a laser field from one beam using a Fresnel mirror with intensity minima due to interference. Bending on part of the mirror produces an accelerated motion of the minima for acceleration of electrons trapped within these minima.

how to produce the field with the moving intensity minima of a laser field [19]. This is shown in Fig. 1b, where the plane wave optical field of a laser is reflected from a Fresnel mirror producing an interference field with minima and maxima. When one of the sections of the kinked mirror is bent, e.g. electronically with piezo ceramics, the minima of this field can be moved with very high speeds.

4. A PARADOX WITH THE MAXWELLIAN STRESS TENSOR

The force density due to electric and magnetic fields, E and H, in a plasma is given by the nonlinear force [9, 20].

$$f_{NL} = \nabla . \{ \underline{T} + (1 - \frac{1}{\omega} \frac{\partial}{\partial t}) [\frac{1 - \tilde{n}^2}{4\pi} \underline{E}\underline{E} - \frac{\chi}{4\pi} \underline{H}\underline{H}] \} + \frac{1}{4\pi c} \underline{E} \times \underline{H} \qquad (1)$$

$$4\pi \underline{T} = \underline{E}\,\underline{E} + \underline{H}\,\underline{H} - \frac{1}{2} (\underline{E}^2 + \underline{H}^2)\,\underline{1} \qquad (1a)$$

where $\underline{T}$ is the Maxwellian Stress tensor in vacuum, ω is the central value of the laser frequency of the temporally finite wave packet, $\tilde{n}$ is the optical refractive index, and χ is the magnetic suceptibility; the latter had to be added to the earlier general transient formula [20] as electrons transmitting a laser beam in vacuum cause optical self focussing and filamentation by magnetic interaction [11]. The nonlinear force of Eq. (1) contains ponderomotive and non-ponderomotive terms as analyzed before [9].

The trapping of electrons within the nodes of standing electromagnetic waves as first derived by Weibel [12] is due to the simplified ponderomotive component of the nonlinear force (1) for one dimensional geometry

$$f_{NL} = - \frac{n_e}{8\pi n_{ec}} \nabla E^2 \qquad (2)$$

where the electron density is n_e and the cut-off density n_{ec} is

$$n_{ec} = \frac{\omega^2 m}{4\pi e^2} \quad , \tag{2a}$$

e is the electron charge and m the electron mass. The single electron motion with the velocity v follows from the force density in Eq. (2) by dividing by the electron density

$$m\frac{dv}{dt} = -\frac{1}{8\pi n_{ec}} \nabla E^2 = \frac{e^2}{2m\omega^2} \nabla E^2 \quad . \tag{3}$$

The result is that free electrons within a standing microwave field are concentrated (trapped) at the nodes of the standing wave. The same happens for plasma where the nonlinear force in a standing wave field pushes plasma to the nodes and causes a density ripple [8].

While the derivation of the confinement of electrons within the minima of E^2 by Weibel and the other authors [13-14] was straight forward and convincing, F.F. Chen mentioned [15] that the nonlinear force (1) for the stationary case of perpendicularly incident electromagnetic radiation in vacuum simplifies to (see [21]

$$f_{NL} = -\nabla(E^2 + H^2)/8\pi \tag{4}$$

Using a standing wave field as shown in Fig. 2, shows that the field expression right hand

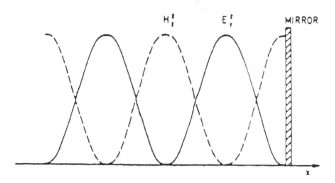

Fig. 2. Electric and magnetic field of a linearly polarized standing wave in vacuum propagating into the x-direction when totally reflected by a mirror M.

side of the del-operator in Eq. 4 is constant. Therefore, no force should be produced. The question is then whether Weibel and the other authors [13-14] are wrong or whether one simply should be cautious by using the Maxwellian stress tensor [15].

This paradox can be solved in the following way. Using the dielectric response of a collisionless plasma given by a refractive index ñ

$$ñ^2 = 1 - \frac{n_e}{n_{ec}} \tag{5}$$

and using the WKB approximation for a medium with spatially varying refractive index where the field amplitudes for plane waves are $E_v/(n)^{1/2}$ and $E_v(n)^{1/2}$ for the electric and magnetic vector respectively, the time average of the force density for a wave propagation into the x-direction is

$$f_{NL} = - \frac{E_v^2}{16\pi} \frac{\partial}{\partial x} \left(\frac{1}{\tilde{n}} + \tilde{n} \right) = \frac{E_v^2}{16\pi} \frac{n_e}{n_{ec}} \frac{\partial}{\partial x} \frac{1}{\tilde{n}} \tag{6}$$

as derived first in 1967 [22]. Taking the same steps as between Eqs. (2) and (3) by expressing the force density $f = mn_e(d/dt)v$ and cancelling on both sides the electron density n_e results in the equation of motion for one single electron, Eq. (3) as used by Weibel [13].

These steps implied dividing by the very large number of the electron density but what is remaining is the equation for one single electron. The action of the Maxwellian stress tensor is therefore including the dielectric response to the one single electron to the field in Eqs. (4) and (5) such that there is no constant (vacuum) field expression to be differentiated in Eq.(4) even if only the presence of one single electron. This is a reminder of the elementary result in electrostatics: If one calculates the forces between two charges from the energy density of the field, one has to take the fields of both charges while the same force results if one takes one charge as a probe (ignoring its own field) when determining the force against the other charge only by working with the electric field of the charge and not the prode.

Being aware of these facts one has to realize that the use of the energy density description of the Maxwellian stress tensor to a charge (or plasma) does include the dielectric response of the charge to the externally given field. In this case, errors will be avoided and the suspicion expressed by F.F. Chen against the Maxwellian stress tensor description can be overcome. The correctness of the results of Weibel and the other authors [13-14] is then clarified.

5. ELECTRON MOTION IN INTENSITY MINIMA

After the clarifications in the preceeding section it is evident that the detailed mechanical description of an electron in the field of a superposition of laser radiations can be described by the equation of motion for the velocity $\underline{v}$ in the finally resulting laser field for $\underline{E}$ and $\underline{H}$ of the well known form

$$\frac{d}{dt} \underline{v} = - \frac{e}{m} \underline{E} - \frac{e}{m} \underline{v} \times \underline{H} \tag{7}$$

If the laser field consists of linearly polarized plane waves moving into (or against) the x-direction with the components E_y and H_z, the differential equations for the velocity components depending on the Cartesian coordinates x, y, and the time t is then

$$\frac{\partial}{\partial t} v_x + \frac{1}{2} \frac{\partial}{\partial x} v_x^2 + v_y \frac{\partial v_x}{\partial y} = - \frac{e}{m} v_y H_z \tag{8a}$$

$$\frac{\partial}{\partial t} v_y + \frac{1}{2} \frac{\partial}{\partial y} v_y^2 + v_x \frac{\partial v_y}{\partial x} = - \frac{e}{m} E_y + \frac{e}{m} v_x H_z \tag{8b}$$

This has to be solved for the superposition of an electromagnetic wave of frequency ω in vacuum propagating to the +x-direction (index 1) and its totally reflected wave (index 2) where the reflection occurs at a mirror moving with a velocity v into the x-direction. Working with a mirror motion in the subrelativistic range the wave field can be approximated even with a time varying v by

$$E_{y1} = E_0 \cos \left(\frac{\omega}{c}x - \omega t\right) \tag{9a}$$

$$E_{y2} = - E_0 \cos\left[\left(\frac{\omega}{c}x + \omega t\right)\left(1 - \frac{v}{c}\right)\right] \tag{9b}$$

$$H_{z1} = E_0 \cos \left(\frac{\omega}{c}x - \omega t\right) \tag{9c}$$

$$H_{z2} = E_0 \cos \left[\left(\frac{\omega}{c}x + \omega t\right)\left(1 - \frac{v}{c}\right)\right] \tag{9d}$$

with the fields after superposition of

$$E_y = E_{y1} + E_{y2} \quad ; \quad H_z = H_{z1} + H_{z2} \tag{10}$$

A further integration of the velocity components arrives at the coordinates x and y of the electron as function of the time t.

For the solution of the nonlinear differential equation system, a numerical integration method was used in conjunction with the necessary Romberg correction method [23], which resulted in a numerically stable solution, reducing the computer time to one twentieth. Nevertheless the computations needed long runs. The results presented in the following are needing more than two hours CPU time on an IBM supercomputer 3090. A special study was necessary to determine the desired initial conditions.

6. RESULTS OF ELECTRON MOTION AND ACCELERATION

We first tested the traces of the electrons in the standing wave patterns at rest (v=0). We used the electromagnetic standing wave field of carbon dioxide laser radiation in vacuum with a wavelength of 10.6 μm. The trapping of an electron starting at the half maximum intensity of the laser standing wave field results in a trace as shown in Fig. 3 for a laser intensity of 1×10^{13} W/cm^2. The initial conditions are chosen in such a way that a small initial velocity (drift) in the positive y-direction is given such that the electron drifts into this direction and the trace is then expanded in Fig.3 for an easier recognition of the details.

As could be expected, the trapping of the electrons to the nodes of the standing wave is not a static one but the electrons between a certain point of the standing wave field oscillates permanently between the corresponding points of the field centered to the minimum of the E-field. These oscillations occur in the same way for each corresponding points such that the concentration of the electrons into the nodes of the standing wave is by time averaging over a large number only of the electrons pending between the corresponding points of the field each.

For the same conditions as in Fig. 3, the trace of an electron has been calculated for the case that the totally reflecting mirror is moving with an acceleration of 1.41×10^{17} m/s^2 from the time zero of the starting of the electron until a time t =7.6 ps from then on the motion is continued at constant velocity (Fig. 4a).It should be mentioned that such an acceleration of a plasma mirror is usual in laser produced plasmas (8).

As in Fig. 3, the electron starts at an area A and is returned at B but (due to the motion of the wave field) is then returned at C, going to D, then to E and F etc. The

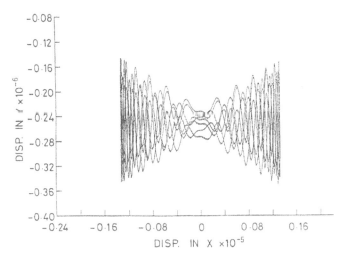

DISP. IN r $\times 10^{-6}$

DISP. IN X $\times 10^{-5}$

Fig. 3. Trace of an electron in a standing wave field of a carbon dioxide laser of intensity 1×10^{13} W/cm^2 starting from the half maximum intensity of the standing wave field.

acceleration into the x-direction is accompanied with such a shift of the phase, that a further acceleration into the negative y-direction occurs. This is typical to the Lorentz-type acceleration as it has been derived from the exact solution of a travelling (not standing) half-wave [30] as shown in the following sections, where the motion into the y-direction is especially pronounced at subrelativistic conditions as applied here.

A more transparent result is seen for the same case when plotting the trace of the velocity components (Fig. 4b). From the parabolic increase of the first part of the oscillations one can see how the electron is accelerated by the laser field. The simple estimation of the increase of the average velocity of the electron can be done from a graphical evaluation of Fig. 4b as follows; the transverse electron oscillation is converted, by the laser field, into a longitudinal (or translative) motion of the electron in which case the electron gains a velocity at the node equivalent to the maximum oscillation energy. This can be seen directly from Fig. 3, where the transverse oscillation energy at the point x=0 is zero and a maximum transverse oscillation at x= $\pm$ 1.325 μm ($\lambda/8$). The evaluation of the velocity at the node from the maximum oscillation energy can be determined from the maximum electric field seen by the electron, this leads to an average nodal velocity of v_{av} =9.88x10^6 m/s^2 ($v_{av}^2 = \dfrac{e^2 E_o^2}{m^2 \omega^2}$

where E_o is the maximum electric field seen by the electron, which is a function of the initial electron position). From the numerical solution of the nonlinear system of differential equations responsible for figure 3, the average electron velocity at the node of the the field intensity is v_{av}=9.89x10^6 m/s^2.

The fact that the velocity oscillation during the acceleration phase is shorter than in the later phase of constant motion of the wave field, can be explained in the following way. The velocity of the longitudinal oscillation is represented by figure 4b, in the first 7.6 ps the

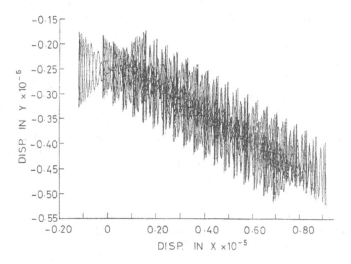

Fig. 4a. Trace of an electron in an accelerated "standing" wave field with subsequent constant velocity.

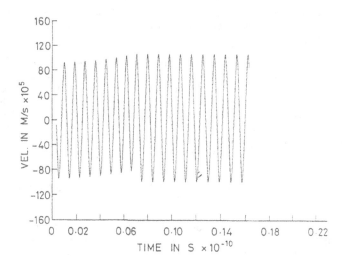

Fig. 4b. Velocity trace of an electron for the same case as in Fig. 4a

mirror is accelerated, in such a way as to increase from zero to 1×10^6 m/s parabolicly then is constant for the remaining time. In the region where the mirror is accelerated the longitudinal oscillation energy is sacrificed in favor of translation energy and the mean velocity is up-shifted. By measuring the up-shift in velocity from figure 4 graphically the average acceleration of the electron is evaluated to be 1.45×10^{17} m/s^2, the average acceleration assigned to the mirror in this region was 1.41×10^{17} m/s^2 which is within the accuracy of the graphical evaluation.

This confirms that the confinement of electrons in the standing wave field of a laser and their concentration at the nodes of the standing wave as phenomenologically explained by Weibel and other authors [13-14] due to the ponderomotive or Miller force (as a special case of the general nonlinear force) is being followed by a single electron motion in the standing wave laser field of a carbon dioxide laser. The basic partial differential equation for the components of the motion is a rather complicated nonlinear system which has been solved numerically. The result indicates that the electrons trapped between the antinodes are still oscillating between corresponding points symmetric to the nodes and the increase of the electron density at the nodes is the result of summing up of the electrons there or by their expectation probability.

When moving the standing wave field, the electrons are following the frame. By accelerating the field, the electrons are accelerated too gaining energy. This is interesting for schemes of laser acceleration of electrons by the intensity minima of superimposed optical fields, in vacuum without plasma conditions [8]. This process may be involved in the experimental discovery of the inelastic Kapitza Dirac effect [24].

The analysis is based on subrelativistic motions of the standing wave field frame only in order to demonstrate and to derive the basic acceleration mechanism of the electrons in all details of the single particle motion in order to support the otherwise identical global result known from the phenomenological theory of the acceleration by the nonlinear force of ponderomotion. A paradox in the theory of the Maxwellian stress tensor description of the nonlinear force has been solved by clarifying how a description of the energy density requires the inclusion of the response field of the test particle (electron) while direct derivation of the force from the given field - as well known in electrostatics - has to neglect the field of the test particle.

It has been shown that the motion of the electrons in the rather simple field of the accelerated "standing" wave field are rather complicated and the solution of the complex nonlinear partial differential equation system needs special numerical techniques and long computer runs. The acceleration of an electron - even in the subrelativistic case only - within the moving or accelerated intensity minima [8,19] of a dynamic Fresnel interference experiment may be even more complicated. But it may be concluded that the traces of the electron may be not basically different to the cases shown here for the electron acceleration using intenisty minima of a standing wave field.

7. EXACT SOLUTION OF ELECTRON ACCELERATION BY TRAVELLING PLANE ELECTROMAGNETIC WAVES IN VACUUM

In the following we present the relativistically exact solution for the acceleration of an electron by linearly polarized, plane (travelling) electromagnetic waves in vacuum. The electromagnetic fields are thought to be produced by laser waves with field strengths up to about 10^{13} V/cm. In this regime the electric and magnetic fields allow a point-mechanical (classical) description of the electron motion. Also the magnetic moment of the electron spin does not need to be included.

In the calculation we study the acceleration of a single electron and neglect its interaction with other ones. Also we do not consider the reaction of the electron on the fields which is a basic assumption for the free electron laser. The amount of energy the electron loses by bremsstrahlung is negligibly small as shown before [25].

Assuming the x-direction as the direction of progression of the transverse wave and the y-direction as the direction of the electric field, we can write the electric and magnetic fields as follows [25]:

$$\mathbf{E} = E_y(x\text{-}ct)\mathbf{e}_y \tag{11a}$$

$$\mathbf{H} = H_z(x\text{-}ct)\mathbf{e}_z \tag{11b}$$

Inserting these fields in the four Maxwellian equations for free fields we find the relation which fulfils these equations:

$$E_y(x\text{-}ct) = H_z(x\text{-}ct) \tag{12}$$

The equations of motion of the electron are given for the individual Cartesian components as:

$$m\frac{d\gamma v_x}{dt} = -\frac{e}{c}v_y E_y \tag{13}$$

$$m\frac{d\gamma v_y}{dt} = -e(1-\frac{v_x}{c})E_y \tag{14}$$

$$m\frac{d\gamma v_z}{dt} = 0 \tag{15}$$

where we have used the relation (12) in replacing H_z by E_y. The factor γ is the relativistic Lorentz-factor defined by

$$\gamma = (1 - \frac{v^2}{c^2})^{-1/2} \tag{16}$$

and m the rest mass of the electron. The last equation of motion, Eq. (15), will not be regarded in the following by assuming no motion of the electron in the z-direction ($v_z = 0$).

A further equation, namely the equation for the time-dependence of the power, can be derived from Eqs (13) to (15):

$$m\frac{d\gamma c^2}{dt} = -ev_y E_y \tag{17}$$

Because the right-hand sides of Eqs (13) and (17) are the same besides a factor c, we obtain an equation between γ and v_x.

$$\frac{d\gamma(1-v_x/c)}{dt} = 0 \tag{18}$$

This equation can be integrated with the initial conditions $\gamma(t=0) = \gamma_0$, $v_x(t=0) = v_{xo}$:

$$\gamma(1-\frac{v_x}{c}) = \gamma_0(1-\frac{v_{xo}}{c}) \tag{19}$$

Next we integrate Eq. (14) with the result

$$\gamma\frac{v_y}{c} = A(x-ct) \tag{20}$$

where

$$A(u) = \gamma_0\frac{v_{yo}}{c} + \frac{1}{mc^2}\int_0^u E_y(u')du' \tag{21}$$

$$u = x-ct \tag{22}$$

In Eq. (20) we used $v_y(t=0) = v_{yo}$ and $x(t=0) =0$ as further initial conditions. Eqs. (19) and (20) are sufficient to express the functions γ, v_x/c and v_y/c as functions of $A(u)$:

$$\gamma = \frac{1+\gamma_0^2(1-v_{xo}/c)^2+A^2(u)}{2\gamma_0(1-\frac{v_{xo}}{c})} \tag{23}$$

$$\frac{v_x}{c} = \frac{1+A^2(u)-\gamma_0^2(1-v_c)^2}{1+A^2(u)+\gamma_0^2(1-v_{xo}/c)^2} \tag{24}$$

$$\frac{v_y}{c} = \frac{2\gamma_0(1-v_{xo}/c)A(u)}{1+A^2(u)+\gamma_0^2(1-v_{xo}/c)^2} \tag{25}$$

where $A(u)$ is given by Eq. (21) and u defined in Eq. (22). Eqs. (23) to (25). Without losing generality we choose the initial conditions $\gamma_0 = 1$, $v_{xo} = v_{yo} = 0$, i.e., $v(t=0)=0$ and get from Eqs. (23) to (25):

$$\gamma = 1 + \frac{1}{2}A^2(u) \tag{26}$$

$$\frac{v_x}{c} = \frac{A^2(u)/2}{1+\frac{1}{2}A^2(u)} \tag{27}$$

$$\frac{v_y}{c} = \frac{A(u)}{1+\frac{1}{2}A^2(u)} \tag{28}$$

where

$$A(u) = \frac{e}{mc^2}\int_0^u E_y(u')du' \tag{29}$$

The coordinates of the location of the electron are obtained by using Eqs. (24) and (25) or (27) and (28), e.g.,

$$\frac{dx}{du} = \frac{dx}{dt}\frac{1}{v_x-c}$$

$$= \frac{1+A^2(u)-\gamma_0^2(1-v_{xo}/c)^2}{-2\gamma_0^2(1-v_{xo}/c)^2} \tag{30}$$

This yields for the initial conditions $x(t=0)=y(t=0)=0$

$$x = -\int_0^u \frac{1+A^2(u') - \gamma_0^2(1-v_{xo}/c)^2}{2\gamma_0^2(1-v_{xo}/c)^2} \, du' \tag{31}$$

$$y = \int_0^u \frac{A(u')}{\gamma_0(1-v_{xo}/c)} \, du' \tag{32}$$

For $\gamma_0 = 1$ the equations reduce to

$$x = -\int_0^u A^2(u')du' \tag{33}$$

$$y = -\int_0^u A(u')du' \tag{34}$$

These are the general solutions of the equations of motion for the electron. Solving the above integrals, one obtains first the various quantities as function of $u = x-ct$. Then one chooses the time, too, as function of u: $ct = x(u) - u$. By this procedure the dependence of the velocities and space coordinates on time becomes known.

8. NET ACCELERATION WITH TRUNCATED WAVE PACKETS

In order to apply the formulas given in the preceding section we assume special types of wave packets, namely harmonically oscillating waves, a constant field strength of finite length, and oscillating waves with only positive amplitudes (rectified waves). Further we consider the case that the electron starts with velocity $v = 0$ from $x=0$ and at $y=0$ of time $t=0$. The case that electrons (or in the same way ions) are accelerated by standard accelerators to relativistic energies and speeded then, up to TeV energies by the laser fields, is a sub-class of the following solutions.

Let us choose the electric field as follows:

$$E_y = -E_0 \sin(kx-\omega t)(\theta(x-ct+l)-\theta(x-ct)) \tag{35}$$

The function $\theta(u)$ is the Heaviside step function. The length of the wave packet is given by the parameter l. We calculate the velocity of the electron after the wave packet has crossed it. We obtain by use of Eqs. (26) to (29):

$$A(u=-l) = \delta \int_0^{-l} (-\sin ku) \, du = \frac{\delta}{k} (\cos kl -1) \tag{36}$$

with

$$\delta = \frac{eE_0}{mc^2} , \tag{37}$$

$$\gamma = 1 + \frac{1}{2}\left(\frac{\delta\gamma}{\pi}\right) \sin^4\left(\frac{\pi l}{\gamma}\right) , \tag{38}$$

$$\frac{v_x}{c} = \frac{1/2(\delta\lambda/\pi)^2 \sin^4(\pi l/\lambda)}{1+1/2 \, (\delta\lambda/\pi)^2 \sin^4(\pi l/\lambda)} \tag{39}$$

$$\frac{v_y}{c} = \frac{-(\delta\lambda/\pi)\sin^2(\pi l, \lambda)}{1+1/2(\delta\lambda/\pi)^2\sin^4(\pi l/\lambda)} \qquad (40)$$

Here, we set $k = 2\pi/\lambda$. These formulas prove clearly that the harmonic wave packet can accelerate an electron if the length is equal to an odd number of half wave lengths: $l = (2n+1)\lambda/2$ with $n=0,1,2...$ For a length equal to an even number of half wave lengths (i.e. any number of complete wave lengths) the electron gets completely stopped as it is known from the cases with electron assembles or plasmas [26,27] or from the lateral injection free electron (or cluster) laser [28], or as mentioned by Sessler [18]: a simple wave packet does not transfer energy into a free electron. This shows that the field amplitude must be in average positive (or negative) in order that a net acceleration occurs. This is achieved e.g. by complete or at least partial rectification of electromagnetic waves. Truncation of a wave packet will be useful too. The action of one half wave would be preferable. It has been shown experimentally [29] that CO_2 laser pulses of less than one wave length can be produced. This kind of rectification should be possible by a short Dugnay shutter [30] as explained in Fig. 5.

Using these half waves, it is important to realize that the electron well follows the lateral acceleration by the transverse electric field, but the generated Lorentz force drives the

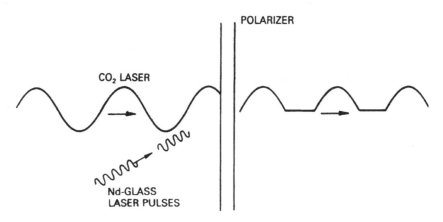

Fig. 5. Semi-rectification of a CO_2 laser pulse by a Duguay shutter [30] using five wavelength Nd-glass laser pulses in a shutter-polariser.

electrons into axial direction. At relativistic conditions $\gamma\gg1$, the lateral end velocity after one half wave length, l, motion even drastically drops against the very high velocity in the propagation direction of the light. Fig. 6 shows the results of Eqs. (39) and (40), modified for a truncated half wave pulse [25]. The velocity in the transverse direction has a maximum of $|v_y/c| = 0.707$ at $\gamma=2$ and drops to $|v_y/c|=0.0014$ at $\gamma=10^6$. This has the consequence that the ratio between the length x and width y of a possible acceleration increases proportional to $(\gamma-1)^{1/2}$. The final coordinates x and $|y|$, and their ratio $|x/y|$ after the acceleration of the electron are depicted in Fig. 7. The accelerator length is $L=x=(\gamma-1)l/3$ where l is the half wave length of the laser beam [25].

The optimised case will be a set of positive half-waves as being produced by a laser wave packet in which the negative amplitudes of E_y and H_z are inverted into positive ones (fully rectified). For simplification we assume a length of the packet given by a multiple N of a half wave length.

$$l = N\frac{\lambda}{2} \qquad (41)$$

The electric fields can be written as:

$$E_y = E_0 \ \sin (kx-\omega t) \ (\theta(x-ct+l)-\theta(x-ct)) \qquad (42)$$

The function A(u), defined in Eq. (29), is obtained as:

$$A(u) = \frac{\delta}{k}(\cos ku -1) [\theta(u+\frac{\lambda}{2}) - \theta(u)]$$

$$+ \frac{\delta}{k}(-\cos ku -3) [\theta(u+2\frac{\lambda}{2}) - \theta(u +\frac{\lambda}{2})]$$

$$+ \frac{\delta}{k}(\cos ku - 5) [\theta(u+3\frac{\lambda}{2}) - \theta(u+2\frac{\lambda}{2})] + ... \qquad (43)$$

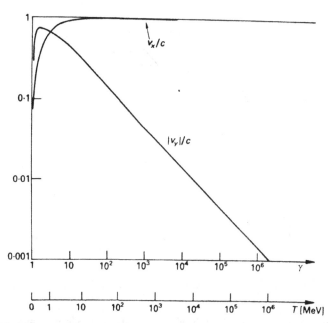

Fig. 6. The final velocities of the electron in the transversal y - and longitudinal x - directions after the acceleration of the electron by a travelling box-like approximated half-wave of length l=λ/2 as functions of γ and the kinetic energy T of the electron [25].

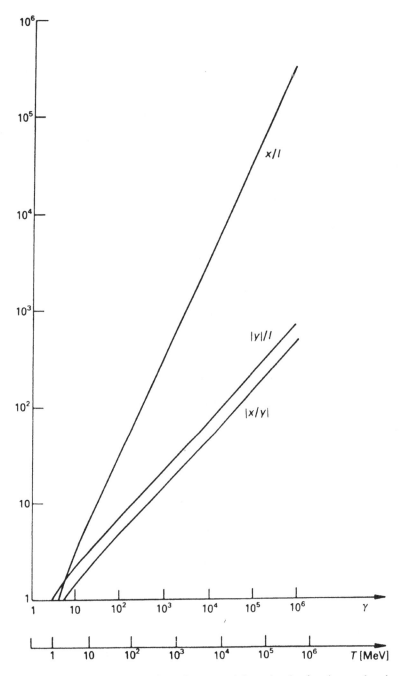

Fig. 7. Final coordinates x and y of the electron and the ratio x/y after the acceleration by a half-wave length l as functions of γ for the case of Fig. 6.

The value of γ is given by:

$$\gamma = 1 + \frac{1}{2}(A(u=-N\frac{\lambda}{2})) = 1 + \frac{1}{2}\left(\frac{N\lambda\delta}{\pi}\right) \tag{44}$$

Next we calculate x and y after the acceleration process:

$$x = -\frac{1}{2}\int_0^{u=-1} [A(u)]^2 \, du$$

$$= \frac{\delta^2\lambda^3}{32\pi^2}(3 + 19 + 51 + 99 \ldots)$$

$$= \frac{\delta^2(N\lambda)^3}{12\pi^2}(1+\frac{1}{8N^2}) \tag{45}$$

$$y = -\int_0^{u=-1} A(u)du = -\frac{\delta(N\lambda)^2}{4\pi} \tag{46}$$

In the next section we shall evaluate the quantities x and y as functions of the intensity $I = \frac{1}{2}(\varepsilon_0/\mu_0)^{1/2}$ E_0^2 (SI units). They are given as follows:

$$\gamma = 1 + \left(\frac{e\lambda}{mc^2\pi}\right)\left(\frac{\mu_0}{\varepsilon_0}\right) I \, N^2 \tag{47}$$

$$x = \frac{1}{6}\left(\frac{e\lambda}{mc^2\pi}\right)\lambda\left(\frac{\mu_0}{\varepsilon_0}\right)IN^3(1+\frac{1}{8N^2}) \tag{48}$$

$$y = -\frac{1}{\sqrt{8}}\frac{e\lambda^2}{mc^2\pi}\sqrt{\frac{\mu_0}{\varepsilon_0}} I \, N^2 \tag{49}$$

9. APPLICATIONS TO NEODYMIUM GLASS AND CO_2 LASERS

For the wavelengths of the Neodymium glass and CO_2 lasers of 1.06 μm and 10.6 μm, respectively, we apply the formulas for wave packets with positive amplitudes derived in the preceding Section. From these calculations we learn about the energy and size of a possible accelerator for electrons.

The formulas (47) to (49) are for λ = 1.06 μm.

$$\gamma = 1 + 1.6425 \times 10^{-22} (I \frac{m^2}{Watt}) N^2, \tag{50}$$

$$x = 2.90168 \times 10^{-29} (I \frac{m^2}{Watt}) N^3 (1+\frac{1}{8N^2})m, \tag{51}$$

$$y = -4.8030 \times 10^{-18} (I \frac{m^2}{Watt}) N^2 \, m, \tag{52}$$

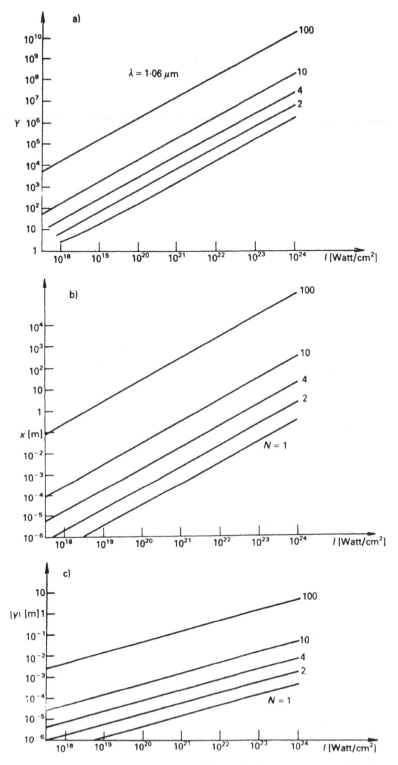

Fig. 8. (a) The relativistic Lorentz-factor γ; (b) the final coordinate x and (c) the final coordinate y after the acceleration of the electron by a wave packet with positive amplitudes and a length of N half wave lengths are shown as functions of the intensity for a wavelength λ = 1.06μm (Neodymium glass laser).

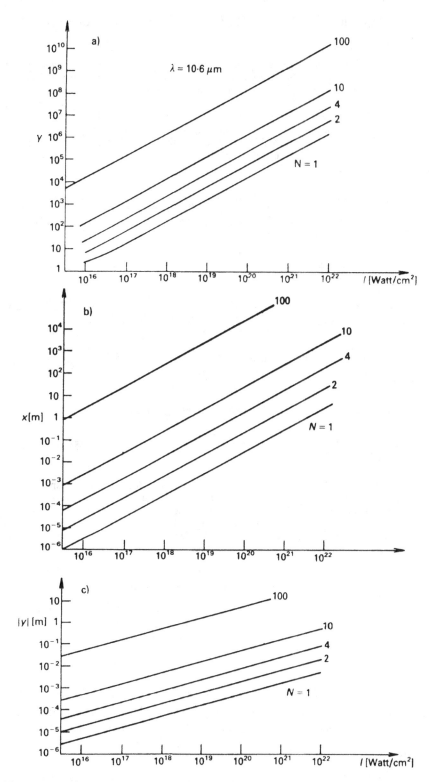

Fig. 9. The same quantities are shown as in Fig. 6 for a wavelength $\lambda = 10.6\mu m$ (CO_2 laser).

For $\lambda = 10.6\mu m$ the power factors are 10^{-20}, 10^{-26} and 10^{-16} in the expressions for γ, x and y, respectively.

In Figs. 8 and 9 we show the γ-factors and the lengths x and y for wavelengths $\lambda = 1.06$ and $10.6\mu m$, respectively. For equal intensities the wave with the larger wavelength needs a longer time to cross the electron and, therefore, the values of γ, x and y are larger. E.g. for $I = 10^{22}$ Watt/cm^2, $\lambda = 1.06\mu m$: $\gamma = 1.6x10^4$, $x = 3.3$mm, $y = -48\mu m$; $\lambda = 10.6\mu m$: $\gamma = 1.6x10^6$, $x = 3.3$m, $y = -4.8$mm (for N = 1). For equal final velocities, i.e, γ-factors, waves with shorter wave lengths need higher intensities, but run shorter distances. E.g., $\gamma = 10^6$ corresponding to 0.511 TeV, $\lambda = 1.06\mu m$: $I = 6.09x10^{23}$ Watt/cm^2, $x = 0.199$m, $y = -0.375$mm; $\lambda = 10.6\mu m$: $I = 6.09x10^{21}$ Watt/cm^2, $x = 1.99$ m, $y = -3.75$mm (for N = 1).

A possible accelerator with a fixed length x and a given wavelength λ has to have a variable power I and has to use wave packets of different lengths $N\lambda/2$ in order to accelerate electrons with different energies. In this case the connection between γ, x, λ and N is given according to Eqs. (47) and (48):

$$\gamma = 1 + \frac{6x}{\lambda N(1+1/(8N^2))} \qquad (53)$$

Fig. 6 shows the γ-factor as a function of the number N of half wave lengths for a given value of x and $\lambda = 1.06 (10.6)\mu m$. The curves demonstrate the effect that with increasing N and fixed value of x the γ-factor drops. As can be seen from Figs. 6 and 7 and by inspecting Eq. (51), the needed power decreases with $1/N^3$. E.g., for an accelerator of length $x = 1$m and wavelength of $\lambda=1.06(10.6)\mu m$ we obtain for N=1: $\gamma = 5.03x10^6(5.03x10^5)$, $I = 3.06x10^{24} (3.06x10^{21})$ Watt/cm^2 and for N = 100: $\gamma=5.66x10^4 (5.66x10^3)$, I= $3.44x10^{18} (3.44x10^{15})$ Watt/cm^2.

The formula (53) can be used to express the energy gain per length:

$$\frac{T}{x} = \frac{6mc^2}{\lambda N(1+1/(8N^2))} \qquad (54)$$

For N = 1 we find 2.57 and 0.257 TeV/m for $\lambda = 1.06$ and $10.6\mu m$ respectively. It is interesting that Eq. (54) depends only on λ and N. This fact stresses the point that lasers with short wave lengths should be applied to accelerate electrons in order to get large energy gains per length. Using the KrF-laser wave length of $0.25\mu m$ and N = 1 (if the rectifying technique could be solved, e.g., by selecting molecules in the polarizers with a response time of half the period of the laser radiation), the gain per accelerator length is then 10.9 TeV/m.

10. STRUCTURE OF THE LASER BEAM IN Y-DIRECTION

The lateral falling off of the laser beam in the transverse direction may lead to high velocities of the electron in the y-direction due to the nonlinear forces [9,31]. In order to estimate this effect we assume the electric field of the packet as following:

$$E_y = f(y) E_y(u) \qquad (55)$$

Here, $f(y)$ is a structure function and E_y the plane packet depending on $u = x-ct$. Further we assume $H_z = E_y$, which is an approximation not solving the Maxwellian equations exactly. An exact solution would also have components of the fields in the x-direction as shown already in literature ([9], Chapter 12.3)[10]. The equations of motion of the electron are now given by:

$$m\frac{d\gamma v_x}{dt} = -\frac{e}{c}v_y f(y) E_y(u) \qquad (56)$$

$$m \frac{d\gamma v_y}{dt} = -e(1 - \frac{v_x}{c}) f(y) E_y(u) \tag{57}$$

These equations can be solved by the same method as discussed in before. Assuming that the coordinate y is also a function of u like x, we obtain the following system of differential equations for y (no initial velocity):

$$\frac{dy}{du} = -A(u) \tag{58}$$

$$\frac{dA}{du} = \frac{e}{mc} f(y) E_y(u) \tag{59}$$

For $f(y) = 1$ we have the solution given in Eq. (34). After inserting Eq. (59) into (58) we can write

$$\frac{d^2y}{du^2} = -\delta f(y) \tag{60}$$

To get some definite result out of Eq. (60) we choose a Gaussian profile for $f(y)$:

$$f(y) = \exp(-y^2/\sigma^2) \tag{61}$$

For this structure function we obtain the following solution of Eq. (62) as an expansion in powers of u:

$$y = -\frac{\delta u^2}{2} (1 - \frac{1}{15} \left(\frac{\delta u^2}{2\sigma}\right) + \frac{19}{1350} \left(\frac{\delta u^2}{2\sigma}\right) + \dots) \tag{62}$$

According to this formula we find the result that the structure function plays only a minor role as long as y remains small compared with the width σ.

For completion it should be mentioned that the exact solution of the acceleration of electrons by plane waves has been studied before by other authors [32] but the here given derivation was specifically performed in order to arrive at detailed evaluations of the acceleration process for the laser intensities which are in the scope of the present and next following level of developments.

11. ACCELERATION OF ELECTRONS TO TeV ENERGIES BY LATERAL INJECTION INTO SHORT LASER PULSES

The just discussed exact solutions of acceleration of electrons by plane electromagnetic pulses of broken wave length duration after rectification [25] are now applied to a discussion of neodymium glass laser beams. As a first step of simplification, laser beams are cut out of a plane wave field corresponding then to a laser beam with nearly rectangular radial intensity profile of a diameter y. It should be reminded that this type of profile has been realized experimentally. The laser pulse profile is explained in Fig. 10 where a relatively small radial decay from the constant interior part of the beam to zero within an increment $(\delta)r$ is used. The laser focus is assumed to be rectangular as shown in Fig. 11 of a width y and a length x.

The electron is assumed to enter the laser beam at a point A where an energy E_{NL} is necessary in order to overcome the nonlinear force for transmitting the gradient of the electrical laser field within the thickness $(\delta)r$. Since we are in the relativistic range only, this energy is given by

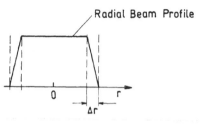

Fig. 10. Intensity profile of the laser pulse on the radius r.

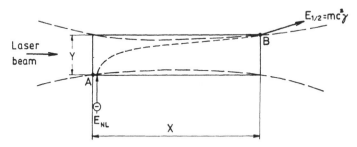

Fig. 11. Rectangular focus area of the laser beam of diameter y and length x with a dashed real beam contour and an input of an electron at point A with an initial energy E_{NL} of radial motion and an energy $E_{1/2}$ when departing the beam at point B. Dotted line corresponds to the relativistic and Maxwellian exact motion of the electron in the rest frame.

$$E_{NL} = mc^2(I/I_{rel})^{1/2} \tag{63}$$

where the relativistic threshold intensity is that where the quiver energy of the electron in the laser field is mc^2. This energy is rather small compared with the energy E gained by the electron when moving within the focus during such a time that just only one half wavelength of the laser beam is acting for the acceleration of the electron according to precedings section 6 to 10, as shown in Fig. 12.

Fig. 10 reproduced the necessary beam widths y focal length x and neodymium glass laser intensities I and power P necessary for accelerating an electron during interaction with only one half wavelength of the laser pulse when entering the Beam on one side A and leaving at the other side B. Any rectification of the laser beam is then not necessary. It is remarkable that the team divergence is then nearly equal to the diffraction limit for the first diffraction minimum of a single laser beam mode.

Table 1 contains the values of the laser intensity, of the laser energy within one half wave length section, the acceleration length x and beam diameter y for producing TeV electrons with lasers of various wave lengths.

It is remarkable that the laser power P increases with the square of the electron energy. Due to the relativistic motion of the electron, a rather long acceleration length of e.g. 38 cm is necessary for arriving at TeV energy until the half wavelength of a neodymium glass laser has overtaken the electron. In order to measure the effect, laser pulses of reasonable magnitudes are nearly available. Restricting to neodymium glass laser only as the most advanced ones for this purposes, in order to arrive at an electron energy of GeV, the laser pulse has to have a power of 2.52×10^{15} Watts and has a single node to be focused to a diameter of 0.016 nm with a focus length of 0.38 mm. The energy in the section of one half wave lengths is then 4.47 Joules. The lateral initial energy of the electron to overcome the nonlinear force - which energy is then returned after leaving the pulse - is 1.21 MeV.

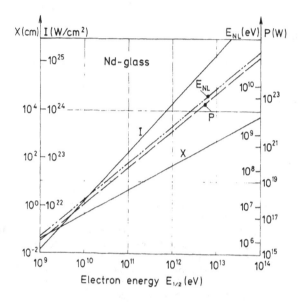

Fig. 12. Laser beam width y, focal length x, laser power P, intensity I, and necessary lateral injection energy E_{NL} for overcoming the nonlinear force at injection into a nearly rectangular beam depending on the energy $E_{1/2}$ which the electron gains when moving with one half wave length in a neodymium glass laser beam.

Such an experiment is close to realization, since laser beams of the type of Nova or Phebus are scheduled to produce pulses of 1 psec duration and 1 KJ energy by end of 1991 [33]. A stationary injection of such 1.21 MeV electron beam parallel to the E-vector of such a pulse with about 2.5 times higher power should immediately lead to about GeV electrons.

The same GeV electrons could be achieved if a 10 fsec pulse of about 25 Joules energy could be produced. 50 fsec KrF laser pulses of 0.1 joule energy have been produced and 100 J energy are being developed [34]. These techniques are well at the stage of being developed [35].

Table 1. Parameters for one laser half wave pulse for acceleration of an electron to 1 TeV energy.

Laser wave length	Laser intensity	Energy in pulse half wave	Acceleration length x	Minimum beam diameter y
[μm]	[W/cm^2]	[MJ]	[cm]	[mm]
10.6	1.91×10^{22}	45.5	389	5.24
2.7	1.836×10^{23}	11.6	99	1.34
1.06	1.191×10^{24}	4.55	38.9	0.524
0.25	2.142×10^{25}	1.07	9.17	0.124

To realize the state of the art, one may summarize the economics of the laser acceleration in the following conceptual framework: the achievement of about 2×10^{15} Watt-1 psec pulses with one Nova laser arm [33] or with Schäfers KrF laser [34] may be considered as present technology, ready for generating GeV electrons. Since one Nova arm is about 1/1000th of the 10 MJ Athena laser [36] which could have been started 1987 to be available 1991, one may be able to conclude that laser acceleration of electrons to 30 GeV is at hand. In order to get TeV 1000 times bigger laser capacities are needed (see Fig. 12). The increase of laser capacity by a factor 1000 needed about 10 years in the past, depending on the continuation of support.

What could be of use for a TeV electron accelerator for the year 2000, is any decision about laser fusion which was delayed too long at present, since it could be shown [37] how present day physics knowledge confirmed an economic (comparable to light water reactors) fusion energy power station, verified technologically within 10 years if a crash program would start now, similar to the realized Apollo-project for flying to the moon or to the not realized Tsongas McCormack bill in the USA 1980 to spend $20 Bill for fusion energy. The attractivity is that laser fusion [37] may provide energy production by multiples cheaper than light water reactors in 2030. A spin-off of this laser development could well be the mentioned TeV or higher energy electron accelerator by lasers, if problems of luminosity and space charge effects of bunches of electrons appear to be of interesting values. Since scientific research is essential for the future standard of living and capacities of defense research have to be re-oriented, the chances for using lasers for solving the greenhouse catastrophy [37] and simultaneously gaining a near direction for accelerator techniques may be in the focus of practical decisions.

For the mechanism described in this section with laser beams, one special quantity has been ignored, the longitudinal electric field of the beam as described in section 2. Using this field for the interaction during one half wavelength only a remarkable forward acceleration of electrons is possible as has been shown by Capers and Jensen [38]. Nevertheless, this mechanism is of linear nature, should the mechanism with the lateral and

Lorentz-like forward acceleration discussed in this section is of quadratic nature. For the range above several GeV energy, the longitudinal field mechanism is much smaller compared with the quadratic mechanism.

12. SUMMARY

The acceleration of electrons in vacuum by lasers to avoid plasma effects or mechanisms of surrounding materials or their dielectric effects, is considered for laser fields with intensity minima which are moved and accelerated. This problem is involved with the lateral motion of laser beams in order to accelerate electrons or by using a Fresnel mirror with turning of a section. In order to demonstrate the basic mechanism, the detailed numerical solution of the motion of an electron in the intensity minima of a CO_2 laser field with standing, moving, or accelerated mirror is evaluated and shows the confinement as predicted by Weibel and expected otherwise from the theory. The subsequent acceleration results in a straightforward way in agreement with the globally assumed theory.

Another acceleration of electrons uses a pulse of a simple penetrating linearly polarized electromagnetic wave packet in vacuum. If a symmetric sequence of waves is in the packet, no net acceleration occurs as known from the basics and preceding detailed analyses. If the pulse has an average electric field different from zero (asymmetric component, rectification, half-wave etc.), as has been demonstrated by CO_2 laser pulses of less than one wavelength, the relativistic exact computation results in an axial acceleration and in a lateral acceleration in the direction of the electric component of the laser field. The lateral motion is very strong and subrelativistic conditions but shrinks the more the relativistic growths. The exact solution shows how the acceleration length by a half-wave length pulse is very large for electrons accelerated to TeV energy. The electron with nearly speed of light needs a large length before it is taken over by the half wave length laser pulse. Due to the lateral motion, a minimum focus diameter is necessary requiring high pulse energies. For a KrF-excimer laser pulse, an energy of MJ is necessary to produce a TeV electron energy. The acceleration length is about 10 cm due to the slow taking over process. Instead of getting half waves by rectification laser pulses can be used with lateral injection such that the accelerated electron leaves the beam after one half wave has transferred the energy.

REFERENCES

[1] B. Richter, in Ref. [4], p.10.
[2] H. Hora, Atomkernenergie-Kerntechnik 48, 185 (1986).
[3] P. Channell (ed.), Laser Acceleration of Particles, AIP Proceedings No. 91 (Am. Inst. Phys. New York, 1982).
[4] C. Joshi and T. Katsouleas (eds.) Laser Acceleration of Particles, AIP Proceedings No. 130 (Am. Inst. Phys. New York, 1985).
[5] C.L. Teng, in Ref. [4] p.1; P. Sprangle , E. Esarey, and A. Ting, The Nonlinear Interaction of Intense Laser Pulses in Plasmas; NRL Memo Rpt. 6545 (Dec. 1989).
[6] B.W. Boreham, and H. Hora, Phys. Rev. Lett. 42, 776 (1979); B.W. Boreham, and B. Luther-Davies, J. Appl. Phys. 50, 2533 (1979).
[7] P.K. Kruit, J.K.B. Kimman, H.G. Müller and M.J. van der Wiel, Phys.Rev. A28, 248 (1983); G. Mainfrey, and C. Manus, Multiphoton Ionization of Atoms, S.L. Chin, and P. Lambro poulos eds. (Adad. Press, New York, 1984); M.D. Perry, O.L. Landen, and E.M. Campbell, Phys. Rev. Lett. 59, 2538 (1987) P.W. Hawkes, ed. Advances in Electronics and Electron Physics (Acad. Press, New York 1987) Vol. 69, p.75.
[8] H. Hora, Laser and Particle Beams 6, 625 (1988).
[9] H. Hora, Physics of Laser Driven Plasmas (John Wiley, New York, 1981).
[10] L. Cicchitelli, Dr. thesis Univ. NSW, 1988.
[11] L. Cicchitelli, H. Hora and R. Postle, Phys. Rev. A41, No. 7 (1990).
[12] H. Hora, J-C. Wang, and P.J. Clark, Laser and Particle Beams 4, 83 (1986).
[13] E. Weibel, J. Electr. Contr. 5, 435 (1958); H.A.H. Boot, S.A. Self, and R.B. Shersby-Harvey, J. Electr. Contr. 22, 434 (1958).
[14] V.A. Gapunov and M.A. Miller, Sov. Phys. JETP 7, 168 (1958).

[15] F.F. Chen, <u>Laser Interaction and Related Plasma Phenomena</u>, H. Schwarz & H. Hora eds. (Plenum, New York, 1974) Vol. 3A, p.291.

[16] H. Hora, Nature <u>333</u>, 337 (1988).

[17] R. Evans, Nature <u>333</u>, 296 (1988).

[18] A.M. Sessler, Phys. Today <u>41</u> (No. 1) 26 (1988).

[19] H. Hora, Opt. Comm. <u>67</u>, 431 (1988). The discussed non-collinear wave field trapping and acceleration of electron discussed in this reference (see Fig. 1a) was evaluated in the following paper: K. Sakai, J. Phys. Soc. Japan <u>58</u>, 2325 (1989).

[20] H. Hora, Phys. Fluids <u>20</u>, 3705 (1985).

[21] H. Hora, Phys. Fluids <u>12</u>, 182 (1969).

[22] H. Hora, D. Pfirsch, and A. Schlüter, Z. Naturforsch <u>22A</u>, 278 (1967).

[23] J.H. Mathews, Numerical Methods (Prentice & Hall, New Jersey 1987), p. 340.

[24] P.H. Bucksbaum et al, Phys. Rev. Lett. 58, 349 (1987); R.E. Freeman, et al., Phys. Rev. Lett. <u>57</u>, 3156 (1986); R.A. Meyers ed. <u>Encyclop. Phys. Sc. & Technol,</u> (Academic Press, New York, 1987) Vol. 7, 315, p.125.

[25] W. Scheid, and H. Hora, Laser and Particle Beams <u>7</u>, 315 (1989).

[26] R. Klima, and V.A. Petrziolka, Cz. J. Phys. <u>B.22</u>, 896 (1972).

[27] R. Dragila, and H. Hora, Phys. Fluids <u>25</u>, 1057 (1982).

[28] P.J. Clark, H. Hora, R.J. Stening, Wang, J.C., Laser and Particle Beams, <u>4</u>, 83 (1986).

[29] C. Roland, and P.B. Corkum, J. Opt. Soc. Am. <u>5B</u>, 642 (1988).

[30] M.A. Duguay, and J.W. Hansen, Appl. Phys. Lett. <u>15</u>, 192 (1969).

[31] A. Christopoulos, H. Hora, R.J. Stening, H. Loeb, and W. Scheid, Nucl. Instr. Meth. <u>A271</u>, 178 (1988).

[32] E.M. McMillan, Phys. Rev. <u>79</u>, 498 (1950).

[33] J. Coutant, Limeil Lab., private com.

[34] J. Hermann and Wilhelmi, <u>Laser for Ultrashort Light Pulses</u>, (Akademic-Verl. Berlin, 1987).

[35] F.P. Schäfer, Private Comm.; ECLIM Conference, Schliersee Jan. 1990; Laser and Particle Beams to be published.

[36] E. Storm, J.D. Lindl, E.M. Campbell, T.P. Bernat, L.W. Coleman, J.L. Emmett, W.J. Hogan, Y.T. Horst, W.F. Krupke, and W.H. Lowderwilk <u>"Progress in Laboratory High Gain ICF: Prospects for the Future"</u> (LLNL, Livermore) Aug. 1988, Rept. No. 47312.

[37] H. Hora, L. Cicchitelli, Gu Min, G.H. Miley, G. Kasotakis, and R.J. Stening, <u>Laser Interaction and Related Plasma Phenomena</u>, H. Hora and G.H. Miley eds. (Plenum, New York, 1990) Vol. 9, p.

[38] F. Caspars, and E. Jensen, <u>Laser Interaction and Related Plasma Phemonena</u>, H. Hora and G.H. Miley eds. (Plenum, New York, 1990) Vol. 9, p.

Note added after finishing this report on the results presented at the conference:

SSC and Laser Accelerators

There may be arguments that the today's technology of laser accelerators are at a comparable level as the SSC to achieve the goals requested by high energy physicists. The present days laser technology may well be able to substitute SSC where the necessary 99.99% reliability of classical helium cooled superconductors is very far away from reality.

The result on the acceleration scheme with electrons injected laterally into a laser beam, Fig. 12, indicated the difficulty that the laser power increases with the square of the electron energy. We evaluated how GeV electrons may be generated by 10^{15} Watt laser beams and that the technology is at hand [33] to operate one arm of the NOVA laser to produce pulses of 1 kJ energy and 1 psec duration.

This has to be compared with the technology of the ATHENA project [36][39] where a 1000 times bigger laser than one NOVA arm can be built for about $700 Million. Using this technology and combining the output beam into the then necessarily large diameter single beam for electron acceleration provides 10^{18} Watt pulses of 1 psec duration resulting in 30 GeV electrons.

Since the costs for increasing laser power by a factor 1000 roughly costs 10 times more, and since this may be considered, e.g., by reducing the pulse length to 100 fsec, 100 MJ pulses (compared with the 10 MJ of ATHENA), we find that the needed pulses of 10^{21} Watts may be in reach for achieving TeV electrons. Apart from the enormous work yet to be done theoretically to evaluate problems of space charge, electron densities per pulse with a psec or 100 fsec laser pulse driving instead of the before discussed half pulses, and apart from the possibilities of other laser types as KrF or of the 80% efficient cluster laser amplifier where pellets with velocities of about Mach 300 are injected laterally into a laser beam and their kinetic energy is converted into optical energy [40], even with the more conservative estimations the new schemes may even reduce the given costs of the well established glass laser technology. On this conservative base one can estimate that the specifications for an accelerator of the type of SSC could be achieved by the present days laser technology. There is no question that there is a development time of the same length of 5 to 10 years as considered for the SSC.

The laser technology seems to be in a more reliable way at hand than the technology of superconductors for the SSC. There is no question that the classical accelerator technology is absolutely correct to be accepted for the SSC, but the superconducting coils - even using the old style metallic superconductors operating with liquid helium - have by far not reached the level of reliability. As estimated [41], the 10,000 superconducting coils have to work properly. Even if each individual magnet is 99.99% reliable, the SSC has a working reliability only of 37%. If each magnet is only 99% reliable, the reliability of the machine is 2×10^{-44}. Knowing that superconducting magnets in tokamaks have a very low reliability, e.g., one only of the eight superconducting magnets of the big tokomak at Cardarache/France broke down and delayed work for about one year despite an explanation that the superconducting magnets there were built with the most refined new technologies, a 99.99% reliability of superconducting magnets for the SSC may be exceedingly questionable.

If the SSC would then technologically be out of reach, we have shown that the aspects of lasers - but only if a very energetic decision is being done for a large scale development immediately - are a strong option for high energy physicists to reach their goals based on an acceleration scheme, which is basically straight forward as explained here.

[39] K.R. Manes, Laser Interaction and Related Plasma Phenomena, H. Hora and G.H. Miley Eds. (Plenum, New York, 1987) Vol. 7, p.21.
[40] H. Hora, Wang Jin-Cheng, P.J. Clark, and R.J. Stening, Laser and Particle Beams 14, 83 (1986).
[41] R.J. Yaes, Nature 344, 188 (1990).

A REVIEW OF BACK LIGHTED THYRATRON PHYSICS

AND APPLICATIONS

Martin A. Gundersen

Department of Electrical Engineering-Electrophysics
University of Southern California
Los Angeles, CA 90089-0484

ABSTRACT

This review discusses a new device, the back-lighted thyratron. The review includes discussion of physics studies of the BLT plasma and cathode, several new applications to the development of practical plasma based devices needed to test and develop concepts for plasma based accelerators, electromagnetic sources, and related devices such as plasma lenses, and the application as a high power switch.

INTRODUCTION

Recently considerable interest has developed in a new class of thyratrons, called the pseudo-spark, and a related, optically switched thyratron (Figure 1) called the back-lighted thyratron (BLT). Research and development of pseudo-sparks has occurred mainly in Europe, with the discovery of the pseudospark occurring in the laboratory of Prof. J. Christiansen of the University of Erlangen[1]. The BLT is optically triggered, has a hollow cathode and a hollow electrode. Optical triggering has been demonstrated with ultraviolet radiation from a variety of sources including 1) a spark, 2) flashlamp, and 3) laser (Kirkman, 1986, 1988). The required optical power is low; it is wavelength dependent, of the order of 1 mJ at 300 nm, and 10 µJ at 220 nm. The conductive phase is initiated by photoemission from the back of the cathode. The recent BLT research results include the following accomplishments: 1) A new cathode, with current density of 10,000 A/cm^2 2) A new plasma based glow switch that operates at up to 100 kA peak, 10^{12} A/sec di/dt 3) The formation of a pulse-repeatable super-dense glow plasma. Included also in this review is a brief discussion of the detailed study of the physics of these devices, including the development of sophisticated in-situ optical diagnostics such as laser induced fluorescence, the development of theoretical models for the plasma electron distribution function and transport properties, and the integration of theory and experiment for practical applications.

The development of new, innovative devices for plasma based accelerator research and technology, and for quantum electronics devices, will provide the basis for major advances in these technologies. Although plasma based devices are expected by some to form the means of advancing accelerator technology beyond the SSC, there are as of this writing very few approaches to the development of these devices that actually address the real device problems. At present, it is very difficult to produce a real, *demonstrated* device with high, uniform plasma density, high current, and other properties described below. Presented in this review also are several new applications to the development of practical plasma based devices needed to test and develop concepts for plasma based accelerators, electromagnetic sources, and related devices such as plasma lenses for enhanced luminosity of relativistic beams.

HIGH POWER SWITCHING

Accelerators, pulsed discharge lasers, such as CO_2 and the excimer families, and laser systems such as pulsed free electron lasers, rail guns, microwave sources, and FMP generators require a pulsed power conditioning system. For an accelerator such as the Stanford Linear Accelerator, pulse shaping, peak current, repetition rate, and current rate of rise are significant issues, and pulse modulators are required for the klystrons and the switching magnets (kickers). These systems, however, are more complex than might be anticipated from a simple consideration of the required components -- a DC power supply, a switch, an energy storage capacitor, and two electrodes. The design and implementation of the system impact the quality, energy, timing, performance and the reliability of the specific device. At the heart of these devices is a switch. Switches in the pulsed power area include spark gaps, semiconductor switches such as the thyristor, the thyratron, and the new pseudospark and BLT. For an accelerator that must operate at a repetition rate of the order of or greater than 100 Hz, the switch most commonly chosen at this time is the thyratron.

Thyratron research peaked at some point following World War II, following the development of the high power hydrogen thyratron by K. Germeshausen. Since the development of the ceramic envelope thyratron in the 1950s there have been a number of innovations, with possibly among the most important being the grounded grid thyratron (EG&G, ≈ 1970) and the hollow anode thyratron (EEV, ~ 1980). The grounded grid thyratron led to improved current rate of rise, and the hollow anode improved reverse current handling capability.

The pseudospark is a closely related device with the main difference being that triggering is electrical, analogous to a conventional thyratron. With the exception of performance characteristics that are specifically related to triggering (such as optical isolation, or jitter associated with a particular triggering design) other characteristics remain the same. Thus pseudospark data is useful in characterizing the expected performance of the switch, particularly in terms of high power, high current, and high repetition rate applications. The isolation possible with the BLT is important for the conceptually new modulator, but is not as critical a consideration as the externally unheated cathode, which is common to both the BLT and pseudospark.

For example, in Erlangen, West Germany, the Kraftwerk Union, a subsidiary of Siemens AG, is testing a multichannel pseudo-spark for excimer laser switching, including test operation[2] of a 20 channel Mechtersheimer-type pseudo-spark at 35 kV, 40 Hz, with a spectacular 100 kA peak current and $2.6 \cdot 10^{12}$ A/sec dI/dt. High peak current operation was demonstrated several years ago at CERN. Single channel pseudo-spark switches were built at CERN under the direction of Dr. H. Riege for a 20 kJ pulse generator which is being used for long-term tests of a plasma lense for the future CERN Antiproton Collector[3]. The switches have been operated at more than 100 kA peak current at 0.3 Hz for ≈ 400,000 shots. In the U.S., an optically triggered switch developed in our laboratory, the back-lighted thyratron, or BLT has been operated under various conditions including as a direct thyratron substitute in a commercial excimer laser. Characteristics including peak current, current rise, etc. support the conclusions of the research in Europe. For example, sub-nanosecond jitter has been achieved in both the pseudo-spark and BLT[4]. The forward drop determines the power that is dissipated in the switch during the steady state. Data indicates that the forward drop at 10 kA peak current is of the order of 100 V, and is less at higher currents. Thus, attractive features of these new switches include improved current rise, improved peak current, improved reverse current, and glow discharge operation with a cold cathode.

A fiber optic waveguide is a simple and practical way to distribute the light and can be designed so that the light is delivered exactly to the edge of the cathode hole - which leads to lower delay and jitter and reduced optical energies required for triggering.

At a wavelength of 308 nm (XeCl excimer laser), *subnanosecond* jitter triggering was found (Braun et al 1988) with as little as 1.5 mJ of laser energy incident in a 10 ns pulse on the electrode. The fiber (either a 1mm diameter PCS (Plastic Coated Silica) good to approximately 250 nm or an all silica fiber 1.5 mm diameter good to 190 nm) was sealed in a glass tube to form a vacuum feedthrough and was positioned relative to the cathode electrode by a small, unobtrusive molybdenum mount. Typically, the light from the fiber is directed immediately around the electrode hole (some light goes through to the anode but this has no significant effect).

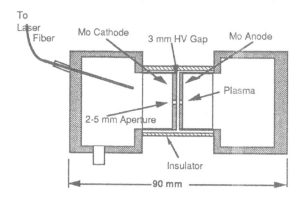

Figure 1. Structure of a BLT switch with optical triggering through an optical fiber. The switch is filled with low pressure (0.01-0.3 Torr) gas and is triggered by unfocused UV light incident on the back of the cathode surface.

The investigation of the cathode emission process is important. Based on existing experimental data on pseudospark performance, scaling to extremely high currents should be fairly straightforward. It should be possible to ultimately produce a device that can switch a MA (10^6 A).

These switches are the result of basic research programs. The anomalously high cold cathode emission (>>hot cathode) and current densities (40,000 A/cm^2, >>normal thyratron plasmas), obtained in a non-arcing mode and with a simple structure, strongly encourage further basic study of plasma devices. The physics of the cathode emission and high current density require study, and further device applications should be sought. The project as envisaged will perform research required to establish feasibility of the new family of high power switches. Versions of these switches under varying operating conditions have switched over 100 kA, operate in a glow discharge mode without electrode degradation from arcing, have demonstrated over $2 \cdot 10^{12}$ A/sec dI/dt, and subnanosecond jitter, using a cold cathode.

THE SUPER-EMISSIVE CATHODE

The cathode emission mechanism is responsible for the high plasma density. This mechanism has been reported in the literature[5]. A fundamental aspect of the cathode is the current density, which is ≈ 10,000 A/cm^2, and extends over a macroscopic area (≈ 1 cm^2). The cathode, and the uniform plasma that it supports, allow a variety of new applications.

As discussed above, recently this laboratory has demonstrated a hollow cathode that operates in thyratron type switches with emission ≈10,000 A/cm^2 -- e.g. much higher currents than dispenser and oxide thermionic cathodes (≈ 100 A/cm^2). The emission is observed to occur over a surface area ≈ 1 cm², and the cathode operates without forming an arc, and hence is useful for thyratrons and other applications[6,7]. This emission is readily achieved in a simple, externally unheated configuration. It appears feasible to extend performance to devices requiring peak currents over 100,000 A. This is an anomalously high -- super-emissive -- regime of operation for a cold cathode operating in a glow discharge plasma. In spite of the considerable phenomenological understanding of gas discharge technology, these data were not predicted, and these results apparently will have important device applications. Although in the past higher but localized current densities have been achieved through the formation of filamentary arcs, devices (e.g. spark gaps) tend to be limited melting, sputtering and cratering of electrode material, and addition of electrode material to the arc plasma. This new cathode supports currents that formerly required arc-type devices, such as spark gaps.

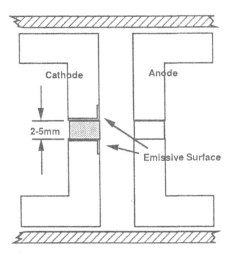

Figure 2. Electrode structure indicating the region of the highly emissive hot cathode surface. The super-emissive area is approximately 1 cm².

The cathode materials used are pure refractory metals. Materials such as Mo and pure W are actually among the best thermionic emitting materials simply because they melt at high temperatures, and thermionic emission is an especially violent function of temperature. These materials by themselves are ordinarily not used for high current emission in devices such as thyratrons, however; typically alkali-oxide coatings, or special preparation of a W matrix that contains Ba, are used, which have much lower work functions and good emission characteristics at temperatures of the order of 800 K.

The cathode emission process is characteristic of the geometry of the device and is distinct from others reported, such as liquid metal plasma cathode or spark gaps. The experimental structure consists of a hollow cathode and anode as shown in Figure 2. The longest path in this device is in the region of the central holes thus the discharge will begin here when the gap is overvolted or triggered by a pulse of electrons in the hollow space behind the cathode. The triggering pulse can be provided by photoemission, electron beam, pulsed glow discharge or other methods described elsewhere[8].

ELECTRON BEAM AND ION BEAM GENERATION

The highly emissive cathode may also be useful to produce an electron beam for additional accelerator applications. Pulsed electron and ion beams have been produced using this structure in low pressure gas. During the initial phase of discharge plasma formation there is an intense electron beam emitted through the anode aperture and an ion beam is emitted through the cathode aperture. The energy of these beams are about equal to the initially applied voltage. Table I shows the electron beam results from several independent laboratories.

Ion beams have also been measured at the University of Erlangen[9] with peak currents of 35A and current density 10^4A/cm^2 produced at a voltage of 50kV in low pressure hydrogen. At the KFK Karlsruhe, West Germany ion beams of 10^6A/cm^2 have been observed in a pseudospark structure operating at over 200kV.

Electron beams can be produced in both single and multi-gap structures (Figure 3). The single gap structures consist of a cathode and anode electrode separated by about 3mm and operate at voltages of 10 to 50 kV. Multi-gap structures consist of several electrode discs with central holes separated by glass, quartz or ceramic insulators and can operate at voltages greater than 200kV. A high voltage pulse of 10 to 50 kV is applied to the single gap structure. When the voltage across the gap nears its maximum UV light from a flashlamp or unfocused laser is incident on the back surface of the cathode electrode releasing electrons through photoemission. The electrons released are accelerated through the gap and anode hole into the drift tube. Some of these electrons collide with neutral gas atoms creating ions that bombard the cathode producing electrons by secondary emission and heating the cathode to a temperature where thermionic emission is possible.

Table I. Electron beam results from several laboratories.

GROUP	peak current	current density	pulse length	energy	gas pressure
	A	A/cm2	ns	keV	
U. Erlangen	10^3	10^5	20	35	< 100 Pa
U. Maryland	10^2	----	20	30	≈ 10 Pa
Dusseldorf	10^3	10^5	0.7-100	50	< 100Pa
USC	10^2	>10^3	>20	15	≈ 10-20 Pa

These electron emission processes produce a high current of electrons that are accelerated to the applied voltage and pass through the anode into the drift tube. This pulse of electrons lasts for 10 to 100ns after which a plasma is formed in the anode cathode gap. After formation of this plasma the voltage on the device is 100-200V allowing only low energy electrons to be produced.

In addition to the pulsed beam during discharge formation there can be a high current low energy beam during the steady state phase of the discharge after plasma formation[10,11]. The heated surface of the cathode is present for many microseconds. The inter-electrode plasma will form a cathode sheath layer at the surface of the cathode with a potential drop of 100 - 500 V. Electrons emitted are accelerated in this cathode sheath before entering the bulk plasma. The mean free path between collisions with electrons and ions of the bulk plasma for 100ev electrons is greater than the typical 3mm electrode spacing, hence many electrons emitted can pass through the anode aperture undisturbed by collisions. The current in this low energy electron beam may be a large fraction of the total discharge current.

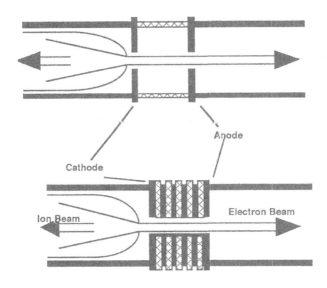

Figure 3. Schematic representation of electron and ion beam formation in single (top) and multi-gap structures.

An experimental apparatus for the study of the pulsed beam is shown in Figure 4. The BLT anode was modified by drilling a three millimeter diameter hole to allow the electron beam to propagate out of the device. The beam was collected by a Faraday cup and the beam current measured with a Rogowski coil. These results suggest the possibility of extracting and accelerating the *low* energy beam. With the addition of an extraction grid and an accelerating gap it should be possible to extract the beam and accelerate it to high energy. Beam generation in the BLT can be studied by observing damage to the metal structures.

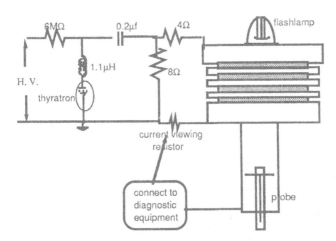

Figure 4. Experimental schematic for initial probe studies of electron beam. Shown is a flashlamp switched BLT. The BLT shown has three gaps in order to achieve higher stand-off voltage relative to a single gap device.

PLASMA BASED ACCELERATORS

The motivation for thinking of plasmas as a possible efficient accelerating medium lies in the fact that a plasma can support very large accelerating fields which scale as SQRT(n) where n is the plasma density. Thus, for a density of 10^{16} cm^{-3}, the maximum longitudinal accelerating field will be on the order of 10 GeV/m , which exceeds by several orders of magnitude the current accelerating gradients. Acceleration due to one of the proposed plasma based accelerator schemes, the plasma wake field accelerator[12], was demonstrated by using a 50 cm long hollow cathode chamber that contained a plasma with a density of 3×10^{13} cm^{-3}. Partly as a result of the low density, accelerating gradients of 1 MeV/m were obtained. Although there will be trade-off issues related to background collisional processes that have yet to be quantitatively determined, it appears that in order for a plasma based accelerator to be realizable, long plasma sources of several meters in length and densities higher than 10^{16} cm^{-3} will have to be developed. These plasmas will likely be pulsed. A high pulse to pulse reproducibility is required for extended periods of time -- requiring the formation of a glow-type plasma with high density and pulse to pulse repeatability of the order of 1%. It is clear from present research efforts in related areas that the plasma *quality* is a key problem.

The pseudospark is a promising high density hydrogen plasma source that satisfies the stability requirements mentioned above to an extent not previously possible. This high quality plasma source, based on the pseudospark or BLT device plasma, occurs because of the very high emission properties of the cathode. Measurements of the plasma electron density and temperature show a very uniform, pulse to pulse repeatable distribution in the transverse and longitudinal directions. The distribution also persists for several microseconds following the current pulse. The electron density has been determined by spatially and temporally resolved measurements of the Stark broadened profile of the H$_\beta$ and other Balmer emission using several spectroscopic methods and a streak camera. Within the limits of the experimental apparatus, no shot-to-shot variation was detected. We propose to develop this plasma source for wakefield accelerator studies.

This approach allows one to produce plasmas *less* dense than most presently proposed schemes and simulations[13,14], but *more* dense than can now be produced while satisfying the constraints listed in the previous paragraph. The plasma density is also more appropriate if one is to reduce background from scattering such as that which produces copious amounts of synchrotron radiation, which will be excessive for beam bunch densities above 10^{10} and at energies in the multi-GeV range.

The densities that have been achieved in reproducible experiments are readily applicable to wakefield experiments -- in order to test concepts, and develop technologically useful devices. We are not aware of competing approaches that have *demonstrated* pulse to pulse reproducibility and spatial uniformity at the densities achieved to date. There are excellent prospects for scaling to higher plasma densities, using the same device principles. In addition, the work will benefit from a mature program that is investigating the physical processes that are responsible for the formation and operation of the plasma.

Approaches that have been suggested include use of a laser for photoionization of a small gas volume, use of a wall stabilized arc, various Θ and Z pinch schemes, and ionization and plasma formation using an electron beam. The use of a laser for multiphoton ionization of hydrogen gas is very inefficient at best, requiring an extremely expensive laser with very high maintenance requirements. The work in this area is important, and laser assisted photoemission will be a technique to be investigated in order to obtain properly shaped pulses for wakefield generation[15]. However, the multiphoton ionization cross sections of hydrogen are small, and because the first excited state of the molecule (and atom) already require more than one UV photon, the process requires a laser with very high intensity at a wavelength that is as short as possible.

501

There is considerable information in the literature describing both the physical processes and the hardware required[16]. Θ and Z pinch plasmas have not been demonstrated to have either the pulse to pulse reproducibility or the longitudinal uniformity required for these experiments. They are inherently limited by their physics -- providing another reason to consider the physics of the plasma sources. If we understand them, maybe we can figure out how to build them. They are also designed for the production of high temperature plasmas, rather than cool, spatially uniform plasmas. The current emission mechanism in the cathode of the BLT is an important intrinsic advantage over the conventional Z or θ type plasma source, in that it allows production of a uniform glow, because it has a large and highly emissive cathode area. As the BLT plasma is scaled to higher densities, it should be possible to achieve uniform, cold plasma densities comparable to that obtained in the pinch device -- without the turbulence, and with reproducible high repetition rate capability.

Approaches to the development of the plasma for wakefield generation also include the successful work of Rosenzweig et.al., which found experimental evidence for wakefield densities in plasmas of the order 10^{13} cm^{-3}. Thus the BLT plasma density is two orders of magnitude higher than that used by Rosenzweig et. al., and is readily implemented in a pulse mode, with simple support requirements.

Various realizations of the pseudospark plasma source are possible (Figures 5 and 6). A multiple gap BLT which is expected to generate a dense plasma column with the length of the plasma determined by the structure, operating conditions, and time during the pulse. A nominal density $> 10^{15}$ cm^{-3} is expected in the version shown in Figure 6. By using a 10^{15} cm^{-3} plasma density column one will be able to increase the accelerating fields reported by Rosenzweig et al. by a factor of 10, operate with a pulse repeatable, simple device, and therefore, scale the dynamics of the beam-plasma interaction to high plasma densities.

PLASMA LENSES FOR FINAL FOCUS

One of the technological issues that the future accelerators are facing, is that in order to obtain luminosities higher than 10^{33} cm^{-2} sec^{-1}, final focusing forces much stronger than the ones provided even with today's superconducting magnets need to be developed.

Plasmas can play an important role in this field too, since a relativistic electron beam traveling through a plasma can excite transverse focusing fields of comparable strength to the longitudinal fields used to accelerate a trailing beam. Thus, a beam going through a dense enough plasma can self focus to a very small radius. The characteristics of the focused beam as well as the lens focal length are a strong function of the plasma density.

The next generation of light particle accelerators will include linear colliders capable of accelerating particles to energies of ≈ 1 TeV. Such high energies require luminosities on the order of 10^{33} cm^{-2} sec^{-1}, which are to be achieved by increasing the machine repetition rate, shaping the beams, increasing the number of particles per bunch, and by developing extremely tight focusing lenses[17]. At present, focusing field gradients on the order of 10 kG/cm can be achieved with superconducting magnets. However, in light of the ability for plasmas to support very large electromagnetic fields, it is possible to create in a plasma medium focusing forces that will exceed those of a superconducting magnet by several orders of magnitude. Such large focusing fields can be created by exciting in the plasma radial wake fields[18]. Therefore, among the candidates for tight focussing lenses are a new class based on plasma engineering.

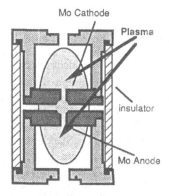

Figure 5. Schematic showing plasma extending from cathode and anode electrodes into the cathode and anode back spaces.

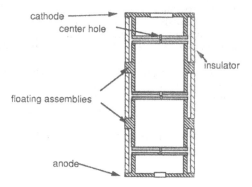

Figure 6. Multiple gap structure designed to produce plasma over an extended length. Propagation of the plasma during and after the current pulse will produce a low temperature plasma in the region between electrodes. Experimental studies are to measure the plasma density in these regions.

A plasma lens, however, developed for a linear collider will have to meet the stringent requirements set by the inherent characteristics of the accelerator. In addition to the high plasma density, these include high repetition rate, high reproducibility from pulse to pulse, device reliability with low maintenance requirements, and low jitter. These constraints place very strong demands on the performance of repetitively pulsed high density plasma sources, which are characteristically unreliable. As described above, the pseudospark and BLT already satisfies most of the requirements listed above.

Measurements of the electron density and temperature of the plasma show a very uniform distribution in the radial direction (figure 7.), which also lasts for several microseconds following the current pulse -- even when the source is an impedance matched, square current pulse. The electron temperature was studied with a spectroscopic investigation of the H_α and H_β emission, and using both a non-equilibrium model and a local thermal equilibrium model for the ratio of the line intensities, which gave a 'low', uniform plasma temperature of the order of one to a few eV. Related physical processes are discussed below. The electron density was spatially and temporally resolved by measuring the Stark broadened profile of the H_β emission. Effects due to Doppler broadening were found to be negligible for hydrogen temperatures of ≈ 1 eV and electron densities $> 1 \times 10^{15}$ cm^{-3}. These data, and streak camera recordings demonstrate that initially during the rising part of the current pulse the electron density is centered around the cathode hole, with a diameter comparable with that of the hole. In the configuration that has been investigated, and not optimized as a plasma lens, the electron density reaches its maximum of 3×10^{15} with a diameter of 1.74 cm. At 300 ns after the peak of the current pulse the plasma density has decayed to 2.8×10^{15} cm^{-3}, reaching radial density scale-lengths of ≈ 1 cm. A plot of the radial plasma density near the peak of the discharge is shown in Fig. 7.

The hollow structure with an axial bore, together with a strong axial current and a high plasma density, make the BLT a very attractive device for focusing of charged particles. First, the axial bore of the BLT allows for an easy injection of the beam into the lens. Second, the high axial current of 5 to 15 kA uniformly distributed over 5 mm in radius, generates radial magnetic field gradients of 4 to 12 kG/cm with field strengths on the order of 2 to 6 kG at a distance of 5 mm from the axis. Such high field gradients can be used to focus relativistic beams on the order of the MeV's to very small spot sizes, and also to perform fine alignment. Third, dense particle beams injected through the cathode bore can create strong focusing forces as a result of the radial wake fields excited in the plasma. These strong focusing forces can focus highly energetic beams of energies 50 GeV and higher to very small spot sizes as needed in order to keep the luminosity high. These two focusing mechanisms make the BLT a versatile lens capable of performing effectively under a variety of experimental conditions.

For example, focussing can be shown to occur assuming a normalized emittance E_n of 8.8×10^{-4} m. rad., relativistic gamma = 40, beam radius at lens location of 0.5 cm and a lens thickness of 1.5 cm. β is the Twiss parameter defined by $\dfrac{E_n}{\gamma} \beta(z) = a^2(z)$, where a is the beam radius and z is the spatial coordinate along the beam axis[19]. These parameters correspond to the beam of the AATF at Argonne National Laboratory. In this case the focusing forces are determined by the lens geometry and are independent of the characteristics of the beam.

For a beam with a radius less than the plasma skin depth l_s ($l_s = \lambda_p/2\pi$), the focusing forces from the wake fields become the dominant mechanism. This corresponds to the beam characteristics of SLAC and also the Advanced Accelerator Test Facility of Argonne National Laboratory. For the 50 GeV (SLAC) case, we have undertaken calculations[20] assuming a long beam limit (beam length ~ 2 mm) in order to estimate the focusing force from a 3×10^{15} cm^{-3} dense plasma by using the linear theory

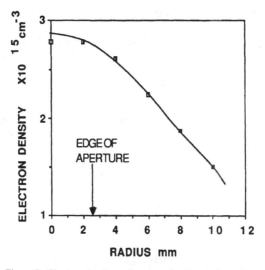

Figure 7. Electron density as function of radius during pulse.

approximation. In this case the BLT acts as thin lens. In the case of energies of 20 MeV (Argonne) nonlinear effects must be considered resulting in higher field gradients than estimated from the linear theory. Under these conditions the plasma acts as a thick lens and a simple calculation of the beam spot size at focus is not possible. The fact that the BLT behaves as a thin lens at 50 GeV and as a thick lens at 20MeV makes it possible to study both configurations with this device.

ELECTROMAGNETIC SOURCES

The propagating electron beam in a pseudospark is known to produce microwaves[21]. Reasonable transmission of microwaves occurs through pyrex glass and alumina, being essentially lossless at microwave frequencies. Detection can be carried out using diode detectors and cutoff waveguides in various bands (L-Band, 1-2 GHz , X-Band, 8-12 GHz,and Ka Band, 27-40 GHz, for example).

Plasmas with a periodic density variation present forbidden bands of frequencies in the electromagnetic dispersion relation[22]. We are also studying the possibility of applying these forbidden bands to the design of efficient "plasma mirrors" of controllable reflectivity in the millimeter wavelength region.

The structure of the multigap device described above has dimensions appropriate for an interaction between an electron beam that can produce mm waves. Injection of a beam into the multigap structure will provide a means for testing beam interactions in a plasma loaded system. This allows higher current densities than previously achievable, because of the space charge neutralization resulting from the (uniform, glow) plasma.

MODELING OF CONDUCTION PROCESSES IN THE BLT DEVICE

The BLT and the Pseudospark are new types of high power glow discharge switches which, compared to thyratrons have advantages such as a self heated emissive cathode and small spatial dimensions. As described above, they are also under consideration as future devices for accelerating or focusing particle beams, and the production of electron and ion beams have also been demonstrated in pseudospark and BLT geometries[23]. For further development and improvement, an improved understanding of the physical principles of the devices should be developed. In this section details of work to understand the basic principles of these devices is described.

Modeling of the devices and the discharge plasma have been pursued at USC in collaboration with experimental work. The electron temperature during the conduction phase is being evaluated from a set of rate equations for a collisional-radiative, partially ionized, and Maxwellian hydrogen "bulk" plasma with electron density of $n=1\text{-}5\times10^{15}$ cm^{-3}. Considering a steady state condition of the population of atomic levels, the comparison of computed and measured line intensity ratios yields a "bulk" electron temperature of about 1eV. The cathode fall voltage was estimated under the assumption, that the high current densities of $j\approx10^4$ A/cm^2 are carried by the "bulk" plasma. The total voltage drop across the BLT gap was calculated from an electrical equivalent circuit. Results are a voltage across the cathode fall of 100-500 V which corresponds to an electric field strength of $E=10^6\text{-}10^7$ V/cm for a cathode fall width of about 10-50 μm. This high electric field accelerates electrons to energies of $\varepsilon=100\text{-}500$ eV and injects them into the "bulk" plasma. From a Fokker-Planck equation it was found that a monoenergetic electron beam with energy $\varepsilon \geq100$ eV can penetrate the gap space filled with a "bulk" plasma, density of $1\text{-}5\times10^{15}$ cm^{-3} at temperature of about 1 eV, before the electrons become thermalized. Impact ionization and excitation of atomic hydrogen by cathode fall generated electrons is to be included in the model of the BLT plasma. Calculations predict an electron beam density $n_b \leq 10^{10}$ cm^{-3} which is comparable to values obtained with electron production due to ion impact at the cathode but has to be compared with values predicted from the analysis for the thermionic electron emission at the cathode.

Electron beams are expected in the commutation phase, $\varepsilon\approx10\text{-}30$ keV, and in the conduction phase, $\varepsilon\approx100\text{-}500$ eV. Calculations with a transport momentum equation are in progress. They include inelastic collisions of electrons with neutrals, which for example are the dominant species in the commutation phase, and collisions with the "bulk" plasma. To support experiments on beam measurements upper limits of the gas pressure, for which beam electrons can penetrate the "bulk" plasma or neutral gas, can be estimated for various gas species. It is necessary to operate at pressures which are high enough to maintain hold-off voltage, high current densities for switching performance, and high ion densities for the production of electrons by ion impact at the cathode, but low enough to allow penetration of particles. Calculations will provide data, which are useful for the extraction of the beam and specification of beam quantities, i.e. energy and density. The principal analysis can also be applied to the case when ions are injected into a plasma-gas mixture. The interaction of beam particles with a "bulk" plasma is also of interest in using the BLT as a particle focusing device or a particle accelerator[24].

A steady state solution can be inappropriate for pulsed discharges, especially if beam electrons are included, which are not at thermal equilibrium. A time dependent solution of the rate equations is in development. Comparison of calculated and measured line intensity ratios will yield "bulk" electron temperature and density, and electron beam energy and density. Further studies of the electron emission from the cathode, heated by ion beam impact, and a more detailed understanding of the plasma sheath in the cathode fall will also be pursued.

These models will be available for investigation of new plasma based devices for many applications, such as those listed above. It is very important to emphasize that these proposed devices are to be based on considerable development, both in theory and devices, and actually represent important and substantial innovation in pulse power technology.

506

EXPERIMENTAL PHYSICS PROGRAM FOR PLASMA BASED DEVICES

A key element of the work has involved spectroscopic studies of the unusually high current plasma that is responsible for conduction in the devices that we have developed over the past several years[25]. Experimental methods that have been used include laser induced fluorescence and Stark broadening measurements. Time resolved knowledge of atomic excited-state populations during the conductive phase is necessary for calculation of electron production and recombination rates, and for the analysis of high power devices that seek to harness plasmas for radically new applications. Excited-state relaxation rates are important in this type of analysis, since any rate equation-based plasma model must take into account collisionally induced transition rates, which are not well known; experimentally obtained rates therefore provide a valuable and needed check on those obtained theoretically.

One of the more serious problems limiting the understanding of plasma properties in high power switches is the lack of accurate data on the transport properties of the plasma. This affects diagnostic capabilities, and also the development of accurate, quantitative models. For example, the electric field, even when small, nevertheless plays a very important role in the plasma because it provides energy; hence small changes may be expected to produce significant effects. It is very difficult to measure the electric field directly using probes because of the complications that occur when probes are introduced into the plasma. There is also difficulty in measuring optogalvanic signals in the presence of high currents, due to the extremely high ionization rates arising from the large Rydberg state cross sections. In previous work relationships were obtained for determination of the electric field in a high-current hydrogen positive column discharge using non-intrusive spectroscopic and current measurements. The conditions considered are typical of the conductive phase of a high-power thyratron. The field is determined as a function of electric current density and the linewidth of atomic hydrogen Balmer emission, quantities that can be accurately measured. Relationships between the field and electron densities and temperatures are also presented. These results provide a method for accurately determining the average electric field and electron temperature in the positive column.

Several parameters of the BLT plasma have been measured to characterize this high current glow discharge plasma. The switch has been observed to operate with an anomalously high current density with a diffuse glow discharge like plasma. Initial measurements estimate a peak current density of $10kA/cm^2$, a voltage of 100 - 200 volts during conduction, a uniform plasma of about $1cm^2$ cross sectional area with electron density of $5x10^{14}$ to $5x10^{15}$ cm^{-3} and a background plasma temperature on the order of 1eV.

A thorough characterization of the plasma should include measurements of the current density, voltage drop, electron density, excited state densities and the species present in the discharge plasma (i.e. electrode material). The results of these measurements can be combined with theoretical modeling to give information about the nature of the electron energy distribution function, plasma conductivity and cathode emission processes.

Streak and framing camera (Figure 8) observations of the discharge plasma give quantitative information of the character of the discharge and its cross sectional area. These measurements along with the total current measured and observation of the region of the cathode utilized by the discharge give an estimate of the plasma current density and the cathode emission current density. Initial streak camera observations show that the discharge has a diffuse character with a cross sectional area on the order of $1cm^2$. A clear distinction between the diffuse mode of operation and occasional arc like events has been demonstrated[26]. These observations were taken in the midplane of the switch gap for a discharge current of 8kA and initial applied voltage of 10kV. At this voltage and current the discharge is normally of the diffuse type at higher currents the number of arc type events increases. Further streak and framing camera observations are needed to characterize the conditions necessary for diffuse operation. Observations at varying peak currents and pulse lengths will be used to determine the maximum current densities and the length of time that the high cathode emission into a glow discharge can be sustained. Observations during reverse conduction will be used to determine peak reverse currents possible without arcing and to compare the cathode processes occurring during forward and reverse currents. Framing camera observations can give information of the plasma character over the entire gap length at a particular time. This data can be used to study the plasma homogeneity and the interactions occurring at the electrode surfaces and between the two electrodes during diffuse and arc type conduction.

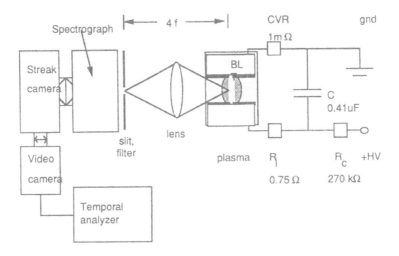

Figure 8. Experimental schematic for plasma diagnostics. Shown is a spectrograph and streak camera to determine plasma densities with mm spatial resolution. This apparatus will be used to study the devices shown in the previous sections, and to investigate regions of the plasma in the longitudinal direction, during and after the excitation pulse.

Accurate measurements of the voltage across the gap during voltage collapse and conduction are required to determine the energy of the ion beams for cathode heating and to investigate the possibility of fast electrons carrying current during conduction. Initial probe and voltage comparison measurements estimate the conduction voltage drop to be about 100v. More accurate measurements will be performed to determine the voltage for several different peak currents, pulse lengths and initial voltages. Voltage dependence on peak current and pulse length can give insight into the cathode emission processes occurring while dependence of the voltage on the initial applied voltage may give information related to the ion bombardment cathode heating mechanism.

Initial measurements of the excited state densities have been completed but more accurate and complete measurements must be performed to compare with the results of theoretical models. The initial measurements with a calibrated optical system using analytical line filters and a photomultiplier tube estimate the hydrogen excited state levels n=3,4,5 to have populations on the order of 10^{10} cm^{-3}. The ratios of the populations have been found to be 0.9 and 0.83 for levels 4/3 and 5/3 respectively indicating a temperature of about 11000 degrees K if LTE is assumed. Measurements of the excited state densities resolved in time and space using line filters and a streak camera (Figure 8) are to be performed.

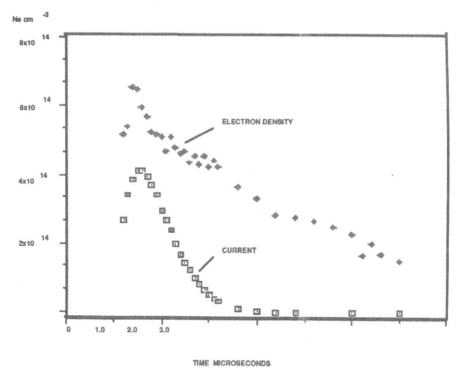

Figure 9. Time resolved electron density measurements shown in comparison with the time dependance of the current. The peak current is 4 kA.

The electron density has been estimated from measurements of the stark broadened linewidth of the hydrogen balmer beta emission. Initial measurement of an 8-10 kA discharge give an estimated electron density $>10^{15}$ cm^{-3} . Higher resolution (< 0.4 Å) time resolved measurements have been made using a 0.75m spectrometer at a peak current of 4kA. The electron density was 7×10^{14} cm^{-3} at current maximum and decayed to 3×10^{14} cm^3 2 μsec after the current maximum (Figure 9). Measurements using a 1m double spectrometer and a linear detector array to obtain the entire line profile in a single discharge are planned to continue to give a complete characterization of the spectral output of the BLT plasma. Time averaged electron densities were measured for several values of peak current using this system. The results show an increasing density with current from 3×10^{14} cm^{-3} at 7.5kA to over 10^{15} cm^{-3} at 40kA.

Initial measurements of the spectral output of the BLT plasma indicate only a small amount of electrode material present in the discharge plasma. Measurements using the 1m spectrometer and detector array will look for the presence of molybdenum atoms and ions in the discharge plasma.

Another diagnostic technique that may be suitable for certain applications involves the use of laser induced fluorescence (LIF) (Figure 10). LIF has been shown to be a useful procedure for obtaining accurate plasma densities with excellent temporal resolution (≈ 5 ns). Use of a picosecond laser will allow greater temporal resolution to be limited only by the laser pulse length.

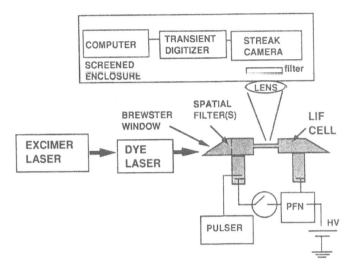

Figure 10. Experimental set-up for LIF study of plasma. Spatial filtering is important in order to reduce background scattering of light.

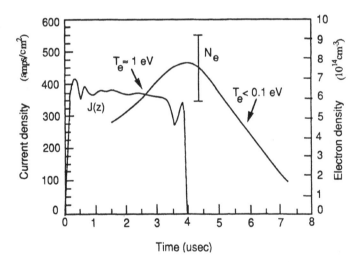

Figure 11. Electron density as function of time for square current pulse produced by an impedance matched circuit and pulse forming network (PFN).

Figure 11 shows data taken using LIF to measure the plasma density. The figure illustrates several aspects of the approach. By using an impedance matched circuit it is possible to control the plasma density temporally to a high degree. This is a very 'clean' pulse over the life of the plasma, and can be compared with the difficulty in maintaining such pulse-to-pulse, repeatable control if using a Z or Θ pinch. Further, the plasma density is very high at the end of, and after, the current pulse. Following an analysis that we have reported[27] -- a time dependent model of the plasma including three body relaxation -- the plasma may be shown to be very cold, as well as uniform. The electron temperature falls to the background gas temperature within approximately 100 ns, and hence the electron temperature is typically between ≈ 0.03 and 0.1 eV. Thus one has a high density, uniform, very cold, pulse repeatable, plasma that may be operated at very high repetition rates.

CONCLUSIONS

At present there are interesting research areas for the BLT device including plasma loaded free electron lasers, generation of microwave radiation for fusion heating, millimeter wave generation through the interaction of the electron beam with the plasma. Other pulse generation applications, such as EMP, excimer and other laser development, will also be affected. The BLT electron and ion beams may be expected to impact microwave, accelerator, and other technology, such as X-ray generation. The prospects for these experiments to produce exciting results are strongly encouraged by recent results -- particularly the remarkable plasma and cathode properties. Research into the physical properties will have important applications, providing a basis for an entirely new technology in plasma devices.

ACKNOWLEDGEMENTS

The work presented here has been supported by the Air Force Office of Scientific Research and the Army Research Office. The plasma mirror concept is due to Dr. Humberto Figueroa. The author would like to acknowledge the work of Dr. C. Braun, Dr. D. Erwin, Dr. G. Kirkman, Mr. H. Bauer and Dr. W. Hartmann, as well as valuable conversations with Prof. J. Christiansen, Dr. H. Riege, and Dr. K. Frank in developing these ideas.

REFERENCES

1. G. Kirkman and M. A. Gundersen, "A Light Initiated Glow Discharge Switch for High Power Applications," Appl. Phys. Lett. **49** (9) 494, (1986). For comprehensive reviews see "High power hollow electrode thyratron-type switches," K. Frank, E. Boggasch, J. Christiansen, A. Goertler, W. Hartmann, C. Kozlik, G. Kirkman, C. G. Braun, V. Dominic, H. Riege and M.A. Gundersen, Proceedings, Sixth IEEE Pulsed Power Conference, June 1987, and "High power hollow electrode thyratron-type switches," K. Frank, E. Boggasch, J. Christiansen, A. Goertler, W. Hartmann, C. Kozlik, G. Kirkman, C. G. Braun, V. Dominic, M.A. Gundersen, H. Riege and G. Mechtersheimer, IEEE Trans Plasma Science PS-16,317, (1988).

2. H.J. Cirkel, "High power excimer laser", Presented at Lasers '87, Lake Tahoe, CA, Dec. 1987.

3. E. Boggasch, V. Brueckner, and H. Riege, "A 400 kA pulse generator with pseudo-spark switches", Proceedings of the 5th IEEE Pulsed Power Conference, Arlington, VA, pp. 820, 1985.

4. C.G. Braun, W. Hartmann, V. Dominic, G. Kirkman, M. Gundersen and G. McDuff,"Fiber optic triggered high-power low-pressure glow discharge switches," IEEE Trans. Electron Devices, Vol. **35**, (4), 559 (1988).

5. W. Hartmann and M. A. Gundersen "Origin of Anomalous Emission in Superdense Glow Discharge", Phys Rev Lett., **60**, 2371 (1988).

6. "Evidence for large area super-emission into a high current glow discharge," W. Hartmann, V. Dominic, G.F. Kirkman, and M.A Gundersen, App. Phys. Lett. **53** (18), 1699 (1988).

7. "A super-emissive self heated cathode for high power applications," W. Hartmann, G. F. Kirkman, V. Dominic, and M.A. Gundersen, IEEE Trans. Elect. Dev. **36** (4), 825 (1989).

8. "An analysis of the anomalous high current cathode emission in pseudo-spark and BLT switches," W. Hartmann, V. Dominic, G. Kirkman, and M. A. Gundersen, J. Appl. Phys. **65** (11), 4388 (1989).

9. See "Workshop on the Fundamentals and Applications of the Pseudospark Discharge", Bad Honnef, West Germany, October 1987.

10. "High current plasma based electron source," H. R. Bauer and M. A. Gundersen, submitted to Appl. Phys. Lett.

11. "A two component model for the electron distribution function in a high current pseudospark or back-lighted thyratron," H. Bauer, G. Kirkman, and M. A. Gundersen, IEEE Trans. Plasma Sci. **18** (2) (1990).

12. J.B. Rosenzweig, D.B. Cline, B. Cole, H. Figueroa, W. Gai, R. Konecny, J. Norem, P. Schoessow, and J. Simpson, Phys. Rev. Lett., **61**, 98 (1988).

13. T. Tajima and J. Dawson, Phys. Rev. Lett. **43**, 267 (1979).

14. P. Chen et. al., Phys. Rev. Lett. **54**, 693 (1985).

15. T. Katsouleas, J.J. Su, W.B. Mori, C. Joshi and J.M. Dawson, "A compact 100 MEV accelerator based on plasma wakefields", presented at OE /LASE, Los Angeles, Jan. 1989.

16. See, for example A.D. Bandrauk, Ed., "Atomic and Molecular Processes with Short Intense Laser Pulses", NATO ASI Series B: Physics, Plenu Press (1987). This provides a current review of multiphoton ionization problems, and several articles specifically address multiphoton ionization in hydrogen.

17. See, for example, S. Turner, Ed., "Proceedings of the workshop on new developments in particle acceleration techniques", Orsay, Fr, 29 June - 4 July 1987, available from CERN, Publication CERN 87-11, ECFA 87/110 (1987).

18. E.B. Forsyth, L.M. Lederman and J. Sunderland, "The Brookhaven-Columbia plasma lens", IEEE Trans. Nucl. Sci. **12**, 872 (1965). See also P. Chen, "A possible final focusing mechanism for linear collider", Particle Accelerators **20**, 171 (1982), P. Chen, J.J. Su, T. Katsouleas and J.M. Dawson, "Plasma focusing for high-energy beams", IEEE Trans. Plasma Sci. **PS-15** (2), 218 (1987).

19. H. Figueroa, W. Gai, R. Konecny, J. Norem, A. Ruggiero, P. Schoessow, and J. Simpson, "Direct measurement of beam-induced fields in accelerating structures," Phys. Rev. Lett. **60**, 2144 (1988).

20. H. Figueroa, G. Kirkman, and M.A. Gundersen, "A plasma lens candidate with highly stable properties," Proceedings of the 1989 Workshop on Advanced Accelerator Concepts, Lake Arrowhead, California, Jan. 9-13, 1989, in press.

21. F. Muller, Reported at the "Workshop on the Fundamentals and Applications of the Pseudospark Discharge," Bad Honnef, West Germany, October 1987.

22. H. Figueroa and C. Joshi, Laser Interaction and Related Plasma Phenomena, 7, 241, Plenum Press (1986).

23. J. Christiansen and Ch. Shultheiss, "Production of High Current Particle Beams by Low Pressure Spark Discharge", Z. Physik A 290, 35 (1979); W. Benker, J. Christiansen, K. Frank, H. Gundel, W. Hartmann, T. Redel, M. Stetter, "Pulsed Intense Electron Beams from Pseudospark Discharges", SPIE OE/LASE Los Angeles, January, 1988; E. Boggasch, T. A. Fine and M. J. Rhee, "Measurement of Pseudospark Produced Electron Beam", Bull. Amer. Phys. Soc., 33, 1951, (1988); G. Kirkman, M. Gundersen, reported at the 1988 IEEE Electron Devices Meeting, Dec. 1988.

24. G. F. Kirkman, H. Figueroa and M. A. Gundersen, "A plasma lens with highly stable properties," 1989 Workshop on Advanced Accelerator Concepts.

25. D.A. Erwin and M.A. Gundersen, "Measurement of excited-state densities during high-current operation of a hydrogen thyratron using laser-induced fluorescence," Appl. Phys. Lett. 48, 1773 (1986). D.A. Erwin, J.A. Kunc, and M.A. Gundersen, "Determination of electric field and electron temperature in the positive column of a high-power thyratron from non-intrusive measurements," Appl. Phys. Lett. 48, 1727 (1986).

26. W. Hartmann V. Dominic, G. Kirkman, and M. A. Gundersen, " Evidence for Large Area Super-Emission into a Glow Discharge," Appl. Phys. Lett. 53, 1699 (1988).

27. C. G. Braun, D. A. Erwin and M.A. Gundersen, "Fundamental processes affecting recovery in hydrogen thyratrons," Appl. Phys. Lett. 50 (19), 1325 (1987).

TWO-DIMENSIONAL CALCULATION OF SEQUENTIAL
ELECTRON LAYER FORMATION BY CROSSED
MICROWAVE BEAMS IN AIR AT LOW PRESSURE*

D. J. Mayhall, J. H. Yee, G. E. Sieger and R. A. Alvarez

Lawrence Livermore National Laboratory
University of California
P. O. Box 808, L-161
Livermore, CA 94550

INTRODUCTION

Laboratory chamber experiments in low-pressure (0.1-10 Torr) air with long-pulse (150-600 ns), 2.856 GHz crossed microwave beams, created by reflection of an incident beam from a metal plate inclined at 45°, have shown that multiple, luminous electron density layers form toward the source of the incident beam.[1] Previous two-dimensional calculations[2] of microwave air breakdown in a rectangular waveguide with full resolution of each cycle of the driving electric field required Cray 1 computation times of about 0.5 hr/ns for runs of up to 20 ns of simulation time. To simulate chamber experiments with pulses of 150-600 ns duration, computation times with full-cycle resolution must be expected to be 75-300 hours per run. Because these are prohibitive run times, a code, which follows only the pulse envelope, has been developed to reduce computation time. This envelope approximation is an extension of the one-dimensional technique for atmospheric microwave propagation reported previously.[3]

TEMPORAL REGIMES

The approach to an envelope solution of the formation of electron density layers in low-pressure air is to divide the formation into two temporal regimes. During the first regime, when the electron density is increasing primarily due to avalanche ionization, the electron density is assumed to have no effect on the driving microwave electric field. This is a precritical state, which occurs before the density reaches the critical value n_c (~1 x 10^{11} cm^{-3} for 2.856 GHz). During the second regime, which starts after the peak electron density reaches a few (1-4) percent of n_c, the electron density is assumed to alter the initial driving electric field pattern. Therefore different sets of equations are solved for the two different temporal regimes.

The First Regime

During the first regime, the electron flux is represented by terms for drift in the transverse electric field and diffusion. The transverse field $\vec{E}_t = \vec{a}_x E_x + \vec{a}_y E_y$ where $\vec{a}_x$ and $\vec{a}_y$ are unit vectors in the x and y directions, and E_x and E_y are field components in those directions. The transverse field is perpendicular to the driving field $\vec{E}_z = \vec{a}_z E_z$, where $\vec{a}_z$ is a unit vector in the z direction and E_z is the associated field component.

*Work done under the auspices of the U.S. Department of Energy by the Lawrence Livermore National Laboratory under Contract No. W-7405-ENG-48.

The curl of the magnetic induction $\vec{B}$ is neglected. The electron conductivity is given by $\sigma = \mu|e|n$, where μ is the electron mobility, e is the electron charge, -4.803×10^{-10} statcoulombs, and n is the electron density.

The equations for the first temporal regime in the CGS system are then

$$\frac{\partial n}{\partial t} = n v_i - \frac{\partial}{\partial x}\left(\mu n E_x\right) - \frac{\partial}{\partial y}\left(\mu n E_y\right) + \nabla_t^2 (Dn) \quad . \tag{1}$$

$$\frac{\partial E_x}{\partial t} = -4\pi\mu|e|n\, E_x - 4\pi|e|\frac{\partial}{\partial x}(Dn) \quad , \tag{2}$$

and

$$\frac{\partial E_y}{\partial t} = -4\pi\mu|e|n\, E_y - 4\pi|e|\frac{\partial}{\partial y}(Dn) \quad , \tag{3}$$

where v_i is the avalanche ionization frequency, $\mu = |e|/(mv_c)$, $\nabla_t^2 = \partial^2/\partial y^2 + \partial^2/\partial x^2$, and D is the diffusion coefficient given by $2\bar{u}/(3mv_c)$. The quantity $\bar{u}$ is the average electron energy, m is the electron mass, and v_c is the collision frequency given by $v_i + v_m$, where v_m is the electron-neutral molecule momentum transfer collision frequency.

The Second Regime

During the second regime, drift, diffusion and all electron losses are neglected in the continuity equation so that Eq. (1) becomes

$$\frac{\partial n}{\partial t} = n v_i \quad . \tag{4}$$

A wave equation for the electric field may be derived from Maxwell's equations and the vector identity, $\nabla \times \nabla \times \vec{E} = \nabla(\nabla \cdot \vec{E}) - \nabla^2\vec{E}$. It is

$$\frac{\partial^2\vec{E}}{\partial t^2} + 4\pi\frac{\partial\vec{J}}{\partial t} = -c^2\nabla^2\left(\nabla \cdot \vec{E}\right) - \nabla^2\vec{E} \quad , \tag{5}$$

where $\vec{J}$ is the current density and c is the speed of light in vacuum.

For a harmonic electric field of frequency w, $\vec{E} = \vec{A}\exp(-iwt)$, where $\vec{A}$ is a slowly varying field envelope and $i = \sqrt{-1}$. With the assumptions of negligible magnetic field and $\vec{J} = \sigma\vec{E}$, the use of Eq. (4) in the electron fluid momentum conservation equation gives the conductivity as

$$\sigma = \frac{ne^2}{m}\frac{v_c + iw}{v_c^2 + w^2} \quad . \tag{6}$$

with the assumption that $\partial v_c/\partial t \approx 0$, the time derivative of the current density becomes

$$\frac{\partial\vec{J}}{\partial t} = \left(-iw\vec{A} + v_i\vec{A} + \frac{\partial\vec{A}}{\partial t}\right)\sigma\exp(-iwt) \quad . \tag{7}$$

With the further assumptions that $\nabla \cdot \vec{E} \approx 0$ and $\partial^2\vec{A}/\partial t^2 \approx 0$, the wave equation becomes

$$\left(\bar{\sigma} - 2iw\right)\frac{\partial\vec{A}}{\partial t} \approx \left[w^2 - \bar{\sigma}(v_i - iw)\right]\vec{A} + c^2\nabla^2\vec{A} \quad , \tag{8}$$

where $\bar{\sigma} = 4\pi\sigma$.

516

Next, $\bar\sigma$ and $\vec A$ are assumed complex, so that $\bar\sigma = \bar\sigma_1 + i\bar\sigma_2$ and $\vec A = \vec A_1 + i\vec A_2$. The Laplacian is specialized to the transverse plane defined by the x and y axes, and the driving field is specialized to the z direction so that $\vec A = \hat a_z A$. With a goodly amount of complex algebra, Eq. (8) becomes

$$\frac{\partial A_1}{\partial t} = aA_1 + bA_2 + f\nabla_t^2 A_1 + d\nabla_t^2 A_2 \ , \tag{9}$$

and

$$\frac{\partial A_2}{\partial t} = -bA_1 + aA_2 - d\nabla_t^2 A_1 + f\nabla_t^2 A_2 \ , \tag{10}$$

where

$$a = \frac{\left[-w^2\bar\sigma_1 - v_i\left(\bar\sigma_1^2 + \bar\sigma_2^2 - 2w\bar\sigma_2\right)\right]}{g} \ , \tag{11}$$

$$b = \frac{-w\left(2w^2 - 3w\bar\sigma_2 + \bar\sigma_1^2 + \bar\sigma_2^2 - 2\bar\sigma_1 v_i\right)}{g} \ , \tag{12}$$

$$d = \frac{-\left(2w - \bar\sigma_2\right)c^2}{g} \ , \tag{13}$$

$$f = \frac{\bar\sigma_1 c^2}{g} \ , \tag{14}$$

and

$$g = \bar\sigma_1^2 + \left(\bar\sigma_2 - 2w\right)^2 \ . \tag{15}$$

Equations (4), (9), and (10) describe approximately the interaction of the avalanching electron density with the slowly time-varying envelope of a harmonic driving field of angular frequency w.

NUMERICAL SOLUTION FOR A SPECIFIC CASE

Computational Geometry

The two-dimensional geometry chosen to model the three-dimensional laboratory experiment is shown in Fig. 1. The coordinate system is an x–y rectangular system. The computational solution space extends from $x = 0$ to x_m and from $y = 0$ to y_m. The dimensions are in centimeters. A perfectly conducting metal plate at 45° inclination to the x axis extends from $(x_1, 0)$ to (x_m, y_1). The space to the right of the plate is ignored. The center of an incident envelope enters the computational

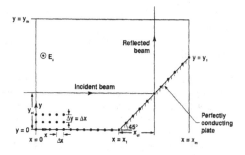

Figure 1. Computational geometry.

space at $(0, y_{cl})$ and strikes the plate at $(x_1 + x_{cl}, y_{cl})$. From this point, the center of a reflected envelope then proceeds upward to leave the space at $(x_1 + x_{cl}, y_m)$. A uniform rectangular grid of spacing $\Delta x = \Delta y$ is superimposed on the computational space. Some of the lower grid points and those along the plate are indicated as dots in Fig. 1.

Initial Electric Field Distribution

At the start of this study, the experimental electric field pattern was not known, so the incident beam was taken to be gaussian in the y direction and given by the complex expression

$$\vec{A}_{in} = \hat{a}_z E_A \exp\left[ik(x - x_1)\right] \exp\left[-(y - y_{cl})^2 / 2\sigma_0^2\right] , \tag{16}$$

where E_A is the beam amplitude, σ_0 is the standard deviation in the y direction, and $k = 2\pi f/c = 0.598$ cm^{-1} is the wave number and $f = 2.856$ GHz.

The reflected beam is given by

$$\vec{A}_r = -\hat{a}_z E_A \exp\left[\frac{-(x - (x_1 + x_{cl}))^2}{2\sigma_0^2}\right] \exp(iky) . \tag{17}$$

The complex initial envelope is given by $\vec{A}_{in} + \vec{A}_r$. Thus the real part of the initial envelope is

$$A_1(0) = E_A \left\{ \cos\left[k(x - x_1)\right] \exp\left[\frac{-(y - y_{cl})^2}{2\sigma_0^2}\right] - \cos(ky) \exp\left[\frac{-(x - (x_1 + x_{cl}))^2}{2\sigma_0^2}\right] \right\} \tag{18}$$

and the imaginary part is

$$A_2(0) = E_A \left\{ \sin\left[k(x - x_1)\right] \exp\left[\frac{-(y - y_{cl})^2}{2\sigma_0^2}\right] - \sin(ky) \exp\left[\frac{-(x - (x_1 + x_{cl}))^2}{2\sigma_0^2}\right] \right\} . \tag{19}$$

In the computation, the initial transit time from the entry of the front of the incident beam into the computational space at $x = 0$ until the emergence of the front of the reflected beam from the computational space at $y = y_m$ is ignored. For a 45 cm x 18.2 cm grid, which is typical, this transit time is less than 2.1 ns. The spatial electric field pattern is thus assumed to be fully formed when the computation starts. For a beam whose amplitude E_A is ramped in time, this assumption may result in some error since the amplitude is assumed to be the same everywhere in the computational space at a given time.

Differenced Equations

The three partial differential equations for each temporal regime are normalized by A_m, the peak magnitude of the incident electric field, and approximated by finite differences with each variable solved at each grid point. The boundary condition that the tangential electric field vanish is applied at the metal plate. In the second temporal regime, the quantities A_1 and A_2 are extrapolated one grid spacing beyond the free space boundaries to approximate the normal Laplacians of the electric field envelope at the free space boundaries.

Two cases of driving field time dependence are considered. The first case is a driving field which increases linearly during the first regime and remains constant throughout the second regime. The second is a driving field, which increases linearly throughout the first regime and continues that behavior into the second regime, where it reaches a plateau value and thereafter remains constant. For the second case, a constant time derivative A_m/t_r, where t_r is the 0-100% rise time, is added to the right sides of Eqs. (9) and (10) to account for the constant increase in the envelope amplitude with time during the ramp.

518

Time Integration of the Differenced Equations

The two sets of differenced equations constitute first order ordinary differential equations (ODEs) in time. In a typical grid of 57 points in the x direction and 25 points in the y direction, there are 3516 equations to be solved. Instead of differencing the left side column vector in time, the entire ODE system is integrated in time with the ODE software package LSODE.[4-6] Since LSODE takes variable time steps, it can slow down or speed up the computation as required by the dominant time scale in the problem. LSODE also has a number of attractive user-selectable options. To reduce code development time, we used the option in which the integrator employs chord iteration with an internally generated banded jacobean matrix of derivatives of the right side of the ODE system. The price for this convenience is a somewhat longer computation time per run than with user specification of the jacobean elements. The stiff system option was also chosen so that any high-frequency, low-amplitude oscillations in the solution will not cause prohibitively small time steps. The code has been implemented on Cray XMP and YMP computers.

CALCULATIONAL RESULTS

Results for the First Driving Field Time Dependence

The calculation is done at 1 Torr air pressure with a uniform initial electron density of 10 electrons/cm^3, which is believed to be a reasonable estimate of the experimental situation. The grid dimensions are 42.4 cm in the x direction and 18.2 cm in the y direction. The first regime equations are integrated until the peak electron density reaches 3.02×10^{15} m^{-3} or about 3% of n_c at 54 ns. At this time, the linearly ramped E_z field, which increases at the rate of 1.99 MV/(m - sec), has reached 0.107 MV/m. The electron density distribution at this time is shown in Fig. 2a. The peak density is located at the driving field maximum, above the incident beam center line which is shown in Fib. 2a, and left of the reflected beam center line. The density peak is at $3\Delta x\sqrt{2} = 3.2$ cm from the plate. The expected distance is at the first maximum in the field pattern or $\lambda/(4 \cos 45°) = 3.71$ cm. A finer resolution may yield a calculated spacing closer to the expected spacing. The magnitude of the driving field is shown in Fig. 2b. The amplitude is 1.97 MV/m.

At 54 ns, the second regime equations are started with the amplitude of the incident field constant in time. As time passes, the elliptic density blob in Fig. 2a slowly grows out along a line parallel to the plate, forming a layer of electron density. The state of this layer at 82 ns is shown in Fig. 3a. The peak value of the electron density, which has migrated downward, is 6.8 n_c. The corresponding E_z magnitude is shown in Fig. 3b, where the field has been excluded from the region of highest electron density near the plate and a standing wave pattern in the x direction with ten nodes has formed toward the source of the incident beam.

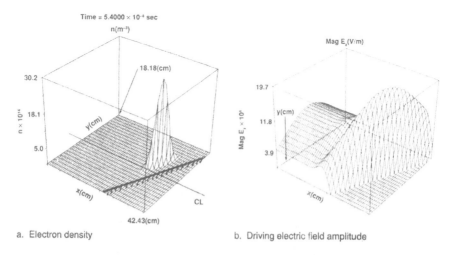

a. Electron density   b. Driving electric field amplitude

Figure 2. Electron density and driving electric field amplitude for the first time dependence at 54 Ns.

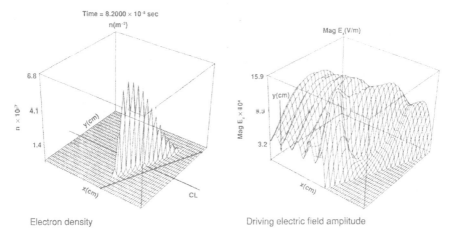

Time = 8.2000 × 10⁻⁸ sec

Electron density

Driving electric field amplitude

Figure 3. Electron density and driving electric field amplitude for the first time dependence at 82 Ns.

As the simulation continues, more electron density layers appear near the standing wave peaks. The electron density at 119.8 ns is shown in Fig. 4 from two different viewpoints. The peak density is 7.3 n_c and occurs in the first layer. Five distinct layers have formed by this time. The orientation of the layer with respect to the plate changes with the distance from the plate. The first two layers are parallel to the plate; whereas, the last three are perpendicular to the incident beam direction.

Experimental streak photography for 1 Torr with an incident waveform, which rises roughly in a linear ramp to 0.159 MV/m at 80 ns, has given times for the appearance of most of the layers.[1] The observed layer appearance times, referred to the incident field initial zero, are shown in Table 1. These values are believed to be good to within ±6%. Experimentally, the third layer appears slightly before the first layer, the fifth layer appears somewhat later than the first layer, then the second layer appears. The fourth layer is not seen.

Calculated layer appearance times have been obtained under the assumption that a layer appears when the peak electron density reaches 0.1 n_c. It is not yet known what value of electron density and temperature correspond to the appearance of light in streak or time-integrated luminosity

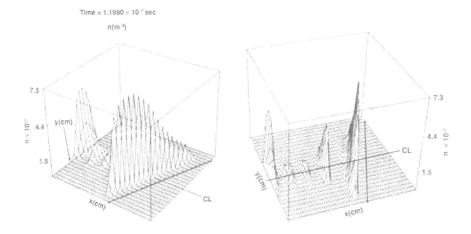

Time = 1.1980 × 10⁻⁷ sec

Figure 4. Electron density for the first time dependence at 119.8 Ns.

Table 1. Experimental and calculational layer appearance times.

Layer number	1	2	3	4	5
Experimental appearance time (ns)	74	99	71	—	87
Calculational appearance time (ns):					
• First waveform (0.107 MV/m peak) to 0.1 n_c	55.2	86.8	103.4	105.2	104.0
% difference from experimental time	-25.4	-12.3	+45.6	—	+19.5
to n_c	58.2	93.6	>120.0	117.6	110.0
• Second waveform (0.159 MV/m peak) to 0.1 n_c	54.8	65.9	66.9	64.7	64.6
% difference from experimental time	-25.9	-33.4	-5.8	—	-25.7
Time to n_c	56.3	67.0	67.3	66.3	65.3
• Third waveform (0.111 MV/m peak) to 0.1 n_c	70.6	83.0	84.9	83.0	82.6
% difference from experimental time	-4.6	-16	+20	—	-5.1
to n_c	72.4	84.5	85.5	85.6	83.5

photography. These calculated layer appearance times are given in Table 1 as times to 0.1 n_c, under the heading "First Waveform." The various layers appear in order, except that the fifth layer shows up 1.2 ns before the fourth layer. The times for the peak density to reach n_c are also shown in Table 1. For the four experimentally observed layers, the calculated times to 0.1 n_c differ from the experimental times by -12.3% for the first layer to +45.6% for the third layer. These differences are also given in Table 1. A negative difference means the calculated time is less than the observed time.

Results for the Second Driving Field Time Dependence

The dimensions of the grid for this calculation are 45 cm in the x direction and 19.3 cm in the y direction. The first regime ends at 53 ns when the peak electron density reaches 0.0123 n_c in a small blob at the maximum of the driving electric field pattern. The second regime then starts and the peak density increases. A second local density peak in the first layer appears at 59.4 ns near the top of the plate. At this time, the first layer consists of two density blobs or local maxima. Thereafter, the peak density of the lower blob remains constant, while that of the upper blob increases. At about 61.2 ns, the two blobs begin to merge, while the peak density in the upper blob continues to increase. At 61.8 ns, the peak density in the upper blob exceeds that in the lower blob and continues to increase until 63.2 ns, when it stops growing. The lower blob then begins to lengthen downward parallel to the plate. At 65 ns, a third detached blob appears in the first layer below the initial blob. The fourth and fifth layers become apparent at this time. The lower third blob in the first layer has begun to coalesce with the initial blob at 66 ns. A blob in the second layer has also appeared near the bottom grid boundary. By 66.6 ns, the peak density in the fifth layer has exceeded that in the first layer, and a small blob in the second layer has appeared near the top grid boundary. The electron density at 67.2 ns is shown in Fig. 5 from two viewpoints. The first layer now consists of three merged density maxima with two discernible, unequal electron density flutes on the side of the layer away from the plate. A blob in the third layer has also appeared near the top of the grid.

By 70 ns, the top detached blob in the second layer denoted by the letter A in Fig. 5, has coalesced with the initial, central blob in the first layer, denoted by the letter B in Fig. 5. This interlayer merging, although different in detail from the calculation, is seen experimentally on time-integrated luminosity photographs. Such a photograph is shown in Fig. 6, where the merging of three local density maxima is indicated. The indication of weak flutes in Fig. 5 on the plate side of the first layer can be argued from the variation of the electron density at the constant normal distance $\sqrt{2} \, \Delta x$ from the plate. Flutes are seen on the surfaces of the first layer in Fig. 6. The flutes on the surface nearest the plate appear to be weaker than those on the other side. This difference in the strength of fluting is also seen in the calculated electron density in Fig. 5. As indicated by the calculated evolution of the first layer, a possible explanation for the flutes in the experimental luminosity is the sequential generation of local density maxima due to migration of electric field maxima after exclusion from regions of previously formed electron density maxima. The details of luminosity photographs vary irreproducibly from shot to shot for the same experimental settings.

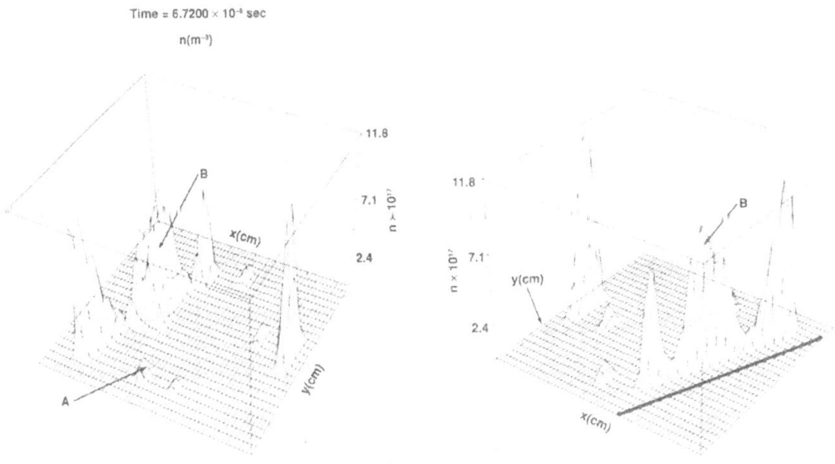

Time = 6.7200 × 10⁻⁸ sec

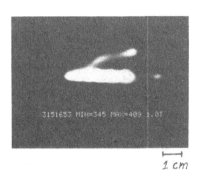

Figure 5. Electron density for the second time dependence at 67.2 Ns.

1 cm

Figure 6. Time-integrated luminosity photograph at 1 Torr.

A contour plot of the electron density at 75 ns is shown in Fig. 7. The metal plate is shown at the right side of the figure. The three local density maxima, denoted B, C, and D, and the bridging density connecting them comprise the first layer. The three detached local maxima, denoted A, E, and F, comprise the second layer. The feature A of the second layer has coalesced into feature B of the first layer. The three detached local maxima, denoted G, H, and I, comprise the third layer. Feature J comprises the fourth layer, which is not very prominent here and was not detected in the experimental streak photography. Feature K comprises the fifth layer. The light lines on Fig. 7 connecting features or running through isolated features indicate the layers and the orientation of their major axes with respect to the x axis. The contour line a of the first layer has the value of 5×10^{16} m^{-3}. The values of the other contour lines are b = 10^{17} m^{-3}, c = 2×10^{17} m^{-3}, d = 4×10^{17} m^{-3}, e = 6×10^{17} m^{-3}, and f = 8×10^{17} m^{-3}.

The camera shutter for the photograph in Fig. 6 was open for the order of a microsecond, which is several times the 270 ns duration of the incident pulse. This photograph corresponds to an incident power waveform, which rises ot its peak value in about 90 ns, remains relatively constant for about 20 ns, then falls in an asymmetric manner more slowly than it rose. The full width at half maximum is about 143 ns. The amplitude of the incident pulse is presently unknown. The length scale for the photograph is shown in the figure.

The length of the main part of the first layer in Fig. 6 is about 11.2 cm; whereas, the total length of the first layer, including the detached upper blob, is about 15.2 cm. The length of the first layer in the contour plot is about 22 cm. The right side flutes on the upper surface of the first layer in

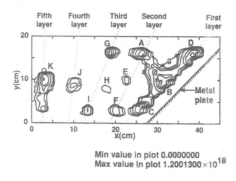

n(m⁻³) at 75 nsec

Min value in plot 0.0000000
Max value in plot 1.2001300×10¹⁸

Figure 7. Electron density contours for the second time
dependence at 75 Ns.

Fig. 6 are about 2.2 cm in extent. In the calculated contour plot, the upper flute in the b contour of the first layer, indicated by the letter g, is about 7 cm in extent. This difference in flute size may be due to a finer scale of local electric field peaks in the experiment. Local maxima with irregular axial (x direction) spacing of the order of 5 cm are seen in homodyne electric field amplitude maps of the experimental chamber without the metal plate. The spacing of the maxima of the interference pattern due to the addition of the plate may influence the size of the luminosity flutes.

The axes of the layers in the contour plot shift from 45° to the x axis to about 63° for the second layer, 69° for the third layer, 88° for the fourth layer, and close to 90° for the fifth layer. The same sort of shift in layer orientation appears in Fig. 6, where the angle between the first and second layers is about 14°. The angle in the contour plot is about 18°.

Calculated layer appearance times have been obtained for the second driving field time dependence. These times are shown in Table 1 under the heading "Second Waveform." All calculated layers appearance times for this case are less than those experimentally measured by 5.8 to 33.4%.

A third simulation has been done with the incident amplitude at 80 ns reduced from 0.159 MV/m by 30% to 0.111 MV/m. The results are very similar to those for the previous simulation, though features develop somewhat more slowly. Calculated layer appearance times are given in Table 1 under the heading "Third Waveform." The percent differences now vary from the experimental values from –4.6 to 20%.

It is highly likely that use of the exact experimental driving field distribution and time dependence in the calculation would give calculated density patterns more like the time-integrated luminosity patterns. It is, however, probably impossible to get perfect agreement since the luminosity photographs are not at all repeatable. Use of the exact, or even an approximate, initial experimental electron density distribution may also increase the agreement. Measurement of the initial electron distribution is presently an unsolved experimental problem. The number of small blobs in the calculated electron density distribution suggests that the distribution may be very sensitive to the details of the driving electric field distribution. Some of the lack of agreement between the calculational and experimental results may also be due to the presence of three-dimensional effects in the experiment.

CONCLUSIONS

A formulation is presented for the approximate calculation of electron density evolution in air in the two-dimensional envelope of an electromagnetic driving field for time periods greater than 20 ns. The formulation is applied numerically to the case of the reflection of a microwave field from a perfectly conducting plate inclined at 45°. The results of three calculations with idealized electric field patterns and time waveforms are compared to experimental results. For the third case, whose waveform more nearly matches the experimental waveform, the electron layers appear nonsequentially. The estimated calculated layer appearance times differ from experimentally measured times by 5 to 20%. Calculated electron density patterns are similar to those of time-integrated luminosity, but differ in detail. More realistic modeling of the experimental conditions may improve the detailed agreement. The results of the present effort indicate that the method of envelope approximation shows promise for the modeling of long-pulse breakdown in gaseous media in complex geometries. A much more intensive and extensive comparison of calculation and experiment is still required for determination of the range of validity of the method.

REFERENCES

1. P. R. Bolton, R. A. Alvarez, and G. E. Sieger, "Temporal Observations of Microwave Generated Air Plasma with a Streak Camera in the Visible Wavelength Region," Paper 3P17, Conference Record—Abstracts, 1988 IEEE International Conference on Plasma Science, Seattle, WA, June 6–8, 83 (1988).

2. D. J. Mayhall, J. H. Yee, R. A. Alvarez, and D. P. Byrne, "Two-Dimensional Electron Fluid Computer Calculations of Nanosecond Pulse Low Pressure Microwave Air Breakdown in a Rectangular Waveguide and Their Verification by Experimental Measurement," in *Laser Interaction and Related Plasma Phenomena*, H. Hora and G. H. Miley, Eds. 8, Plenum Press, New York, 121–137 (1988).

3. G. E. Sieger, D. J. Mayhall, and J. H. Yee, "Numerical Simulation of the Propagation and Absorption Due to Air Breakdown of Long Microwave Pulses," in *Laser Interaction and Related Plasma Phenomena*, H. Hora, and G. H. Miley, Eds. 8, Plenum Press, New York, 139–148 (1988).

4. A. C. Hindmarsh, "ODEPACK, A Systemized Collection of ODE Solvers," *Scientific Computing*, R. E. Stepleman, Ed. 1, North Holland Publishing Company, Amsterdam, 55–64 (1983).

5. G. D. Byrne and A. C. Hindmarsh, "Stiff ODE Solvers: A Review of Current and Coming Attractions," *J. Comp. Phys.* 70:1 (1987).

6. A. C. Hindmarsh and L. R. Petzold, "Numerical Methods for Solving Ordinary Differential Equations and Differential/Albegraic Equations," *Energy and Technology Review*, Lawrence Livermore National Laboratory, Livermore, CA, 23-36, September (1988).

SHOCK WAVE DECAY AND SPALLATION IN LASER-MATTER INTERACTION

S. Eliezer, Y. Gazit and I. Gilath

Plasma Physics Department
Soreq Nuclear Research Center, Yavne 70600, Israel

ABSTRACT

Spall at ultra high strain rate (10^7 sec^{-1}) was investigated in a variety of materials using short pulsed laser induced shock waves. The intensities of our 3.5 nsec Nd: Glass laser were in the range 10^{10}-10^{12} W/cm^2, and the foil thicknesses in the 0.01-0.1 cm range. The laser generated shock wave pressure was in the domain of a few hundred kilobars. The controlled stepwise increase in laser energies allowed us to find the stages of damage evolution from incipient to complete perforation of the target foils.

A new experimental method was developed in order to calculate the decay of the laser generated shock waves. This technique enabled us to evaluate the laser induced spall pressure in different materials. In particular, experiments were performed on aluminum, copper and unidirectional carbon fiber epoxy composites. For the composites, the tensor character of the spall strength was demonstrated.

1. INTRODUCTION

Shock loads can be generated by high speed impact, explosives or by intense short time energy deposition. While in most impact experiments the impact time is in the microsecond time scale, the laser impact time is in the nanosecond regime.

There is a considerable interest in determining the response of composite materials to shock loads since they are increasingly being used in applications where shock loading is significant. It is important to know the stages of material failure, in particular the conditions of incipient damage and the pressure responsible for this damage as well as the pressure attenuation in the sample.

Our work is based on our previous experience on spall and dynamic fracture in metals[1-3]. The advantage of our method as compared to other impact experiments is the study of thermomechanical damage at strain rates of the order of 10^7 sec^{-1}, corresponding to hypervelocity impact conditions. Other advantages are the small quantity of material involved and the close control of experimental conditions.

2. LASER INDUCED SHOCK WAVES

The interaction physics between high irradiance lasers ($\geq 10^{10}$ W/cm^2) and matter, leading to the creation of a shock wave, can be summarized schematically by the following intercorrelated processes.

LASER ABSORPTION → ENERGY TRANSPORT → SHOCK WAVE

The irradiated target creates a plasma (corona) instanteneously ($< 10^{-12}$ sec), so that during the laser pulse duration (of the order of nanoseconds in our case) the irradiated matter consists of a plasma medium in front of the dense target (solid density). The laser radiation entering the plasma encounters increasing density of plasma and it is absorbed up to the critical density region where the plasma frequency equals that of the laser light. The plasma refractive index becomes zero at the critical density (where the resonant frequency of the plasma electrons in the oscillating electric field is equal to the laser frequency) so that reflection of the laser light occurs in this region. The critical density n_c is given by $n_c = 10^{21}/\lambda^2$ electrons cm^{-3} where the laser wavelength λ is given in microns. (Our laser wavelength is about 1 micron). For a fully ionized plasma (where the atomic mass A ~ 2Z) the critical density corresponds to a mass density of $\rho_c = 3.3 \times 10^{-3}$ λ^{-2} g/cm^3, so that in general the critical mass density is small relative to solid density. In our regime of irradiance ($10^{10} - 10^{12}$ W/cm^2) the laser absorption is through the inverse bremsstrahlung mechanism in which an electron in binary collision with an ion absorbs a photon.

The energy is carried away from the point of absorption by the diffusion process. The energy absorbed by the electrons up to the critical density is transported outward into the expanding plasma and inwards into the plasma of greater than critical density. The inward transport of energy reaches the ablation surface where the plasma is created.

In laser-matter interaction under consideration the inward momentum is imparted to the target by the ablation of the hot plasma rather than by the momentum of the laser photon themselves. The heated material is blown off the target and the ablation drives a shock wave into the target. At the final stages of the laser pulse the ablation pressure at the surface drops and a rarefaction wave propagates into the target, following the shock wave. This pressure wave is approximately of a triangular shape and it propagates into the target. After reaching the back surface of the target it is reflected and a tension (negative pressure) is created. For a triangular pressure pulse with a maximum pressure P_O reaching the free surface of the target, a spall is formed at a distances Δ, by the spall pressure $-P_s$ (the negative pressure required for spall.) There is a minimum value, $P_O = P_S$, for spall to occur. For higher values of P_O, multiple spall layers can be obtained in brittle materials while for ductile material the energy is dissipated in plastic deformation.

The values of ablation induced shock pressure can be calculated by sophisticated hydrodynamic simulation codes. However, to a good approximation for our laser irradiances an analytic quantitative estimate of the ablation pressure is given by the following formula[4].

$$P_{kb} = 230 \quad A^{7/6} \quad z^{-9/16} \quad \lambda^{-1/4} \quad \tau^{-1/8} \quad \left[\frac{I_L}{10^{12}} \right]^{3/4} \tag{1}$$

where, the pressure is given in kilobars, A is the atomic weight, Z the ionization number, λ the laser wavelength in μm, τ the laser pulse time in nanoseconds and I is the laser intensity in W/cm^2.

Using our laser induced shock waves, spall in metalic targets was investigated by us and all stages of damage or dynamic failure were identified[1-3]. The intensities of the 3.5 nsec Nd:Glass laser were in the range of 10^{10}-10^{12} W/cm^2, and the foil thickness was in the 100-600 μm range. The laser-generated shock wave pressure was in the range of a few hundred kilobars (kb). The shock wave traversed the foils in a few tens of nsec. The controlled stepwise increase in laser energies allowed us to find the stages of damage evolution from incipient to complete perforation of the target foils. The incipient spall for ductile metals was identified from the level of separate voids in aluminium, while brittle metals from the level of cracks. Those voids or cracks coalesce resulting in a continuous spall layer. For higher laser intensities the spall layer break away and target penetration is observed for specific high intensities.

3. ESTIMATION OF THE STRAIN RATE

Strain rate estimates were done for aluminium using experimental and simulation results.

The strain rate in ballistic experiments is estimated[5] as $\dot{\varepsilon} = \Delta\varepsilon/\Delta t$, where $\Delta\varepsilon = 2U_p/C$, where U_p is the particle velocity and C is the speed of sound. This gives only a very rough estimate of the strain rate, as Grady[5] himself comments: "Variations in strain rate among experiments are a consequence of differing amounts of dispersion before wave interaction at the spall plane". The exact strain rate associated with spall for impact times of a few nanoseconds is difficult to determine accurately. Using the analogy of laser experiments for strain rate calculations, Δt, can be substituted by τ_L, the laser pulse time. The strain for laser experiments can be written in the same way, and the rough estimate for the strain rate is accordingly:

$$\dot{\varepsilon} = \frac{2\ U_p}{C\ \tau_L}\ sec^{-1} \tag{2}$$

By substituting values from equation of state for aluminum in our experimental range of a few hundred kb, we obtain the following strain rate:

$$\dot{\varepsilon} = (3-4) \times 10^7\ sec^{-1} \tag{3}$$

The computer simulation of the phenomena of laser-solid interactions and the material behaviour involves the simultaneous solution of the three conservation equations (of mass, momentum and energy), along with the equation of state (EOS) of the material. As long as the size of the laser spot is larger by a factor of 3 than the foil thickness, one can employ a one dimensional computer code. In our simulation[3,6] we use a simple spall model in which the material is spalled at the cell whose (negative) pressure attains a value more negative than the 'spall pressure'.

The strain rate can be calculated from the simulation as following:

$$\dot{\varepsilon} = \frac{-\dot{\rho}}{\rho} = -\frac{1}{\rho}\frac{\partial\rho}{\partial t} = -\frac{1}{\rho}\frac{\partial\rho}{\partial x}\,C \qquad (4)$$

where $\partial\rho/\partial x$ is the density gradient obtained from the simulations at time and position right before the spall occurs, and C is the speed of sound. The density gradient is laser intensity dependent. In our simulations for aluminium in the domain of laser irradiance 10^{10} – 10^{12} W/cm^2 the density gradient varies in the range of -5×10^6 to -2×10^7 kg/m^4. Therefore the strain rate will be in the range of $(1-4) \times 10^7$ sec^{-1}, in agreement with the strain rate calculated above.

The strain rates we obtained are about one order of magnitude larger than those obtained by the other experimental methods.

4. THE EXPERIMENTAL METHOD

The Advanced Laser Amplifier Design Integration System is capable of delivering 3-10 nsec, up to 100 Joule pulses in the 1.06 µm wavelength. The system comprises three stages: the pulse forming and pre-amplifying stage, the filtering stage and the main amplifying stage.

The first stage contains a Q-switched oscillator which delivers Gaussian shaped pulses with variable FWHM between 3-10 nsec. This Gaussian shape can be modified by steepening its edges for shock wave production by using a fast electro-optical shutter.

The second stage includes spatial filters defining the appropriate divergence needed for the third stage. Optical isolation is also included to avoid back reflection of the amplified beam.

The third stage includes a triple pass amplifier with its associated optics and a double pass amplifier.

The amplified beam is focused in the target chamber through a f/20 lense. For our experiments the beam intensity obtained is in the range 10^{10} – 10^{12} W/cm^2. For spall experiments, where lower intensities and larger spots are required, 1-3 mm diameter spots can be obtained by moving the target out of focus.

The diagnostics include routine monitoring of beam characteristics, such as energy, temporal and spatial shape, by means of a calorimeter, optical streak camera, various X-ray diagnostics etc. The laser system is described in more detail in ref [7].

The experimental work consisted of determining the irradiance and energy density conditions required to produce spall in one dimensional shock wave geometry.

Very close energy density steps can be obtained at the amplifier stages, enabling to monitor all failure stages.

For damage evaluation, metallurgical or scanning electron microscopy was used. For internal damage evaluation, the samples were sectioned, polished, and examined by optical microscopy.

5. DAMAGE IN CARBON EPOXY UNIDIRECTIONAL COMPOSITE AS COMPARED TO ALUMINUM

A continuous replacement of conventional metals by composite materials is observed in space applications. It is due to their superior mechanical properties, high strength to low weight ratio and high chemical resistivity.

Carbon fibers (CF) are among the most promising reinforcing materials for composites having very high modulus and strength, high stiffness, low density, low coefficient of thermal expansion, good fatigue and creep resistance, dimensional stability and transparency to X-rays. Carbon fibers are chemically inert, show good resistance to stress corrosion and other environmental degradation.

A comparison will be made between the impact behaviour at ultra high strain rate of CF/epoxy and aluminum in order to evaluate their relative strength for short pulsed laser induced shock waves.

Two impact directions will be used for CF/epoxy composite: parallel and perpendicular to fiber directions.

The high strain rate impact experiments were performed with the Nd:glass laser. The pulse length was 7.5×10^{-9} sec and irradiances of the order of $10^{10} - 10^{11}$ Watt/cm^2 were used to obtain threshold for spall in samples of different thicknesses.

Laser energies were changed stepwise to ensure damage evolution measurements. Laser spots were three times greater than sample thicknesses to ensure planar shock waves.

Scanning electron microscopy was performed on a JEOL ISM-1300 instrument at 25 KV on gold coated samples.

Mechanical sectioning and polishing was used to reveal internal damage.

The spall damage was observed as fiber and matrix breaking at the back surface of the samples, the damage increasing with increasing laser energy densities[8]. The threshold energy density and corresponding irradiances and plasma ablation pressure to produce spall in composites, aluminum and iron are presented in table 1. Plasma ablation pressure was calculated using eq. 1.

Impact experiments parallel (along the fiber direction) revealed a greater resistance than in the perpendicular direction, as expected.

A comparative study was presented for dynamic spall development in carbon epoxy unidirectional fiber composites and pure aluminum. The very brittle fracture mode of the composites was observed at very low energy densities. However, the fracture was localized at the spall zone, and no internal delaminations or crack propagation were observed through the sample as an effect of shock wave propagation for impacts perpendicular to fiber direction.

When impacts were along the fiber direction, the cracks propagated through the sample because the fibers did not act as crack arresters. No crack branching was observed on the sectioned samples.

The intensive heat pulse ablated the matrix on the front, while tensile (spall) fracture without heating was evident at the back surface.

The unidirectional composite is stronger along the fiber direction than in the perpendicular direction. Quantitative comparison will be presented in the next section.

Table 1. Threshold energy for spall, irradiance and plasma ablation
pressure for carbon/epoxy unidirectional composite and aluminum.

Sample thickness [mm]	Laser energy threshold for spall $[J/cm^2]$	Laser irradiance threshold for spall $[Watt/cm^2]$	Corresponding plasma ablation pressure, [kbar]
	Aluminum		
0.100	100	0×10^{10}	29.7
0.175	350	1×10^{11}	33.9
0.275	460	1.3×10^{11}	41.4
0.600	700	2×10^{11}	57.2
	Carbon Epoxy Composite (perpendicular impact)		
0.2	25	3.3×10^9	2.2
0.5	195	2.6×10^{10}	8.6
0.9	325	4.3×10^{10}	13.8
2.0	650	8.6×10^{10}	25.2
	Carbon Epoxy Compositee (parallel impact)		
0.2	225	6.4×10^{10}	24.9
0.4	460	1.3×10^{11}	43.0
0.6	700	2.0×10^{11}	60.0
0.8	1150	3.4×10^{11}	86.8

6. AN EXPERIMENTAL ESTIMATION OF THE LASER INDUCED SHOCK WAVE ATTENUATION IN DIFFERENT MATERIALS

High irradiance lasers can produce very strong shock waves in targets. The pressure pulse duration is comparable to the laser pulse length and consequently the shock attenuation is rapid. Diagnostics of high pressures in the range of kbars-Mbars on a nanosecond time scale is still a challenge, therefore calculations of these effects are achieved by large laser-matter hydrodynamic codes. These codes require a full understanding of many parameters and are therefore a major effort.

Analytic models have been developed[9,10] for high irradiance lasers to calculate the formation and decay of laser generated shock waves. These models find that the pressure drops by more than a factor of two traversing a 50 μm aluminum slab for laser irradiances in the domain of $5.10^{13} - 5.10^{14}$ W/cm^2.

Dynamic fracture and spall is a result of very short tensile stresses exceeding the dynamic strength of the material. Spall pressure is impact time dependent, however for very short transient times a reasonable prediction of dynamic spall strength (pressure) is needed. In our experiments, we have very short transient times and therefore the evaluation of the instantaneous dynamic spall strength is important.

A new approach is presented here to estimate experimentally the pressure gradient using the effect of spall in targets by laser induced shock waves. At threshold spall, we have the same spall pressure and the same shock velocity at the rear surface for different foil thicknesses. Using targets with decreasing thicknesses, the laser induced plasma ablation pressure P_{th} represents the real spall stress (P_s) and a value corresponding to pressure dispersion through the target thickness ΔX.

$$P_{th} = P_s + \frac{\partial P}{\partial x} \Delta X + \cdots \qquad (5)$$

By extrapolating the experimental values for zero thickness $\Delta X = 0$, the limiting value will represent the dynamic spall pressure. For small values of ΔX the Taylor expansion can be approximated by the first two terms given in eq. (5). In this case the slope will represent the pressure gradient. Experiments were performed to evaluate the pressure decay and spall pressure for isotropic and anisotropic materials using laser induces shock waves.

The experimental work consisted of determining the irradiance and energy density conditions required to produce spall in one dimensional shock wave geometry threshold conditions, see table (1).

The threshold pressure necessary to obtain incipient spallation of the samples was calculated from the plasma ablation pressure using the equation (1).

A linear dependence was obtained experimentally for the pressure as a function of target thickness for isotropic materials such as aluminum and copper. Similar experiments were performed on carbon fiber epoxy unidirectional composites for two impact geometries: one for perpendicular impact vs fiber direction (denoted $\perp$ in fig. 1) and the other along the fiber direction (denoted $\parallel$ in fig. 1.) See table 1 and fig. 1.

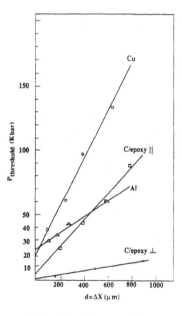

Fig. 1. Plasma threshold ablation pressure for spall as a function of target thickness.

Table 2. The spall pressure and pressure gradient in different materials.

Material	Pspall [kb]	Pgradient [kb/mm]
Copper	20 ± 2	180 ± 20
Aluminum	26 ± 2	57 ± 6
CF/Epoxy (perpendicular to fiber)	0.3 ± 0.2	16 ± 2
CF/epoxy (parallel to fiber)	7 ± 0.5	103 ± 10

Table 3. Dynamic tensile strength of unidirectional carbon or graphite expoxy composites (Impact configuration 0° is along fiber direction while 90° is perpendicular direction).

Material	Impact Config- uration	Dynamic Method	Tensile Strength kb	Ref.
Graphite- epoxy	0° 90°	Instron	17.2 0.4	11
Graphite- epoxy	0° 90°	Explosive pressure pulse in ring specimen	10-12 1.2-1.6	12
Carbon epoxy	0° 90°	Drop weight	17 1	13
Carbon epoxy	0°	Hopkinson bar	12	14
Carbon epoxy	0° 90°	Hopkinson bar	11.5 3.4	15
Carbon epoxy	0° 90°	Laser induced shock waves	7 ± 0.5 0.3 ± 0.2	Our Results

Our experiments yield directly the pressure gradient values. The values for Pspall were obtained by extrapolating to zero foil thickness (see table 2). It is interesting to point out the ablation pressure calculated by eq. 1 for aluminum for a laser irradiance of 1 x 10^{11} W/cm^2 is the same as the value obtained from our simulation, which is 34 Kbar. The pressure gradient simulation is 0.135 (Kb/μm) while the experimental value for aluminum pressure gradient is 0.057 (Kb/μm). This is about a factor of two less than the simulation values. The simulation results are less accurate for the lower pressure regions that prevail at the back free surface of the sample, because our simulation is a fluid code and therefore does not take into account material strength. It is therefore probable that our experimental results are more reliable than our simulation results in the domain of low pressure.

In table 3 a compilation (see references) of impact tensile strength for carbon and graphite epoxy undirectional composites is made. We see that our method fit well with other published data. By analyzing the results for impact along fiber direction we see that for low strain rate (drop might, instron) the tensile strength is higher than for higher strain rates (Hopkinson bar or explosive pressure pulse). Our experimental results correspond to the highest strain rate and therefore are consistent with the compilation of the experimental results, suggesting that Pspall decreases with increasing strain rate for composite materials. However, it might be possible that the different values of Pspall are due to specimen preparation.

7. DISCUSSION

A new experimental method was described to estimate the pressure gradient for laser induced shock waves in thin slab targets. The following values were obtained: aluminum has 57 ± 6 Kb/mm, copper has 180 ± 20 Kb/mm, for carbon fiber/epoxy composite it is 16 ± 2 Kb/mm when the impact is perpendicular to the fiber direction and 103 ± 10 Kb/mm when the impact is parallel (or along) the fiber direction. The large difference between the pressure gradient for perpendicular or parallel impact in unidirectional CF/epoxy is an experimental confirmation that the dispersion of the shock wave is significantly higher in the parallel direction. These results emphasize the tensor characteristics of anisotropic materials.

The dynamic spall pressure for very short impacts was obtained by extrapolating the threshold values for zero target thickness to eliminate the contribution of the pressure dispersion. The following values were obtained: aluminum 26 ± 2 Kbar, copper 20 ± 2 kbar, carbon/epoxy composite perpendicular to fiber direction is 0.3 ± 0.2 Kb and 7 ± 0.5 for impact parallel to the fiber direction.

Acknowledgement

The research reported herein has been sponsored in part by the United States Army through its European Research Office. We are grateful for this support.

The authors would also like to acknowledge the collaboration of Dr. S. Skholnik and the devoted technical assistance of Mr. Y. Sapir, S. Maman and M. Gur.

REFERENCES

[1] .I. Gilath, S. Eliezer, M.P. Dariel, L. Kornblit: "Total elongation at fracture at ultra-high strain rates", J. Mat. Sci. Letters, 7, 915 (1988).

[2] .I. Gilath, S. Eliezer, M.P. Dariel, L. Kornblit: "Brittle to ductile transition in laser induced spall at ultra high strain rate", Appl. Phys. Letters, 52, 1207 (1988).

[3] .I. Gilath, S. Eliezer, M.P. Dariel, L. Kornblit, T. Bar Noy, "Laser induced spall in aluminum and copper", J. de Physique C3, 49, 191 (1988).

[4] .M.H. Key: "The physics of the superdense region" in Laser-Plasma Interactions edited by R.A. Cairns and J.J. Sanderson, Publ. SUSSP, p. 219, (1980).

[5] .D.E. Grady: "The spall strength of condensed matter", J. Mech. Phys. Solids, 36, 353 (1988).

[6] .A.D. Krümbein, D. Salzmann, H. Szichmann, S. Eliezer, Israel Atomic Energy Commission Report, IA-1396 (1985).

[7] .S. Jackel, M. Givon, a. Ludmirsky, S. Eliezer, J.L. Borowitz, B. Arad, A. Zigler Y. Gazit; Laser and Particle Beams 5, 115 (1987).

[8] .I. Gilath, S. Eliezer, S. Shkolnik, "Spall behaviour of carbon epoxy unidirectional composites as compared to aluminum and iron of composites", J. of Composite Mat. (to be published).

[9]. R.J. Trainor, Y.T. Lee: "Analytic models for design of laser generated shock-wave experiments". Phys. Fluids 25, 1898 (1982).

[10]. A. Loeb and S. Eliezer: "Analytical model for creation and decay of strong shock waves caused by a trapezoidal laser pulse", Phys. Fluids 28, 1197 (1985).

[11]. P.N. Kousiounelos and J.H. Williams, "Dynamic fracture of unidirectional graphite fiber composite strips", Int. J. of fracture 20, 47 (1982).

[12]. I.M. Daniel, R.H. La Bedz and T. Liber, "New method for testing composites at very high strain rates", Experimental mechanics, 71 Febr. (1981).

[13]. P.T. Curtis, "An investigation of the mechanical properties of improved carbon fibre composite materials" J of Composite Mat., 21, 1118 (1987).

[14]. J. Harding, L.M. Welsh, "A tensile testing technique for fibre-reinforced composites at impact rates of strain" J. of Mat. Sci. 18, 1810 (1983).

[15]. C. Cazeneuve and J.C. Maile: "Etude du comportement des composites a fibres de carbon sons differente vitesses de deformation", J. de Physique, 46, C5-551, (1985).

HIGH-GAIN DIRECT-DRIVE TARGET DESIGN FOR ICF

G. Velarde, J.M. Aragonés, L. Gámez, C. González,
J.J. Honrubia, J.M. Martínez-Val, E. Mínguez,
J.M. Perlado, M. Piera, U. Schröder and P. Velarde

Instituto de Fusion Nuclear (DENIM)
Universidad Politécnica de Madrid
José Gutiérrez Abascal, 2
28006 Madrid, Spain

INTRODUCTION

The study of the efficiency of high-gain direct-drive targets for ICF has been increased during last years because of the significant progress achieved in laser experiments. Two major key issues have been observed. First, main advances on the uniform illumination to implode targets, due to the techniques developed at NRL[1], Rochester[2] and Osaka[3]. Second, lower growth rates for the hydrodynamic instabilitites up to a 30% of the clasical value by using short wavelength lasers.

It has been also assumed[4] that using a driver delivering about 10 MJ in a 10 ns pulse, density compressions about 200 g/cm^3 and temperatures higher than 3 KeV can be obtained in cryogenic DT targets, achieving gains about 100.

Following these typical target performances, the main goal of this work is to analyze what driver energies and pulse shapes should be needed to obtain ignition conditions in a new family of ICF targets, under different illumination conditions. We have considered a 1 mg DT target driven by 0.35 µm laser light, 30 MeV Li$^+$ (or, equivalently, 4 MeV proton beams) and 10 GeV Bi$^+$ beams. It has also been assumed that the beams impinge with a perfect spherical symmetry on the target, and the study has been focussed to the implosion process as well as on the ignition conditions.

Two one-dimensional codes, NORMA[5] and SARA[6] have been used to simulate the high gain targets driven by laser, light ion and heavy ion beams. The NORMA and SARA treats the target evolution with a 3-temperatures and multigroup radiation transport schemes, respectively.

In previous works[5,7] we characterized the evolution of multilayered HIBALL-like targets with moderate aspect ratio and 1 or 2 milligrams of DT. Those targets included an external high-Z tamper and an internal high-Z radiation shield, that also acts as a pusher to compress the DT. The numerical simulation of those targets showed a low hydrodynamic efficiency, roughly 7%, in such a manner that driver energies higher than 2-3 MJ were needed to obtain high gain. In addition, two major disavantages were observed. First, the interface between the high-Z pusher and the DT is highly Rayleigh-Taylor unstable, degrading the fuel compression process, and second, the high-Z layer produces high energy photons that could preheat the fuel, producing in addition high radiation losses.

Because of these reasons, a family of simpler targets to obtain high gain with less pulse energy has been proposed[8]. These targets satisfy some implosion stability criteria found by comparing 1-D simulations with experiments. The composition of the targets is a low-Z material layer, to absorb and convert the beams into hydro-motion, and a layer of 1 mg of cryogenic DT (fig. 1). The low-Z absorber avoids the conversion of the beams energy into X-rays, that could preheat the fuel. The initial aspect ratio of the targets is moderately high to obtain a good hydrodynamic efficiency. The mass of the absorber material for each target has been selected following several physical reasons, such as prevent the preheating in the fuel, and also to optimize the fuel compression (Table 1).

To implode this class of targets without degrading the symmetry along implosion process, the pulse should have a smooth variation in time and a low intensity, avoiding strong shock waves generation in the ablation surface and high values of the in-flight aspect ratio (IFAR). Under these conditions, the DT fuel, and eventually the payload, is accelerated smoothly and the velocity profile in the fuel is essentially flat. In addition, the low intensity regime avoids plasma instabilities in the corona region for the laser target.

Basically, the evolution of these targets is as follows, although some differences exist among the different kind of driver beams used. First, when the illumination begins, a moderate shock wave generated in the deposition zones goes toward the fuel, increasing the DT density and locating it in a high-density isentrope. Thus, in a first stage the fuel and part of the absorber is compressed, increasing the IFAR by at least, one order of magnitude. Second, the ablation of the absorber takes place smoothly, occuring a maximum density of the absorber equal or less than the density of the compressed DT. Third, the ablation pressure is almost constant, accelerating uniformly the fuel.

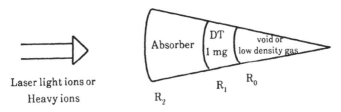

Figure 1. New family of direct driven targets

Table 1. Masses of the absorber materials for each target.

Driver	Absorber	Mass (mg)
Laser	CH	4
Light ions	Li	32
Heavy ions	Li	376

The criteria used to assess the stability of the implosion have been to keep the IFAR[9] below 200 and the convergence ratio below 30-40. The capsules have been designed to obtain implosion velocities close to 4.10^7 cm/s (with a flat profile), appropriate to obtain a good ignition propagation, and a high compression of the fuel. The requirements along the ignition and burn phases have been to have a stagnation free compression, that can be achieved with the already mentioned implosion velocities, and to obtain the ignition propagation before the shock wave produced by the center collapse is reflected by the fuel-pusher surface.

LASER TARGETS

The laser target used in the simulations has been 4 mg of CH as the absorber layer and 1 mg of cryogenic DT. The absorber mass has been fixed to prevent the external fuel preheating produced by electron conduction. The initial fuel aspect ratio has been decided to have 50 or 100. The deposited driver energy in the target has been selected to 2 and 3 MJ.

Two pulse shapes have been considered in the simulations. The first one is a single Gaussian pulse with a maximum intensity of 200 TW/cm^2 and 5 ns of FWHM. The second one has a Gaussian ramping as prepulse followed by a constant intensity pulse, which maximum intensity depends of the total pulse duration, the prepulse duration and the deposited driver energy.

The simple Gaussian pulse leads to a lower compression of the DT ($\rho R \sim 1.4$ g/cm^2), because the time between the void closure and the maximum compression is too large, avoiding an efficient compression of the fuel.

The second pulse shaping, conceptually similar to that found by others authors[10], has been selected for doing a parametric analysis. In this, for each deposited driver energy, the time duration of the total pulse and the prepulse have been changed, obtaining differents peak intensitites (TW/cm^2).

So, the analysis has been focused to know how the implosion conditions (ρR and T), the hydrodynamic efficiency, the implosion velocities and the hydrodynamic stability are changed, due to the driver energy, pulse shaping and initial fuel aspect ratio.

When the initial fuel aspect ratio is fixed at 100, it has been observed from the simulation that good implosion conditions ($\rho R = 4.3$ g/cm^2 and T=4 keV), can be obtained, using 2 MJ deposited in 10 ns, with 3.7 ns of prepulse. However, when the same case is calculated depositing the same energy in 13 ns, with 4.8 ns of prepulse, the implosion conditions are reduced until 3.5 g/cm^2 for ρR and 2.2 keV for T. This reduction is because the maximum intensity, while in the first case is

Table 2. Implosion conditions at several pulse shapes
(Initial Fuel Aspect Ratio=50)

E_{driver} (MJ)	Pulse shaping	ρR(g. cm^{-2})	T (keV)	n_H%
2	194 ⋯⋯ 4.8 13	2.3	1.3	7
2	207 ⋯⋯ 6 13	2,4	1.4	7
2	300 ⋯⋯ 6 10	2.3	2.2	7.5
2	328 ⋯⋯ 7 10	2.2	2.3	8.5
3	311 ⋯⋯ 6 13	3	2.7	7
3	300 ⋯⋯ 3.7 10	2.7	5.1	7.5
3	328 ⋯⋯ 6 10	3.2	4.9	8

300 TW, in the second case is reduced for 194 TW. The hydrodynamic efficiency is 11.7% for the first case and 9% for the second one, maintaining high implosion velocities after the void closure ($\sim 4 \times 10^7$ cm $\cdot$ s^{-1}).

In targets with high fuel aspect ratios hydrodynamic instabilities,can appear because the time-dependent in-flight aspect ratio (IFAR)[9] takes values higher than 300 during large periods of time. So, an analysis of 2-D should be done to know the grade of stability of these targets.

Using an initial fuel aspect ratio of 50, several important changes have been observed, comparing with the above outlined case. First, depositing 2 MJ in the target, the implosion conditions are poor, but using shorter pulses better conditions are found, because the maximum intensity grows. The prepulse duration is very important, because for larger prepulse durations best implosion conditions and lower IFAR values.are observes However, cases with 2 MJ give $\rho R \sim 2.3$ g/cm^2 and T$\sim$2.4 keV which are very low to get high-gains.

With the same fuel aspect ratio (50) but depositing 3 MJ better implosion conditions have been obtained. In this way, a case with a maximum intensity of 378 TW, during 10 ns with 3.7 ns prepulse, gives $\rho R=2.7$.g/cm^{-2} and T=5.1 keV. While, doing a large prepulse until 6 ns, the ρR of the fuel grows until 3.2.g/cm^{-2}, maintaining the same temeperature. But, if the same energy is deposited in 13 ns, the implosion conditions are reduced until $\rho R=3$ g/cm^2 and T=2.7 keV, because the maximum intensity is now 310 TW. (Tables 2 and 3).

The hydrodynamic efficiency for these targets, with lower fuel aspect ratio is around 7%, and the hydrodynamic stability is in the limit of 200.

Table 3. Implosion conditions at several pulse shapes
(Initial Fuel Aspect Ratio=100)

E$_{driver}$ (MJ)	Pulse shaping	ρR(g. cm^{-2})	T (keV)	n$_H$
2	300 ...... (3.7 10)	4.3	4.1	11.7
2	194 ...... (4.8 13)	3.52	2.25	9%

LIGHT ION FUSION TARGETS

The target considered has a lithium layer and 1 mg of cryogenic DT fuel layer. The mass of the lithium layer has been considered as a parameter to optimize the fuel compression. Because the implosion velocity is determined by this mass, the ignition performance (ρR and T) depends to a large extent on it. We have found that the optimum value is 32 mg, that corresponds to the range of light ions at the beginning of the illumination. The range shortening effect reduces its penetration when the lithium layer becomes ionized, forming a payload of 3-4 mg that acts as a fuel pusher. The initial aspect ratio of the target is 5 and the aspect ratio of the fuel layer 50.

The 4 MeV proton beam pulse has a Gaussian shape with a FWHM of 12 ns and a maximum intensity of 110 TW/cm^2, relative to the initial outer surface area of the target (the maximum intensity that impinges on the target is, roughly, 50 TW/cm^2). To have realistic energy deposition profiles, we have considered an energy spread of the beam of 10% (the proton energy is distributed from 4.4 to 3.6 MeV according to a Gaussian profile centered in 4 MeV) and a Gaussian profile in angle, which spread is obtained from the ratio of the beam radius (initially 0.324 cm). Thus, when the target expands, the angular spread of the beam relative to the external surface is reduced. The total energy deposited in the target is 1.12 MJ.

The evolution of the LIF target is similar to that before outlined. It is worth pointing out that the IFAR increases quickly to high values (>200) when a moderate shock wave compress the fuel at the beginning of the pulse (~3.5 ns). However, this high value of the IFAR lasts only a short time, because the maximum density switchs very fast to the lithium layer, which density increases as a consequence of the range shortening (that is small because of the low-Z absorber considered). The IFAR is then reduced to 60-70. If a LIF target with higher initial aspect ratio (~10) is considered, the IFAR goes over 200 to almost the end of the pulse.

The hydrodynamic efficiency achieved is, roughly, 25% and the implosion kinetic energy of the fuel is almost 1/3 of the total (fuel+pusher). Thus,the fuel is compressed by the pusher, but because the densities are much lower than those found with high-Z pusher, the pusher-fuel interface is not so sensitive to the Rayleigh-Taylor instability.

The maximum values of ρR and temperature obtained in the pure hydro calculation were 4.5 g/cm^2 and 2.5 keV, respectively, with a convergence ratio of almost 35. These conditions are enough to burn a ~40% of the fuel to have a gain of 120, as given by the complete calculation including detailed transport of the fusion alphas and neutrons.

HEAVY ION FUSION TARGET

The HIF target consists of a lithium layer of different thickness according to the energy of the heavy ions and a cryogenic DT layer with 1 mg and aspect ratio of 100. The mass of the lithium layer has been taken in such a manner that initially the ions penetrate until the lithium-pusher interface. Thus, for 10 GeV Bi$^+$ ions the lithium mass is 376 mg and the total aspect ratio of the capsule is 1.3.

Targets were illuminated by centrally focused Bi^+ ions of energies between 1 and 10 GeV (fixed energy in each case). The time profile was very simple: a box without prepulse, without voltage ramping and without power variation within the pulse. Specifically, for the 10 GeV target the pulse has an intensity of 45 TW/cm^2 (200 TW) and 10 ns length, that corresponds to a total energy deposited in the target of 2 MJ.

The hydrodynamic evolution is characterized by low temperatures, almost constant ablation pressure and uniform acceleration of the fuel. The range shortening of the beams plays an important role, generating a massive payload, that reduces the implosion velocity to 2.5 10^7 cm/s in this case. The range shortening is due to the effective charge of heavy ions, that increases with temperature, rather than the ionization of the low-Z absorber. The hydrodynamic efficiency is almost 22%, but a large fraction of the implosion kinetic energy (0.92) is carried out by the payload. Nevertheless, the fuel efficiency, defined as the fuel internal energy at maximum ρR over the total beam energy, is about or higher than 5%. The IFAR observed in the calculations were much less than 100.

The ignition conditions reached in the fuel are similar to those formed for the LIF targets, high enough to burn on 40% of the fuel (gain ~79). The maximum values of ρR and T obtained in this particular case has been 4.5 g/cm^2 and 2.5 keV with a convergence ratio of 47. However, the fuel compression time (from the void closure to the maximum compression) is significantly greater than that corresponding to the LIF case. Hence, the fuel compression is more sensitive to hydrodynamic instabilities, unless the implosion velocity and pulse energy be increased.

From the analysis of a set of HIF targets with different ion beam energy and different pulse intensities, it has been found that the efficient compression of the fuel depends mainly on the speed profile within the fuel. If the innermost fuel meshpoints have been accelerated much more than the outer part, the shock reflected from the center crashes against the outer fuel and the fuel becomes isothermalized. So to speak, a fraction of the kinetic energy of the inner fuel is used to stop the kinetic energy of the outer fuel, so reducing dramatically the compression degree. On the contrary, if the speed profile is uniform, the shock reflexion does not slow down significantly the outer fuel. In this case, a hot spark (not an ideal, but a physical one) is formed and the compression achieved is very high.

The main conclusion of our HIF analysis is that target performance of directly driven ICF targets can achieve ignition conditions with moderate driver energies of 2 MJ, although optimum performances (higher fuel efficiencies and higher gains) could be obtained with energies between 3 and 4 MJ. Figure 2 shows the results obtained at the end of the compression phase. The fuel efficiency η is measured as the fuel maximum internal energy over the total driver input. It is noted that this simulations were done assuming perfect illumination, recognizing the larger effect of non-uniformities in this HIF cases, which could give more restrictive requirements.

Cases of figure 2 have different power and pulse lengths for the same energy, and are only valid for Bi^+ ions between 6 and 10 GeV. For lower ion energies, the range shorthening during the beam deposition is so strong that the fuel hydrodynamic efficiency is much lower and the target performance is poorer.

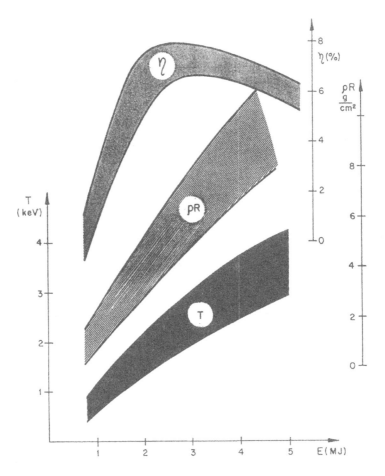

Figure 2. Temperature, ρR and fuel efficiency (η) of HIF targets (η=fuel internal energy at maximum ρR over total beam energy).

Table 4. Summary of results

	Laser		LIF	HIF
Initial fuel aspect ratio	50	100	50	100
Intensity (TW/cm^2)	400	204	110	45
Energy (MJ)	3	2	1.1	2
Hydro-efficiency	.075	.11	.25	.216
Implosion velocity (cm/s)	3*10^7	4.5 10^7	3.6 10^7	2.5 10^7
IFAR	> 200	> 200	< 100	< 100
$(\rho R)_{máx}$	2.7	4.3	4.5	4.5
$T_{máx}$	5.1	4.1	3	2.5
Convergence ratio	28	35	35	47
Gain	—	—	120 - 130	60 - 70

CONCLUSIONS

The major conclusion of this work is the similar energies needed to have ignition conditions in so different targets under dramatically different illumination conditions (Table 4). In any case, high gain could be obtained with energies of a few MJ, that is low enough to make atractive the direct drive targets. Previous works about the indirect drive target have shown that at least 4-5 MJ are needed.

In future work we will further study the high-gain targets for ICF reactor as well as low-gain targets for breakeven experiments.

REFERENCES

1. R. Lemburg, A. Schmitt and S. Bodner, J. Appl. Phys. 62, 2680 (1987).
2. R.L. McCrory et al., Proc. of "12th International Conference on Plasma Physics and Controlled Nuclear Fusion Research", paper IAEA-CN-50/B-1-2, Nice, France (October, 1988).
3. Y. Kato et al., Phys. Rev. Lett. 53, 1057 (1984).
4. J. D. Lindl, Fusion Technology Vol. 15, 227 (1989).
5. G. Velarde et al. Laser and Particle Beams, 4, 239 (1986).
6. G. Velarde et al., DENIM Annual Research Report 1987,
7. G. Velarde et al. Nucl. Instr. and Methods in Phys. Re., A278, 105 (1989).
8. G. Velarde et al., "High-gain direct-drive capsule design for ICF" Proceedings of the ICENES-89, July 1989. To be published by World Scientific Publishing.
9. J. H. Gardner and S. E. Bodner, Phys. of Fluids 29, 8 (1986).
10. K. Mima, H. Takabe and S. Nakai, Laser and Particle Beams 7, Part 2 (1989).

ADIABATIC COMPRESSION OF FUEL IN ICF TARGET

Keishiro Niu and Takayuki Aoki

Department of Energy Sciences, Graduate School at Nagatsuta
Tokyo Institute of Technology
Midori-ku, Yokohama 227, Japan

INTRODUCTION

Although a proposal of light ion beam (LIB) fusion power plant has been given (Niu, Aoki and Takeda 1988), again proposal of power plant of LIB by inertial confinement fusion (ICF) is given here as a reasonable one from technical and economical points of view, including improvement for propagation of beam and target structure in which a spherical fuel compression will be performed adiabatically in a spherically symmetric way.

PULSED POWER AND REACTOR

The power supply system and fusion reactor are similar to those proposed before. Here proton beam is proposed to be used extracted from the diode with a particle energy of 8MeV. In order to extract the proton beam from a diode with a particle energy of 8MeV (4MeV for propagation, 3MeV for rotation and 1MeV for thermal oscillation), the output voltage of Marx generator is 16MV. This value seems the upper limit for ordinary Marx generator. For LIB impinging a spherical target, spherically symmetric irradiation of LIB on a target may be difficult to be realised. Therefore, the indirect driven target is proposed to be used (Niu 1989). In this case, the beam energy required to release the fusion energy of 3GJ from a target is 12MJ. Here, 12 power supply systems are prepared. Each two of them combined together extracts one proton beam of the beam energy of 2MJ. The power supply system is as follows:

Table 1. Power supply system to extract proton beam.

Marx Generator;	12 modules
charging voltage;	200kV
capacitance of a bank;	$2.5\mu F$
number of capacitor banks;	80

Table 1.(cont.)

stored energy;	4MJ
output voltage;	16MV
Cylindrical Intermediate Storage Capacitor;	12 modules
insulator;	water
inner (anode) radius;	3m
outer (cathode) radius;	5m
length;	3.5m
charging time;	155ns
Pulse Forming Line;	12 modules
input voltage;	16MV
output voltage;	8MV
length;	0.67m
pulse width;	$\tau_b = 30$ns

Once the power in a Marx generator is shifted to an intermediate storage capacitor, which is a coaxial cylinder filled with water as the insulator. The inner (anode) radius is 3m, the outer (cathode) radius is 5m, and the length is 3.5m. The charging time is estimated to be 155ns.

The pulse forming line has the length of 0.67m for the pulse width of 30ns. The input voltage is 16MV and the output voltage is 8MV.

To construct the total power supply systems of 12 modules, the cost is estimated to be US$120M, which is less than 10% of the construction cost of light water reactor of 1GW. Economically and technically, LIB fusion system seems reasonable.

In order to neutralised the charge of the proton beam during propagation in the reactor cavity, the electron beam is launched from a triode to the target at the same time with the launching of proton beam. To give the same velocity to the electron beam with the that of proton beam which has the propagation energy of 4MeV, the cathode voltage of triode to accelerate the electron is -6.20kV. This voltage is supplied by a kind of Marx generator, whose stored energy is 3.93×10^2J. The total energy stored in the Marx generators of 6 modules to extract the six electron beams is only 2.38kJ. To the grid of triode, the voltage of 168kV is supplied through a capacitor bank of 354pF where 5kJ is initially stored. Thus the total stored energy for six capacitor banks of grid circuits is only 30kJ. Required energy to extract six electron beams is negligible in comparison with that for proton beams.

The reactor to be used is just the same with that proposed by Niu and Kawata (1987). The flibe is proposed to be used as the coolant of the rotating reactor. The flibe is chemically stable and the neutron with the energy of 14.1MeV is thermalised in the flibe layer with the thickness of 30cm (Naramoto and Niu, 1989). No damage is expected to be accepted on the structural reactor wall by the spattering, blistering and swelling.

DIODE FOR PROTON BEAM AND TRIODE FOR ELECTRON BEAM

The diode for proton beam and triode for electron beam is schematically shown in Fig. 1. Two powers from the two pulse forming lines are supplied to a diode to extract a proton beam of 2MJ. The parameters of diode is tabulated as follows:

Table 2. Diode for proton beam

anode inner radius;	r_i=5.0cm
anode outer radius;	r_0=23.6cm
(anode area;	S_A=1.67×10³cm²)
electrode gap distance;	d=9.51mm
magnetic field for insulation of electron current	
in the radial direction (average);	B_r=0.921T
in the azimuthal direction (average);	B_θ=3.90T
(total intensity;	B_{tot}=4.55T)
proton current intensity on the anode surface;	j_b=5kA/cm²
(total proton current;	I_b=$j_b S_A$=8.33MA,
	$I_{b\|\|}$ =4.17MA)

By the action of the Lorentz force due to B_r, the proton particle is accelerated in the azimuthal direction. Thus the proton beam has the particle energy as follows:

the total particle energy;	e_p=8MeV,
the average propagation energy;	e_z=$mv_z^2/2$=4MeV
the average rotation energy;	e_θ=$mv_\theta^2/2$=3MeV
the average thermal energy;	e_r=$mv_r^2/2$=1MeV

The proton beam is confined in a small radius of 5mm by the azimuthal magnetic field B_θ which is induced by the beam current $I_{b\|\|}$ =4.17MA. This beam propagation is stabilised by the large thermal motion e_r and by the axial magnetic field B_z which is induced by the beam rotation. All proton particles rotate around the axis extracted from a diode shown in Fig.1, the proton beam forms a hollow cylinder with the inner radius of 2mm and the outer radius of 5mm.

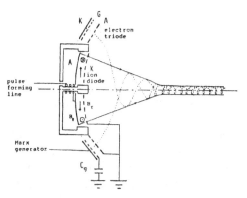

Fig. 1. Diode for proton beam and triode for electrode beam.

At the leading and trailing edges of a strong proton beam, there remains the particle charges unneutralised and they induce a strong electrostatic field (Kaneda and Niu 1989) and cause to diverge the beam propagation. To delete these unneutralised ion charge, the proposal is given for launching the electron beam with the proton beam as follows:

Table 3. Triode for electron beam.

cathode inner radius;	r_{in}=25cm
cathode outer radius;	r_{out}=32cm
(cathode area;	S_K=1.06×10³cm²)
cathode voltage;	Φ_K=-6.20kV
distance between cathode and anode;	d_{KA}=1cm
grid voltage;	Φ_G=168kV→84kV
grid capacitor bank;	C_G=354pF
grid current;	I_G=800kA
distance between cathode and grid;	d_{KG}=2.0mm
electron current density on the cathod surface;	j_e=8kA/cm²
(total electron current;	I_{be}=$j_e S_K$=4.17MA)

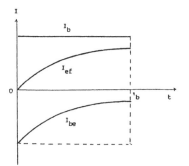

Fig. 2. Proton current density I_b, electron current density I_{be}, and effective current density I_{ef} versus time t.

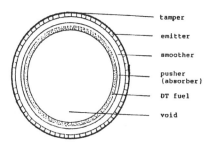

Fig. 3. Schematic view of indirect driven target.

The velocity of electron accelerated by the voltage of Φ_K=-6.20kV is v_e=2.12×10⁷m/s which is equal to the proton velocity v_z with the propagation energy of e_z=4MeV. Since the partial electron current of 800kA flows in the grid circuit, the grid voltage decreases from the initial value of 168kV to the final value of 84kV during 30ns because the grid capacitor bank is 354pF in which the initial stored energy is 5kJ. Thus the electron current I_{be} decreases from the initial value of 4.17MA to 1.48MA with time, although electron velocity remains constant due to the constant cathode voltage. Figure 2 shows that the proton current $I_{b\parallel}$, the electron current I_e and the effective current $I_{ef}=I_{b\parallel}-I_{be}$ versus the time t. Thus the electron beam neutralises the proton charge at the leading edge of the proton beam and the effective current I_{ef} is expected to confine the beam in a small radius.

Table 4. Indirect driven target for beam energy E_b=12MJ and pulse width τ=30ns.

target radius;	r_t=8.716mm
tamper (lead)	
density;	ρ_{Pb}=11.3g/cm^3
thickness;	δ_{Pb}=23.4μm
mass;	M_{Pb}=120mg
rate of beam energy deposition;	C_{Pb}=20%
temperature;	T_{Pb}=8K$\rightarrow$500keV
leaking out radiation energy loss;	E_{rl}=1.5MJ
radiator (lead)	
density;	ρ_{ra}=2.13g/cm^3
thickness;	δ_{ra}=690μm
mass;	M_{ra}=1.49g
rate of beam energy deposition;	C_{ra}=80%
temperature;	T_{ra}=1.76keV (after 2.4ns)
expansion velocity;	V_{ra}=1.61$\times$10^{5}m/s
inward radiation intensity;	I_{ra}=4.06$\times$10^{13}W/cm^2
radiation gap (smoother, vacuum)	
thickness;	δ_{sm}=2mm
pusher (absorber, aluminium)	
density;	ρ_{Al}=2.7g/cm^2
thickness;	δ_{Al}=217μm
mass;	M_{Al}=264mg
temperature;	T_{Al}=8K$\rightarrow$200eV
propagation velocity of hot surface;	V_h=5.42$\times$10^{3}m/s
(transparent time;	δ_{Al}/V_h=40ns)
DT fuel	
density;	ρ_{DT}=0.19g/cm^3
thickness;	δ_{DT}=286μm
mass;	M_{DT}=23mg
inner void	
radius;	r_v=5.5mm
saturated vapour pressure;	p_v=7$\times$10^{7}Pa

INDIRECT DRIVEN TARGET

The indirect driven target which is proposed to be used is schematically shown in Fig. 3 (Niu 1989). The parameters are summarised in Table 4.

ADIABATIC COMPRESSION OF FUEL

When the target is irradiated by proton beams of 12MJ, 80% of the beam energy is deposited in the radiator layer, and the temperature of radiator layer increases to T_{ra}=1.76keV 2.4ns after the start of beam irradiation. At this temperature, 60% of the deposited beam energy is converted to the radiation energy.

About 12% of the radiation energy leaks out from the target through the tamper layer. Thus the average inward radiation (soft x-ray) intensity is $I_{ra}=4.06\times10^5$m/s. The radiation gap of thickness of 2mm smoothes out the nonuniformity of the radiation intensity on the outer surface of absorber. This radiation gap closes 20ns after the beam irradiation starts, by the expansion of radiator and absorber layers. The outer surface of aluminium absorber layer absorbs the x-ray from the radiator layer, and the temperature increases to $T_{Al}=200$eV (the pressure reaches $p_{Al}=7.15\times10^{12}$Pa). Since the temperature of inner absorber layer is so small as 8K that the radiative mean free path is very short. The outer layer of radiator absorbs x-ray and becomes hot to be transparent for the x-ray. Thus the hot region advances in the absorber layer from the outer layer to the inner layer with the velocity

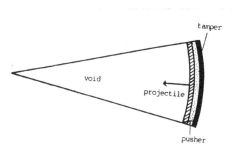

Fig. 4. Projectile, accelerated by pusher, flies with supersonic velocity in a converging nozzle.

$V_h=5.42\times10^3$m/s. This velocity is lower than the sound velocity $(s_{Al}=5\times10^4$m/s) in the solid aluminium. Thus the pressure p_{Al} at the boundary between the aluminium absorber and DT fuel increases gradually to $p_{Al}=7.15\times10^{12}$Pa during 7.5ns and it saturates at that value because of the expansion of absorber layer. In the case that this pressure increase is not too steep, the fuel acceleration is done in an adiabatic way (Morreeuw and Saillard 1978, Kawata, Abe and Niu, 1981). The acceleration α of the fuel reaches $\alpha=p_{Al}S_{DT}/M_{DT}=1.15\times10^{14}$m/s^2, where S_{DT} means the outer surface of DT layer. Thus the fuel implosion velocity u arrives at $u=\alpha\tau=3\times10^5$m/s for $\tau=2.6$ns. As Fig. 4 shows a spherical hollow target forms a supersonic converging nozzle for the DT fuel (Niu and Aoki 1989). To compress the fuel to 2000 times the solid density , the adiabatic compression is very important. In a supersonic converging nozzle, the fuel is compressed adiabatically, and by the action of small pusher pressure $p_{Al}=7.15\times10^{12}$Pa the fuel obtains the final high pressure of $P_{DT}=10^{17}$Pa. According to another adiabatic compression method proposed by Morreeuw and Saillard (1978) or by Kidder (1974), always pusher pressure should be larger than the fuel pressure, and the DT fuel cannot be compressed 2000 times the solid density. In the supersonic converging nozzle, the fuel density ρ_s at the sonic surface r_s is given by

$$\frac{\rho_s}{\rho_0}=\left\{\frac{2}{\gamma+1}\left[1+\frac{(\gamma-1)M_0^2}{2}\right]\right\}^{1/(\gamma-1)},\tag{1}$$

where ρ_0 is the initial (solid) density, and M_0 is the initial Mach number. The sonic surface r_s is given by

$$\frac{r_s}{r_0}=M_0^{1/2}\left\{\frac{2}{\gamma+1}\left[1+\frac{(\gamma-1)M_0^2}{2}\right]\right\}^{-(\gamma+1)/4(\gamma-1)}.\tag{2}$$

The initial fuel position is $r_0=5\times10^{-3}$m, and the sound velocity s_0 of the fuel is $s_0=2.34\times10^5$m/s. Therefore the initial Mach number of fuel is $M_0=12.8$. Equations (1) and (2) leads to $\rho_s=2.69\times10^2\rho_0=5.12\times10^4$kg/m^3 and $r_s=2.63\times10^{-6}$m. In the nozzle inner than r_s, the flow is subsonic and flow-choking occurs. The strong shock wave propagates outward in the fuel. Thus again the fuel is compressed to 2000 times the solid density and especially heated to 4keV. The most important process for ICF is the adiabatic compression of fuel in the supersonic converging nozzle.

REFERENCES

Kaneda T. and Niu K., 1989, Laser and Particle Beams, 7, 207.

Kawata S, Abe T and Niu K, 1981, J. Phys. Soc. Jpn., 50, 3497.

Kidder R. E., 1974, Nucl. Fusion, 14, 53.

Morreeuw J. P. and Saillard Y., 1978, Nucl. Fusion, 18, 1263.

Naramoto H. and Niu K., 1989, Inst. Plasma Phys. Rept., IPPJ-900, 12.

Niu K., 1989, Laser and Particle Beams, 7, 505.

Niu K. and Aoki T., 1988, Fluid Dyn. Res., 4, 195.

Niu K., Aoki T. and Takeda H, Laser Interaction and Related Plasma Phys., 8, 605.

Niu K. and Kawata S., 1987, Fusion Tech., 11, 365.

X–RAY CONVERSION IN HIGH GAIN RADIATION DRIVE ICF

X.T. He and T.Q. Chang

Center for Nonlinear Studies
Institute of Applied Physics
and Computational Mathematics
P.O.Box 8009, Beijing

M. Yu

China National Nuclear Corporation, Beijing

ABSTRACT

In this paper, the conversion mechanism of soft X–ray emission, efficiency and the related physics for high gain ICF are analysed; the radiation temperature in hohlraum is estimated. Finally, the numerical results of laser cavity–target coupling for typical models are given and discussed

I. Introduction

Recent years, indirect drive studies for high gain ICF have achieved an important progress[1],[2]. Indirect drive, i.e., radiation drive, relies on first converting the laser light into X rays that are contained inside a hohlraum which is designed to provide a temporal and spatially uniform radiation surrounding capsule for implosion.

Physical research for such a laser–target coupling system may be mainly divided into fields: hohlraum physics and implosion dynamics.

Hohlraum physics is to study processes of laser light absorbed, converted into soft X rays, and X ray transport etc. How further to raise their efficiencies for economization of laser energy E_L is an essential goal of hohlraum physics, though it has been shown[1] that the absorption efficiency η_a may be great than 90% and the conversion efficiency η_x great than 70% . In this paper, we shall lay emphasis on some physical characteristics in hohlraum. Perhaps we may divide such a laser–plasma interaction system into four ranges.

(i)The subcritical range where the plasma density $\rho < \rho_c$ (critical density) and high temperature T_e . The laser for short wavelength, such as $\lambda_0 = 0.35\mu m$, is essentially absorbed through inverse bremsstrahlung, and is greatly converted into X rays nearby critical surface. Hot electrons are produced due to anomalous absorption. Plasma in high temperature is rapidly expanding.

(ii)The supercritical range where $\rho \geqslant \rho_c$. This is an electron heat conduction region where conversion may be able to be neglected for the laser light at short wave–

length, but important for longer wavelength laser, such as $\lambda_0 = 1.06\mu m$.

(iii)The radiative ablation range where radiation conduction plays a decisive role.

(iv)The shock heat range happened in front of the radiation wave where temperature is low, which will not be discussed here.

This paper is emphasized on the X ray conversion, and is organized as follows:section Ⅱ for hohlraum environment; Ⅲ for X ray conversion; Ⅳ for numerical results and discussions.

The units adopted in this paper, unless otherwise stated, are as follows:time t:ns; laser wavelength λ_0:μm; laser intensity I_a:10^{14}w/cm^2;temperature T:10^6 ^{0}K; electron(ion) number density $N_e(N_i)$:1/cm^3; fluid density ρ;g/cm^3; energy:joules.

Below the following parameters for high gain ICF are adopted: Laser energy $E_L = 10^7$, pluse width~10, intensity $I_a \sim 1-10$, $\lambda_0 = 0.35$.

Ⅱ. Hohlraum Environment

In this section, we shall briefly discuss the hohlraum environment with laser light absorbed, including the parameters of the critical surface, the radiative temperature and plasma states inside hohlraum. For convenience, it is assumed that the hohlraum case is made up of gold, and the capsule ablator inside hohlraum is the low Z matter, such as Al.

1. Critical surface parameters

The critical number density for electrons is expressed from $ck_0 = \omega_e$

$$N_e = \frac{1.11 \times 10^{21}}{\lambda_0^2} \tag{2.1}$$

where $k_0 = \frac{2\pi}{\lambda_0}$;$\omega_e = (\frac{4\pi n_e e^2}{m_e})^{\frac{1}{2}}$ electron plasma frequency; e,m_e charge and mass for electrons respectively. Thus, the critical fluid density is in the form

$$\rho_e = \frac{0.364}{\lambda_0^2 z'} \tag{2.2}$$

where z' is the effective charge of the ions in the plasma, and

$$z' = 31.32 T_e^{0.1489} \tag{2.3}$$

for Au target and $\lambda_0 = 0.35$.

The critical temperature for electrons is given by ref.[3]

$$T_{ec} = 26.7 I_a^{0.45} \lambda_0^{0.85} \tag{2.4}$$

For $\lambda_0 = 0.35, I_a = 1 \sim 10$, we have

$$T_{ec} = 10.9 \sim 30.8; \quad \rho_e = 0.066 \sim 0.057; \quad z' = 44.7 \sim 52. \tag{2.5}$$

2. Radiative ablation region

Although to produce a temporal and spatially uniform temperature T_R by radiation conversion in hohlraum is very complex and it is necessary to have a precisely numerical computation, we can estimate T_R in a simple and believable way.

Because the laser energy for $\lambda_0 = 0.35$ and high gain is converted into X rays in high efficiency, we can discuss the equation for radiation heat conduction by neglecting the hydrodynamic effect, and give a effective result. We write the equation for one dimension as follows

$$\rho \frac{\partial}{\partial t}(C_v T) = \frac{\partial}{\partial x}(K \frac{\partial}{\partial x} T) \tag{2.6}$$

where the radiative conduction coefficient

$$K = K_0 T^{n+3} \rho^m, \qquad\qquad K_0 = \text{constant}. \tag{2.7}$$

The specific heat

$$C_v \propto T^l \tag{2.8}$$

Assume that the boundary temperature T_0 is a constant, Eq.(2.6) has a self-simlar solution

$$T(x,t) = T_0 (f(z))^{\frac{1}{l+1}} \tag{2.9}$$

where

$$z = \frac{X}{(\frac{K}{(N+1)\rho C_v}(\frac{T_0}{T})^{N-l} t)^{\frac{1}{2}}} \tag{2.10}$$

$$N = n + 3$$

From Eq.(2.6), we have approximately

$$f(z) = (1 - \frac{z}{z_0})^{\frac{1+l}{N-1}} \tag{2.11}$$

where

$$z_0 = (\frac{2(N+1)}{N-l})^{\frac{1}{2}} \tag{2.12}$$

then, we obtain the ablation depth as follows

$$X_R = \left(\frac{2K}{(N+1)\rho C_v}(\frac{T_0}{T})^{N-l} t\right)^{\frac{1}{2}} \tag{2.13}$$

and the ablation mass ρX_R.

Thus, the radiation energy entering ablator by the heat conduction is

$$E_R = \int \rho C_v T dx = \frac{N-1}{N+1}\left(\frac{C_v}{T^l}\right) T_0^{l+1} \rho X_R \tag{2.14}$$

The expression for ρX_R, to a great extent, is independent of ρ due to $X_R \propto \rho^{-\frac{(1-m)}{2}}$, where $m \approx -1$. Thus, the self–similar variable z and the radiative temperature T(x,t), which are the function of ρX, are independent of ρ as well.

Assume that the ratio between the Au–Al surface area absorbed energy equals to 10, by using laser parameters $E_L = 10^7$, pulse $t = 10$, and $\eta_a = 0.9$, $\eta_x = 0.8$, we have the upper limit temperature from Eq.(2.14)

$$T_{max} \approx 3.5 \tag{2.15}$$

On the other hand, if the main fuel is compressed under ideal condition, and $\rho = 200$, then, taking the hydrodynamic efficiency $\eta_h = 10\%$, we have the lower limit temperature

$$T_{min} = 2.5 \tag{2.16}$$

When $T = 3$, the ablation mass is $\rho X_R \approx 0.015 g / cm^2$ for Au.

3. States between both sides of the critical surface

The density and the temperature for electron and equilibrium radiation in both sides of the critical surface are schematically plotted in Fig.1, where the position of the critical surface is, indeed, unceasingly moving forward during the laser light irradiation.

In the subcritical region, T_e nearby critical surface rapidly decreases due to X ray emission and the heat conduction to the supercritical region, and $T_e \sim$ constant varying with X at $\rho \ll \rho_e$ region. Whereas ρ rapidly decreases due to plasma expansion toward approximate vacuum.

Nearby the critical surface, there are

$$T_e \approx T_c \sim 10 T_c$$

$$\rho \approx 0.01 \rho_e \sim \rho_e \tag{2.17}$$

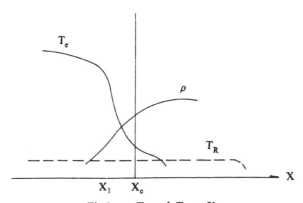

Fig.1 ρ, T_e and T_R vs X

In $\rho \ll \rho_e$ region, the fluid velocity

$$u = C_T \ln\frac{\rho_1}{\rho}.$$ (2.18)

where C_T for the isothermic acoustic speed, $\sim 280 T_e^{2/3}\left(\dfrac{\mu m}{ns}\right)$ $\rho_1 = \rho(x_1)$,

$$\rho = \rho_1 exp\left\{\frac{(x-x_c)}{C_T t}\right\}, \quad \text{for } t = 0, X = X_c.$$ (2.19)

In the supercritical region, T_e raised is caused by the heat conduction for electrons, where $T_e < T_c$ due to X ray emitted rapidlly. The electron conduction depth from the critical surface is very short, where energy by the electron conduction across the critical surface is essntially balanced by X ray emission. However, the density in this region decreses, i.e., $\rho_e \leqslant \rho \leqslant \rho_0$ (19.3 g/cm^3 for Au) essentially by the radiation ablation.

III. X Ray Conversion

Laser light for short wavelength is essentially absorbed by electrona through the inverse bremsstrahlung process in the subcritical region, where the X ray are converted, especially, near critical surface, and the electrons are heated to high temperature T_e accompanied plasma rapid expansion. In this system there exist a lot of free electrons, however, many electrons are still bounded on ions, and the bremsstrahlung emission is unimportant for conversion.

The contribution for X ray conversion is mainly from bound electrons on ions. The interactions of free and bound electrons are of an essential role in the process of X ray emission. When the bound electrons are in equilibrium distribution with temperature T_e same as free electron's, i.e., occupation probability satisfies the Fermi–Dirac distribution, both positive and inverse processes of free and bound electron interactions are not cancelled out each other.

However, in our case, the distribution of the bound electrons is non–equilibrium due to the non–equilibrium distribution of X rays which are emitted in an optically thin region. They satisfy a rate equation of occupation probability P_n for the nth energy levels, thus, the positive and inverse processes of free and bound electron interactions are not cancelled out. It determines the characteristic of X ray conversion.

The non–equilibrium photons, i.e., X rays, greatly escape from corona region except for some of them to be trapped, and transport into the supercritical region and the radiation ablation region, then they are strongly absorbed in optically thich regions where the spontaneous emission produces the equilibrium X rays.

In order to discuss the conversion processes in detail, we write the rate equation for P_n as follows:

$$\frac{dP_n}{dt} = [N_e \beta_{\gamma n}(1-P_n) - \alpha_{\gamma n}P_n] + [N_e^2 \beta_{en}(1-P_n) - N_e \alpha_{en}P_n]$$
$$+ (1-P_n)N_e\left[\sum_{h>n}\frac{C_h}{C_n}\beta_{hn}P_n + \sum_{l<n}\frac{C_l}{C_n}\alpha_{ln}P_l\right]$$
$$- P_n N_e\left[\sum_{h>n}(1-P_n)\alpha_{nh} + \sum_{l<n}(1-P_l)\beta_{nl}\right]$$
$$+ (1-P_n)\left[\sum_{h>n}\frac{C_h}{C_n}A_{hn}P_n + \sum_{l<n}A'_{nl}P_l\right]$$
$$- P_n\left[\sum_{h>n}\frac{C_h}{C_n}A'_{hn}(1-P_n) + \sum_{l<n}A_{nl}(1-P_l)\right]$$ (3.1)

The meaning of various terms in the right hand side of Eq.(3.1) is as follows:

The first square bracket indicates the photoionization recombination and its inversion processes. The second for the three—body recombination and its inversion—collisional ionization; the thirt and fourth square brackets for the collisional deexcitation and emission; and the last terms for the spectral line absorption(trapping) and emission. The other processes, such as Auger effect and dielectronic recombination etc., have been neglected in Eq.(3.1).

We now give the expressions for the rate(1 / ns unit) of above mentioned processes under a hydrogen—like approximation in partly ionized Au plasma.

A. Electron—electron interaction

1. Collisional ionization rate for nth level

$$\alpha_n^{(e)} = N_e \alpha_{en} \approx 0.167 \times 10^5 \frac{\rho Z'}{T_e^{\frac{3}{2}}} G\left(\frac{I_n}{T_e}\right), \tag{3.2}$$

where I_n is ionization energy.

$$G(x) = \frac{e^{-x}}{x} - E_1(x)$$

$$E_1 = \int_x^\infty \frac{e^{-x}}{x} dx \tag{3.3}$$

2. Three—body recombination rate

$$\beta_n^{(e)} = N_e^2 \beta_{en} = 0.633 \times 10^{-3} \frac{\rho Z'}{T_e^{\frac{3}{2}}} e^{\frac{I_n}{T_e}} \alpha_n^{(e)} \tag{3.4}$$

3. Collisional excitation rate for n→h, n <h

$$\alpha_{nh}^{(e)} = N_e \alpha_{nh} \approx 0.106 \times 10^6 \rho Z' \frac{I_{nh}}{T_e^{\frac{3}{2}}} \frac{|\vec{R}_{nh}|^2}{Z_{nh}^{*2}} G\left(\frac{I_{nh}}{T_e}\right), \tag{3.5}$$

where $I_{nh} = I_n - I_h$, I_n for ionization energy of the nth level, $Z_{nh}^* = \frac{Z_n + Z_h}{2}$ for effective nuclear charge. $|\vec{R}_{nh}|^2$ is the transition probability from n to h states, and $C_n |\vec{R}_{hn}|^2 = C_h |\vec{R}_{nh}|^2$, thus $C_n \alpha_{nh}^{(e)} = C_h \alpha_{hn}^{(e)}$, $C_n = 2n^2$.

4. Collisional deexcitation rate(h→n, h > n)

$$\beta_{hn}^{(e)} = N_e \beta_{hn} = \left(\frac{n}{h}\right)^2 e^{\frac{I_{nh}}{T_e}} \alpha_{nh}^{(e)} \tag{3.6}$$

B. X Ray emission and absorption

1. Photoionization recombination rate

$$\beta_n^{(v)} = N_e \beta_{vn} = \frac{1.10\rho Z' I_n^2}{C_n T_e^{\frac{3}{2}}} \frac{I_n}{e^{T_e}} E_1\left(\frac{I_n}{T_e}\right) \tag{3.7}$$

2. Spontaneous emission rate for spectral line(h→n, h⟩ n)

$$A_{hn} = 0.68 \times 10^3 \frac{|\overrightarrow{R}_{hn}|^2}{Z_{hn}^{*2}} \int_0^\infty v^3(1 + f_v) b_{hn}(v) dv \tag{3.8}$$

where v_{hn} is the X ray energy emitted in h→n transition, in 10^6 °K unit.

3. The resonant absorption rate of the spectral line(n→h, h⟩ n)

$$A'_{hn} = 0.68 \times 10^3 \frac{|\overrightarrow{R}_{nh}|^2}{Z_{nh}^{*2}} \int_0^\infty v^3 f_v b_{hn}(v) dv \tag{3.9}$$

where $C_n A'_{nh} = C_h A'_{hn}$. However, the photoionization effect(α_{vn}) may be unimportant to contribute p_n relative to above mentioned processes for n⟩ 1 shells in the optically thin system.

From Eq.(3.1) and these rate coefficients, we can say something as follows:

a. The three–body recombination relative to the collisional ionization is important only in a high ρ and low T_e region.

b. The colliosnal deexcitaton rate is faster than the collisional excitation for higher levels, and the higher the levels, the faster collisional excitation and deexcitation happened.

c. The spectral line emission of the higher energy levels is dominant processes for X ray conversion, however, the resonant absorption of the spectral line(trapped) may be important in the conversion region.

d. Free electrons unceasingly collide with bound electrons until free electrons are getting a lower T_e state.

In Eq.(3.1), f_v is the photon number on a quantum state, which is included in the terms related the processes of X ray absorption and emission and satisfies the following equation:

$$\frac{\partial f_v}{\partial t} - \nabla \cdot K_v \nabla f_v = j_{ff}(v)(1 + f_v) - C\mu_{ff}(v)f_v + j_{bf}(v)(1 + f_v) - C\mu_{bf}(v)f_v$$
$$+ j_{bb}(v)(1 + f_v) - C\mu_{bb}(v)f_v \tag{3.10}$$

where v in 10^6 °K uint, j_{ff}, j_{bf} and j_{bb} are free–free, free–bound and bound–bound emission rates, respectively, and $c\mu_{ff}$, $c\mu_{bf}$ and $c\mu_{bb}$ are correspondingly their absorption rates. $K_v = \frac{l_v C}{3}$ is a diffusive coefficient and l_v is the mean free path for photon of energy v, where

$$j_{bb}(v) = \sum_{l=1}^{n_0} \sum_{n=l+1}^{n_0} j_{nl}^{(b)}(v), \quad \left(\frac{1}{ns}\right), \tag{3.11}$$

$$j_{nl}^{(b)}(v) = 0.806 \times 10^{-16} N_i C_n P_n (1 - P_l) \frac{v_{ln}^3}{v^2} \frac{|\vec{R}_{nl}|^2}{z_{nl}^2} b_{nl}(v) \qquad (3.12)$$

$$b_{nl} = \begin{cases} \delta(v - v_{ln}) & \text{for line width} \\[2mm] \dfrac{1}{\sqrt{2\pi}\,\gamma_{ln}} exp\left(-0.5\dfrac{(v_{ln} - v)^2}{\gamma_{ln}^2} \right) & \text{for Gaussion broad.} \end{cases}$$

$$(3.13)$$

γ_{ln} is the width. N_i is the ion density;

$$C\mu_{bb} = \sum_{l=1}^{n_0} \sum_{n=l+1}^{n_0} C_{ln}^{(b)}(v) \; (\frac{1}{ns}) \qquad (3.14)$$

$$C_{ln}^{(b)}(v) = 0.806 \times 10^{-16} N_i C_l P_l (1 - P_n) \frac{v_{ln}^3}{v^2} \frac{|\vec{R}_{ln}|^2}{z_{ln}^2} b_{ln}(v) \qquad (3.15)$$

For the laser parameters which we are interested in, i.e., $I_a = 1 \sim 10, \lambda_0 = 0.35$, from Eqs.(2.2)–(2.4), we have:

$$T_e \approx 1 \sim 10, \; \rho \approx 0.0057 \sim 0.066, \; z' \approx 45 \sim 52 \qquad (3.16)$$

In the conversion region, $T_r > T_e$, generally speaking, the ionic shells for $n = 3$ are partly or fully filled, and the $n > 4$ levels are greatly ionized. Therefore, the collisional excitation, the line transitions and trapping effect happened between $n = 3$ and $n = 4$ are the most important processes. We can discuss them based on two levels as shown in Fig.2.

Now, Eq.(3.1) and (3.10) can be considerably simplified. The equations for P_n and f_v are rewritten as follows:

$$\frac{dP_4}{dt} = P_3(1 - P_4)A_{43}'$$
$$- P_4(1 - P_3)A_{43} + P_3(1 - P_4)\alpha_{43}^{(e)} \qquad (3.17)$$

$$\frac{\partial f_v}{\partial t} = \frac{\partial}{\partial x} K_v \frac{\partial}{\partial x} f_v + j_{bb}(v)(1 + f_v) - C\mu_{bb}(v)f_v \qquad (3.18)$$

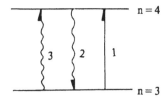

Fig.2.   Two–level diagram
1.   Collisional excitation
2.   Spectral line emission
3.   Trapping process

The deexcitation process in Eq.(3.17) has been ignored due to $P_4 \ll P_3$ and $e^{\frac{l_{34}}{T_e}} \sim O(1)$. On our interesting time scale, let $\dfrac{dP_4}{dt} = \dfrac{\partial f_\nu}{\partial t} = 0$, we have:

$$\frac{P_4(1-P_3)}{P_3(1-P_4)} = \frac{\alpha_{43}^{(e)}}{A_{43}} + \frac{1}{\zeta_3}\int\left(\frac{\nu}{\nu_{34}}\right)^3 b_{43}(\nu)f_\nu d\nu \tag{3.19}$$

and

$$f_\nu = \frac{j_{bb}(\nu)}{C\mu_{bb}(\nu) - \frac{1}{f_\nu}\frac{\partial}{\partial x}K_\nu\frac{\partial f_\nu}{\partial x}} \tag{3.20}$$

where the induced emission has neglected in Eqs.(3.17) and (3.18). ζ_i
$\equiv \int\left(\dfrac{\nu}{\nu_{34}}\right)^i b_{43}(\nu)d\nu$. We now approximately express the flux term of Eq.(3.17) in the form,

$$\frac{1}{f_\nu}\frac{\partial}{\partial x}\left(K_\nu\frac{\partial f_\nu}{\partial x}\right) \approx - \begin{cases} \alpha_0^2\dfrac{l_\nu C}{3\triangle^2}, & \text{for diffusion process} \\[3mm] \alpha_1\dfrac{C}{\triangle}, & \text{for straight transport} \end{cases} \tag{3.21}$$

where $\triangle$ is a thickness of the conversion region. α_0^2, α_1 are the adjustable parameters, and arc of the order of unit.

Thus, in a trapping region, we have

$$f_\nu = \frac{P_4(1-P_3)}{P_3(1-P_4)}\left(1 - \frac{1}{3}(\frac{l_\nu}{\triangle_0})^2\right) \qquad \text{for } l_\nu \ll \triangle_0, \tag{3.22}$$

where $\triangle_0 \equiv \dfrac{\triangle}{\alpha_0}$

$$l_\nu^{-1}(\mu m) = \mu_{bb}(\nu) = 0.269 \times 10^{-21} N_i C_4 P_3(1-P_4)\frac{\nu_{34}^3}{\nu^2}\frac{|\vec{R}_{34}|^2}{z_{34}^2}b_{43}(\nu) \tag{3.23}$$

Obviously, $f_\nu \ll 1$ from Eq.(3.22) due to $P_4 \ll P_3$. Substituting Eq.(3.22) into Eq.(3.19), we obtain

$$\frac{1}{3}\left(\frac{l_{\nu_{34}}}{\triangle_0}\right)^2 = \frac{P_3(1-P_4)\zeta_3\alpha_{43}^{(e)}}{P_4(1-P_3)\zeta_7 A_{43}} \tag{3.24}$$

where $\zeta \equiv \int\left(\dfrac{\nu}{\nu_{34}}\right)^i\dfrac{1}{\sqrt{\pi}\,\gamma_{34}}exp\left(-0.5\dfrac{(\nu_{34}-\nu)^2}{\gamma_{34}^2}\right)d\nu$, $\quad l_{\nu_{34}} \equiv l_{\nu=\nu_{34}}$. Eq.(3.24) shows that the photon flux in the trapping region is determined by the collisional excitation.

We now write the expression for conversion intensity as follows

$$W_{bb} = \frac{8\pi h}{C^3}\int \nu^3(j_{bb}(\nu) - C\mu_{bb}(\nu)f_\nu)d\nu \tag{3.25}$$

h: Planck constant.

Substituting into Eq.(3.22) and (3.24), we get:

$$\frac{W_{bb}}{\rho_{\cdot}}\left(\frac{J}{g\cdot ns}\right) = 0.515\times10^{9}\,\xi_{3}\frac{\zeta_{5}}{\zeta_{7}}P_{3}(1-P_{4})v_{34}^{4}\frac{\left|\overline{R}_{34}\right|^{2}}{Z_{3}^{*2}}\frac{\alpha_{43}^{(e)}}{A_{43}} =$$

$$0.417\times10^{13}\frac{\zeta_{5}}{\zeta_{7}}P_{3}(1-P_{4})v_{34}^{2}\frac{\rho Z'}{T_{e}^{\frac{3}{2}}Z_{34}^{*2}}G\left(\frac{I_{34}}{T_{e}}\right). \tag{3.26}$$

where $\left|\overline{R}_{34}\right|^{2}\approx51.97$. Finally, conversion energy is written in the form

$$E_{bb} = \int\frac{W_{bb}}{\rho_{\cdot}}\,dm\,dt. \tag{3.27}$$

where m is the fluid mass.

Eqs.(3.25) or (3.26) show that conversion energy is determined by the collisional excitation, i.e., the radiation flux as shown in Eq.(3.18). Thus, physical picture for the short wavelength laser converting in X rays is approximately described as follows:

The bound electrons on the nth levels collide with the free electrons, and are excited to the hth($>$ n+1) levels due to trapping effect and collision excitation, then immediately jump back to the nth levels, and thus X rays are emitted.

Eq.(3.5) is only related to slowly varying parameters p_3, p_4, T_e and ρ, whereas z' satisfies the equation

$$\frac{dZ'}{dt} = 18\alpha_{3}^{(e)}P_{3} + 32\alpha_{4}^{(e)}P_{4} \tag{3.28}$$

where the fast atomic processes have been cancelled out in Eq.(3.28), and $\alpha_{n}^{(e)}$ is the collisional ionization rate as shown in Eq.(3.2). As a result, the conversion energy can be approximately obtained from solving the simplified hydrodynamic equations coupled with electron and ion temperature equations.

IV. Numerical Results and Discussion

We use 1D three–temperature hydrodynamic code to compute the conversion energy by laser light, and the hohlraum environment with no X ray transport. In numerical simulation, laser light with the parameters($E_L=10^7$, $\Delta t=10$, $\lambda=0.35$, $I_a=2.4$) in supergaussion time distribution is injected into an one–dimensional cylindric hohlraum of radius $10^4\mu m$. The area effect of capsule is equivalently considered,

The results show that although we considered the principal quantum number for atomic energy levels up to 9, the X ray energy emission is essentially from the levels n = 3 and 4 as shown in Fig.3. The conversion efficiency is about 70%, and the conversion processes dominantly occur in the subcritical region nearby the critical surface about range $(3.9-5.8)\times10^{-3}$g / cm^2, where $\rho_0=19.3$ g / cm^3 as shown in Fig.4, where T_e rapidly decreasing is mainly caused by the strong X ray conversion.

The input energy and the conversion energy vary with time as shown in Figs.5 and 6.

The radiation temperature in hohlraum is shown in Fig.7, the maximum value approaches $T_R=3$.

Our results also show that the conversion mechanism and the characteristics can be simply explained by the two–level modelas proposed in section III. In addition, the radiative temperaturecan be rough predicted by the process of the radiative heat conduction without hydrodynamics.

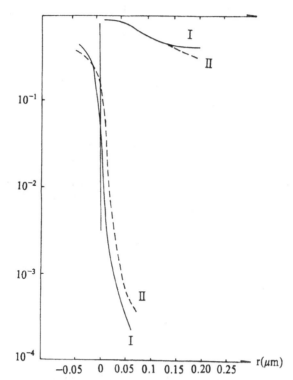

Fig.3. P_3 and P_4 in both sides of critical surface $(r = 0)$
curves   I for t = 5 ns
         II for t = 8 ns

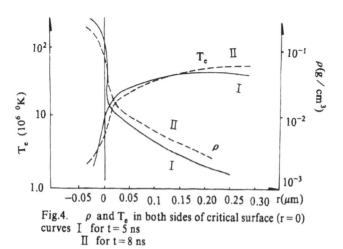

Fig.4. ρ and T_e in both sides of critical surface $(r = 0)$
curves   I for t = 5 ns
         II for t = 8 ns

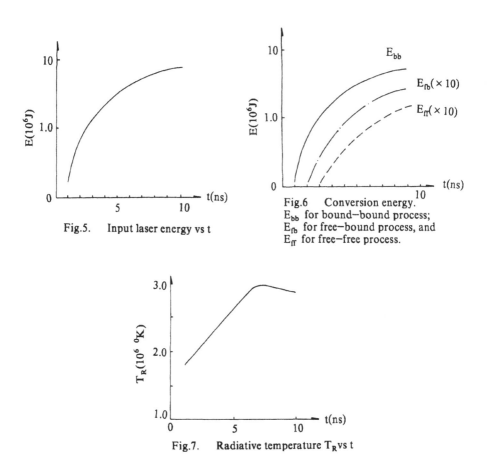

Fig.5.   Input laser energy vs t

Fig.6   Conversion energy.
E_{bb} for bound–bound process;
E_{fb} for free–bound process, and
E_{ff} for free–free process.

Fig.7.   Radiative temperature T_R vs t

Acknowledgements

    This work was performed in collaboration with Dr.G.N.Chen and Dr.C.S.Sui who participated in the numerical computation.

References

    1.E.Storm, J.D.Lindl, E.M.Campbell et al., Progress in Laboratory High Gain ICF, Prospects for the Future, the paper on the Nuclear War, Erice Sicily, Italy, (1988)
    2.E.Storm, The Proclamation on the Eight session of the International Seminars on Nuclear War, Erice Sicily, Italy, (1988).
    3.R.P.Drake, Comments On Plasma Phys. Controlled Fusion, 12, 181(1989).

IMPLOSION CHARACTERISTICS OF RADIATION-DRIVEN

HIGH GAIN LASER FUSION

T.Q. Chang and X.T. He

Centre for Nonlinear Studies
Institute of Applied Physics and
Computational Mathematics
P.O.Box 8009 , Beijing , China

M. Yu

China National Nuclear Corporation, Beijing

ABSTRACT

This paper theoretically discusses the implosion characteristics of radiation driven high gain laser fusion. The authers have estimated some important characteristic quantities, including efficiencies, fuel compression and hot spot ignition etc. In addition, the paper emphasizes some major harmful factors which have to be controlled for successful performance.

INTRODUCTION

Since the early years of 1970's the ICF (Inertial Confinement Fusion) research, especially driven by high power laser, has been a key subject to which many scientists devoted[1]. Its final goal is the clean energy source for next era and it has many important applications in science and other fields . Nowadays, there are two major approaches to the research, direct drive and indirect drive. The former is directly to impinge a target, in which the DT fuel is filled, with many laser beams as uniform as possible and the produced high pressure would compress the fuel to high density and required temperature to ignite the thermonuclear reaction. In contrast to the direct drive, the indirect drive fashion requires the laser light convert into soft X ray at first and then, the later drives the DT pellet to implode. As well known, the main advantage of indirect drive is relatively easier to satisfy the strict demand of uniform implosion of the pellet. Morever, the indirect drive needs no so many laser beams as in direct drive.

In recent years the indirect drive has made many important progresses [2,3,4]. It has been shown that a proper radiation environment (radiation temperature and its spatial uniformity), which is near to the requirement for high gain laser fusion, has been created in some scaling down laboratory experiments (Nova, 0.35 μm laser) and the results are consistent with the predictions of one dimensional calculations. Of course, the ignition could not be realized in suck scaling down experiments. However, it has been also pointed out that, energy gain ~ 100 will be achieved if $\sim 10^7$ J laser energy is used.

This paper is to discuss the physics of indirect drive laser fusion of high gain target, trying to give a comparatively clear physical picture. Generally speaking, the concerned physics can be divided into two parts, hohlraum physics and radiation driven implosion dynamics. The former includes the target absorption of laser light, X ray conversion and radiation transport, etc. However, this part is not the emphasis of this paper, we will discuss it elsewhere. In this paper we are to concentrate our attention to the later, implosion dynamics.

The second section is devoted to briefly discussing the fundamental demands of high gain fusion and some characteristics of the related target. The third section discusses the problems of implosion dynamics, such as the radiative ablation, environment temperature of pellet, hydrodynamic efficiency of the implosion, fuel compression and hot spot, etc. The two dimensional effects which must be controlled during the implosion will be talked over in section four.

MAIN DEMANDS FOR HIGH GAIN LASER FUSION

There are two prerequisites as the starting points of our discussion. The first is the laser energy which can be used for the high gain laser fusion and the second is the energy gain which must be achieved for a real reactor. By the end of this century, from the economic and technical point of view, the attainable standards of laser facility will be about $10^6 \sim 10^7$ Joules^3 (Nd glass laser, $3\omega_0$) . For the purpose in question, its pulsewidth should be about ~ 10ns, hence the power is $\sim 10^{15}$W. We will take these values for the following discussion.

As a practical energy source, high gain is necessary, but it is limited by some actual factors. Now, we discuss the problem. The energy gain is defined as follows.

$$G = \frac{\text{released fusion energy}}{\text{input laser energy}}$$

$$= \frac{m_{DT}\eta_b \times 3.5 \times 10^{11}(J/g)}{E_L} \tag{1}$$

where η_b is burn up faction of DT fuel. The fusion energy will be used to produce electric power. We denote the corresponding efficiency by η_e. However a part of the generated power, say, β has to be backfeeded to the laser facility. In addition, we assume the efficiency of the laser facility is η_L, therefore

$$GE_L \eta_e \beta \eta_L = E_L$$

$$G = 1/n_e \beta \eta_L \tag{2}$$

In order to suppress the demand for G, we should increase the values of η_e, β and η_L as far as possible. However, in general β_e could hardly be greater than 0.5 . As a power plant β should not be too high, we suppose it is about $1/5 - 1/3$. η_L is also difficult to exceed 0.1 , so that the demand for G is

$$G > 60 - 100 \tag{3}$$

Suppose a million KW power plant is to be built, then

$$E_L G\eta_e (1 - \beta)N = 10^9 J \qquad (4)$$

Where N is the laser shot frequency. Substituting $E = 10^7$ J and the other related quantities, we have

$$N > 2 \quad / sec \qquad (5)$$

In order to realize the radiation—driven laser fusion, a proper target is needed. As well—known, the main difference of this kind target from the one for direct is the need of a radiation case[5], at the middle of which a fusion pellet is installed. The laser beams will enter the case from the entrance holes located at two ends of the target. The target should absorb the laser energy as much as possible and convert it into soft X—ray with high efficiency. The case can prevent the X—ray from escaping and form a proper radiation environment around the pellet. In addition, the laser should be avoided directly impinging upon the pellet since the target is designed for indirect drive. Besides this, the laser produced plasma must also be avoided stricking the pellet because the plasma behaviour could hardly be controlled otherwise it will affect the spherically symmetric implosion of the pellet.

For the purpose of increasing the efficiencies of laser absorption and X—ray conversion, the radiation case should be made of heavy material for example Au and the short wave length laser must be used ($\lambda \leqslant 0.35\mu m$). Making use of the short wave length laser is one of the most important progresses in laser fusion in recent years. For $0.35\mu m$ laser light the absorption efficiency of Au target could reach > 90 % , at the same time , the X—ray conversion efficiency,60—80 %. It is such a high efficiencies that make the high gain laser fusion become possible.

Being different from planner target, in cavity target (the case) the converted radiation will be re—absorbed by its wall. Because of the relatively large special heat of wall material the radiation temperature is much lower than the one of laser produced plasma. Furthermore, for the sake of having sufficiently high radiation temperature and low enough temperature gradient in the environment of fusion pellet, the radiation transport must be fluent. It is the radiation that drives the pellet and passes its partial energy to the DT fuel during the implosion.

According to the above brief description, the whole implosion efficiency of the target can be defined as follows

$$\eta = \frac{fuel - obtained \ energy \ in \ implosion \ process}{input \ laser \ energy} \qquad (6)$$

and

$$\eta = \eta_a \eta_x \eta_d \eta_h \qquad (7)$$

η_a is the laser absorption efficiency of the target defined by $\eta_a = E_a / E_L$(E_a, absorbed laser energy) , η_x the X—ray conversion efficiency $\eta_x = E_x / E_a$, $\eta_d = E_{xp} / E_x$, the distribution efficiency of radiation energy (E_{xp} is the part of radiation energy which could be absorbed by the pellet) and $\eta_h = E_{DT} / E_{xp}$ is the implosion hydrodynamic efficiency. (E_{DT}, the energy obtained by DT fuel during the implosion) .

The problems of the radiation energy distribution and the hydrodynamic efficiency will be discussed in the following section. Depending upon different design, their possible , reasonable values are η_d:0.3–0.5, η_h:0.07–0.15.

Now we discuss some general characteristics of this kind of high gain target.

Central Ignition

Ignition needs some conditions, one of them is the high fuel temperature of $T > 60$ (10^6K) (for detail see the discussion later) . According to the temperature requirement and formula(3) we can discuss the ignition feature of the pellet. If we assume that DT as a whole should reach such a high temperature at the end of implosion, then

$$m_{DT}C_vT = E_L\eta \tag{8}$$
$$C_v = 10^7 J / g(10^6 K)$$

Substituting it into the expression for G . We obtain

$$G = \frac{\eta\eta_b}{T} 3.5 \times 10^4 \tag{9}$$

from $G > 60–100,$ we have

$$\eta_b\eta > 0.1 - 0.17 \tag{10}$$

This is not realistic because η is only about a few percent, η_b is $1/3-1/2$ at most.Therefore. the ignition takes place simultaneously in the whole DT fuel is impossible. The ignition has no alternative but to take place in a small part of the fuel, that is, central ignition.   The energitic particles, which are generated in the central region as the products of DT thermonuclear reactions, then heat the surrounding DT medium, gradually burning up to a considerable η_b. It is obvious that the central ignition would greatly reduce the required laser energy (10–20 times less).

Compression of DT Fuel

Besides the small part of DT which is for the ignition , the most part of the fuel should keep its temperature as low as possible during the implosion, at the meantime being compressed up to a rather high density, so that

$$\eta E_L = m_h c_v T + m_c \varepsilon \tag{11}$$
$$m_c = m_{DT} - m_h \approx m_{DT}$$

ε is internal energy of unit mass of DT.   In a perfect cold compression,   after considered the quantum effect, the Thomas–Fermi equation of state gives

$$\varepsilon = 2.3 \times 10^5 \rho^{2/3} (J / g) \tag{12}$$

It is the high density DT that contributes the majority of thermonuclear energy and its burn–up fraction can be expressed by

$$\eta_b = \rho R / (6 + \rho R) \tag{13}$$

where ρ and R are the density and radius respectively,after the compression is completed

$$\rho R = (\frac{3}{4\pi} \rho^2 . m_{DT})^{1/3} \qquad (14)$$

$\eta_b \sim 0.33$–0.5 means $\rho R \sim 3$–6. Considering the needed ρR value and some nonideal factors happened in a real implosion, we can estimate the fuel mass and the density that DT must be achieved in the implosion. There are many nonideal factors, for example, η could not reach the expected value, two dimensional effects, the effect of preheat which would damage the compression and so on. In a nutshell, these factors are equivalent to a reduction of laser energy. So that in (11), ηE_L should be replaced by $(\eta E_L)_{eff}$. In following estimates, $(\eta E_L)_{eff}$ takes the value $1 \times 10^5 J$.

We will consider the hot and cold regions respectively, with assuming half of the energy would be used to ignite the central region and the remain to compress the cold region.

$$\frac{1}{2}(\eta E_L)_{eff} = m_{DT}\varepsilon = 2.3 \times 10^5 m_{DT}\rho^{2/3} = 3.71 \times 10^5 \rho R(m_{DT})^{2/3} \qquad (15)$$

$$m_{DT} = (\frac{(\eta E_L)_{eff}}{7.42\rho R \times 10^5})^{3/2} \qquad (16)$$

Substituting the required ρR

$$m_{DT} \langle 9.5 - 10mg; \qquad \rho \rangle 110g / cm^3 \qquad (17)$$

From the point of view of the required E_L, we hope to enhance ρ, with keeping ρR fixed so as to reduce the m_{DT}, for example, when $\rho \sim 200$ g / cm^3, m_{DT} may decrease down to $\sim$ 3mg. However, on the other hand, there is also a permited upper limit. Substituting (15) into (1), we obtain

$$\rho = [\frac{1.52 \times 10^5 \eta \eta_b}{G}]^{1.5} \qquad (18)$$

G has a lowest value 60, η, η_b has a highest value about $0.07 \times 0.5 = 0.035$, so

$$\rho \langle 850g / cm^3 \qquad (19)$$

Apparently, high density compression is also a necessary means for reducing the required laser energy.

Hot Spot

Using similar procedure, we can discuss the hot spot. From the energy relation

$$\frac{1}{2}(\eta E_L)_{eff} = m_h c_v T \qquad (20)$$

$\rho r \rangle 0.3$ (see following section)

$T = 60$ $c_v = 10^7 J / (g 10^6 K)$

the obtained results are

$$m_h \sim 0.1mg \quad , \qquad \rho_h \rangle 34g / cm^3 \qquad (21)$$

Imagine that a fusion pellet is placed in an environment and driven by it. According to the radiation hydrodynamics, we can obtain

$$E_T(t) - E_T(0) = -\int_0^t dt \int_{s(t)} d\vec{s}\, \vec{v} P - \int_0^t dt \int_{s(t)} d\vec{s}\, \vec{F_k} = W(t) + F_R(t) \qquad (22)$$

where $E_T(t)$ is the total energy of the pellet at time t, $S(t)$ is its outer surface area at t, w(t) is the work on the pellet acted by the environment, and F_R is radiation energy flow from the environment. It is obvious that w(t) is mainly contributed by the plasma surrounding the pellet and it should be avoided as far as possible for controlling the asymmetry of the implosion.So that the radiation drive can be defined as

$$W(t) \ll F_R(t) \qquad (23)$$

Radiation Ablation

A pellet which is embeded into a radiation environment will be heated by propagating a radiation wave to its interior. The heated part of the pellet will isothermally expand into the environment. At the same time, a presure wave will propagate in oppesite direction. This is the radiation—ablative process. Because of the coexistance of radiation, pressure and rarefaction waves and their interactions, the picture is very complicated. The pellet mainly consists of an outer shell and the fuel DT. It has been known that DT is in liquid state and is absorbed in a porous plastic foam as in fig 1. When the pressure wave reaches the boundary between the shell and DT, another rarefaction wave will produced and propagate backwards into the shell, that makes the whole picture more complicated.

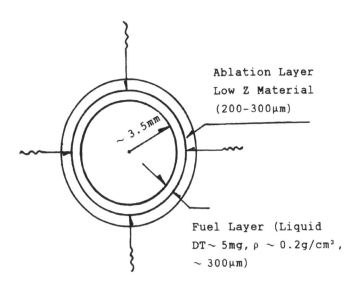

Ablation Layer
Low Z Material
(200–300µm)

$\sim 3.5mm$

Fuel Layer (Liquid
DT $\sim$ 5mg, $\rho \sim 0.2g/cm^3$,
$\sim$ 300µm)

Fig.1.   Fusion pellet.

The exact description of the problem must appeal to computer simulation. However, here we'd like to consider a simplified situation so as to get a clear physical picture and, furthermore, some numerical estimations. For the purpose, we will ignore the hydrodynamic motion induced by the radiation wave. In fact, although the hydrodynamic motion consumes a part of radiation energy, the reduction of density and radiation temperature arisen from the motion will, in turn, lead to increasing the radiation flow. Therefore this simplified model may be used to describe the radiation ablation.

The radiation transfer equation is

$$\rho \frac{\partial}{\partial t}(C_v T) = \frac{\partial}{\partial x} K \frac{\partial}{\partial x} T \; ; \qquad K = \frac{4aT^3 c}{3} l_R \; ; \qquad l_R = A\rho^m T^n \tag{24}$$

where l_R is the mean free path for radiation, a is the Stefan–Boltzman constant. c light speed. Suppose the radiation temperature at boundary T_0 doesn't change with time, there is a self similarity solution

$$T(x,t) = T_0 [f(Z)]^{\frac{1}{1+l}} \tag{25}$$

where z is the similarity variable. z is the similarity variable

$$Z = X / \sqrt{K'T_0^{N-l} t} \tag{26}$$

$$K' = \frac{4}{3} \frac{ac}{(N+1)\rho} (\frac{l_R}{T^n})(\frac{C_v}{T^l})^{-1}$$

$$N = n + 3, C_v \propto T^l$$

T_0 is the environmental temperature, $f(z)$ can be expressed as

$$f(Z) = (1 - \frac{Z}{Z_0})^{\frac{1+l}{N-l}} \; ; \qquad Z_0 = [\frac{2(N+1)}{N-l}]^{1/2} \tag{27}$$

As a result, the ablation depth is

$$X_R = [\frac{8}{3} \frac{ac}{(N-l)\rho} (\frac{l_R}{T^n})(\frac{C_v}{T^l})^{-1} T_0^{N-l} t]^{1/2} \tag{28}$$

(the ablation mass is ρx_R), meanwhile the transfered radiation energy is

$$E_R = \int C_v T\rho dx = \frac{N-l}{N+1}(C_v / T^l)T_0^{1+l}\rho X_R \tag{29}$$

Although above solution is obtained under the assumption of constant density, the obtained ρx_R is also approximately suitable to the ρ varied with space case and it is an exact one if m = −1, whereas in many cases, m is not very different from −1.

Selection of Outer Shell Material of the Pellet and the Ideal Value of η_d

From (28) and (29) it can be seen that E_R may be very different for different materials, even if they exist in the same radiation environment. It is the property that makes one have the liberty of selecting the materials of the cavity wall and the outer shall of the pellet so as to increase η_d . We have mentioned that the cavity wall should be made of high Z element. Comparing with gold, the relative abilities of Fe , Al and Be for absorbing the radiation are

$$(E_R)_{Fe} / (E_R)_{Au} \approx 1.2T^{0.35}$$
$$(E_R)_{Al} / (E_R)_{Au} \approx 6.1T^{0.23}$$
$$(E_R)_{Be} / (E_R)_{Au} \approx 7.1T^{1.13} \tag{30}$$

If T takes the value 3, they are about 1, 8, 8 and 25 respectively.We assume that the area ratio of the pellet surface and cavity inner surface is about 0.1, selecting Fe as pellet material is not proper if you want η_d reaching or above 0.5, you must select the lower Z material,for example Al or other still lower Z elements.

Environmental Radiation Temperature of Pellet

Increasing the environmental radiation temperature is an important consideration for indirect drive laser fusion, since high radiation temperature would enhance the pellet−absorbed radiation energy, and therefore,the radiation presure and the hydrodynamic efficiency of the implosion. However , the temperature is upperly limited by the laser energy. We assume the outer shell is made of Al, the cavity Au, the surface area ratio still is 0.1, the radius of the pellet is 0.4 cm, $E_L = 10^7 J$, laser pulsewith is 10ns. $\eta_a = 0.9$, $\eta_x = 0.8$. From (29), we easily obtain the upper limit of radiation temperature

$$T_{max} \sim 3.5(10^6 K) \tag{31}$$

We can estimate the lower limit of the temperature that must be reached as follow: from the minimum energy needed for DT fuel and the possible hydrodynamic efficiency η_h the required radiation energy absorption for pellet could be estimated, and hence the required radiation temperature. Assuming the whole mass of DT is 5 mg. Hot spot mass is 0.1 mg.Hot spot temperature is 60, the final DT density reached during compression is 200 g / cm^3, and η_h is 10% we have the minimum radiation temperature required for high gain

$$T_{min} \sim 2.5 \tag{32}$$

Radiation Pressure and Radiation−Driven Hydrodynamic Efficiency

The radiation pressure here we mean is the pressure at the front of radiation wave. $P = \Gamma \rho T$, Γ is approximately a constant, $T \sim T_0$. Therefore we should determine the density at that front. It seems difficult to obtain the exact $\rho(t)$ at that place, so we appeal to the characteristic method to estimate it. Because of the ablative mass obtained previously is reliable we use Lagrangian coordinates (m,t), with $m = \int \rho dx$ being mass, the hydrodynamic equations are

$$\frac{\partial}{\partial t} ln\rho = -\rho \frac{\partial}{\partial m} u$$
$$\frac{\partial}{\partial t} u = -\rho \frac{\partial}{\partial m} ln\rho . \tag{33}$$

where ρ, u is the dimensionless quantities ρ / ρ_0, u / c_s (c_s is the isothermal sound speed) and in eq.(33) m and t have been stood for m / ρ_0 and tc. respectively. The equations (33) can be rewritten as

$$\frac{\partial}{\partial t}[ln\rho \pm u] \pm \rho\frac{\partial}{\partial m}(ln\rho \pm u) = 0 \qquad (34)$$

There are two characteristic lines in the space (m,t)

$$
\begin{aligned}
c_+ : \quad & m_\alpha = \rho t_\alpha \\
c_- : \quad & m_\beta = -\rho t_\beta
\end{aligned}
\qquad (35)
$$

and the quantities $u+ln\rho$ and $u-ln\rho$ are the invarainces along the c_+ and c_- respectively. (35) indicates that $\pm \rho$ are just the slopes of these characteristics.

The characteristics and radiation–ablative curve have been shown in fig.2. Comparing the slopes of c_+ and the ablative curve, we have

$$\frac{1}{\rho} < \frac{1}{D} \qquad (36)$$

where D is the ablative velocity. $D = \dfrac{d}{dt}(X_R / c_s)$. As a result,

$$p = \Gamma\rho T > \rho_0 c_s \frac{dX_R}{dt} = \frac{\rho_0}{2} c_s X_R t^{-1} \qquad (37)$$

Introducing a phenomenological factor $\xi > 1$.

$$p = \frac{\rho_0}{2} \xi c_s X_R t^{-1} \qquad (38)$$

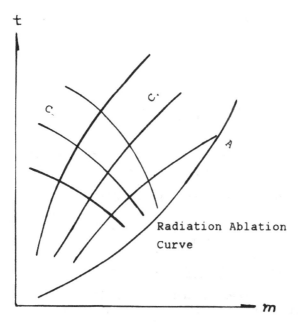

Fig.2. Radiation ablation curve and characteristics

In the following we will take $\xi = 1.5$ for Al, which is estimated by comparing eq.(38) with the result of computer simulation. So that

$$p = 5.25 T^{3.0} t^{-\frac{1}{2}}$$ (39)

Substituting $t = 10$ ns, $P \sim 100$ Mb when $T = 3$.

Now we estimate the hydrodynamic efficiency. It can be obtained by using solid state model and , morever, for simplifying , we use planar geometry ,which is approximately valid for the period in which the DT gains its major kinetic energy . By $m_0 = m_s + m_{DT}$ denoting the total mass to be driven, and by $m(t)$ the ablated mass at time t, we have

$$[m_0 - m(t)]\frac{du}{dt} = p(t), \qquad\qquad m(t) = \rho X_R$$ (40)

Substituting eq.(28) and eq.(38) for $m(t)$ and $P(t)$ we have a solution

$$u = 6.56 \times 10^{-3} \xi T^{0.5} ln(\frac{1}{1-y})$$ (41)

with $y = m(t) / m_0$. The hydrodynamic efficiency is defined as

$$\eta_h = \frac{1}{2} m_{DT} u^2 / S_t E_R$$ (42)

where S_t is the outer surface area of the pellet, so

$$\eta_h = \frac{1}{2}\frac{N+1}{N-1}\frac{\xi^2 c_s^2}{c_v T}\frac{m_{DT}}{M_0} f(y) = 1.7 \times 10^{-2} T^{-0.39} \xi^2 f(y)$$ (43)

$$f(y) = [ln(1-y)]^2 / y$$ (44)

The flight distance is

$$|R_0 - R(t)| = \int_0^t u dt = 2c_s \xi \left(\frac{M_0 \sqrt{t}}{m(t)}\right)^2 g(y)$$ (45)

$$g(y) = \frac{3}{4} + \frac{(1-y)}{2}[ln(1-y)(1+y) - \frac{1}{2}(3+y)]$$ (46)

From these expressions it can been seen that the hydrodynamic efficiency is sensitive to the ablated mass fraction y. Seting the thickness of Al shell is 0.025 cm, $m_{DT} = 5$ mg,$\xi = 1.5$, $T = 3$ the hydrodynamic efficiency could reach 10–15% if 85–90% of the shell has been ablated. The corresponding velocity is $30–40 \times 10^{-3}$ cm / ns and the flight distance is ~ 0.2 cm. From the needed fraction of ablation mass for high η_h , we can also estimate the corresponding ablation time, it is about 11 ns. This is also the required width of the main laser pulse.

Critical Isoentropic Line for Implosion Compression

The required energy for high compression of DT fuel is dependent on compressive precess, more precisely speaking, the entropy enhancement during the process. Assuming DT is ideal gas

$$d\varepsilon = Tds - pdv$$

$$c_v dT = Tds - \Gamma \rho T dv$$

$$\frac{dT}{T} = \frac{ds}{c_v} + \frac{2}{3}\frac{d\rho}{\rho}$$

$$\frac{T}{T_0} = \left(\frac{\rho}{\rho_0}\right)^{\frac{2}{3}} e^{\frac{\Delta s}{c_v}}$$

$$\frac{\varepsilon}{\varepsilon_0} = \left(\frac{\rho}{\rho_0}\right)^{\frac{2}{3}} e^{\frac{\Delta s}{c_v}} \tag{47}$$

where $\triangle s$ is the whole increasing of entropy during the compression. By using the real equation of state of DT we can obtain $\varepsilon \sim \rho$ curve corresponding different entropy values, see fig.3 . Once given the energy which can be supplied to DT by the radiation ablation and the density that you want by the end of compression, a special isoentropic line, which passes the given ε, ρ in fig.3, could be obtained . It may be called critical isoentropic line. Its meaning is that the DT state could never transcend the line at any time during the whole comprssion process. Otherwise, the energy supplied is not sufficient for the fuel to be compressed to the value you expected .

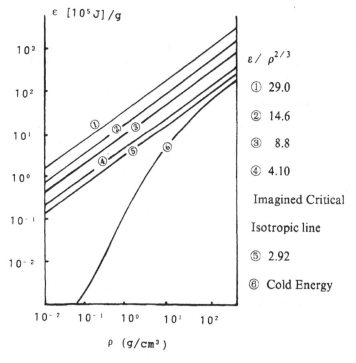

Fig.3. Isoentropic lines for DT

Entropy Increasing of Implosion

Several factors would lead to the entropy–increasing of DT during implosion. The first is radiation transfer. Therefore the direct heating of DT induced by radiation must be avoided in the whole implosion precess. We have seen from above that , in order to increase the hydrodynamic efficiency the majority of pellet shell should be ablated, the remain of the shell must be very thin. If the target design is not acurate or the very thin unablated shell has been broken down due to hydrodynamic instability occured in the implosion, the unpermissable entropy increasing will take place. This is one of the most difficult problems in implosion dynamics research. If the heating happens, the corresponding entropy increasing is

$$\Delta s = \frac{\Delta Q}{T} \tag{48}$$

Where ΔQ just is the heat transfered to the fuel from radiation.

Besides the process, there are two main mechanisms which would bring about the entropy–increasing during the implosion, they are shock wave and the preheating due to hot electrons.

Shock Wave Induced Entropy–Increasing. When the shock wave, which is produced by the radiation ablation of pellet ,propagates to the boundary between the shell and DT fuel, a rarefaction wave will be reflected , meanwhile the fuel will be compressed. Because of the large difference in density of shell and the fuel, shock wave presure drop will take place. According to uncontinuety analysis,

$$\frac{p_f}{p} = \frac{20}{9}\left(\frac{2}{\bar{\gamma}-1}\right)^2 \frac{1}{\bar{\gamma}} \frac{\rho_{DT}}{\rho_0}\left[\left(\frac{\rho_0}{\rho}\right)^{\frac{\bar{\gamma}-1}{2\bar{\gamma}}} - 1\right]^2 \tag{49}$$

Where p_f , p are the shock wave presures in DT and shell , respectively ,ρ_{DT} is the DT density ahead of the shock, whereas ρ_0 is shell density behind of the shock. γ is the effective adiabatic index of the shell material (for DT has been taken to be 5/3). For Al shell the pressure drop will be about 5–10 times. For certainty, we will assume it is 5 in following discussion.

In order to control the shock wave induced entropy–increasing the compression should be proceed slowly and gradually. Therefore, the pulse for driving implosion should be either continuously increasing in strength or divided into a series of step by step strengthend ones ,the formers are for pre–compressions and the last one supplies the main required energy for compressing DT.

For ideal gas , the entropy–increasing produced by shock wave is[6]

$$\frac{\Delta s}{c_v} = \ln(p_1 \rho_0^r / p_0 \rho_1^r) \tag{50}$$

Subscripts "0", "1" denote ahead and behind of shock. Setting $p_1 / p_0 = \omega$, $\rho_1 / \rho_0 = \sigma$. There exist following relation between ω and σ[7]

$$\omega = \frac{\sigma - \mu^2}{1 - \mu^2 \sigma} \tag{51}$$

$$\mu^2 = (\gamma - 1)/(\gamma + 1)$$

The corresponding energy increasing factor is

$$e^{\frac{\Delta s}{c_v}} = \omega\sigma^{-\gamma} \tag{52}$$

For decreasing the entropy–increasing we must control the shock intensity ω. The shock induced first pulse is often strong, because the pressure,temperature and density ahead of the shock are rather low. The density behind of the first shock will be near $\rho_0(\gamma+1)/ (\gamma-1)=0.8 \text{ g}/\text{cm}^3(\gamma=5/3)$. In this case, the factor for pressure–increasing is rather large,but the p_1 is still limited owing to small p_0.

Following the first compressing there are a series of succesive shock waves to compress DT. For simplifying the description we suppose the pressure of given shock wave always is ω times as that of the previous one, ω is a constant. If there are overall m times compressions, we have

$$e^{\frac{\Delta s}{c_v}} = [\omega\sigma^{-\gamma}]^m \tag{53}$$

Now we can estimate the restriction of the pressure p_1 induced by first shock in DT fuel. Suppose that the critical isoentropic line corresponds to 2 times of the energy needed for adiabatically compressing the DT up to $\sim 200\text{g}/\text{cm}^3$, then $\varepsilon_c \sim 1.5\times 10^7(\text{J}/\text{g})$, from eq.(47),

$$\varepsilon_1\left(\frac{200}{0.8}\right)^{\frac{2}{3}}[\omega\sigma^{-\gamma}]^m = 1.5\times 10^7 \tag{54}$$

or

$$p_1[\omega\sigma^{-\gamma}]^m = 2\times 10^5 \tag{55}$$

where p_1 is just the pressure of the first shock wave (J/cm^3) in DT. If $\omega = 4$, $m = 3$,

$$p_1 \sim 1.35(Mb) \tag{56}$$

The corresponding energy and temperature are

$$\begin{aligned}\varepsilon_1 &\sim 2.5\times 10^5 \quad (\text{J}/\text{g})\\ T_1 &\sim 0.025 \quad (10^6 K)\end{aligned} \tag{57}$$

In fact, p_1 should be restricted more strictly than above value. The reason is that, above discussion has implicitly assumed that the shock series would coincide at the inner boundary of DT medium, without considering the rarefaction of DT in the intervals of succesive shocks. Nevertheless, this assumption is what we should advoid in design, because the too early coincidence of the succesive shocks would lead to a very strong wave, hence a unduly hot spot would be formed in the followed centripetal motion. If we considered the rarefactions, the entropy–increasing would be larger, since the same pressure would bring about a larger entropy in a rarer medium. So that the p_1 should be lower than what we estimated. By the same reason the DT should be appropriately precompressed before the main shock wave has arrived, for example x g/cm^3.

Therefore the whole compression of DT may be roughly divided into three stages: precompression, main acceleration stage and followed convergent compression. In the last stage the kinetic energy of DT converts into high density of the material and, generally speaking, in the stage the entropy–increasing is not severe.

In order to realize the above described compressive process with the entropy–increasing as samll as possible, the time behavier of the drive pulses must be carefully designed and adjusted. It is another difficult problem of the implosion hydrodynamic research.

Hot Electron Preheating. If hot electrons deposit some energy $\triangle Q$ into the DT fuel, the entropy increasing will be $\triangle s = \triangle Q / T$. It is obvious that the $\triangle s$ would be more severe at earlier stage since at that time the temperature of DT is lower. From the $\rho \sim \varepsilon$ curve in fig.3, the preheating–increased temperature should be less 0.01 at $\rho = 0.2$.

The outer shell of the pellet is a good shield layer for preventing the fuel from being severely preheated. Again assume that the shell is made of Al. In that material the electron range is

$$\bar{R}(cm) = 0.412 E^n (mev) / \rho(g / cm^3) \tag{58}$$

$$n = 1.265 - 0.0954 ln E(mev)$$

If the thickness of the shell is 250 μm, only the electron whose energy is greater than 270 Kev can penetrate into the DT. We can estimate the tolerable hot electron energy deposited into DT as follows.

The deposited electron energy in unit mass is

$$\frac{dQ}{dm} = \int J(t, E_e, \bar{r}) \frac{-dE_e}{\rho dx} dE_e dt \tag{59}$$

where J is the hot electrons flux. We suppose that the hot electron is uniform everywhere and that the electron energy would be fully absorbed once it has entered into the pellet. Let $E_L \eta'_{he}$ denotes the hot electron energy which can be deposited into the fuel. Upon considering that the number of electrons which have penetrated into a specified part of the target should be proportional to its surface area exposed to the hot electron irradiation.

$$\eta'_{he} < \frac{m_{DT} c_v T_p}{E_L (S_p (\Delta Q)_{DT} / S \sum_{Al+DT} [\Delta Q])} \tag{60}$$

where S_p is the surface area of the pellet and S is the whole area being exposed by hot electrons in target, and $\Delta Q = \int (dQ / dm) dm$. T_p is the tolerable temperature for the preheating. Setting $E_L = 10^7 J$; $(S_T / S) \sim 0.1$; $T_p \sim 0.01$ and

$$-\frac{dE_e}{\rho dx} = \frac{2\pi e^4 n_e}{m_e v^2} \left[ln \frac{m_e v^2 K}{2 I^2 (1 - \beta^2)} - (\sqrt{1 - \beta^2} - 1 + \beta^2) ln 2 + 1 - \beta^2 \right.$$
$$\left. + \frac{1}{8} (1 - \sqrt{1 - \beta^2}) \right] \tag{61}$$

where n_e is the number density per mass, v is electron velocity, K is its kinetic energy, I is average ionization energy of material $\beta = v / c$, we obtain

$$\eta'_{he} < 0.014 \qquad\qquad (62)$$

As to the restriction of whole hot electron energy, it depends upon the effective temperature T_h of hot electrons. When T_h is lower, the preheating is not important. In addition, short wavelength laser is conducive to suppressing the generation of the hot electrons.[8]

Hot Spot

The design requirements for hot spot are $(\rho r) > 0.3$ and $T > 60$, the former is the critical ignition condition. The critical ignition means that the energy released by nuclear reaction in a medium could heat the medium itself up to the temperature needed for proceeding the reaction further. The condition also implicates that the range of α particle produced by D+T reaction is less than the size of the hot spot. In fact, in this case, α's range is about $\rho R_a \sim 0.1$. The condition of hot spot temperature comes from the requirement: the energy released by the nuclear reaction should at least be equal to or greater than the spontenious bremsstrahlung energy loss. With these two necessary conditions having satisfied, the mass and temperature, hence the hot spot energy, should be strictly controlled. The hot spot is formed in the convergent motion of the fuel. The first reflection of the shock at the center could hardly reach the required density $\rho \sim 34$ g / cm^3 . However, after the reflection at center the temperature formed in the central part of DT is much higher than the surrounding cold DT, the sound speed $(\alpha \sqrt{T})$ is large , so that the shock can quickly reflect several times between the center and the convergently moved cold DT, resulting in satisfying the density requirement.

We have mentioned that, in the really designed pellet, the DT is absorbed into a porous plastic. Therefore the mixing of some plastic into the hot spot may be unavoidable and that will increase the temperature needed for ignition. Because of the rate of bremsstrahlung in plasma is

$$p_b = 1.57 \times 10^{-31} n_e \sum_i (n_i Z_i^2) \sqrt{T} \qquad (W / cm^3) \qquad\qquad (63)$$

p_h is sensitive to Z of element. Suppose some atoms of carbon have mixed into DT, and setting $\varphi = n_c / n_D$, by make p_b and reaction rate be equal, we can see the required temperatures are 72 and 87 for $\varphi = 0.05$ and 0.1, respectively. The detail data can be found in fig.4.

Because the temperature of hot spot is very high and the surrounding medium is much colder, the electron conduction will decrease the temperature of hot spot and enlarge its size. Of course, it would also raise the energy requirement.

Large Cavity Structure of the Pellet

The DT compression and the hot spot formation, in the final analysis, rely on the amount of work acted by the shell of pellet. The work can be approximately expressed as

$$P_f v = E_{DT} \qquad\qquad (64)$$

where P_f is the pressure in DT fuel, v is roughly the DT initially occupied volume and E_{DT} is the needed energy. The radiation ablation pressure in pellet shell is $\sim$ 100Mb, so that the $P_f \sim$ 20Mb. If E_{DT} takes two values 2×10^5J and 4×10^5J, (the required lowest energy for DT compression and central ignition is $\sim 10^5$J, if $m_{DT} \sim 5$ mg,

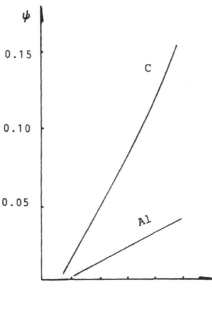

ψ

0.15

0.10

0.05

C

Al

T (10^6K)

Fig.4. Ignition temperature changes
with impurity fraction

and its final density is about 200 g / cm^3),the corresponding radii are R$_{DT}$ = 0.3, 0.36cm, respectively. We can also explain the problem as follows. Suppose that the interier of the pellet is initially fully filled with DT and its initial density still is 0.2 g / cm^3. For the same whole DT mass the radius of DT will be ~ 0.18cm, after precompression, say $\rho \approx 1$, the radius will be ~ 0.11cm.Setting E$_{DT}$ = 2 × 10^5, we obtain P$_f$~400Mb, the presure in shell should be 2000 Mb. So large! It is not possible for realistic radiation ablation.

TWO DIMENSIONAL EFFECTS

Above discussion is based on one dimensional consideration that is assuming the implosion is perfectly spherically symmetric and we have seen there are several severe difficulties in the design. However, the greatest risk for successful performance come from some two dimensional effects. If these effects could not be controlled, the high gain laser fusion would be difficult to be realized. Those are implosion symmetry, breakdown of ablation layer and medium mixing. We will briefly discuss them.

Implosion Symmetry

This is the indispensable condition for high compression and central ignition. It is the point that makes the indirect drive manifest advantageous.As you have seen that the radius of hot spot is about 90μm, 40 times smaller than its initial value. In addition, density,temperature and mass have to be strictly controlled . These harsh terms would not be realized under so large convergent ratio untill the design is acurate enough and the spherical implosion is highly guaranteed. Similarly, the high compression of DT fuel also requires this summetry. Numerical simulation indicates that the average density would largely decrease if there is a small drive nonuniformity. A simple estimate gives that the nonuniformity of drive pressure should be strictly controlled, $\Delta p / p < 1-2\%$

Breakdown of Ablation Layer

We have pointed out above that the high hydrodynamic efficiency demands very thin remaining shell of the pellet, so that the hydrodynamic instability easyly makes it break down. However, this is never permitted to happen, a sharp contrdiction !

From the classical theory for hydrodynamic instability, a disturbance will develop according to

$$e^n = e^{\int dt \sqrt{ak}} \tag{65}$$

where a is acceleration, $k = 2\pi / \lambda$, λ is the wave length of the disturbance. From the previous estimation (41) a(t) can be obtained, and for a fixed k, n can be integrated. The most dangeous λ may be the thickness of the remaining ablation layer. Setting $\lambda = 0.025 \times 0.15$ cm, which corresponds to 85 percent ablation of 250 μm thick outer shell of the pellet, the integration gives $n \sim 10$. That is intolerable. Fortunately, the radiation ablation can play a role of stabilizing the hydrodynamic instability. After considered this effect, [9]

$$n \sim \int dt \sqrt{ak} - \int dt k v_a \tag{66}$$

where v_a is about the ablation velocity and $\int dt v_a$, ablation depth. In the same condition as above, $\int dt k v_a \sim 35$. Of course, these figures are not exact. But, any way,the ablation is a very important factor for stabilizing the hydrodynamic instability.

There are also other mechanizms for stabilizing the instability, for example, the density gradient. It may be plays a considerable role, especially when the rarefaction wave, which is reflected from the boundary between the pellet shell and fuel, propagates to the ablation front in the shell, with smoothing the density gradient.

Mixing

Another problem brought about by hydrodynamic instability is the mixing which happens between the shell material and the fuel, and between hot and cold fuels. The later could happen since the formation of hot spot needs shock reflections between centre and cold DT, therefore the boundary of cold and hot DT plasma is some times accelerated and some times decelerated The another reason for mixing is microjets which may be produced at boundary due to shock wave. Perhaps, these mixings could hardly be completely avoided, but must be controlled as far as possible. Severe mixing would seriously affect the ignition and the burning, even make them fail.

BRIEF SUMMARY

This paper mainly discuss the physical processes happened in radiation—driven laser fusion. We have estimated the principal characteristics in it. In order to outstand the physics, regularity and the key points, we mainly appealed to analytical means. The more detail picture must rely on numerical simulation.

Using short wavelength laser makes the high gain laser fusion become possible.But, it also puts forward some difficult problems to the designer, especially the 2—dimensional effects. However, the achievements which have obtained up to now show that these difficulties are not impassable.

REFERENCES

1. J.W.Nuckolls et al.,Nature (London)239 ,139(1972).
2. E.Storm, Progress of Inertial Confinement Fusion at LLNL, IAEA 11th Conference on Plasma Physics and Controlled Nuclear Fusion Research, Kyato, Japan, 13–20 Nov. 1986.
3. E.Storm, J.D.Lindl, E.M.Campbell et al., Progress in Laboratory HighGain ICF, Prospects for the Future, the Paper on the Eighth Session of the International Seminars on Nuclear War, Erice Sicily, Italy (1988).
4. E.Storm, Speech on the Eighth Session of the International Seminars on Nuclear War, Erice Sicily, Italy (1988).
5. Alexander J.Glass, Laser Focus, Vol. 17 No.4, 20(1981).
6. H.Motz, The Physics of Laser Fusion, Academic press, London. New York. San Francisco, 181(1979).
7. R.Courant and K.O.Friedrichs, Supersonic Flow and Shock Waves Interscience Publishers, Inc. New York, 148(1958).
8. R.D.Drake, Laser and Particle Beams, 6, part 2, 235(1988).
9. J.O.Kilkenny et al., Bulletin of American Physical Society, 32, 1769(1987).

FISSION-INDUCED INERTIAL CONFINEMENT HOT FUSION

AND COLD FUSION WITH ELECTROLYSIS

Yeong E. Kim

Department of Physics
Purdue University
West Lafayette, IN 47907

INTRODUCTION

Conventional theoretical estimates for deuterium-deuterium (D-D) fusion can not explain tritium production and excess heat generation above that due to the electrode reaction observed by Fleischman, Pons, and Hawkins (FPH) [1] and others [2-5] in their electrolysis experiments with a palladium cathode immersed in heavy water (with 0.1M LiOD), since the estimated D-D fusion cross-sections and rates are too small at room temperature. Recent experimental results indicate that the extrapolation method is not valid at low energies. Plausible nuclear theory explanations are discussed. Experimental measurements of the D-D fusion cross-sections and branching ratios at very low energies are suggested. In order to explain the FPH effect, a surface reaction mechanism is proposed for the cold D-D fusion with electrolysis. Experimental tests of the proposed mechanism for cold fusion are discussed. Other nuclear reactions involving neutron-induced reaction processes, which may occur subsequent to and concurrent with the D-D fusion, are also discussed.

One of the neutron-induced reaction processes suggested as a plausible mechanism for cold fusion is used to design pellets for neutron-induced inertial confinement fission-fusion for large-scale power generation, which is different from conventional approach to fusion. The process involves chain reactions: (1) $n + {}^6Li \rightarrow {}^4He + T$ and/or $n + {}^7Li \rightarrow {}^4He + T + n$, and (2) $T + D \rightarrow {}^4He + n$, in $Li-D$ pellets surrounded by Li and other blankets. Several pellet designs are described to achieve a plasma density of $\lesssim 10^{19} cm^{-3}$ and a confinement time $\tau \gtrsim 10^{-5} sec$ for the neutron-induced fission-fusion.

SURFACE REACTION MECHANISM FOR COLD D-D FUSION

Because of the complexity of four-nucleon system, no rigorous theoretical calculations of the D-D fusion rates and branching ratios have been carried out.

With $E \gtrsim 0$ appropriate for a surface reaction mechanism for the electrolysis experiments [1-5], two dominant channels for the D-D fusion are

$$D + D \rightarrow {}^3H(1.01MeV) + p(3.02MeV) \qquad (1a)$$

and

$$D + D \rightarrow {}^3He(0.82MeV) + n(2.45MeV). \qquad (1b)$$

Reaction (1a) is not a "real" fusion but a neutron-transfer reaction, while reaction (1b) is a fusion reaction in which two protons are fused to form a ${}^3He$ nucleus. The reaction rate, R_{DD}, for (1a) or (1b) is given by [6]

$$R_{DD} = \frac{n_D n_D}{2} <\sigma v> (cm^3 sec^{-1)})$$

with

$$<\sigma v> = \frac{(8/\pi)^{1/2}}{M^{1/2}(kT)^{3/2}} \int \sigma(E) E e^{-E/kt} \, dE$$

where the cross-section, $\sigma(E)$, is parameterized as (E is in the c.m.)

$$\sigma(E) = \frac{S(E)}{E} e^{-(E_G/E)^{1/2}} \qquad (2)$$

in the conventional estimates assuming that non-resonant charged particle reactions for reactions (1a) and (1b). E_G is the "Gamow energy" given by $E_G=(2\pi\alpha Z_D Z_D)^2 Mc^2/2$ or $E_G^{1/2} \approx 31.28(keV)^{1/2}$ with the reduced mass $M \approx M_D/2$. The extrapolated values $(S(E \approx 0) \approx 55keV$ - barn) of the S-factors for both reactions (1a) and (1b) are nearly equal at $E \approx 0$, although S(E) for reaction (1b) is slightly larger than S(E) for reaction (1a) for $E \gtrsim 20keV$ [7].

Rigorous tests of eq.(2) at $E \lesssim 3keV$ are yet to be carried out by future experimental measurements. Justification of eq.(2) has not been demonstrated by theoretical calculations based on a dynamical four-nucleon theory for reactions (1a) and (1b).

If there is no resonance behavior for $\sigma(E \lesssim 3keV)$ and the above relation (2) between $\sigma(E)$ and S(E) turns out to be valid for $0 < E \lesssim 3keV$, then a minimum value of the deuterium kinetic energy (c.m), $E \approx 100eV$, is required to obtain the claimed value of $\Lambda \approx 10^{-10} sec^{-1}$ [1, 3] for reaction (1a), since $R_{DD} = (n_D^2/2) < \sigma v > \approx 1.8 \times 10^{14} cm^{-3} sec^{-1}$ (or $\Lambda \approx 0.3 \times 10^{-10} sec^{-1}$ per D-D pair) with $<\sigma v> \approx 10^{-31} cm^3 sec^{-1}$ and $n_D \approx 6 \times 10^{22} cm^{-3}$ in the Pd cathode or in heavy water [1, 3]. However, there are no currently known mechanisms which enable deuterium atoms to gain average kinetic energies of $\gtrsim 100eV$ and at the same time suppress reaction (1b) in the electrolysis experiments with the Pd cathode [1, 3, 5]. For the claimed value of $\Lambda \approx 10^{-23} sec^{-1}$ [8] for reaction (1b), deuteron kinetic energy (c.m.) of $E \approx 25eV$ is required.

Conventional theoretical estimates of the cold D-D fusion rate and branching ratio are arbitrary and may not be valid since they are based on an extrapolation of the reaction cross sections at higher energies ($\gtrsim 4keV$) to lower energies where no direct measurements exist, except the indirect measurements of FPH and others. Recent experimental results of Beuhler, Friedlander and Friedman [9] at $\sim 0.3keV$ show that both the fusion rate and the branching ratio for $(^3H - p)$ channel are extremely large as in the case of the FPH effect compared to the conventional estimates, thus indicating that the extrapolation method is not valid at low energies. For $E = 0.15$ keV in the c.m. (corresponding the case of Beuhler et al. [9] with 0.3 keV in the laboratory system), the use of eq. (2) gives $\sigma_{calc}(E \approx 0.15keV) \approx 3 \times 10^{-57}cm^2$ for both reactions (1a) and (1b), while the experimental result of Beuhler et al. is $\sigma_{exp}(E \approx 0.15keV) \approx 0.67 \times 10^{-31}cm^2$ [10] which is 25 orders of magnitude larger than $\sigma_{calc}(E \approx 0.15keV) \approx 3 \times 10^{-57}cm^2$ obtained from eq. (2) [10].

The results of $\sigma_{exp}(E \approx 0.15keV) \approx 0.67 \times 10^{-31}cm^2$ extracted from the experiment by Beuhler et al. [9] can be regarded as the first direct measurement of $\sigma_{exp}(E)$ at $E \approx 0.15keV$ for reaction (1a), and also as an experimental test invalidating the conventional extrapolation method based on eq. (2) at low energies. It also supports the recent results of enhanced D-D fusion rates observed in electrolysis experiments by Fleischmann et al. [1] and others [3, 5, 8].

Another significance of the experimental results of Beuhler et al. [9] is that the reaction rate for (1b) appears to be substantially suppressed compared with that of reaction (1a) in their experiment, contrary to the conventional assumption of nearly equal rate for both (1a) and (1b). The suppression of reaction (1b) and enhancement of reaction (1a) at low energies have also been observed in the recent electrolysis experiments by Wolf et al.[3], Iyengar [5], and others. One possible explanation is that there may be a broad resonance behavior in $\sigma(E)$ for reaction (1a) but not in $\sigma(E)$ for reaction (1b), which is plausible since the final state Coulomb interaction is present in reaction (1a) but not in reaction (1b). If $\sigma(E)$ happens to have resonance behavior near $E \approx 0$, the extrapolation may yield erroneous values of $\sigma(E \approx 0)$ or $S(E \approx 0)$, since the non-resonant relation (2) is not applicable to resonance reactions. Therefore, it is very important to investigate the possibility of resonance behavior for $\sigma(E)$ near $E \approx 0$ theoretically, and also to measure $\sigma(E \approx 0)$ directly with precision experiments.

Recent observation of neutron bursts at $-30°C$ at a rate of $\Lambda \approx 10^{-23}sec^{-1}$ reported by Menlove et al. [11] may be interpreted as the existence of a sharp resonance in the reaction channel (1b).

At present, there are no direct experimental measurements and theoretical calculations of the branching ratios for reactions (1a) and (1b) below $E \lesssim 3keV$. One would expect the branching ratio of reaction (1a) to be larger than that of reaction (1b), since reaction (1b) involves a fusion of two protons to form ^3He while reaction (1a) does not fuse two protons but merely transfers a neutron from one deuteron to another to form ^3H (known as the deuteron stripping reaction in nuclear physics). Theoretical calculations of the reaction cross-sections and branching ratios of reactions (1a) and (1b) are being carried out based on four-nucleon scattering theory [12] using nucleon-nucleon forces and the Coulomb interaction.

Direct experimental tests of the above interpretation of the results of the electrolysis experiments [1-5] and Beuhler et al. [9] can be carried out by measuring $\sigma(E)$ for both reactions (1a) and (1b) at the energy range of $0 \lesssim E \lesssim 0.3keV$, using the inverse reactions, $p(^3H, D)D$ and $^3H(p, D)D$ for (1a) and $^3He(n, D)D$ for (1b) in addition to using the direct reactions (1a) and (1b).

If the experimental result of Beuhler et al., $\sigma_{exp}(0.15keV) \approx 10^{-13}b$, is conclusively confirmed in future experiments, smaller average deuterium kinetic energies of $E_D \approx 1 \sim 10eV$ are needed to obtain the reaction rates of $\Lambda \approx 10^{-10} \sim 10^{-23}sec^{-1}$ for reactions (1a) and (1b). There are no known mechanisms which can provide $E_D \approx 1 \sim 10eV$ for deuterium atoms imbedded in Pd lattice sites in the electrolysis experiments. However, it is possible to generate the average kinetic energies of $E_D \approx 1 \sim 10eV$ in the surface zone outside the Pd cathode as described below.

It is known that whiskers of Pd deuteride and/or LiD form on the cathode surface in electrolysis experiments in which the Pd cathode is immersed in D_2O with 0.1 M $LiOD$ electrolyte. These whiskers are known to occupy the surface zone of $\gtrsim 10\mu m$ thickness, where most of D_2 gas are formed from the dissociation of D_2O. Depending on electrolysis conditions, many spherical and hemispherical D_2 gas bubbles of varying sizes (radii ranging from few μm to few mm) will be produced continuously in the surface whisker zone and will stay there for certain time durations before they move out of the electrolysis cell. Although the potential difference across a given D_2 gas bubble is expected to be a fraction of the applied potential across the electrolysis cell, high potential differences can occur in nonequilibrium situations in which D_2 gas bubbles can form into a layer that virtually covers the Pd cathode, thereby allowing the formation of a high double-layer electric field. Most of these D_2 gas bubbles in the surface whisker zone will have whiskers protruding into the bubbles creating field emission potentials around the tips of whiskers. The average potential in each D_2 bubble is expected to be approximately that of the applied potential of the electrolysis cell, but the electric field near the whisker tips can be several orders of magnitude larger than the average field, as is well known from field emission studies. Due to this electric field, D^+ ions in the bubble will gain kinetic energies with a statistical distribution which depends on the bubble size and values of the varying electric field inside the bubble. The average kinetic energy of the D^+ ions in each bubble is expected to be $1 \sim 10eV$, when the applied potential is $1 \sim 10V$.

The surface reaction mechanism described above is consistent with the recent observation reported by Appleby et al. [2] that the use of NaOD does not produce excess heat during electrolysis, since whiskers may not form with NaOD electrolyte (without Li) during electrolysis.

The surface reaction mechanism for cold D-D fusion can be tested by measuring the production rates of fusion products of reactions (1a) and (1b) from field emission experiments in which metal deuteride tips (Pd deuteride, Ti deuteride, etc.) are surrounded by D_2 gas at different pressures in a container with different values ($1 \sim 10V$) of applied electric potential.

NEUTRON-INDUCED CHAIN-REACTION HYPOTHESES

Neutron-induced chain-reaction hypotheses [13-16] has been proposed, which involves radiative neutron cature in the Pd cathode,[13] and/or neutron-induced fission-fusion process [14-16]. The first stage of the fission-fusion process is ignited by neutrons from reaction (1b), which produces tritium (^3H or T) via the following fission reaction,

$$n + {}^6Li \rightarrow {}^4He(2.05MeV) + T(2.73MeV), \tag{3}$$

with kinetic energies indicated in parentheses. The cross-section for reaction (3) is very large at thermal energies ($949 \times 10^{-24}cm^2$) [17]. The reaction rate (cross-section times velocity, σv) for reaction (3) is also very large, $\sigma_{nL_i}v_n = (2.1 - 1.3) \times 10^{-16}cm^3s^{-1}$, for a range or neutron energies up to 14 MeV [18]. The natural abundances of Li are 7.5% ^6Li and 92.5% ^7Li.

The second stage of the process is T-D fusion with T(2.73 MeV) generated from the first stage (3) and/or from reaction (1a),

$$T + D \rightarrow {}^4He(3.52MeV) + n(14.07MeV). \tag{4}$$

The T-D fusion cross-section is maximum ($\sim 10^{-23}cm^2$) at a T kinetic energy of $\sim$ 100 keV and is nearly three orders of magnitude larger than the D-D fusion cross-section for the same D kinetic energy. The T-D fusion reaction rate for reaction (4) is large [19], $\sigma v \stackrel{>}{\sim} 10^{-16}cm^3s^{-1}$, for T(10 keV $\sim$ 10 MeV).

Both chain-reaction hypotheses (3) $\rightarrow$ (4) for the FPH effect lead to a set of predictions which can be tested experimentally. However, very low reported values [3, 20-22] of the neutron production rate ($\stackrel{<}{\sim} 10^{-23}sec^{-1}$ per D-D pair) compared with the reported rates of tritium production ($\sim$ 6 hour burst, $10^{-10}sec^{-1}$) [3] and excess heat generation ($\sim$ days to weeks, with an inferred D-D fusion rate of $10^{-10}sec^{-1}$) [1, 2, 4] are inconsistent with the chain-reaction hypothesis, (3) $\rightarrow$ (4), at present.

HYBRID INERTIAL CONFINEMENT FISSION-FUSION

In designing pellets for T-D fusion ignited by neutron flux, the following advantages will be utilized: (a) the fusion ignition may be accomplished without an enormous electromagnetic energy input; (b) the neutron flux is capable of heating $Li/D/T$ fuel selectively or preferentially while laser or charged particle beams cannot; (c) an external neutron flux can be generated economically from (i) conventional fission reactors, (ii) radioactive wastes product of fission reactors, and (iii) thermonuclear explosive stockpiles when atomic and hydrogen bombs are disarmed. In the following, several pellet designs are described for neutron-induced inertial confinement fission-fusion (pellet designs 1(a) and 2(b)) and for magnetically insulated confinement fission-fusion (pellet designs 2(a) and 2(b)).

The design 1(a), as shown in Fig. 1(a), consists of a set of concentric shells with a void inner space. The outer shell is a cannon-ball type shell made of metals or ceramics which have very small neutron activation cross-sections. The intermediate shell is a Li blanket (Li oxide or other Li compounds). The inner shell consists of a Li/D mixture (LiD etc.) or a $Li/D/T$ mixture (LiD/LiT, etc.). The inner void core has a sufficiently large volume ratio with respect to the inner shell to maintain the D/T plasma density ($n_D \approx n_T$) at less than a critical value n_c, when the solid D/T fuel coated on the inner shell is ablated as plasma into the void inner space.

The design 1(b), as shown in Fig. 1(b), is the same as design 1(a), except there is an extra larger outer shell made of materials similar to those used for the outer shell of 1(a). This outermost shell can provide additional containment if the smaller shell inside ruptures. It can also act as a imploding pusher shell when both neutron flux and laser beams are used.

For pellet designs 2(a) and 2(b), shown in Figs. 2(a) and 2(b), we use the magnetically insulated and inertially confined fusion (MICF) scheme with a pulsed laser beam developed by Hasegawa et al.[23,24]. For the new scheme, a neutron flux is used instead of a laser beam (design 2(a)) or both a neutron flux and laser beam (design

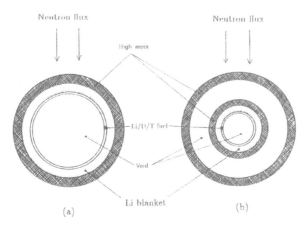

Fig. 1 Pellet designs for hybrid confinement fission-fusion. High
mass material is a metal or ceramic neutron reflector with a
very small neutron activation cross-section. .

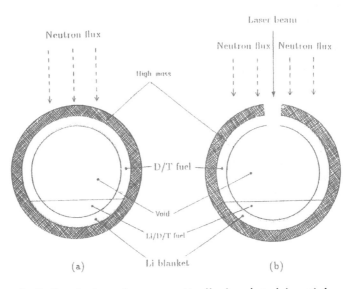

Fig. 2 Pellet designs for magnetically insulated inertial confinement fission-fusion. High mass material is a metal or
ceramic neutron reflector with very small neutron activation cross-section.

2(b)) are used. In MICF, the heated fuel plasma is insulated by a self-generated magnetic field. The magnetic field is produced by ablation due to reaction (3) and/or by direct impact of a laser beam incident on the solid $Li/D/T$ fuel which is coated on the inner surface of the pellet shell as shown in Figs. 2(a) and 2(b). Hasegawa et al.[23,24] have demonstrated that a plasma density $\gtrsim 10^{20}cm^{-3}$ can be contained in a pellet shell due to the insulation of its heat by a self-magnetic field of $\gtrsim 100T$ when a 100 J CO_2 laser is incident on a parylene ($C_8H_7C\ell$) inner shell of $15\mu m$ thickness. With a confinement time of $\tau \approx 8$ n sec and a density of $n \approx 6 \times 10^{20}cm^{-3}$, a preliminary result of $n\tau \approx 5 \times 10^{12}cm^{-3}sec$ at $kT \approx 0.5keV$ has been achieved [23,24]. With the use of an external neutron flux, it should be possible to improve the above result by raising the plasma temperature and confinement time τ.

For all designs, 1(a), 1(b), 2(a), and 2(b), the thickness of each shell is to be optimized by theoretical calculations and experimental tests. Typical values for pellet and shell dimensions are given below.

The amounts of Li blanket and $Li/D/T$ mixture in the pellet are adjusted so that they transform into plasma with a density of less than n_c in order to prevent rupture of the outer shell for a desired time duration. The critical pressure, P_c, of mechanical stress that any metallic container can sustain is $P_c \lesssim 10^5$ atmospheres. The plasma pressure exceeds the critical pressure P_c when the plasma density exceeds $n_c \approx 10^{19}cm^{-3}$ at a plasma temperature of $\sim 10keV$.

For LiD, the density is $\sim 0.923g/cm^3$ and $n_{solid} \approx 6 \times 10^{22}cm^{-3}$. Hence, when $0.923 \times 10^{-4}g$ of solid LiD becomes plasma in a unit volume, we obtain the plasma density of $n_c \cong 6 \times 10^{18}cm^{-3}$. Therefore, the ratio of volumes of the LiD shell and the inner void space has to be kept $\sim 10^{-4}$ in order to maintain the $Li - D$ plasma density at less than or equal to $n_c \approx 6 \times 10^{18}cm^{-3}$. For the total volume, $\frac{4\pi}{3}r_b^3$, of the LiD shell and the inner void space, we require the LiD $(or\ Li/D/T)$ shell volume $\frac{4\pi}{3}(r_b^3 - r_a^3)$ to be $10^{-4} \times \frac{4\pi}{3}r_b^3$, i.e. $\frac{4\pi}{3}(r_b^3 - r_a^3) \approx 10^{-4} \times \frac{4\pi}{3}r_b^3$ or require the LiD $(or\ Li/D/T)$ shell thickness to be $(r_b - r_a) \approx \frac{1}{3} \times 10^{-4}r_b$.

Since reaction (4) for T-D fusion releases $17.6MeV \approx 2.82 \times 10^{-12}J(1MeV = 1.6 \times 10^{-13}J)$ per reaction, 1g of D/T fuel, which contains $\sim 1.2 \times 10^{23}$ D/T atoms, can release a total energy of $\sim 3 \times 10^5$ MJ which is equivalent to 75 tons of TNT. Since the largest explosive energy, which can be expected to be practically handled, is 100 MJ (equivalent to 25 kg of TNT), the practical amount of D/T fuel is limited to $\lesssim 0.3$ mg per pellet.

If we assume $r_b \approx 1.3cm$, then $(r_b - r_a) \approx 0.43 \times 10^{-4}cm = 0.43\mu m$ so that the total volume $\approx 9.2cm^3$, and the LiD $(or\ Li/D/T)$ shell volume $\approx 9.2 \times 10^{-4}cm^3$. The pellet then contains $\sim 0.85mg$ of LiD $(or Li/D/T)$ or $\sim 0.3mg$ of D (or D/T).

An estimate of the appropriate amount of Li to be included in the pellet and blanket is made assuming a thermal neutron flux of $\Phi_n \approx 10^{15}cm^{-2}sec^{-1}$, which may not currently be available. When the neutron flux is incident on the pellet (1(a) or 1(b)), it will not heat the outer shell because of the very small value of its neutron activation cross-section but will heat both the Li blanket and the $Li/D/T$ shell via reaction (3) preferentially. The rate of T(2.7 MeV) production is

$$R_T = n_{Li}\Phi_n\sigma_{nL_i} \approx n_{Li} \times (0.949 \times 10^{-6})sec^{-1}$$

$$\cong 5.7 \times 10^{16}cm^{-3}sec^{-1},$$

with $n_{Li} \approx 6 \times 10^{22} cm^{-3}$ in $LiD(\rho \approx 0.923 g/cm^3)$. The corresponding power production by reaction (3) is then

$$P_{Li}(0.923g) = R_T Q_{nL_i} \approx 4.4 \times 10^4 W\, cm^{-3} \approx 44 KW\, cm^{-3}$$

using $Q_{nL_i} \approx 7.68 \times 10^{-13} J/reaction$. Note that the total energy released by burning 0.92g (1 cm^3) of LiD is $E_{total} \approx n_{Li} Q_{nL_i} \approx 4.6 \times 10^{10} J = 46 TJ$.

Even though the amount of Li in the LiD shell is limited to $\sim$ 6mg due to the above considerations, we can increase the input power due to reaction (3) by inserting an intermediate shell containing a Li blanket. The amount of Li in the blanket is adjusted to provide sufficient input power in the pellet for the TD fusion ignition. If the TD fusion ignition cannot be achieved by the neutron flux alone, pulsed laser beams can be used to cause implosion of the outer shell with the inner shell preheated by the neutron flux. For this latter situation, design 1(b) may be more suitable than designs 1(a) and 2(a).

When the incident neutron flux produces T(2.7 MeV) via reaction (3) it can either enter the inner void space as a positive ion or be stopped within the LiD or Li blanket. The range of T(2.7 MeV) in LiD is $\sim 6.4 \times 10^{-3} cm$. The probability, P_{TD}, that T(2.7 MeV) will interact with D in solid LiD before slowing to rest is given by

$$P_{TD} = \int dx\ \sigma_{TD}\, n_D = \int dE\ \sigma_{TD}(E)\, n_D/(dE/dx)$$

$$\approx 1.8 \times 10^{-4},$$

with $n_D \approx 6 \times 10^{22} cm^{-3}$ in LiD. The rate of the T-D fusion in LiD is then

$$R_{TD} = R_T P_{TD} \approx 0.84 \times 10^{13} cm^{-3} sec^{-1}.$$

However, P_{TD} may increase substantially in $Li-D$ plasma. Both T(2.7 MeV) and ^4He (2.1 MeV) will heat up the LiD and Li blanket, ablating electrons as well as Li and D ions into the inner core space, thereby creating a hot plasma of electrons, Li, D, and T ions. If the confinement time τ (before the outer shell is ruptured by heat conduction and plasma pressure) is of sufficient duration for Lawson's criteria for T-D fusion ignition, then T-D fusion will be ignited and D originating from the LiD fuel will be burned to release energy via reaction (4). Note that 0.9 mg of LiD contains $\sim 6 \times 10^{19}$ D atoms and will release a total energy of 169 MJ.

Since Lawson's criteria for TD fusion ignition is $n\tau \gtrsim 3 \times 10^{14} cm^{-3} sec$ at $kT \approx 10 keV$, the required confinement time is $\tau \approx 3 \times 10^{-5} sec$ for $n = 10^{19} cm^{-3}$, and $\tau \approx 1 sec$ for $n = 3 \times 10^{14} cm^{-3}$. The outer shell of the pellet must be able to contain hot TD plasma inside the pellet for the confinement time τ in order to achieve the TD fusion ignition. The above parameters and dimensions for design 1(a) are based on the assumed neutron flux of $\Phi_n \simeq 10^{15} cm^{-2} sec^{-1}$. Other possible parameters and dimensions for design 1(a) with varying values of the neutron flux, $\Phi_n < 10^{15} cm^{-2} sec^{-1}$, are to be investigated to determine the lower limit of Φ_n for which the T-D fusion can be ignited for design 1(a) as well as for designs 1(b), 2(a), and 2(b). If a neutron flux of $\Phi_n \approx 10^8 cm^{-2} sec^{-1}$ is shown to be capable of igniting T-D fusion for some specific designs of the $Li/D/T$ pellet, it has an important and practical consequence in that the radioactive wastes (such as in the form of a $Am-Be$ capsule, for example) can be utilized as a neutron source for T-D fusion ignition.

The ignition of TD fusion by neutron-induced fission is expected to be more involved for magnetic plasma confinement because of the need for more complex designs required. The following magnetic confinement devices need to be investigated for the proposed process: (a) tokamak; (b) reversed field pinch; (c) magnetic mirror; and (d) ion trap. Because of design complexicity, T-D fusion ignition by neutron flux in a tokamak reactor is expected to be the most difficult.

The neutron-induced fission-fusion reactor with the $Li/D/T$ pellet described above is a realistic alternative to the conventional fusion reactors, designed for magnetic plasma or inertial confinement fusions, which have not yet achieved a break-even stage for a sufficient time duration in spite of enormous scientific and engineering efforts invested during the last two decades. Therefore, it is important and urgent to investigate the feasibility of achieving an ignition stage in new prototype fission-fusion reactors with the use of an external neutron flux, as described in this paper.

REFERENCES

1. M. Fleischmann, S. Pons, and M. Hawkins, Journal of Electroanalytic Chemistry, **261**, 301 (1989); and errata, **263**, 187 (1989).

2. A. J. Appleby, S. Srinivasan, Y. J. Kim, O. J. Murphy, and C. R. Martin, "Evidence for excess heat generation rates during electrolysis of D_2O in LiOD using a palladium cathode - a microcalorimetric study", a talk presented by Appleby at Workshop on Cold fusion Phenomena, May 23-25, 1989, Santa Fe, New Mexico; "Anomalous calorimetric results during long-term evolution of deuterium on palladium from lithium deuteroxide electrolyte" submitted to Nature; A. J. Appleby, Y. J. Kim, O. J. Murphy, and S. Srinavasan, "Specific effect of lithium ion on anomalous calorimetric results during long-term evolution of deuterium on palladium electrodes", submitted to Nature.

3. K. L. Wolf, N. J. C. Packham, D. R. Lawson, J. Shoemaker, F. Cheng, and J. C. Wass, "Neutron emission and the tritium content associated with deuterium loaded palladium and titanium metals", in the Proceedings of Workshop on Cold Fusion Phenomena, May 23-25, 1989, Santa Fe, New Mexico, to be published in J. Fusion Energy.

4. A. Belzner, U. Bischler, G. Crouch-Baker, T. M. Gur, G. Lucier, M. Schreiber, and R. Huggins "Two fast mixed-conductor systems: deuterium and hydrogen in palladium-thermal measurements and experimental considerations", presented by R. A. Huggins, in the Proceedings of Workshop on Cold Fusion Phenomena, May 23-25, 1989, Santa Fe, New Mexico, to be published in J. Fusion Energy.

5. P. K. Iyengar, "Cold Fusion Results in BARC Experiment", in the Proceedings of the 5th International Conference on Emerging Nuclear Energy Systems (ICENES V), Karlsruhe, West Germany, July 3-6, 1989.

6. W. A. Fowler, G. R. Caughlan, and B. A. Zimmermann, Ann. Rev. Astr. Astrophys. 5, 525-570 (1967).

7. N. Jarmie and N. E. Brown, Nucl. Inst. Method **B10/11**, 405-410 (1985); A. Krauss, H. W. Becker, H. P. Trautvetter, and C. Rolfs, Nucl. Phys. **A 465**, 150 (1987).

8. S. E. Jones et al., Nature **338**, 737 (27 April 1989).

9. R. J. Beuhler, G. Friedlander, and L. Friedman, Phys. Rev. Lett. **63**, 1292 (1989).

10. Y. E. Kim, "Comment on Cluster-Impact Fusion", PNTG-89-11 (October 1989).

11. H. O. Menlove et al., "The measurements of neutron emissions from Ti plus D_2 gas", in the Proceedings of Workshop on Cold Fusion Phenomena, May 23-25, 1989, Santa Fe, New Mexico, to be published in J. Fusion Energy.

12. O. A. Yakubovsky, Yad. Fiz. **5**, 1312 (1967) [Sov. J. Nucl. Phys. **5**, 937 (1967)]; P. Grassberger and W. Sandhas, Nucl. Phys. **B2**, 181 (1967); E. O. Alt, P. Grassberger and W. Sandhas, J.I.N.R. Report No. E4-6688 (1972).

13. Y. E. Kim, "Neutron-induced photonuclear chain-reaction process in palladium deuteride", PNTG-89-7 (July 1989).

14. Y. E. Kim, "New cold nuclear fusion theory and experimental tests", an extended summary of a talk presented at Workshop on Cold Fusion Phenomena, May 23-25, 1989, Santa Fe, New Mexico, to be published in J. Fusion Energy.

15. Y. E. Kim, "Hybrid inertial confinement fission-fusion for large scale power generation", Purdue Nuclear Theory Group Report PNTG-89-10 (September 1989).

16. Y. E. Kim, "Neutron-induced tritium-deuterium fusion in metal hydrides", Purdue Nuclear Theory Group Report PNTG-89-4 (April 14, 1989); "Fission-induced tritium-deuterium in metal deuterides", PNTG-89-5 (June, 1989).

17. T. Lauritsen, and F. Ajzenberg-Selove, Nucl. Phys. **78**, 39 (1966).

18. H. Goldstein, and M. H. Kalos, "An Index to the Literature on Microscopic Neutron Data", NDL-SP-10, Nuclear Defense Laboratory, Edgewood Arsenal, MD (1964); Nuclear Data Tables.

19. D. L. Book, "NRL Plasma Formulary", Naval Research Laboratory Publication 0084-4040 (Revised 1987).

20. R. D. Petrasso, et al., Nature **339**, 183 (1989); and erratum, **339**, 264 (1989).

21. M. Fleischmann, S. Pons, and R. J. Hoffmann, Nature **339**, 667 (1989).

22. R. D. Petrasso, et al., Nature **339**, 667 (1989).

23. A. Hasegawa, et al., Phys. Rev. Lett. **56**, 139 (1986).

24. A. Hasegawa, et al., Nuclear Fusion **28**, 369 (1988).

ION BEAM-PLASMA INTERACTION FOR PARTICLE DRIVEN FUSION

C. Deutsch and G. Maynard
L.P.G.P.[+], Université Paris Sud, 91405 Orsay Cedex, France

R. Bimbot, D. Gardes, M. Dumail, B. Kubica, A. Richard,
M.F. Rivet and S. Servajean
I.P.N., B.P. 1, 91406 Orsay Cedex, France

C. Fleurier and M. Hong
GREMI, CNRS, Université d'Orléans, 45067 Orleans, France

D.H.H. Hoffmann, K. Weyrich and H. Wahl
MPQ-Garching and GSI, Postfach, 110552, D-6100 Darmstadt, FRG

ABSTRACT

The interaction of energetic multicharged ion beams with separately produced target plasmas is investigated within a projectile ion-target electron binary pattern. The validity conditions of the relevant standard model are asserted. A few extensions are suggested.

The corresponding stopping results are contrasted to recent experimental results obtained by producing a fully ionized plasma on accelerator beam lines with the SPQR2 setup.

1. INTRODUCTION

As already advocated several times[1], the stopping of intense and energetic ion beams in fully ionized plasmas may be correctly understood at zero order of the beam-plasma interaction, through the behaviour of a single projectile within the target. Simple order of magnitude estimates demonstrate that in the most intense beams of ICF interest[1,2], the average ion-ion distance remains two orders of magnitude larger than the mean electron screening length in cold target. This fortunate occurrence, often referred to as a reduction principle, provides us with two fundamental consequences.

First, it allows us to investigate plasma stopping properties within the theoretical framework already worked out for cold matter.

This latter is essentially patterned on the ion projectile-target electron encounter. In this approach, the excitation of one-electron orbitals bound to the target, plays a crucial role. So, it should be carefully evaluated.

[+] associé au C.N.R.S.

Next, the reduction principle opens the way to a novel stopping metrology of ICF interest[1], which makes use of already existing accelerating facilities, provided the target plasma is separately ignited. The latter simulates a coronal plasma arising in a compression process through intense beams.

We are thus led to investigate at length the stopping of energetic ions in dense and hot matter. The resulting energy losses and charge exchange cross-sections will then be of a great help in assessing the performances of ICF driven by intense ion beams compressing and heating the pellet.

The present approach relies on the neglect of any significant collective beam-plasma effect[3], as already suggested several times[1,2,4], by analytical estimates as well as numerical simulations demonstrating that intense ion beams ($E/A \sim 50$ MEV) are unlikely to be significantly deflected by beam-target coronal instabilities.

In addition, even in a rather turbulent situation, the binary projectile-target electron pattern is expected to build up the beginning of a more complete understanding.

Sec. 2 is devoted to a systematic extrapolation of plasma stopping in the cold matter approach.

We thus explain the content of the various assumptions underlying the so-called Stopping Standard Model (SSM) implicitly used, up to now, in computing the stopping of intense ion beams within multilayered ICF targets[1,2,4].

Basic characteristics of plasma stopping and projectile effective charge are discussed within the SSM.

A few obvious extensions beyond the SMM are examined in Sec. 3.

The experimental setups currently developed to check out the present assertions are briefly displayed in Sec. 4. An emphasis is laid on the SPQR2 program which makes use of a dense plasma column inserted on the linear-accelerator beam line. Preliminary results are discussed. Future programs are sketched out.

2. STOPPING STANDARD MODEL (SSM)

The basic content of our standard knowledge about cold matter stopping is validated by a few assumptions and scaling laws underlying the use of the well-known 3B (Bohr-Bethe-Bloch) formula for swift particle stopping (see Table 1 and eq. (1)). The very content of the Stopping Standard Model (SSM) for hot matter, essentially consists in applying the same guidelines for ion-plasma interaction.

This is not an arbitrary assumption. It simply amounts to the most economical formalization of our present understanding of this topic[1,5], while allowing us to focus on the specific plasma effects. These latter, as detailed below, points out to a enhanced stopping capability of a plasma with a given linear electron density $n_e \ell$ (ℓ = range of projectiles into the medium) when compared to its cold gas equivalent (same $n_e \ell$). The various items listed in Table 1 may be systematically questioned. However, except very near the end of range, they can be checked out as highly robust statements.

Table 1 - Stopping standard model: Basic facts and assumptions

- Intense ion beams appear **dilute** in target
- rectilinear trajectoires
- pointlike projectiles
- Nonrelativistic regime ($\beta \leq 0.35$)

- $n_e \ell$-scaling $\quad -\dfrac{\Delta E}{E} \simeq \dfrac{n_e \ell}{E^2}$

- prefactor $\quad \dfrac{4\pi Z_1^2 e^4 n}{m_e V_1^2} \qquad$ dominant for E/A between 1 Mev/a.m.u. and
1 Gev/a.m.u.

- log terms dominant at very small (end of range) and very large projectile velocity.

As mentioned above, the assumptions listed in Table 1 essentially sustain the validity range of the 3B formula for the stopping of nonrelativistic point charges by isolated atom or cold matter which may be straightforwardly extended to hot and partially ionized material. Then, bound and free electrons, and also to a very limited extent, plasma ions, contribute to the projectile stopping. Restricting our attention to the overwhelming electron contribution, one thus writes

$$ -\frac{dE}{dx} = \frac{4\pi(Z_1 e^2)^2 \, n}{m_e V_1^2} \times [\,(Z_T - \bar{Z})\, \ell n\, \Lambda_B + \bar{Z}\, \ell n\, \Lambda_F\,] \quad , \tag{1} $$

$\bar{Z}$ denotes target average ionization (number of free electrons/nucleus). For high (with respect to target electron velocities but not yet relativistic) velocities,

$$ \Lambda_B = \frac{2m_e V_1^2}{I_{av}} \tag{2} $$

with I_{av} = mean excitation energy, while

$$ \Lambda_F = \frac{m_e V_1^2}{\hbar\omega_p} \tag{3} $$

where ω_p = target plasma frequency.

Actually, Λ_F is a high T limit of a more general expression derived from a complete RPA dielectric function for the electron jellium.

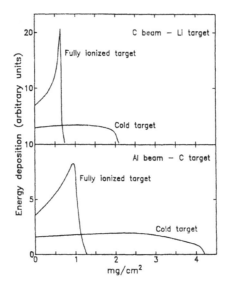

Figure 1- Energy deposition profiles. Top, 12 MeV carbon beam in a
fully ionized and in cold lithium target. Bottom, 54 MeV Al
beam in a fully ionized and in a cold C target (Nardi and
Zinamon[5]).

A/ Enhanced stopping (Fig. 1)

The first basic discrepancy between stopping in a heated target and
in the cold equivalent with the same $n_e\ell$, arises from the enhanced
stopping capabilities of the corresponding plasma. As long as the Born
RPA approximation works, the small relative projectile ion-target
electron velocity allows for an efficient energy exchange. Therefore,
the highest electron orbitals are expected to respond in the most
versatile fashion to the incoming electrostatic field. One thus gets the
approximate rule of thumb, according to which free electron orbitals
show up the most efficient stopping, while highly excited bound orbitals
are more effective than deeply bound ones.

This latter statement may be illustrated by the range shortening
arising from the bound electrons contribution.

For this purpose, let us remark that the isolated ion $\ell n \Lambda_B$ may still
be used up to a 10 ev target temperature[8] (eq. (2)). Above 10 eV, plasma
corrections systematically lower Λ_B values, thus demonstrating the above
statement.

B/ Enhanced projectile effective charge ($Z_1(V_1)$)

The basic physics of the projectile effective charge is driven by the
observation that it is much more difficult for the travelling ion to
pick up bound orbitals out of plane waves than from target bound
states[5].

This specific behaviour explains that once the ionization fate of the
projectile is settled at the target entrance (Fig. 2), it then remains
nearly unchanged excepted very close to the end of range, where it does

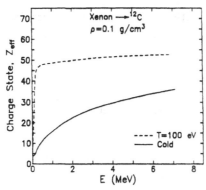

Figure 2 - The charge Z_{eff} of a xenon ion as a function of its energy as slowed in a fully ionized carbon target (Nardi and Zinamon[5]).

recombine. Such a feature is at variance with projectile charge state in cold target which provides easily a lot of prepared bound orbitals, ready for pick up, so to speak. As a final result, we expect from the enhanced $Z_1(V_1)$ an additional contribution to the range shortening already considered in A/.

It is often a rather substantial one (see (Fig. 1).

3. BEYOND THE S.S.M.

The modelling developed up to now, seems quite adequate in investigating quantitatively ion stopping on most part of its range. However, the SSM is likely to break down near the end of range, where the beam-driver energy conversion takes place. Actually, the Coulomb logarithms (2) and (3), thus fall below unity, and the projectile behaves as a brownian particle assaulted by the target electrons. The stopping is then proportional to V_1. On the other hand, when the projectile kinetic energy is smaller than 100 kev/a.m.u., Coulomb ion-ion interactions are no longer negligible.

The very basic SSM assumption, according to which the beam appears dilute in target, might also be questioned in the presence of beam filamentation instabilities[1]. Then, clusters of beamlets are strongly compressed transversally. So, when the given projectiles flow with parallel velocities, they experience strong correlations. As a consequence, temperature extensions of cluster stopping currently used to determine the vicinage effect in solids, might be useful in addressing such questions.

4. BEAM-PLASMA EXPERIMENTS

A/ Target plasma

The theoretical predictions of the SSM for ion energy losses and effective charge in target can be checked out by firing a plasma located on the beam line of a conventional linear accelerating structure[1].

These beam/target interaction experiments can be easily scaled under the form (see Table 1)

$$-\frac{\Delta E}{E} \sim \frac{n_e\ell}{E^2} \quad , \tag{4}$$

where $n_e\ell$ is the product of the target electron number density with the projectile range in the target given by the linear part of the range expression

$$\int_{E_0}^{E_0/10} \frac{dE}{dE/dx}$$

Table 2 – Target plasma currently used in probing heavy ion beam stopping

Set-up (location)	electron linear density $n_e\ell$ (cm^{-2}) in target	target electron temperature ev	projectile (energy in Mev/a.m.u.)
Laser ablated plate synchronized with a tandem accelerator (SPQR1) (Bruyère le Chatel)	10^{18}	80-200	Cu^{5+}, Si^{12+} at 0.5 Mev/a.m.u.
Dense hydrogen discharge on a beam line (SPQR2) (Orsay, IPN)	a few 10^{19}	2-5	C^{4+}, S^{7+} at 2 Mev/a.m.u.
GSI (Darmstadt)			Ar^{18+}, Ge^{20+} Ca^{13+}, U^{33+} at 1.4 Mev/a.m.u.
Z-Pinch (Aachen and GSI)	10^{21}	15	heavy ions available on UNILAC line $(\frac{E}{A} \geq 1.4$ Mev/a.m.u.)

The content of eq. (4) essentially amounts to saying that an optimum energy loss corresponds to a given line density $n_e\ell$. For instance, it is of little interest to fire too energetic ions in a moderately dense plasma. The corresponding E is likely to be too small. In these respects the projectile energy envisoned for SPQR3 lies near the ''real game'' value, i.e., 10 to 50 MeV/amu.

These considerations motivate the SPQR2 program with projectile energies in the 1 to 4 MeV/amu range and a target $n_e\ell \geq 10^{19}$/cm^2, described below.

Table 2 displays several projectile-target plasma pairs listed in order of increasing $n_e\ell$.

We detail a little bit more the SPQR2 setup which has been extensively used, very recently.

Our basic goal is to simulate an ablation corona resulting from an implosion conducted through intense ion beams. This is achieved with the discharge tube pictured on Fig. 3. Low pressure hydrogen (10 Torr) is introduced in a quartz capsule (neck diameter = 25 cm, length = 40 cm). High power electric discharge (20 kA) ignites a plasma. A free electron density ~ 4-5×10^{17} e-cm^{-3} is then established for 100 µsec. An axial and steady 2 kG magnetic field prevents the onset of Rayleigh-Taylor and other instabilities. The interelectrode distance is 40 cm.

Plasma parameters (n_e, T) are checked out through laser interferometry and line broadening (Stark effect) of H_α, H_β lines emitted by excited neutral species. Along the beam line, we can still rely on a neutral density $\sim 10^{15}$ cm^{-3}, sufficient for diagnostic purposes.

Electrodes are hollowed in their center to allow the ion beam through. Moreover, there are no solid plasma boundaries. An efficient differential pumping (Fig. 3) secures a swift circulation of helium or hydrogen.

SPQR 2 $n\ell \approx 10^{19}$ e cm^{-2}

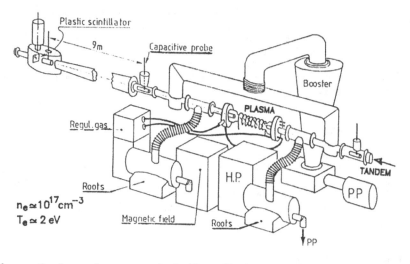

Figure 3- General set-up including discharge, differential pumping and time-of-flight. (Bimbot et al.,[11]).

The primary pump works at 17 ℓ/sec. The roots at 140 ℓ/sec, while the booster operates at 2000 ℓ/sec in air or 6000 ℓ/sec in H and He.

The whole set-up is located, altogether with the capacitor bank in a Faraday cage. In Figure 4, we display theoretical results closely related to the SPQR2 interaction experiment. One can see how the effective charge results from a dynamic balance between ionization and recombination. However, the latter is much less likely to occur in a plasma.

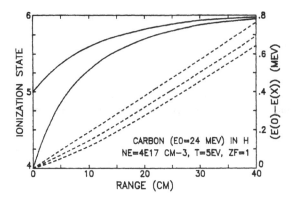

Figure 4 – Effective charges and energy losses of C^{4+} ions with 2 MeV/a.m.u. in fully ionized hydrogen

The important point explains a marked discrepancy between the effective charge in plasmas and in cold matter, where the usual expression

$$Z_{eff} = Z[1-1.034 \times exp[-(v/2.19x10^8 \, cm/s)]Z^{-0.688}]$$

applies. On the beam line (see Fig. 3) the three capicitive phase probes pick up the ion velocity before and after interaction with the plasma. The last two are disposed on both sides of a 9 m long time-of-flight which allows for a 4% accurate energy loss. When the beam current is too low, multichannel plates disposed at the end of the beam line can provide an alternative beam probe, provided the corresponding time structure is stable enough.

5. MEASURED ENERGY LOSS

The SPQR2 setup is particularly instructive, because it afford to contrast the behaviour of light ions to the heaviest ones, when flowing through the same plasma. As a rule of thumb, ions with the smallest ionicity/mass ratio are the easiest to detect after interaction with the plasma. They are less likely to be suddenly deflected by bursts of instabilities or other collective effects.

A/ Light ion stopping[12]

The energy losses of ^{12}C and ^{32}S beams in 9 torr H_2, deduced from time of flight measurements with and without gas, are found to be 0.61 $\pm$ 0.05 MeV and 3.11 $\pm$ 0.15 MeV respectively. These results are in fair agreement with semi-empirical tabulations for cold matter stopping.

The energy losses in plasma have been measured for ^{12}C and ^{32}S projectiles, and for two time windows corresponding to the beam intensity maximum mentioned above. The results are displayed in Fig. 5 for ^{32}S ions, and for a 13 kV discharge in 9 torr H_2. (A similar

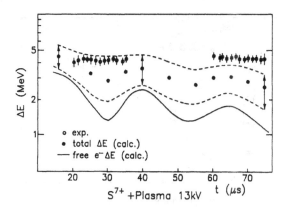

Figure 5 – Experimental and calculated energy loss in a fully
hydrogen plasma (see text) for a S^{7+} ion beam
(2MeV/a.m.u.) (Gardès et al.[12])

behaviour is observed for ^{12}C projectiles, the energy losses being
reduced by a factor 5, consistent with the ratio of effective charges of
C and S ions). The experimental energy loss ΔE (open symbols) is
compared in Fig. 5 to the calculated value.

B/ Heavy ion stopping

Figure 6 shows the energy loss of 333 MeV ^{238}U ions in neutral
hydrogen gas at 6.5 mbar (dashed line) and the energy loss during the
discharge time of the capacitor bank. Due to the oscillating current in
the plasma discharge circuit we find a corresponding oscillation in the
energy loss of the ions. The peaks in the energy loss occur when the
plasma parameters temperature and density reach maximum values. After
about 70 µs the energy loss drops below the value measured in neutral
hydrogen. At this time a rarefaction wave reaches the axis of the plasma
tube and the density of the target gas is reduced.

We compare the experimental results with theoretical precdictions for
both the well known case of ions penetrating cold hydrogen gas and the
plasma case. Figure 6 shows the evaluation of the SSM taking into
account the charge state development of the ions passing through the
target. The main difference between the cold gas case and the plasma
stems however from the difference in the Coulomb logarithms in eq. (1).
The enhanced energy loss of heavy ions in a plasma environment predicted
by this theory is thus well documented by the experimental data.

6. FUTURE PROSPECTS

The success of the above SPQR2 program leads us to consider more
dense plasma targets with higher $n_e\ell$ values. This is the purpose of our
next step making use of a Z-pinch target, presently under development in
Aachen, and displayed on the UNILAC beam line at GSI. One thus expects a
two order of magnitude increase in $n_e\ell$, with a temperature in the 15-20
ev range for a highly ionized hydrogen discharge.

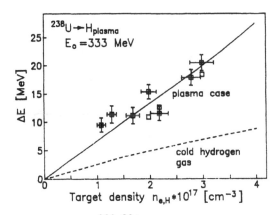

Figure 6 - Energy loss of $^{238}U^{33+}$ ions in a fully ionized hydrogen plasma and in its cold gas equivalent (Hoffmann et al.[13]) at the same pressure.

REFERENCES

1) C. Deutsch, Ann. Phys. Fr **11** (1986) 1.
2) R.C. Arnold, J. Meyer-Ter-Vehn, Rep. Prog. Phys. **50** (1987) 559.
3) C. Deutsch, P. Fromy, X. Garbet and G. Maynard, Fus. Techn. **13** (1988) 362.
4) R.O. Bangerter, Fus. Tech. **13** (1988) 348.
5) E. Nardi and Z. Zinamon, J. Physique **C8** (1983) 93.
6) R. Dei-Cas, Report CEA-R-5119 (1981).
7) G. Maynard and C. Deutsch, J. Physique **46** (1985) 1113.
8) X. Garbet and C. Deutsch, Europhys. Lett. **2** (1986) 761.
9) C. Deutsch and G. Maynard, Euro. Phys. Lett. **7** (1988) 31, and Phys. Rev. A **40** (1989) 3209.
10) G. Maynard, Thesis, Orsay, 1987.
11) R. Bimbot, S. Della-Negra, D. Gardès, M.F. Rivet, C. Fleurier, B. Dumax, D.H.H. Hoffmann, K. Weyrich, C. Deutsch, G. Maynard, to be published IN Laser Interactions and Related Plasma Phenomena, ed. H. Hora, Vol. 8 (1988) 665.
12) D. Gardès, R. Bimbot, S. Della-Negra, M. Dumail, B. Kubica, A. Richard, M.F. Rivet, A. Servajean, C. Fleurier, A. Sanba, C. Deutsch, G. Maynard, D.H.H. Hoffmann, K. Weyrich, H. Wahl, Europhys. Lett. **8** (1988) 701.
13) D.H.H. Hoffmann, K. Wevrich, H. Wahl, T. Peter, J. Jacoby, R. Bimbot, D. Gardès, S. Della-Negra, M.F. Rivet, M. Dumail, C. Fleurier, A. Sanba, C. Deutsch, G. Maynard, R. Noll, R. Haas, R. Arnold, and S. Maurmann, Z. Phys. A **330** (1988) 339.
14) D.H.H. Hoffmann et al., letter of intend, GSI (1987) and C. Fleurier et al., J. Phys. (Paris) **C7** (1988) 141.

TIGHT FOCUSING OF PROTON BEAM

AND ITS INTERACTION WITH TARGETS

Weihua Jiang, Katsumi Masugata and Kiyoshi Yatsui

Laboratory of Beam Technology
Nagaoka University of Technology
Niigata 940-21, Japan

ABSTRACT

A high-bright, self-magnetically insulated, coaxial ion beam diode, "Plasma Focus Diode" (PFD), was successfully developed and studied system-atically. Using such the tightly focused beam, we investigated the ion beam energy deposition in the intense beam-target interaction. Experi-mental results on various diagnostics are reported briefly involving diode operation, ion-current density, beam-power density and energy spectra. The beam-target interaction experiment is described in detail, where the proton energy loss was measured by a time-resolvable Thomson-parabola spectrometer (TPS). In comparison with the energy loss for cold target (calculated by Bethe equation), enhanced energy deposition was observed for aluminium targets of 7 μm thick with maximum enhancement ratio of $\sim$ 1.5. Theoretical efforts involve diode behaviour analysis and simulations of beam-target interaction. We analyzed diode behaviour through calcula-tions of field distribution, electron and ion densities and diode imped-ance. In the simulation of beam-target interaction, we applied bound electron and free electron stopping power model, thermal equilibrium model, radiation and conduction model and hydrodynamic expansion model. The simulated results of proton energy loss in aluminium are in good agreement with the experimental data. Following above experimental and theoretical studies, physical understanding was obtained on PFD and the associated interaction with targets.

Ⅰ. Introduction

Intense light ion-beam (LIB) is considered as a hopeful candidate for inertial confinement fusion (ICF) driver.[1] Two of the major problems of LIB-fusion are beam power concentration and beam-target interaction. For beam power concentration, the expected beam power density is more than 100 TW/cm^2 on a small spherical target of diameter of 5 $\sim$ 10 mm. To obtain such a high power density, it is very important for us to achieve large solid angle of irradiation and small divergence angle of the LIB.[2] We have successfully developed a new and simple type of self-magnetically insulated diode named "Plasma Focus Diode" (PFD).[3-5] The ion beam in PFD focuses two-dimensionally (line focusing) with an axial length of 40 mm. Since the diode parameter diagnostics were reported elsewhere,[6-9] we here present the experimental results of diode operation, ion-beam current density, beam-power density at the focusing and the energy spectra of proton. We compared these data with the analytical calculation of diode behavior by using a simple model of laminar electron flow and space-charge limited ion emission.

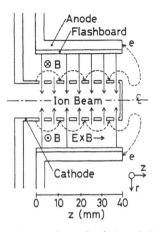

Fig. 1   Basic principle of PFD.

For studing the physical processes of compressing and heating the target in LIB-ICF studies, it is necessary to understand clearly the energy deposition mechanism in high-power beam interaction with targets. Recently, enhanced energy deposition due to target ionization has been considered of importance in the interactions. Both experimental and theoretical works were carried out to study this phenomenon.[10-15] For observing the enhanced ion energy deposition, it is important to heat the target to a sufficiently high temperature. Considering the wide range of incident angle of the ion beam, we used PFD to study the enhanced energy deposition in targets, even with the beam power density of ~ 0.1 TW/cm² which is smaller compared with that used by others.[10-12] In this paper, we will report our experimental results of proton energy loss in aluminium targets and give some discussions about the enhanced stopping power by comparing the experimental energy loss with both the calculation of cold target and the simulation of target plasma.

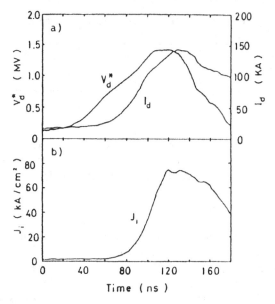

Fig. 2   Typical waveforms of a) $V_d{}^*$ and I_d, and b) J_i.

Ⅱ. Plasma Focus Diode

a) Experimental Diagnostics

Figure 1 shows the principle of PFD schematically. It consists of two coaxial cylindrical electrodes: anode (outer cylinder, 35 mm in ID) and cathode (inner cylinder, 22 mm in OD). The axial length is 40 mm. Epoxy was used as the ion source for the flashboard anode. When the pulse power is fed to PFD, electrons produced at the cathode initialy irradiate the anode. The diode current produces a self-magnetic field (B_θ). When the diode current exceeds a certain critical value, the flow of electrons is self-insulated by B_θ. The ion beam generated from the anode is two-dimensionally focused (line focusing) toward the central axis through the perforated cathode which is drilled with holes (1 mm in diameter) with transparency of $\sim$ 40 %.

The experiment was carried out on the pulse-power generator "ETIGO-II".[16,17]   Figure 2 shows the typical waveforms of (a) inductively-calibrated diode voltage ($V_d{}^*$), diode current (I_d) and (b) ion-current density (J_i) at the focusing region.   The ion-current density was measured by a   biased-ion collector (BIC) at r = 11 mm and z = 5 mm.

By using a three-channel BIC and a Rutherford-scattering pinhole camera, we have measured the ion-current density (J_a) on the anode surface and the focusing radius of ion beam (r^*) at three positions along the axial direction of PFD, z = 7, 20 and 33 mm.   Very tight focusing has been obtained to be 0.36 mm in diameter toward the focusing axis at z = 7 mm.[6-9]   Thus the ion-beam power density was estimated as:

$$P_f = (r_a/r^*)J_aV_d{}^*. \tag{1}$$

Table I summarizes the data for three axial positions. From Table I, the

Table I. Beam power density (P_f) estimated for three axial positions.

z (mm)	r* (mm)	J_a (kA/cm^2)	P_f (TW/cm^2)
7	0.18	1.4	0.18
20	0.23	1.7	0.17
33	0.25	1.9	0.18

power density is seen to be approximately same at the three positions, which is due to the fact that the focusing radius increases as the ion-current density increases toward the top of the diode.

b) Analytical Calculation

Figure 3 shows the conceptual structure of our calculation model. In Fig. 3, r_c and r_a are the cathode and anode radii where the electrical potentials are 0 and V_d*, respectively. Here, r_s is the surface radius of the electron sheath which carries the current I_e. The ion-current density is J_i and there is a current I_c in the cathode. The radius r_s divides the A-K gap into two parts and we used space-charge limited ion emission model for region A and the laminar electron flow model for region B:

$$\nabla \cdot \mathbf{E} = -e\,n_i / \varepsilon_0, \quad \text{(for region A)}$$
$$-\mathbf{E} = \mathbf{v}_e \times \mathbf{B}, \quad \text{(for region B)} \quad (2)$$
$$\nabla \cdot \mathbf{J}_i = 0,$$

where $\mathbf{E}$, $\mathbf{B}$, $\mathbf{J}_i$ and $\mathbf{v}_e$ are electric field, magnetic field, ion current density and electron velocity, respectively. We supposed that electrons move only in the z direction. The boundary condition of continuity of electrical potential and field was used between A and B.

Figure 4 shows the calculated results for potential (ϕ), ion density (n_i) and electron density (n_e) distribution in the r direction, where we supposed that $V_d* = 1.5$ MV, $I_c = 60$ kA, $r_c = 11$ mm and $r_a = 17.5$ mm. By considering the continuity of parameters, we obtained the distribution of r_s, J_a ($= J_i(r_a)$) and I_e in the z direction, which is shown in Fig. 5. From Fig. 5, we got I_d (total diode current) ~ 120 kA, when $V_d* = 1.5$ MV. Thus the diode impedance is nearly 12.5 Ω.

From the experimental waveforms of V_d* and I_d (cf. Fig.2), we get the time evolution of impedance. To fit the above model with the experiental impedance which falls rapidly, we have to suppose the A-K gap closing velocity of ~ 6 cm/μs. This velocity seems to be too large for us to find its mechanism only with electrode plasma expansion.

From Fig. 5, furthermore, we have got J_a, which has the maximum of 0.25 kA/cm^2 at z = 40 mm. Even if we consider the gap closure, the calculated J_a is ~ 0.46 kA/cm^2 which is much smaller than the experimental result ($J_a = (1.4 \sim 1.9)$ kA/cm^2, cf. Table I). Therefore, the real ion emission is larger than that predicted by the space-charge limited emission model.

The above two problems are similar to those for applied-B ion diode,[18] which will be studied in more detail on PFD.

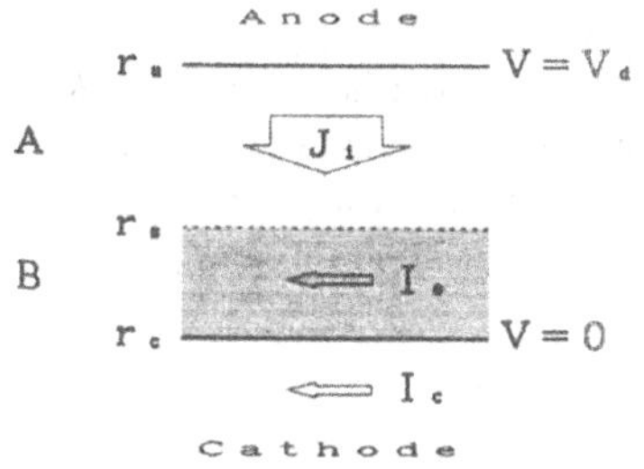

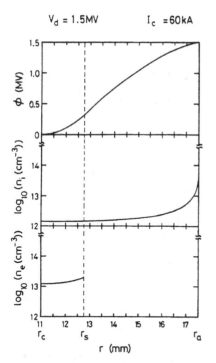

Fig. 3   Conceptual   structure of calculation model of PFD.

Fig. 4   Calculated result of
radial distribution of ϕ,
n_i and n_e.

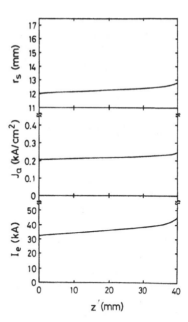

Fig. 5   Calculated result of
axial distribution of r_s,
J_a and I_e.

III. Ion Beam-Target Interaction

a) Experimental arrangement and diagnostics

Figure 6 shows the experimental arrangement of the diagnostic system for beam-target interaction.   In PFD, the target foil was inserted on the focusing line of the diode with an angle of 45° to the axis, which was also used as the Rutherford-scattering foil.   As shown in Fig. 6, ions scattered by the target enter the first pinhole after penetrating the target foil with an effective length of $\sqrt{2}d$, where d is the target thickness.

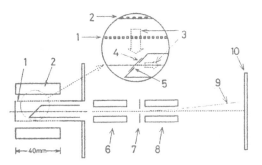

Fig. 6   Schematic of PFD and TPS.   1-cathode, 2-anode, 3-ion beam, 4-scatterer, 5-first pinhole, 6-magnetic deflector, 7-second pinhole, 8-electric deflector, 9-typical proton orbit, 10-detector (CR-39).

The scattered ion beam is collimated by the pinholes and then analyzed by a time-resolvable Thomson-parabola spectrometer (TPS) as shown in Fig. 6.   Principally, it is similar to that used in Ref. 11.   We prepared the second pinhole of the collimator between the magnetic and the electric deflectors reducing the spot-size of the beamlet from 1.7 mm to 0.96 mm. The former size corresponds to the situation that both pinholes are in the same side of the magnetic deflector.   As a result, the time- and energy-resolutions are significantly improved.   We used an appropriate time-varying electric field to carry out the time-resolved detection of the ion energy.   Experimentally, the deflecting electric field was in the oscillating waveform with the peak value of 25 kV/cm and the period of ∼ 300 ns, which is generated by a C-L oscillator triggered by the delayed pulser.   The intensity of the magnetic field generated by a permanent magnet is 0.78 T.   Both pinholes of the collimator have the diameter of 0.6 mm.   The ion traces are recorded on a CR-39 track-recording plastic film, which can be read out after etching in a solution of sodium hydroxide for at least two hours.   From the traces of the ions, we calculated their energy and time dependence after taking the time of flight into account.

In order to determine the proton energy loss correctly, we must find the incident energy of the protons. We carried out it in two respects: first we made the inductive correction of the voltage waveform very carefully with accurate calculation of the total inductance, and then tested the corrected waveform by using a thin gold scatterer, since the energy loss of protons in this scatterer is very small. Then the proton energy loss is calculated from the difference between the detected energy and V_d* at the corresponding instant. The scatterer is made of 0.22-μm thick gold coated on a 2-μm mylar film.

b) Experimental results

Figure 7 shows the experimental results of energy spectrum of the protons in the absence of a target, where the energy loss in the scatterer has been considered. The error bar shown in Fig. 7 mainly comes from the

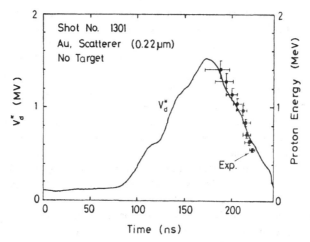

Fig. 7   Experimental result of energy spectrum
of protons in the absence of target.

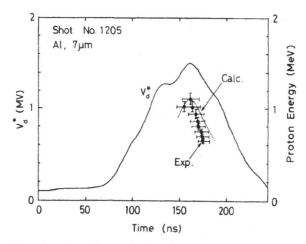

Fig. 8   Experimental (Exp.) and calculated (Calc.)
(cold-target) proton energy spectrum together
with the waveform of V_d* for 7-μm Al target.

uncertainty caused by the sizes of the two pinholes. The proton energy spectrum is found to be in good agreement with the waveform of V_d*.

Proton energy loss measurement has been carried out for aluminium target with thickness of 7 μm, the data of which is presented in Fig. 8. In Fig. 8, for comparison, we also plotted the proton energy calculated from the waveform of V_d* by subtracting the cold-target energy loss and the loss due to nuclear elastic scattering. Here we have used the data of cold target stopping power given by Andersen et al.[19] From Fig. 8, we can clearly see the enhanced energy loss which is indicated by the difference between E_{exp} and E_{calc}. We also noted that the enhancement ratio (defined as the ratio of energy loss in plasma to that in cold-target) increases with time from 1 to ~ 1.5.

c) Simulation Results

In the interested regions of ion energy and plasma temperature, the energy loss rate of the ions can be expressed by:

$$(dE/dx)_{total} = (dE/dx)_{bound} + (dE/dx)_{free} + (dE/dx)_{ion}, \qquad (3)$$

where the first term of the right-hand side is due to the bound electrons of plasma ions or neutral atoms, the second term is due to the free electrons in the plasma, and the third term is due to the ions in the plasma. The third term is usually very small compared to the other two terms. To calculate the energy loss due to bound electrons, one can use the Bethe equation after determining carefully the average ionization potential.[14] Alternatively, a model called generalized oscillator strength (GOS) was used by McGuire et al[20] to give the tabulated proton stopping powers for Al^{+n} ($0 \leq n \leq 11$). The free-electron stopping power should be described by binary collision theory[21,22] within a Debye radius coupled with collective plasma wave excitation outside the Debye radius.[23] Comparison of calculated results of different methods was made by Mehlhorn[13] showing that the simple binary model gives good approximation for the plasma electron stopping power.

Because of the difference between the experimental result and the calculated result for cold-target, we have to investigate the process of beam-target interaction in more detail. A one-dimensional hydrodynamic calculation code is used to do this. The hydrodynamic equations are the mass equation, momentum equation, energy equation and the equation of state, where the equation of energy included several terms due to energy deposition (ε_d), ionization (ε_i) and energy flux (ε_f):

$$\frac{\partial \rho}{\partial t} + \rho \frac{\partial v}{\partial x} + v \frac{\partial \rho}{\partial x} = 0$$

$$\frac{\partial v}{\partial t} + v \frac{\partial v}{\partial x} + \frac{1}{\rho} \frac{\partial p}{\partial x} = 0$$

$$\frac{\partial e}{\partial t} + v \frac{\partial e}{\partial x} + \frac{p}{\rho} \frac{\partial v}{\partial x} = \varepsilon_d + \varepsilon_i + \varepsilon_f \qquad (4)$$

$$p = (\gamma - 1) \rho e$$

Here, ρ, v, p, e, x, t and γ are mass density, velocity, pressure,

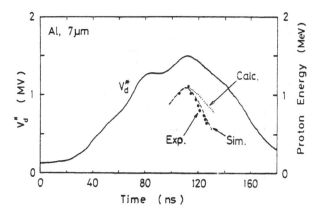

Fig. 9   Simulation (Sim.) of proton energy loss for
          7-μm aluminium target in comparison with
          experimental (Exp.) and calculated cold-
          target energy loss (Calc.).

specific energy, length coordinate, time and isentropic exponent, respectively.   The ion beam is supposed to be parallel to the incident angle
of 45°.   Beam composition is supposed to be 80% protons and 20% carbon
ions. For different shots the same beam characterization was used, where
the ion-beam current density is experimentally measured waveform similar
to Fig.2, and the individual voltage history was used.   The stopping
power of bound electrons was calculated using the tabulated data of
McGuire et al[20] by means of fitting.   Free electron stopping power was
calculated by simple binary collision model[21].   The population of each
ionization degree was calculated from Saha equation under the assumption
of local thermodynamic equilibrium (LTE).   The energy flux term (ε_f) of
eq. (4) involves radiation transport and electron thermal conduction.
Radiation heat transfer was treated in the conduction approximation and
the radia-   tion conductivity was calculated according with Ref. 24.
The electron conductivity was calculated according with Ref. 22.
Boundary conditions were determined by blackbody radiation.
      The simulation result corresponding to Fig. 8 is shown in Fig. 9.
For comparison, the calculated energy spectrum based on cold-target model
is also shown in Fig. 9.   This cold target energy spectrum is calculated
from the cold-limit of GOS data[20] which is ∼ 5 % smaller than that given
by Ref. 19 for 1-MeV proton.   Comparing the simulation with the experiment, good agreement is evidently obtained.   Therefore, by considering
the bound and free electron stopping powers individually, our model has
been proved to be practical for the aluminium target.
      Figure 10 shows the simulation results of time evolution of tempera-
ture, density and average ion- ization for 7-μm aluminium target, where t
= 0 corresponds to the starting time of the ion beam which is at ∼ 95 ns
on Fig. 9.   At t = 30 ns, for example, we see the target temperature ∼ 25
eV, target density ∼ (10^{19} ∼ 10^{20}) cm^{-3}, and average iniza- tion ∼ (5 ∼
6).   Hence, the maximum electron density is close to ∼ 10^{21} cm^{-3}, where the
degree of ionization is ∼ 40 %.

611

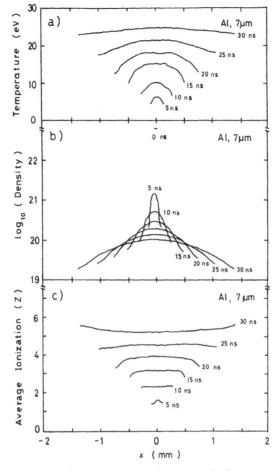

Fig. 10 Simulation result for 7-µm Al of a) temperature (eV), b) density (cm⁻³) and c) average ionization (Z).

IV. Concluding remarks

From the above experimental and theoretical studies, we summarize the following concluding remarks.

1) By using PFD, tightly focused (line focused) ion beam was generated along the central axis. The focusing radius is 0.18 ~ 0.25 mm, and the beam-power density ~ 0.1 TW/cm² at ion energy of ~ 1.5 MeV.

2) Comparing with calculation results, the experimental ion-current density on the anode (1.4 ~ 1.9 kA/cm²) is found to be a factor of 3 larger than that given by the model of space charge limited emission (0.46 kA/cm²).

3) Using such the tightly focued PFD, the enhanced proton energy deposition was observed in aluminium targets with the maximum enhancement ratio ~ 1.5.

4) According to hydrocode results, the target in PFD can be heated to temperature of ~ 25 eV with the degree of ionization of ~ 40 % at the electron density of ~ 10²¹ cm⁻³.

5) Theoretical model incorporating free and bound electron stopping terms is suitable to describe the energy loss of protons in aluminium

plasma. Thus, the enhanced deposition process can be understood by plasma effect.

Acknowledgments

This work was partly supported by a Grant-in-Aid from the Ministry of Education, Science and Culture of Japan. Experimental support given by M. Matsumoto, K. Aga, H. Isobe and N. Yumino of Lab. of Beam Technology are acknowledged with thanks.

References

1) S. Humphries, Jr.: Nuclear Fusion, 20, 1549(1980).
2) J. P. VanDevender, J. A. Swegle, D. J. Johnson, K. W. Bieg, E. J. T. Burns, J. W. Poukey, P. A. Miller, J. N. Olsen and G. Yonas: Laser and Particle Beams 3, 93 (1985).
3) K. Masugata, T. Yoshikawa, A. Takahashi, K. Aga, Y. Araki, M. Ito and K. Yatsui: Proc. 6th Int'l Conf. High-Power Particle Beams, Kobe, 1986, ed. by C. Yamanaka (Inst. Laser Eng., Osaka Univ.), 152(1986).
4) K. Masugata, K. Aga, A. Takahashi and K. Yatsui: Proc. of 2nd Int'l Top. Symp. ICF Res. by High-power Particle Beams, Nagaoka, 1986, ed. by K. Yatsui (Lab. Beam Tech., Tech. Univ. of Nagaoka), 81 (1986).
5) K. Yatsui, Y. Shimotori, Y. Araki, K. Masugata, S. Kawata and M. Murayama: Proc. 11th Int'l Conf. Plasma Phys. & Controlled Nucl. Fusion Res., Kyoto, IAEA-CN-47/B-Ⅲ-9 (1986).
6) K. Yatsui, K. Masugata and S. Kawata: Proc. 8th Int'l Workshop on Laser Interaction and Related Plasma Phenomena, Monterey, USA (1987); also, Laser Interaction and Related Plasma Phenomena (Plenum Press, New York and London), 8, 653 (1988).
7) K. Yatsui, Y. Shomotori, H. Isobe, W. Jiang and K. Masugata: Proc. 12th Int'l Conf. Plasma Phys. & Controlled Nuclear Fusion Research, Nice, France, IAEA-CN-50/B-4-2 (1988).
8) K. Yatsui, K. Masugata, Y. Shimotori, K. Imanari, M. Murayama, M. Yokoyama and T. Takaai: Proc. 7th Int'l Conf. High-Power Particle Beams, Karlsruhe, West Germany, I, 522 (1988).
9) K. Masugata, H. Isobe, K. Aga, M. Matsumoto, S. Kawata, W. Jiang and K. Yatsui: Laser and Particle Beams 7, 287 (1989).
10) F. C. Young, D. Mosher, S. J. Stephanakis, S. A. Goldstein and T. A. Mehlhorn: Phys. Rev. Lett. 49, 549 (1982).
11) J. N. Olsen, T. A. Mehlhorn, J. Maenchen and D. J. Johnson: J. Appl. Phys. 58, 2958 (1985).
12) H. Bluhm, B. Goel, P. Hoppe, H. U. Karow and D. Rusch: Proc. 7th Int'l Conf. High-Power Particle Beams, Karlsruhe, West Germany, I, 381 (1988).
13) T. A. Mehlhorn: J. Appl. Phys. 52, 6522 (1981).
14) E. Nardi, E. Peleg and Z. Zinamon: Appl. Phys. Lett. 39, 46 (1981).
15) K. A. Long and N. A. Tahir: Phys. Fluids, 29, 4204 (1986).
16) K. Yatsui, Y. Araki, K. Masugata, M. Murayama, M. Ito, E. Sai, M. Ikeda, Y. Shimotori, A. Takahashi and T. Tanabe: in Ref. 3, 329 (1986).

17) A. Tokuchi, N. Nakamura, T. Kunimatsu, N. Ninomiya, M. Den. Y. Araki, K. Masugata and K. Yatsui: in Ref. 4, 430 (1986).

18) P. A. Miller and C. W. Mendel, Jr., J. Appl. Phys. 61, 529 (1987).

19) H. H. Andersen and J. F. Ziegler: Hydrogen-Stopping Powers and Ranges in All Elements (Pergamon Press, New York) (1977).

20) E. J. McGuire, J. M. Peek, and L. C. Pitchford: Phys. Rev., A26, 1318 (1982).

21) J. D. Jackson: Classical Electrodynamics (John Wiley & Sons, New York), 643 (1975).

22) L. Spitzer, Jr.: Physics of Fully Ionized Gases (John Wiley & Sons, New York), 128 (1960).

23) D. Pines and D. Bohm: Phys. Rev. 85, 338 (1952).

24) Ya. B. Zeldovich and Yu. P. Raizer: Physics of Shock Waves and High Temperature Hydrodynamic Phenomena (Academic Press, New York, 1967), I, Chap. II; II, Chap. X.

THEORETICAL ANALYSIS OF CHARGE NEUTRALIZATION OF

THE INTENSE LIGHT ION BEAM

Takeshi Kaneda and Keishiro Niu

Department of Energy Sciences, Graduate School at Nagatsuta
Tokyo Institute of Technology
Midori-ku, Yokohama 227, Japan

INTRODUCTION

In this paper, electromagnetic fields appearing around the intense light ion beam (LIB) with the leading and tailing edges are obtained on the basis of the steady Vlasov equation for LIB, the Krook equation for background electron and the steady Maxwell equations for electromagnetic fields. The result shows that strong electrostatic fields remains uncanceled at the both edges of the beam due to the delay by electron inertia to neutralize the beam charge, although the charge in the main part of the intense LIB are completely neutralize[1]. It is turned out that the beam current is scarcely neutralized because the strong magnetic field suppresses the electrons from following the beam-ion particles[2]. The strong electrostatic field appearing at the leading edge causes the beam to diverge. The charge at the leading edge is inevitably required to be neutralized. A following method is proposed here to neutralize the beam charge at the leading edge in order that the beam avoids diverging. The beam passes through a plasma with a high density and high temperature[3]. In such a case, the electric fields appearing at the leading edge will trap electrons in the region of dense plasma. Physical parameters with which the beam traps electrons in the high dense plasma are made clear in this paper. It is shown that building up such a high dense plasma is not so difficult to be realized from the technical point of view.

FUNDAMENTAL EQUATIONS

The governing equations used here are the Vlasov equation for beam particles, and the Krook equation for the background electrons;

$$\frac{\partial f_b}{\partial t} + \mathbf{v} \cdot \frac{\partial f_b}{\partial \mathbf{r}} + \frac{e_b}{m_b} (\mathbf{E} + \mathbf{v} \times \mathbf{B}) \cdot \frac{\partial f_b}{\partial \mathbf{v}} = 0 \quad , \tag{1}$$

$$\frac{\partial f_e}{\partial t} + \mathbf{v} \cdot \frac{\partial f_e}{\partial \mathbf{r}} + \frac{e_e}{m_e} (\mathbf{E} + \mathbf{v} \times \mathbf{B}) \cdot \frac{\partial f_e}{\partial \mathbf{v}} = -(\nu_{ee} + \nu_{ei}) (f_e - f_{eM}) \quad . \tag{2}$$

Here, b,e refer to the beam and electron respectively, and f_{eM} refers to the Maxwellian distribution function for background electron;

$$f_{eM} = n_{e0} \left(\frac{m_e}{2\pi k_B T_e} \right)^{3/2} exp \left(- \frac{\frac{1}{2}m_e v_e^2 - e\Phi_0(r)}{k_B T_e} \right) \quad . \tag{3}$$

Electron charge e_e is related to e ($= 1.6 \times 10^{-19}$ C) by $e_e = -e$ and beam charge is $e_b = e$. The electromagnetic fields appearing in eqs.(1) and (2) satisfy the Maxwell equations,

$$\nabla \times \mathbf{H} = \mathbf{j} + \frac{\partial \mathbf{D}}{\partial t} \quad , \tag{4}$$

$$\nabla \times \mathbf{E}_m = -\frac{\partial \mathbf{B}}{\partial t} \quad , \tag{5}$$

$$\nabla \cdot \mathbf{D}_s = \rho_e, \tag{6}$$

$$\nabla \cdot \mathbf{B} = 0 \quad , \tag{7}$$

where $\mathbf{j}$ is the current density and ρ_e is the charge density. Subscripts m and s refer to the electromagnetic and electrostatic fields, respectively. The current and charge densities are expressed respectively by

$$\mathbf{j} = \int (e_b \mathbf{v} f_b + e_e \mathbf{v} f_e) \, d\mathbf{v} \quad , \tag{8}$$

$$\rho_e = \int (e_b f_b + e_e f_e) \, d\mathbf{v} + e n_i \quad . \tag{9}$$

Here, i refers to the background ion.

STEADY SOLUTION OF BEAM

The solution to the steady (and one-dimensional in space) Vlasov equation for beam particles is

$$f_{bo} = f_{b0}(H, P_\theta, P_z) \quad . \tag{10}$$

Here, the total energy H, canonical axial momentum P_z and canonical angular momentum P_θ are given respectively by

$$H = \frac{1}{2} m_b v^2 + e_b \Phi \quad ,$$

$$P_z = m_b v_z + e_b A_z \quad ,$$

$$P_\theta = m_b r v_\theta + e_b r A_\theta \quad .$$

Equation (10) is the exact expression for the distribution function of the beam which has infinite length.

SOLUTION OF BEAM WITH TWO EDGES

The unsteady solution for the beam which has two edges is chosen in the form as

$$f_b = f_{b0}(H, P_\theta, P_z)\, g_b(Z) \quad , \tag{11}$$

where

$$g_{bA}(Z) = \frac{1}{z_1}Z + 1 + \frac{l}{2z_1} \quad , \tag{12}$$

$$g_{bB}(Z) = 1 \quad , \tag{13}$$

$$g_{bC}(Z) = -\frac{1}{z_1}Z + 1 + \frac{l}{2z_1} \quad , \tag{14}$$

and

$$Z = z - V_{bz}t \quad . \tag{15}$$

Here Z is a coordinate in the frame moving with beam velocity V_{bz} ($Z=0$ is the central point of the beam). The subscript A refers to the tailing edge of the beam, B to the main part and C to the leading edge. The solution (11) expresses that the gradients of number density are linear at the two edges. In this simple form, the number density of the beam particles is obtained by

$$n_b(r,z,t) = \int f_b \, dv = g_b(Z) \int f_{b0} \, dv = g_b(Z) n_b'(r) \quad . \tag{16}$$

In the above equation, the fact that function n_b' of r can be chosen arbitrarily, is equivalent to the arbitrariness of the functional form f_{b0} regarding H, P_θ and P_z in eq.(10). The number density is chosen here as constant within a beam radius, for simplicity,

$$n_b'(r) = n_{b0} \quad . \tag{17}$$

The profiles of the beam number density is shown in Fig.1.

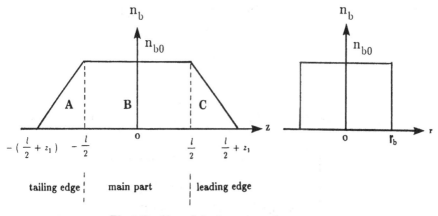

Fig.1 Profiles of the beam number density

In this simple case, the electromagnetic fields can be calculated analytically.

●Magnetic fields

Magnetic fields inside and outside the beam in the azimuthal direction are obtained by Ampere's law and Biot-Savart's law.

$$B_{\theta in}(r,Z) = \frac{1}{2}\mu_0 e V_{bz} n_b(Z) r \quad , \tag{18}$$

$$B_{\theta out}(r,Z) = -\frac{\mu_0}{4r} e r_b^2 V_{bz} n_{b0} \left[\{(\frac{Z}{z_1} + 1 + \frac{l}{2z_1})(\cos\theta_2 - \cos\theta_1) + \frac{r}{z_1}(\sin\theta_2 - \sin\theta_1)\} \right.$$
$$\left. + (\cos\theta_3 - \cos\theta_2) + \{(-\frac{Z}{z_1} + 1 + \frac{l}{2z_1})(\cos\theta_4 - \cos\theta_3) - \frac{r}{z_1}(\sin\theta_4 - \sin\theta_3)\} \right] \quad , \tag{19}$$

where,

$$\sin\theta_n = \frac{r}{\sqrt{(Z + b_n)^2 + r^2}} \quad , \tag{20}$$

$$\cos\theta_n = \frac{Z + b_n}{\sqrt{(Z + b_n)^2 + r^2}} \quad , \tag{21}$$

$$(n = 1,2,3,4)$$

$$b_1 = \frac{l}{2} + z_1 \quad b_2 = \frac{l}{2} \quad b_3 = -b_2 \quad b_4 = -b_1 \quad . \tag{22}$$

●Induced electric fields

Induced electric fields in the axial direction can be obtained by the Maxwell equation (eq.(5)) together with eq.(15),(18) and (19) ,as follows,

$$E_{mzout}(r,Z) = -V_{bz} \int_{\infty}^{r} \frac{\partial B_{\theta out}}{\partial Z} dr$$

$$= \frac{\mu_0}{4z_1} e r_b^2 V_{bz}^2 n_{b0} \left[\frac{Z+b_2}{|Z+b_2|} \log \left| \frac{r}{|Z+b_2| + \sqrt{r^2+(Z+b_2)^2}} \right| \right.$$

$$+ \frac{Z+b_3}{|Z+b_3|} \log \left| \frac{r}{|Z+b_3| + \sqrt{r^2+(Z+b_3)^2}} \right|$$

$$- \frac{Z+b_1}{|Z+b_1|} \log \left| \frac{r}{|Z+b_1| + \sqrt{r^2+(Z+b_1)^2}} \right|$$

$$\left. - \frac{Z+b_4}{|Z+b_4|} \log \left| \frac{r}{|Z+b_4| + \sqrt{r^2+(Z+b_4)^2}} \right| \right] \quad , \tag{23}$$

$$E_{mzin}(r,Z) = E_{mzout}(r_b,Z) - V_{bz} \int_{r_b}^{r} \frac{\partial B_{\theta in}}{\partial Z} dr = E_{mzout}(r_b,Z)$$

$$+ \begin{cases} -\frac{1}{4}\mu_0 e V_{bz}^2 \frac{n_{b0}}{z_1}(r^2 - r_b^2) & \text{(Region A)}, \\ 0 & \text{(Region B)}, \\ \frac{1}{4}\mu_0 e V_{bz}^2 \frac{n_{b0}}{z_1}(r^2 - r_b^2) & \text{(Region C)}, \end{cases} \tag{24}$$

where b_n (n=1,2,3,4) is given by (22).

● Electrostatic fields

In the case of LIB, the beam number density is very high ($n_b \sim 10^{22} /m^3$). Due to this high density and high propagation velocity of the beam, charges near two edges are not neutralized, because the neutralization of the beam charge by the background electrons is delayed due to their inertial effects. If we take the number density as is shown in Fig.1, the number density of unneutralized beam particles at the leading edge can be expressed approximately as $\bar{n}_b = \frac{1}{z_1} V_{bz} \tau n_{b0}$. Here, z_1 is the length of the leading edge (see Fig.1), τ is the time delay of charge neutralization with the approximation, the number density of unneutralized beam particles has a constant value throughout the leading part. The charge in the leading part is positive. The time delay τ of charge neutralization is approximated as inversely proportional to the electron plasma frequency ν_{pe} of the background plasma. The z-component of the electrostatic field has the maximum value on the axis of the beam, and are obtained from eq.(6) for each region at r=0 as follows;

$$\text{for} \quad Z > \frac{l}{2} + z_1 \ \ or \ \ Z < -(\frac{l}{2} + z_1) \ : \ E_{sz}(Z) = \frac{\bar{n}_b e}{2\varepsilon_0} f(Z) \ , \tag{25}$$

$$\text{for} \quad Region \ A \ : \ E_{sz}(Z) = -\frac{\bar{n}_b e}{2\varepsilon_0} \{2Z + l + 2z_1 - f(Z)\} \ , \tag{26}$$

$$\text{for} \quad Region \ B \ : \ E_{sz}(Z) = \frac{\bar{n}_b e}{2\varepsilon_0} \{ -2z_1 + f(Z)\} \ , \tag{27}$$

$$\text{for} \quad Region \ C \ : \ E_{sz}(Z) = \frac{\bar{n}_b e}{2\varepsilon_0} \{2Z - l - 2z_1 + f(Z)\} \ , \tag{28}$$

where,

$$f(Z) = \sqrt{(Z + \frac{l}{2} + z_1)^2 + r_b^2} + \sqrt{(Z - \frac{l}{2} - z_1)^2 + r_b^2}$$
$$- \sqrt{(Z + \frac{l}{2})^2 + r_b^2} - \sqrt{(Z - \frac{l}{2})^2 + r_b^2} \ . \tag{29}$$

The r-components of the electrostatic fields are obtained as follows;

$$E_{srin}(r) = \frac{e\bar{n}_b r}{2\varepsilon_0} \ , \tag{30}$$

$$E_{srout}(r,Z) = \frac{\sigma}{4\pi\varepsilon_0 r} (\cos\theta_1 - \cos\theta_2 + \cos\theta_3 + \cos\theta_4) \ , \tag{31}$$

where $\sigma = \pi r_b^2 \bar{n}_b e$ and the $\cos\theta_n$ is given by eq.(21).

● Profiles of the electromagnetic fields

The typical profiles of the magnetic field B_θ, and the induced electric field E_{mz} as well as the electrostatic field E_{sz} are shown in Fig.2 versus Z for several r. The maximum electrostatic fields E_{sz} and E_{sr} are of order of 10^8V/m. Even in the main part of the beam, the beam current is not neutralized[1] and induces the strong magnetic field in the azimuthal direction.

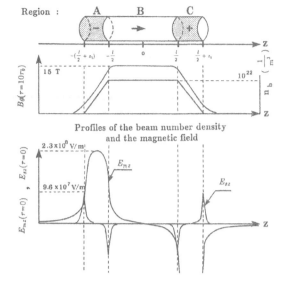

Profiles of the beam number density
and the magnetic field

Fig.2 Profiles of the Electromagnetic Fields

DETAILED ANALYSIS FOR MOTION OF BACKGROUND ELECTRON

In this section, the velocity distribution function of the background electrons is analyzed accompanied with the temporal development of the velocity distribution function of the beam. The function is chosen here as follows,

$$f_e(\mathbf{r}, \mathbf{v}, t) = f_{eM}(r,z,v_r,v_\theta,v_z)\, g_e(z,v_z,t) \quad . \tag{32}$$

Initially, the distribution function should be the Maxwellian with the temperature
T_e.

$$f_e(t=0) = f_{eM} \tag{33}$$

If (32) is substituted into eq.(2), the equation for g_e is derived as

$$\frac{\partial g_e}{\partial t} + \dot{z}\,\frac{\partial g_e}{\partial z} + \frac{e_e}{m_e}(E_{mz} + E_{s0z} + E_{s1z} + v_r B_\theta)\frac{\partial g_e}{\partial \dot{z}}$$

$$= \Big[\frac{e_e}{k_B T_e}\big\{ E_{s1r}\dot{r} + (E_{mz} + E_{s1z})\dot{z} \big\} - (\nu_{ee} + \nu_{ei}) \Big]g_e + (\nu_{ee} + \nu_{ei}) \quad . \tag{34}$$

Here it is provided that f_{eM} satisfies the steady form of eq.(2). Equation (34) is
so called Lagrange's equation and the characteristic equations are given by

$$\frac{dt}{1} = \frac{dz}{\dot{z}} = \frac{d\dot{z}}{\dfrac{e_e}{m_e}\Big(E_{mz} + E_{s0z} + E_{s1z} + v_r B_\theta\Big)}$$

$$= \frac{dg_e}{\Big[\dfrac{e_e}{k_B T_e}\big\{ E_{s1r}\dot{r} + (E_{mz} + E_{s1z})\dot{z}\big\} - (\nu_{ee} + \nu_{ei}) \Big]g_e + (\nu_{ee} + \nu_{ei})} \quad . \tag{35}$$

Then we obtain the following three independent equations,

$$z = \dot{z}\, t + z_0 \ ,$$
(36)

$$\int F_z \, dz = \frac{1}{2} m_e \dot{z}^2 + \xi \ ,$$
(37)

$$g_e = \left[\frac{m_e}{F_z} (\nu_{ee} + \nu_{ei}) \int exp \left[-\frac{m_e}{F_z} \int f(\dot{z}) \, d\dot{z} \right] d\dot{z} + C \right] exp \left[\frac{m_e}{F_z} \int f(\dot{z}) \, d\dot{z} \right] \ ,$$
(38)

where

$$f(\dot{z}) = \frac{e_e}{k_B T_e} \left\{ E_{s1r} \dot{r} + (E_{mz} + E_{s1z}) \dot{z} \right\} - (\nu_{ee} + \nu_{ei}) \ ,$$
(39)

and

$$F_z = e_e (E_{mz} + E_{s0z} + E_{s1z} + v_r B_\theta) \ .$$
(40)

From (36),(37) and (38), the peak of Maxwellian shifts along the $\dot{z}$ axis as a function of t, and the number densities of electrons captured by the beam edge can be obtained. If the region of high density of the background plasma is located in the beam path, the leading edge of the beam traps electrons. The solution g_e for the case of collisionless ($\nu_{ee} + \nu_{ei} = 0$) and constant field quantities is obtained by eq.(38) as,

$$g_e = \eta \ exp \left\{ \frac{e_e \int (E_{mz} + E_{s1z}) \ t \ d\dot{z} - \xi}{k_B T_e} \right\} \ .$$
(41)

Here η is determined by the initial condition,

$$g_e(t=0) = 1 \ .$$
(42)

It is obvious from eq.(42) that η is unity. Then the distribution function for background electron can be obtained as,

$$f_e = f_{eM} \, g_e$$

$$= n_{e0} \left(\frac{m_e}{2\pi k_B T_e} \right)^{3/2} exp \left(- \frac{\frac{1}{2} m_e \{ \dot{r}^2 + (r\dot{\theta})^2 + \dot{z}^2 \} - e\Phi_0 - e(E_{mz} + E_{s1z}) \, \dot{z} t}{k_B T_e} \right)$$
(43)

The peak of the distribution function shifts along the z axis with velocity

$$V_{peak} = \frac{e_e(E_{mz} + E_{s1z})}{m_e} t \ .$$
(44)

In order to neutralize the charge in the leading part of the beam, high temperature and high density will be required to exist in the beam path. In order to neutralize the beam charge of the leading edge, the required number density in the high density region is shown in Table 1 as a function of electron temperature. This table shows that these parameters are technically reasonable to build up the background plasma.

Table 1

n_{e0}	$k_B T_e$
$10^{22} / m^3$	$231 \, eV$
10^{23}	$186 \, eV$
10^{24}	$156 \, eV$
10^{25}	$134 \, eV$

REFERENCES

1) T.Kaneda and K.Niu , Laser & Particle Beams **7** (1989), part2, p.207.

2) K.Niu and S.Kawata , Fusion Tech. **11** (1987) p.365.

3) T.Kaneda and K.Niu , IPPJ-900, Mar.(1989) p.154.

AN IMPLICIT FLUID-PARTICLE MODEL
FOR ION BEAM-PLASMA INTERACTIONS†

D.B. Kothe

Los Alamos National Laboratory

C.K. Choi

School of Nuclear Engineering
Purdue University

INTRODUCTION

A new method for modeling the hydrodynamic behavior of collisional plasmas is presented. The numerical model uses finite-sized "particles" in a *full* particle-in-cell (PIC) representation of the plasma, where *full* is used to emphasize that a complete characterization of the plasma is obtained solely with the properties carried by the particles. Plasma motion is modeled by integrating the equations of motion for each particle implicitly in time on an arbitrarily-adaptive computational grid. The full PIC method is a conservative scheme that has no numerical diffusion and is ideally suited for modeling distorted and unstable hydrodynamic fluid flow. Computational results of a shock tube problem, the Rayleigh-Taylor (R-T) instability, and planar heavy-ion-driven ICF target implosions illustrate the properties of the method.

Two-dimensional compressible hydrodynamic motion of plasmas and fluids in general can be numerically modeled in a variety of ways [1]–[14]. The Lagrangian method [15], in which the computational mesh is embedded in the fluid, is usually the most frequent choice [10]–[14], especially in one-dimensional calculations. The Lagrangian approach allows tracking and resolution of material interfaces but cannot accurately represent distorted interfaces and shear flows. The use of triangular [16] or arbitrary [17] zones has somewhat alleviated the Lagrangian mesh constraints. The Eulerian method, in which the computational mesh stays fixed relative to the fluid, has been employed as well [1]. The Eulerian approach does not have problems with shearing flows, but it is inherently diffusive and material interface resolution is difficult. The mixed or coupled Eulerian-Lagrangian method [18, 19] has also been applied to study the motion of beam-heated material [3, 4]. The particle-in-cell (PIC) method [20]–[26], which in its original form (*classical* PIC) represents the fluid as "particles" (Lagrangian mass points) moving through a fixed (Eulerian) mesh, has also proven useful in modeling the hydrodynamic response of material irradiated by ion and laser beams [2],[5]–[8].

The PIC method described below is chosen over all other options for modeling two-dimensional plasma hydrodynamic behavior because of its computational ability to accurately treat distorted flows, conserve all fluid quantities (including angular momentum [63]), arbitrarily resolve specified flow gradients, and minimize numerical diffusion. This choice is due to

†Work performed under the auspices of the United States Department of Energy

a recent advance in PIC algorithms, known as FLIP (FLuid Implicit Particle) [27, 28], which has virtually eliminated problems of diffusion, multistreaming, resolution, and statistical noise. The name FLIP is also attached to the present work, which is called FLIP-PHD (Plasma HydroDynamics) [29], because it treats the particles and hydrodynamic finite-difference equations in the same fashion. After a brief review of the physical model in FLIP-PHD, the particle hydrodynamics and the difference schemes are detailed. The applicability of FLIP-PHD is then illustrated with several numerical examples.

PHYSICAL MODEL

A brief description of the physical model in FLIP-PHD is itemized below, with more details available in references [29] and [44].

1. The plasma is assumed to be a charge-neutral fluid of electrons and ions co-moving with a mean velocity $\vec{V}$, characterized by thermodynamic state variables such as specific volume, pressure, and internal energy. The hydrodynamic equations of motion therefore apply (equations (22) and (26)), describing the dynamic evolution of the plasma. The fluid is assumed to be inviscid except for artificial viscosities necessary to resolve shocks.

2. Energy is transported in the fluid by hydrodynamic compression or expansion, thermal conduction, collisional equilibration between electrons and ions, artificial viscous dissipation, and an arbitrary energy deposition source. Transport of energy via thermal X-rays is neglected. Both electron and ion energy transport equations are solved (equations (27) and (28)), as the fluid is assumed to carry separate electron and ion temperatures.

3. The physics associated with thermonuclear burn of DT fuel is neglected. This includes burn product (i.e., neutrons and alpha particles) transport and energy deposition. Gamma ray emission and absorption associated with thermonuclear burn is also neglected.

4. Local thermodynamic equilibrium (LTE) atomic physics is assumed in the calculation of the average material charge state. The further assumption that all particles obey Boltzmann statistics results in the Saha relation for an estimation of average charge state.

5. The thermodynamic relations of the electrons and ions are governed by realistic equations of state (EOS). The necessary relations are obtained from the Los Alamos National Laboratory Sesame EOS data library in tabular form [30]. Multi-material regions are assumed to be uniformly mixed and governed by equilibrium thermodynamics. Transport coefficients such as the equilibration coefficient and the electron and ion thermal conductivities are computed for arbitrary EOS and material charge state [31]. The coefficients reduce to those given by Braginskii for ideal gas EOS [32].

6. If ion beams interact with the plasma, ion beam energy is deposited in the plasma as follows:

   (a) Energy loss to bound electrons is calculated using the modified Bethe theory at high projectile energies [33, 34] and the LSS theory at low projectile energies [35, 36]. Mean excitation potentials are estimated with the scaling relation suggested by Mehlhorn [37]. Tabular stopping power data are available for cold aluminum and gold background materials.

   (b) Energy loss to nuclei is calculated using a least-squares fit to Linhard's theory [35] as given by Steward and Wallace [38] and used in the same fashion by Mehlhorn [37].

624

(c) Energy loss to plasma ions is calculated using either a binary Coulomb collision theory [39] or a unified theory [40].

(d) Energy loss to free electrons is calculated using either a collective binary-wave theory [37, 39] or a unified theory [40].

(e) Ion beam charge state in cold material is calculated using either the empirical expression of Brown and Moak [41] or the least-squares fit of Steward [42] to range-energy data given by Northcliffe [43]. Charge states in ionized material are estimated by scaling the previously mentioned expressions with the relative velocity between the beam ions and free electrons [37].

PARTICLE HYDRODYNAMICS

Particle hydrodynamics refers to the use of finite "particles" in the numerical representation of fluid flow [45]. The Lagrangian equations of motion are integrated on numerous globs of fluid called "particles" having properties such as mass, momentum, and energy. The particle method has been surprisingly successful in modeling highly distorted flow in two and three dimensions. Its robustness has been exploited in a wide range of physics to model phenomena such as hypervelocity impact [24, 26], [46]–[48], atmospheric effects [25, 49], shock dynamics [23], hypersonic shear flow [23], chemically reacting flow and detonation [26, 50, 51], supersonic wake flow [26, 52], relativistic fluid flow [53], high speed jets [26, 54], hydrodynamic instabilities [55], strength of materials [46]–[48], MHD [56], and, finally, the hydrodynamic response of ion- and laser-heated material [2],[5]–[8].

Background

The PIC computing method, originally developed by Harlow and co-workers at the Los Alamos National Laboratory in the 1950's [20]–[26], was devised to treat higher-dimension hydrodynamic problems having multi-material regions with strong shear flow. The contemporary Eulerian and Lagrangian finite-difference methods had severe difficulties with problems of this type. Eulerian methods gave poor representation of material interfaces. Accurate treatment of the convective terms in the conservation equations often led to unreasonable numerical diffusion and instability. Lagrangian methods, on the other hand, gave low accuracy in regions of strong shear because of the quadrilateral shapes given to the fluid. The *classical* PIC method attempted to eliminate these inherent difficulties by simulating the convective terms in the conservation equations through the movement of fluid particles relative to a fixed Eulerian mesh. According to this prescription, grid distortion is eliminated with a fixed Eulerian mesh, convective derivatives are treated with movement of the particles relative to the grid, and material interfaces are easily tracked by following the particles.

In Harlow's *classical* PIC method [25], particle properties are first assigned to the grid, typically using a nearest grid point (NGP) interpolation (Figure 1A), so that particles in a cell are therefore the only contributors to the cell's properties. The Lagrangian equations of motion are then solved on an Eulerian grid. Cell- and grid-based quantities are updated with these tentative solutions. The particles are next moved with a velocity that is linearly-interpolated (Figure 1B, known as area weighting in two dimensions) from the tentative velocities of neighboring cells. A particle, upon changing cells, carries temporary mass-proportioned energy and momentum from the cell it exits into the cell it enters. Finally, grid quantities are reassigned by again summing over all particles in a given cell (NGP interpolation). The advantages of the classical PIC method have already been mentioned, but there are also disadvantages, some of which include (1) statistical noise caused by the finite number of particles and lower order interpolation, (2) limited spatial resolution due to the use of an Eulerian grid to calculate gradients, (3) high diffusion in all the primitive variables (except mass) due to the manner in which fluid quantities are convected by the particles from cell to cell [28], (4) an inability to solve low speed (incompressible) flow problems, (5) a lack of translational and rotational invariance, and (6) computation time and memory storage typically double that of equivalent Eulerian or Lagrangian methods [26].

Many attempts have since been made to overcome the difficulties of the classical PIC method itemized above [6, 27],[56]–[58]. The first approach, as in classical PIC, only allows the particles to carry mass and position in an effort to minimize computation time and memory storage. For example, the second order algorithm developed by Nishiguchi and Yabe [57] was devised to increase accuracy and decrease diffusion (items (1) and (3) above), with further improvements allowing for a nonuniform grid to increase resolution (item (2) above) [61]. The second approach, known as *full* PIC, borrows from plasma PIC simulation methodology [59, 60] by allowing each particle to carry all the necessary fluid information, which results in increased computational time and memory storage. The full PIC formulation has been adopted by McCrory, et al. [6], Leboeuf, et al. [56], Marder [58], and FLIP in its original form by Brackbill and Ruppel [27]. Allowing the particles to carry momentum and energy virtually eliminates the diffusion problem encountered in classical PIC (item (3) above), but introduces a particle multistreaming possibility due to the collisionless nature of particle motion. A natural representation of a fluid with a many body system, the particles, is obtained with FLIP by distinguishing between the momentum a particle carries and the velocity with which it moves through the mesh [27]. This technique eliminates multistreaming by assuring a single-valued displacement field.

Many features distinguish FLIP from *classical* PIC: (1) a *full* PIC formulation in which particles carry all the permanent fluid information, thereby reducing the grid to a computational convenience; (2) higher order (quadratic) interpolation; (3) differentiating between a particle's momentum and the velocity with which it moves through the grid; (4) arbitrary spatial resolution with a grid which can be Eulerian, Lagrangian, adaptive to user-specified data [62], or a combination thereof; and, (5) an implicit time-differencing scheme for all of the conservation equations.

Particle Kinematics

Finite-sized particles carrying a position $\vec{r}_p$, color c_p (material identifier), mass m_p, velocity $\vec{v}_p$, electron energy e_{ep}, and ion energy e_{ip} represent the plasma fluid in FLIP-PHD. The mapping of the particle's physical position in the domain (consisting of a grid of convex quadrilaterals with physical vertex positions $\vec{r}_v$) is defined to be

$$\vec{r}_p = \sum_v \vec{r}_v S(\vec{r}_{vn} - \vec{r}_{pn}), \tag{1}$$

where S is an interpolating kernel which will hereafter be called a shape function. The summation is over all vertices included in the support of S. The value of the shape function above (a particle-to-vertex mapping) depends upon the difference between the natural coordinates of the particles, $\vec{r}_{pn}(\xi, \eta)$, and the grid vertices $\vec{r}_{vn}(i, j)$. Cell-centered mapping depends upon the difference between the natural coordinates of the particles and the cell centers, $\vec{r}_{cn}(i + 1/2, j + 1/2)$. Information in natural coordinate space (denoted by a subscript n) is preferred since the coordinate system simplifies to a mesh of unit squares. The shape function for a particle-to-vertex (denoted by pv) mapping can be written

$$S(\vec{r}_{pn} - \vec{r}_{vn}) = S(\xi - i, \eta - j) \equiv S_{pv}(\xi', \eta'), \tag{2}$$

while a particle-to-cell center (denoted by pc) mapping takes the form

$$S(\vec{r}_{pn} - \vec{r}_{cn}) = S(\xi - i - 1/2, \eta - j - 1/2) \equiv S_{pc}(\xi', \eta'). \tag{3}$$

For the particle and grid motion to be equivalent during the Lagrangian phase of the calculation, the shape function S must be a Lagrangian invariant,

$$\frac{dS}{dt} = 0. \tag{4}$$

The function must also be normalized to one,

$$J^{-1} \int S(\vec{r}) \, d^3\vec{r} = \int \int S(\xi, \eta) d\xi d\eta = 1 \,, \tag{5}$$

where $J \equiv \partial(x,y)/\partial(\xi,\eta)$. The integral becomes a summation

$$\sum_{ij} S(\xi'(i), \eta'(j)) = 1 \tag{6}$$

in the PIC method. If S is a spline, the summation reproduces the exact value of the integral for linear functions [45]. Normalization, together with bounded support, guarantees mass conservation on the grid because constants such as individual particle masses are interpolated correctly [45].

The possibilities for shape functions include a set of interpolating kernels called the central B-splines [45]. A central B-spline exactly reproduces linear functions through ordinary or smoothing interpolation. The first three B-splines, shown in Figure 1, are the nearest grid point interpolation,

$$S_1(\xi', \eta') = \begin{cases} 1; & \text{if } |\xi'|, |\eta'| \le \frac{1}{2}, \\ 0; & \text{otherwise,} \end{cases} \tag{7}$$

linear interpolation,

$$S_2(\xi', \eta') = \begin{cases} (1 - |\xi'|)(1 - |\eta'|); & \text{if } |\xi'|, |\eta'| \le 1, \\ 0; & \text{otherwise,} \end{cases} \tag{8}$$

and quadratic (TSC) interpolation,

$$S_3(\xi', \eta') = \begin{cases} (\frac{3}{4} - |\xi'|^2)(\frac{3}{4} - |\eta'|^2); & \text{if } |\xi'|, |\eta'| \le \frac{1}{2}, \\ \frac{1}{2}(\frac{3}{2} - |\xi'|)^2(\frac{3}{4} - |\eta'|^2); & \text{if } \frac{1}{2} \le |\xi'| \le \frac{3}{2}, |\eta'| \le \frac{1}{2}, \\ \frac{1}{2}(\frac{3}{4} - |\xi'|^2)(\frac{3}{2} - |\eta'|)^2; & \text{if } |\xi'| \le \frac{1}{2}, \frac{1}{2} \le |\eta'| \le \frac{3}{2}, \\ \frac{1}{4}(\frac{3}{2} - |\xi'|)^2(\frac{3}{2} - |\eta'|)^2; & \text{if } \frac{1}{2} \le |\xi'|, |\eta'| \le \frac{3}{2}, \\ 0; & \text{otherwise.} \end{cases} \tag{9}$$

Quadratic interpolation is the first of the B-splines to smooth the data and give a continuous first derivative [45]. The FLIP algorithm uses linear interpolation to obtain vertex mass and momentum from the particles (S_{2pv}) as well as particle quantities such as position, advancement velocity, and acceleration from the grid vertices (S_{2vp}). Quadratic interpolation is used to obtain cell-centered mass, electron energy, and ion energy from the particles (S_{3pc}). Since the data points (particles) are typically disordered, quadratic interpolation onto cell centers is used because it is a smoothing interpolation formula designed for interpolation from noisy data [45]. The support of a bi-quadratic (2-D) interpolation is nine cells, enabling one to use fewer particles which do not have to be fluctuation-averaged.

The particle equations of motion become, upon differentiating equation (1) with respect to time,

$$\frac{d\vec{r}_p}{dt} = \sum_v \vec{V}_v S_{2vp} \,, \tag{10}$$

and

$$\frac{d\vec{v}_p}{dt} = \sum_v \frac{d\vec{V}_v}{dt} S_{2vp} \,. \tag{11}$$

Grid accelerations $d\vec{V}_v/dt$ are calculated using any finite-difference method such as that discussed in the following section. The vertex momentum,

$$\vec{P}_v = \sum_p m_p \vec{v}_p S_{2pv} \,, \tag{12}$$

627

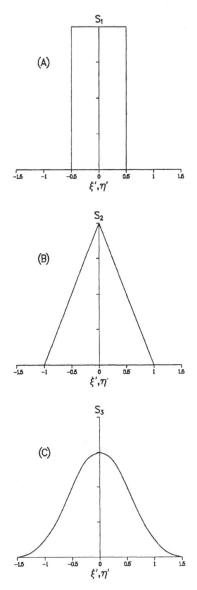

Figure 1. Examples of basis splines which can be used for the shape function S in particle-grid interpolation: (A) nearest grid point (S_1), (B) linear (S_2), and (C) quadratic (S_3).

divided by the vertex mass,

$$M_v = \sum_p m_p S_{2pv}, \qquad (13)$$

results in a center-of-mass or fluid velocity defined to be the vertex velocity,

$$\vec{V}_v = \frac{\vec{P}_v}{M_v}, \qquad (14)$$

where the summation is over all particles p included in the support of S. Defining the particle velocity to be the fluid velocity at the point $\vec{r}_p$ (equation (10)) [27] is a crucial step toward modeling collisional fluids. Individual particle velocities $\vec{v}_p$ are only used to find the vertex momenta $\vec{P}_v$. The $\vec{v}_p$ are not used for particle advancement as in the methods of Marder [58] and Leboeuf et al. [56]. Velocities and accelerations, hence momenta, are imparted to particles according to equations (10) and (11), respectively. The inherent collisional features of FLIP had to be performed artificially by Leboeuf et al. (using a Krook-type drag term) [56] and Marder (using a particle velocity smoothing procedure) [58]. McCrory et al., in their PAL method [6], also advanced particles according to equation (10) but imparted momentum to the particles using a classical PIC mass-weighted cell momentum instead of equation (11). This technique required two references to the particles in a given cycle, as opposed to only one needed in FLIP.

As mentioned earlier, cell-centered quantities are obtained with quadratic interpolation (S_{3pc}). The quantities needed are cell mass,

$$M_c = \sum_p m_p S_{3pc}, \qquad (15)$$

cell specific electron and ion internal energies,

$$\varepsilon_{ec} = \frac{\sum_p e_{ep} S_{3pc}}{M_c}; \qquad \varepsilon_{ic} = \frac{\sum_p e_{ip} S_{3pc}}{M_c}, \qquad (16)$$

and cell density,

$$\rho_c = \frac{M_c}{V_c}, \qquad (17)$$

where V_c is the cell volume. Other cell-centered quantities such as electron and ion temperatures, electron and ion pressures, total pressure, sound speeds, specific heats, average charge states, etc., are found from EOS relations.

Particle Dynamics

Once solutions to the equations of motion are obtained on the grid (discussed in the following section), they are transferred to the particles, giving the particles a new position, momentum, and energy. Using the vertex velocities and accelerations at time t^θ $(t^0 + \theta \Delta t,$ where $0 \le \theta \le 1)$, linear interpolation is used to advance the particle positions,

$$\vec{r}_p^1 = \vec{r}_p^0 + \Delta t \sum_v \vec{V}_v^\theta S_{2vp}, \qquad (18)$$

and find new particle velocities,

$$\vec{v}_p^1 = \vec{v}_p^0 + \theta^{-1} \sum_v \left(\vec{V}_v^\theta - \vec{V}_v^0 \right) S_{2vp}. \qquad (19)$$

Energy is quadratically interpolated to the particles (S_{3cp}) and normalized according to particle mass, not particle energy as in the early version of FLIP [27]. By summing over all cells surrounding the particle within the support of S_{3cp}, the change in particle electron energy is found to be

$$e_{ep}^1 = e_{ep}^0 + m_p \sum_c \left[(\Delta\varepsilon_e)_c - \Delta t\, p_{ec} \left(\nabla \cdot \vec{V}\right)_c^\theta \rho_c^{-1} \right] S_{3cp}, \tag{20}$$

where the first term $(\Delta\varepsilon_e)$ is a cell-centered energy change due to sources, thermal conduction, and thermal equilibration [29], and the second the PdV work. Similarly, the change in particle ion energy is given by

$$e_{ip}^1 = e_{ip}^0 + m_p \sum_c \left[(\Delta\varepsilon_i)_c + \Delta t \left(\sigma_{kl}^{art}\right)_c \left(\frac{\partial V_l}{\partial x_k}\right)_c^\theta \rho_c^{-1} - \Delta t\, p_{ic} \left(\nabla \cdot \vec{V}\right)_c^\theta \rho_c^{-1} \right] S_{3cp}$$

$$+ m_p \theta^{-1} \sum_v \left[\vec{V}_v^\theta \cdot \left(\vec{V}_v^\theta - \vec{V}_v^0 \right) \right] S_{2vp} - \frac{1}{2} m_p \left\{ (\vec{v}_p^1)^2 - (\vec{v}_p^0)^2 \right\}, \tag{21}$$

where the first term $(\Delta\varepsilon_i)$ is the cell-centered energy change due to sources, thermal conduction, and thermal equilibration [29], the second the viscous heating (with an artificial viscosity tensor σ_{kl}^{art}), and the third the PdV work. The fourth and fifth terms represent the energy dissipated due to backwards Euler differencing and interpolation [27].

DIFFERENCE SCHEMES

In the Lagrangian phase of the calculation, the grid and particles do not move relative to each other. Finite difference approximations to the conservation equations are solved on the grid, instead of directly on the particles as in other methods such as SPH [64]. This is a reasonable substitution in FLIP-PHD since there are typically many more particles than grid points. The finite-difference equations must be consistent with the particle representation and conserve important quantities such as mass, momentum (linear and angular), and energy. Exact details of the implicit difference schemes used in FLIP-PHD for momentum and energy transport are described elsewhere [29], so what follows represents a very brief overview.

Momentum Transport

An attractive feature of the FLIP-PHD model for momentum transport is an implicit solution for the total pressure. An implicit treatment obviously eliminates linear stability constraints such as the Courant condition, enabling arbitrarily large time steps. (Except for supersonic or slightly subsonic problems, where the Courant stability condition is no longer a critical restriction.) Numerical tests have shown the implicit pressure solution to be robust and smooth over a wide variety of problems. Implementation is easy and relatively inexpensive since the matrix is usually diagonally dominant, therefore easily invertible with an ICCG solver [65].

In the Lagrangian frame, the continuity equation is written

$$\frac{d\rho}{dt} + \rho(\nabla \cdot \vec{V}) = 0, \tag{22}$$

where ρ and $\vec{V}$ are the density and velocity of the fluid, respectively. The continuity equation is consistent with the particle representation, so its solution is automatically satisfied. During this phase of the calculation, the fluid is assumed to be isentropic, so one may write the electron and ion pressures as functions of densities and their respective entropies (s_e and s_i):

$$p_e = p_e(\rho, s_e), \qquad p_i = p_i(\rho, s_i). \tag{23}$$

A relationship between the total pressure $(p_e + p_i)$ and density can therefore be written

$$\frac{dp}{dt} = c_s^2 \frac{d\rho}{dt}; \qquad c_s^2 = \left(\frac{\partial p_e}{\partial \rho}\right)\Big|_{s_e} + \left(\frac{\partial p_i}{\partial \rho}\right)\Big|_{s_i}, \tag{24}$$

where c_s is the sound speed. The continuity equation then becomes

$$\frac{dp}{dt} + c_s^2 \rho (\nabla \cdot \vec{V}) = 0 \, . \tag{25}$$

Discontinuous solutions are not allowed, so the artificial viscosity method is used to resolve discontinuities such as shocks in the flow. The Lagrangian momentum equation takes the form (using (k, ℓ) tensor notation)

$$\rho \frac{dV_k}{dt} + \frac{\partial p}{\partial x_k} = \frac{\partial \sigma_{k\ell}^{\mathrm{art}}}{\partial x_\ell} \, , \tag{26}$$

where $\sigma_{k\ell}^{\mathrm{art}}$ is an artificial viscosity tensor [29]. When equations (25) and (26) are written in finite-difference form, that resulting system of two equations and two unknowns (an implicit pressure and velocity) is combined into one equation for an implicit pressure [29], similar to the method of Casulli and Greenspan [66].

Energy Transport

The Lagrangian time rate of change of the specific internal energy can be written as

$$\rho \frac{d\varepsilon_i}{dt} = \nabla \cdot [\kappa_i \nabla T_i] + \omega_{ei}(T_e - T_i) - p_i \nabla \cdot \vec{V} + \sigma_{k\ell}^{\mathrm{art}} \left(\frac{\partial V_\ell}{\partial x_k} \right) + S_i \tag{27}$$

for the ions, and

$$\rho \frac{d\varepsilon_e}{dt} = \nabla \cdot [\kappa_e \nabla T_e] - \omega_{ei}(T_e - T_i) - p_e \nabla \cdot \vec{V} + S_e \tag{28}$$

for the electrons. Changes in specific internal energy result from thermal diffusion (κ_e, κ_i), collisional energy transfer (ω_{ei}), PdV work, artificial viscous dissipation ($\sigma_{k\ell}^{\mathrm{art}}$), and an energy source (S_e and S_i). Equations (28) and (27) can be written in finite-difference form as

$$\varepsilon_e^1 = \alpha^0 \nabla \cdot [\kappa_e \nabla T_e^1] + \Delta t \, \omega_{ei}(T_i^1 - T_e^1) - \alpha^0 p_e^0 \left(\nabla \cdot \vec{V} \right)^\theta + \alpha^0 S_e^0 + \varepsilon_e^0 \, , \tag{29}$$

for the electrons, and

$$\varepsilon_i^1 = \alpha^0 \nabla \cdot [\kappa_i \nabla T_i^1] + \Delta t \, \omega_{ei}(T_e^1 - T_i^1) - \alpha^0 p_i^0 \left(\nabla \cdot \vec{V} \right)^\theta$$
$$+ \left(\sigma_{k\ell}^{\mathrm{art}} \right)^\theta \left(\frac{\partial V_k}{\partial x_\ell} \right)^\theta + \alpha^0 S_i^0 + \varepsilon_i^0 \, , \tag{30}$$

for the ions, where $\alpha^0 = \Delta t / \rho^0$. The above expressions contain thermal equilibration and conduction terms written in fully implicit form. This is a desirable system of equations to solve since explicit linear stability constraints are avoided, and solution accuracy is of order $\Delta t + \Delta x^2$ [67]. Higher accuracy in time (order Δt^2), if desired, can be obtained by writing the temperature terms in the above equations in Crank-Nicholson fashion [67]. The cross-relaxation terms and the nonlinear dependence of the thermal conductivities present a problem, however, because an implicit treatment usually requires an expensive iteration procedure [68]. A time-splitting formulation, however, circumvents the expense and complexity of a fully implicit, iterative calculation. By accumulating effects separately within a time step, the difference equations are simplified into a non-iterative form with accuracy comparable to fully implicit methods.

Time-Step Splitting

The finite-difference forms of the electron and ion energy transport equations are formulated with a time-step splitting method in which source and sink contributions to the specific internal energies of ions and electrons are computed at a hierarchy of time levels. Each contribution follows an assumed energy transport time scale ordering [29]. Ion beam energy deposition changes the ion and electron specific internal energies from ε_i^0 and ε_e^0 at time t^0 to ε_i^* and

ε_e^* at time t^* $(t^0 < t^* < t^0 + \Delta t)$. Thermal equilibration effects are next calculated, altering ε_i^* and ε_e^* at time t^* to ε_i^{**} and ε_e^{**} at time t^{**} $(t^* < t^{**} < t^0 + \Delta t)$. Finally, thermal conduction effects are considered, which results in a further change to ε_i^{***} and ε_e^{***} at time t^{***} $(t^{**} < t^{***} < t^0 + \Delta t)$. Since these are strictly thermal processes resulting from collisions, the density dependence in EOS relations is neglected (i.e., temperatures can be calculated from internal energies with specific heats). The change in ion thermal energy, $\Delta\varepsilon_i = \varepsilon_i^{***} - \varepsilon_i^0$, and electron thermal energy, $\Delta\varepsilon_e = \varepsilon_e^{***} - \varepsilon_e^0$, are then interpolated, along with PdV work, in a mass-weighted fashion onto the particles (equations (20) and (21)). This process, very similar to a fully implicit formulation of the energy equations, is advantageous because it is simple and modular. Iterative procedures are not required and computational expense is minimized without the use of oversimplifying assumptions. The order or deletion of physical processes does not change the overall algorithm.

Ion Energy Deposition

Ion-beam energy deposition is modeled in FLIP-PHD with a 3-D ray trace algorithm [12]–[14]. The beam is represented by an arbitrary number of rays, each carrying a specified direction and beam power fraction. A ray's two properties, direction and power, enable the representation of ion beams with detailed intensity profiles and illumination geometries. Since rays are initialized with powers, intensities are varied by changing either the target or beam cross-sectional area. Resolution and detail in beam deposition is governed both by the number of rays per beam and the number of computational zones representing the target. Three-dimensional effects are simulated in a cylindrically symmetric geometry by rotating the computational grid in the (r, z) plane around the axis of symmetry (z-axis). A ray traveling in a straight line through such a cylinder would trace out a hyperbolic path on the computational grid. Each ray is consequently forced to follow a hyperbolic path on the computational grid, $r^2 - r_0^2 = (z - z_0)^2 \tan^2\Theta$, where (r_0, z_0) is the point of closest approach and Θ is the incident angle relative to the z-axis. The ray trace algorithm [69] traces each ray through the computational grid in a cell-by-cell fashion until the ray thermalizes with the background plasma (Figure 2A) or leaves the problem (Figure 2B). High accuracy in ion range is obtained by solving a predictor-corrector loop within each cell [70]. The rays are "launched" from some outer extremity in vacuum and the tracing begins when a ray hits a material boundary. Unlike Lagrangian fluid codes where a given cell follows the same material throughout the calculation, any cell in FLIP-PHD can have at any time a mixture of materials or no material at all. The material boundaries of a problem must therefore be located prior to tracing rays. Reflecting boundaries reflect rays by altering either the point of closest approach or directional angle in its hyperbolic path. The energy deposited in a cell is therefore based upon the sum of energies deposited by each ray passing through that cell. The electron source term S_e results from losses to background electrons (bound and free) and the ion source term S_i results from losses to nuclei and plasma ions. For a regular, rectilinear mesh there is no need for more than one ray per cell, but a fully Lagrangian mesh may need many rays per cell for an accurate beam representation.

Ion beam energy deposition to cells with mixed materials is treated with either an average atom model or Bragg superposition. In the average atom model, an average density, temperature, charge state, and atomic number is assigned to the mixture so that only one energy loss calculation is necessary in each cell. With Bragg superposition, the total energy loss in a cell ij having a mixture of materials is equal to the sum of energy losses to each individual k^{th} component, i.e., $S(\text{ij}) = \sum_k S_k(\text{ij})$. This expression holds for deposition to both electrons and ions. Bragg superposition is the more accurate but more expensive alternative to beam energy deposition. If mixed regions are limited to finely resolved areas, then the average atom model may give adequate accuracy.

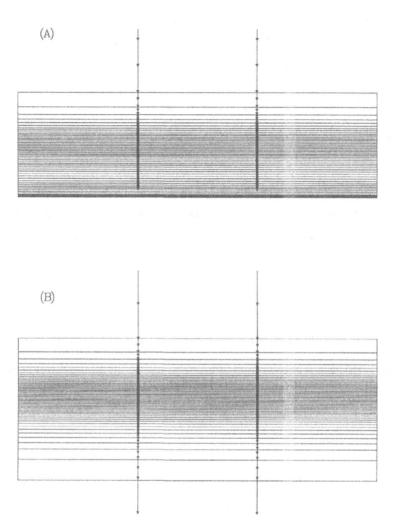

Figure 2. Illustration of the use of rays in FLIP-PHD to represent ion beams streaming through the computational mesh. In (A) the beam thermalizes in the plasma and in (B) the beam has enough energy to completely penetrate the plasma.

The diffusion or conduction of heat in ion and energy energy transport leads to an equation of the form

$$\frac{\partial \phi}{\partial t} = A_0 \nabla \cdot (\kappa \nabla \phi) \,, \tag{31}$$

where A_0 is a constant, ϕ is the electron (T_e) or ion (T_i) temperature, and κ is the electron or ion thermal conductivity. The thermal conductivity remains inside the divergence since it is a function of spatially-dependent plasma thermodynamic variables. The finite-difference operator for the diffusion of thermal flux, $\nabla \cdot (\kappa \nabla \phi)$, must be formulated to be accurate to at least second order in space as well as mimicking properties such as symmetry and energy conservation possessed by the actual differential operator [71].

The physical coordinate system in FLIP-PHD can be highly irregular due to the use of a Lagrangian and/or adaptive mesh, so differentiation is simplified by mapping physical coordinates (x, y) into natural coordinates (ξ, η), which are linear quadrilaterals of unit length. Transformation of the diffusion term into the boundary-fitted coordinates (ξ, η) results in

$$\nabla \cdot (\kappa \nabla \phi) = \frac{1}{J x^\delta} \left[x^\delta \alpha \kappa \phi_\xi - x^\delta \beta \kappa \phi_\eta \right]_\xi + \frac{1}{J x^\delta} \left[x^\delta \gamma \kappa \phi_\eta - x^\delta \beta \kappa \phi_\xi \right]_\eta \,, \tag{32}$$

where partial derivatives are denoted by subscripts, J is the Jacobian of the transformation,

$$J \equiv \frac{\partial(x, y)}{\partial(\xi, \eta)} = x_\xi y_\eta - x_\eta y_\xi \,, \tag{33}$$

with α, β, and γ defined by

$$\alpha \equiv \frac{1}{J}(x_\eta^2 + y_\eta^2), \qquad \beta \equiv \frac{1}{J}(x_\xi x_\eta + y_\xi y_\eta), \qquad \text{and} \qquad \gamma \equiv \frac{1}{J}(x_\xi^2 + y_\xi^2). \tag{34}$$

The superscript δ on the variable x is a constant equal to 1 in cylindrical (where x is the r coordinate) and 0 in cartesian geometry. A differential volume element is $dV = x^\delta J d\xi d\eta$ per radian, which defines a zone volume to be $x^\delta J$ per radian. By finite-differencing equation (32), an operator matrix M can be written such that the thermal diffusion equation becomes $M\phi^1 = \phi^0$, where ϕ^0 and ϕ^1 are temperatures at t and $t + \Delta t$, respectively. The matrix M is symmetric and usually diagonally dominant, enabling an easy and fast inversion with an ICCG method [65].

Choice of the Computational Grid

A unique feature of FLIP is the freedom with which one can use the computational grid to gain increased accuracy and resolution, which stems from the fact that the particles carry all the fluid information. A global rezone of a problem does not cause numerical diffusion or result in loss of information, rather it simply requires relocating the particles on the new grid, which can be done in a systematic fashion [27]. A typical computation can easily employ a grid which is Eulerian (fixed), Lagrangian (moving with the fluid), adaptive (which attempts to resolve user-specified data) [62], or some combination thereof. For example, many of the planar heavy-ion-driven ICF target implosions computed with FLIP-PHD [44] exploit this capability.

The structure of FLIP-PHD can be reviewed with the help of the flow chart in Figure 3 showing the sequence of tasks performed every time step. A computational cycle consists of particle kinematics (interpolating particle data onto the grid), formulating EOS for the plasma, solving the equations of motion on the grid, relocating the grid, and particle dynamics, in which particles are moved after they have received changes in momenta and energy from the grid.

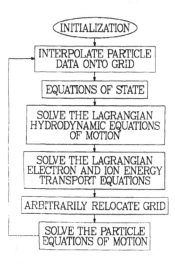

Figure 3. Flow chart of the tasks performed every time step in FLIP-PHD.

NUMERICAL EXAMPLES

The numerical examples that follow illustrate the ability of FLIP-PHD to model a wide variety of fluid flow problems, spanning several decades of fluid density-temperature space, from incompressible hydrodynamics (the Rayleigh-Taylor (R-T) instability) to compressible gas dynamics (the shock tube), and finally, to the hot, dense ablation plasmas imploding in a planar version of the heavy-ion-driven HIBALL ICF target [73, 74].

Rayleigh-Taylor Instability

The evolution of a single-mode, 30 cm wavelength perturbation separating heavy and light densities of air at standard temperature and pressure is followed for 6τ, where τ is the R-T e-fold, equal to $(gk\alpha)^{-1/2}$, with g the gravitational field, k the wave number, and α the Atwood number. Interface growth is computed with FLIP-PHD for 4 Atwood numbers equal to 0.2, 0.4, 0.6, and 0.8, corresponding to density ratios of 1.5, 2.3, 4.0, and 9.0, respectively. An initial surface perturbation amplitude of 1.5 cm is imposed, giving an initial amplitude to wavelength ratio of 0.05. The evolution of a one-half wavelength portion of the interface is computed with periodic boundary conditions on all boundaries. A constant gravitational force ($g_y = -980.67$ cm/s^2) is applied downward throughout the calculation. A 15×120 computational mesh is used with zone spacing in both the x (lateral to the interface) and y (perpendicular to the interface) directions equal to 1 cm. The R-T instability is implicitly computed by FLIP-PHD with a Courant number ($u\Delta t/\Delta x$) of 10. A very low Mach number of about 0.01 verifies the virtual incompressibility of the problem, which in the past has been difficult for compressible PIC algorithms to solve. Figure 4 displays interface density contours for the four Atwood numbers at $t = 5\tau$. Daly [75] and others [76, 77] found that the K-H (Kelvin-Helmholtz) instability and tip widening on the spike in the nonlinear stage of the R-T instability are more pronounced at low Atwood numbers. This is also the general conclusion obtained here, but for the above-mentioned zoning, FLIP-PHD did not resolve the interface well enough at low Atwood numbers (0.20 and 0.40) to obtain a pronounced K-H roll-up. The

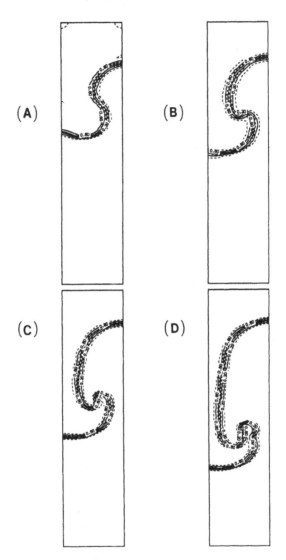

Figure 4. Density contours of the interface in a FLIP-PHD calculation of the R-T instability at an e-fold time of five for Atwood numbers of (A) 0.20, (B) 0.40, (C) 0.60, and (D) 0.80.

tendency, however, toward asymmetry in spike and bubble shapes for larger Atwood numbers is apparent. In all cases the computed R-T growth rate compares very well with linear theory up to an amplitude-to-wavelength ratio of about 0.4 [29].

Shock Tube

The shock tube consists of a diaphragm separating two regions of an ideal gas ($\gamma = 1.4$) initially at rest. One region has higher pressure and density than the other, so that when the diaphragm is removed the two uniform initial states separate into 3 constituent nonlinear waves (shock, rarefaction, and contact discontinuity). The shock propagates ahead of the contact discontinuity through the less dense region, while the rarefaction propagates through the more dense region. The initial conditions are the dimensionless ones used by Sod [78], who compared several finite-difference methods with the shock tube problem solved here. For $0.0 \leq x \leq 0.5$, labeled as region 1, $p_1 = 1.0$, $\rho_1 = 1.0$, $u_1 = 0.0$, while for $0.5 \leq x \leq 1.0$, labeled as region 5, $p_5 = 0.10$, $\rho_5 = 0.125$, and $u_5 = 0.0$. The system can be reduced to four equations and four unknowns [72], yielding an analytic solution for the rarefaction region (region 2), behind the contact discontinuity (region 3), and in front of the contact discontinuity (region 4). The computational mesh consists of 100 grid points from $0.0 \leq x \leq 1.0$ with reflecting boundary conditions at $x = 0.0$ and $x = 1.0$. The diaphragm is located at $x = 0.5$. Sixteen particles per cell are loaded initially and the grid is allowed to adapt to pressure gradients in the x direction. The problem is calculated up to $t = 1.0$ to allow comparison with the solution of Winkler, et al. [79, 80].

Figure 5 shows the numerical solution (individual grid points) compared with the analytic solution (solid line) at a time of 0.2013. Fluid quantities in front of and behind the rarefaction and shock are almost exact (within 0.1% of theory), but the structures of the rarefaction and shock are not as well resolved as other finite-difference algorithms [78]–[80]. The corners at the endpoints of the rarefaction are rounded off, which is typical of PIC results in rarefied regions where smaller particle densities lead to greater statistical errors. The shock transition occupies a fairly broad region of approximately 5 zones and has a slight overshoot. The slight overshoot at the shock front can be dampened with a higher artificial viscosity without sacrificing much in the way of shock thickness. Quadratic interpolation, artificial viscosity, and backward Euler differencing are the primary reasons for the broad shock structure. Higher order interpolation introduces a quadratic particle shape which limits shock thicknesses to greater than 3 zones. This consequence does have its advantages, however, in that one is given more freedom in adjusting the artificial viscosity [29] compared with some finite-difference algorithms [78]. Diffuseness, oscillatory behavior, and overshoot around shocks can be alleviated somewhat with finer resolution made possible by the adaptive grid. The contact discontinuity, on the other hand, is very steep, occupying only one to two zones, just the opposite of typical results from finite-difference calculations [78]. In conventional finite-difference schemes, numerical diffusion builds up with each time step as mass is advected, causing the contact discontinuity to be smeared out. This does not occur in FLIP-PHD, since sharply resolved contact discontinuities are automatically preserved by the presence of the particles carrying all the fluid information.

Planar HIBALL Target Implosion

FLIP-PHD has been used to study the implosion dynamics of the planar HIBALL target [44], which represents the single-shell multilayer target employed in the HIBALL reactor study [73, 74] with an infinite radius. The planar target consists of a 140 μm tamper layer of lead (Pb), a 500 μm absorber/pusher layer of lithium seeded with lead (LiPb), and a 150 μm layer of cryogenic deuterium-tritium (DT) fuel. The target, driven by a 10-GeV Bi$^+$ heavy-ion beam penetrating deep into the LiPb layer for a period of about 30 ns, expands into vacuum on the beam-irradiation side (Pb) and as well as on its backside (DT). The hydrodynamic response of this target during and subsequent to irradiation by the high-energy Bi$^+$ beam has been studied in 1-D [29] and in 2-D [44]. In the 2-D implosions the target is subjected to a nonuniformity

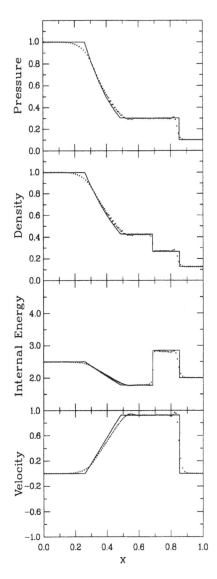

Figure 5. FLIP-PHD solution of Sod's shock tube problem at a time of 0.2013. Lines represent the exact analytic solution and grid points are shown individually.

in the form of a single-mode perturbation given either to the beam intensity or a particular material interface.

An initial grid spacing in the y-direction parallel to the beam penetration is 15 μm, and lateral (x-direction) fluid flow is resolved for the 2-D calculations with 20 zones per half wavelength (the wavelength λ referring to either the beam intensity or surface perturbation wavelength). In the 2-D calculations, the grid is relocated after every cycle in one of two fashions: (i) a semi-Lagrangian fashion, in which lateral (x) movement of the grid is restricted but parallel (y) movement takes place at the local y-component of the Lagrangian fluid velocity; and, (ii) in a Lagrangian mode that is weighted with body-fitted coordinates computed by a grid generator [62] which evenly distributes the grid throughout the domain. Approximately 10,000 total particles comprise the HIBALL target in the 2-D calculations, i.e., an initial particle density of nine per computational cell. The Pb/vacuum and DT/vacuum interfaces are treated as Lagrangian boundaries. Simple free-slip conditions along the left and right boundaries are used so that end effects of the target can be ignored. The 10-GeV Bi$^+$ ion beam is represented with 20 rays (1 per zone) for uniform intensities and 100 rays (5 per zone) for nonuniform intensities. Multiple rays per zone are needed only when the grid distorts in a direction lateral to the beam propagation. A spacing of 5 rays per zone adequately resolves any nonuniform beam intensity profile. The calculations are initialized at room temperature (0.025 eV).

1-D Hydrodynamic Response

Some of the important dynamical properties of the 1-D planar HIBALL target which can be modeled with FLIP-PHD are illustrated in Figure 6. Profiles of stopping power, electron temperature, pressure, and density are displayed in 1-D slices along the beam direction at times of 21.7, 24.8, and 31 ns. It is apparent from Figure 6 that FLIP-PHD does very well in resolving steep gradients driving the implosion, i.e., density, temperature, and pressure. These 1-D calculations are very useful in studying beam-coupled hydrodynamics, and have been a vital tool in deriving intensity scaling relations and understanding target optimization [29].

2-D Hydrodynamic Response

Using FLIP-PHD as a tool, a recent symmetry and stability analysis of the 2-D HIBALL target [44] led to the identification of an ion-beam-driven hydrodynamic instability and a better overall understanding of the complex coupling of ion beams to the fluid response of ablation plasmas in ion-driven ICF targets. The ability of FLIP-PHD to accurately model the resulting distorted flow was of vital importance in recognizing that ion-driven ICF targets are sensitive to spatial perturbations at the target interface and in beam intensity.

CONCLUSIONS AND SUMMARY

In summary, the full PIC representation in FLIP-PHD has a number of superior numerical features which result in a more accurate representation of plasma fluid behavior. Some of these features include (1) the lack of numerical diffusion in convective transport, (2) sharp resolution of contact discontinuities and material interfaces, (3) an arbitrarily-adaptive ALE-type (arbitrary Lagrangian-Eulerian) computational grid, (4) implicit time-differencing of the equations of motion, (5) second-order conservation of all relevant fluid quantities, i.e., mass, momentum (including angular), and energy; (6) FLIP equations of motion that exploit recent advancements in PIC hydrodynamics [27], and (7) a fully self-consistent, state-of-the-art physical model. FLIP-PHD has been extensively tested and successfully applied on a number of diverse problems [28].

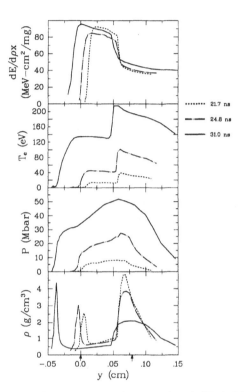

Figure 6. Profiles of $dE/d(\rho x)$, T_e, p, and ρ computed with FLIP-PHD at three different times within a 1-D planar HIBALL target driven by a 10-GeV Bi^+ beam with a reactor pulse shape. Left and right arrows on the bottom axis point to initial DT/vacuum ($y = 0.0$ cm) and Pb/vacuum ($y = 0.079$ cm) interface locations, respectively, with the beam entering from the right.

ACKNOWLEDGEMENTS

The authors would like to acknowledge the assistance of J. U. Brackbill during the development and application of FLIP-PHD.

REFERENCES

1. G. Buchwald, et al., Laser and Part. Beams 1, 335 (1983).
2. S. Kawata and K. Niu, J. Phys. Soc. Jap. 53, 3416 (1984).
3. I. Yamagushi, S. Kawata, and K. Niu, IPPJ-725, Institute of Plasma Physics, Nagoya University (1985).
4. S. Nakajima and K. Niu, IPPJ-769, Institute of Plasma Physics, Nagoya University (1986).
5. R. C. Arnold and J. Meyer-ter-Vehn, MPQ-113, Max Planck Institut Für Quantenoptik (1986).
6. R. L. McCrory, R. L. Morse, and K. A. Taggart, Nucl. Sci. Engr. 64, 163 (1977).
7. Z. Zinamon, E. Nardi, and E. Peleg, Phys. Rev. Lett. 34, 1262 (1975).
8. D. Lackner-Russo and P. Mulser, PLF-32, Max Planck Gesellschaft zur Foerderung der Wissenschaften, F. R. Germany (1980).
9. M. Tamba, et al., Laser and Part. Beams 1, 121 (1983).
10. P. C. Thompson and P. D. Roberts, Laser and Part. Beams 2, 13 (1984).
11. S. Atzeni, Comp. Phys. Commun. 43, 107 (1986).
12. G. B. Zimmerman, UCRL-74811, Lawrence Livermore National Laboratory (1973).
13. G. B. Zimmerman and W. L. Kruer, Comm. Plasma Phys. 2, 51 (1975).
14. J. Norton, Private Communication, Los Alamos National Laboratory (1987).
15. W. D. Schulz, Meth. Comp. Phys. 3, 1 (1964).
16. C. P. Verdon, R. L. McCrory, R. L. Morse, G. R. Baker, D. I. Meiron, and S. A. Orszag, Phys. Fluids 25, 1653 (1982).
17. M. J. Fritts, W. P. Crowley, and H. Trease, eds., **The Free-Lagrange Method**, (Springer-Verlag, New York, 1985), Proc. First Inter. Conf. on Free-Lagrange Methods, 1985.
18. W. F. Noh, Meth. Comp. Phys. 3, 117 (1964).
19. R. M. Frank and R. B. Lazarus, Meth. Comp. Phys. 3, 47 (1964).
20. F. H. Harlow and M. Evans, LAMS-1956, Los Alamos National Laboratory (1955).
21. F. H. Harlow, J. Assoc. Comput. Mach. 4, 137 (1957).
22. M. W. Evans and F. H. Harlow, LA-2139, Los Alamos National Laboratory (1957).
23. F. H. Harlow, et al., LA-2301, Los Alamos National Laboratory (1959).
24. F. H. Harlow, Proc. Symp. in Appl. Math., **Experimental, Arithmetic, High Speed Computing and Mathematics**, vol. XV, pp. 269–288, 1963.
25. F. H. Harlow, Meth. Comp. Phys. 3, 319 (1964).
26. A. A. Amsden, LA-3466, Los Alamos National Laboratory (1969).
27. J. U. Brackbill and H. M. Ruppel, J. Comp. Phys. 65, 314 (1986).
28. J. U. Brackbill, D. B. Kothe, and H. M. Ruppel, Comp. Phys. Commun. 48, 25 (1988).
29. D. B. Kothe, Ph.D. Thesis, Purdue University (1987).
30. K. S. Holian, Ed., LA-10160-MS, Los Alamos National Laboratory (1984).
31. C. W. Cranfill, Private Communication, Los Alamos National Laboratory, 1986–1987.
32. S. I. Braginskii, in **Reviews of Plasma Physics**, (Consultants Bureau, New York, 1965), Vol. 1, pp. 205–311.
33. H. Bethe, Ann. Physik. 5, 325 (1930).
34. H. Bethe, Z Physik. 76, 293 (1932).
35. J. Linhard, M. Scharff, and H. E. Schiøtt, Mat. Fys. Medd. Dan. Vid. Selsk. 33(14), 1 (1963).
36. J. Linhard and M. Scharff, Phys. Rev. 124, 128 (1964).
37. T. A. Mehlhorn, J. Appl. Phys. 52, 6522 (1981).
38. P. G. Steward and R. W. Wallace, UCRL-19128, Lawrence Livermore National Laboratory (1970).

39. J. D. Jackson, **Classical Electrodynamics**, (John Wiley and Sons, New York, 1975), 2nd Ed., ch 13.

40. C. K. Choi and M.-Y. Hsiao, Nucl. Tech./Fus. **3**, 273 (1983).

41. M. D. Brown and C. D. Moak, Phys. Rev. B **6**, 90 (1972).

42. P. G. Steward, UCRL-18127, Lawrence Radiation Laboratory, Ph. D. Thesis, (1968).

43. L. C. Northcliffe, in **Studies in Penetration of Charged Particles in Matter**, (National Academy of Sciences, National Research Council, Pub. 1133, Washington, D. C., 1964), p. 173.

44. D. B. Kothe, C. K. Choi, and J. U. Brackbill, in **Laser Interaction and Related Plasma Phenomena**, (Plenum Press, New York, 1988), vol. 8, pp. 701–721.

45. J. J. Monaghan, Comp. Phys. Rep. **3**, 71 (1985).

46. R. L. Bjork, RAND-P-1662, Rand Corporation Report (1958).

47. T. D. Riney, R62SD95, General Electric Space Sciences Laboratory Report (1962).

48. T. D. Riney, R64SD13, General Electric Space Sciences Laboratory Report (1964).

49. F. H. Harlow and B. D. Meixner, LAMS-2770, Los Alamos National Laboratory (1962).

50. C. Mader, LA-3077, Los Alamos National Laboratory (1964).

51. W. R. Gage and C. L. Mader, LA-3422, Los Alamos National Laboratory (1965).

52. A. A. Amsden and F. H. Harlow, AIAA J. **3**, 2081 (1965).

53. F. H. Harlow, A. A. Amsden, and J. R. Nix, J. Comp. Phys. **20**, 119 (1976).

54. F. H. Harlow and W. E. Pracht, Phys. Fluids **9**, 1951 (1966).

55. F. H. Harlow and D. O. Dickman, LA-2256, Los Alamos National Laboratory (1959).

56. J. N. Leboeuf, T. Tajima, and J. M. Dawson, J. Comp. Phys. **31**, 379 (1979).

57. A. Nishiguchi and T. Yabe, J. Comp. Phys. **52**, 390 (1983).

58. B. M. Marder, Math. Comp. **29**, 434 (1975).

59. J. W. Dawson, Rev. Mod. Phys. **55**, 403 (1983).

60. J. W. Eastwood, Comp. Phys. Commun. **43**, 89 (1986).

61. A. Nishiguchi and T. Yabe, J. Comp. Phys. **47**, 297 (1982).

62. J. U. Brackbill and J. S. Saltzman, J. Comp. Phys. **46**, 342 (1982).

63. J. U. Brackbill, Comp. Phys. Commun. **47**, 1 (1987).

64. J. J. Monaghan and R. A. Gingold, J. Comp. Phys. **52**, 374 (1983).

65. D. S. Kershaw, J. Comp. Phys. **26**, 43 (1978).

66. V. Casulli and D. Greenspan, Int. J. Num. Methods Fluids **4**, 1001 (1984).

67. R. D. Richtmyer and K. W. Morton, **Difference Methods for Initial-Value Problems**, (Interscience Publishers, New York, 1967), 2nd Ed., ch. 8.

68. J. J. Duderstadt and G. A. Moses, **Inertial Confinement Fusion**, (John Wiley and Sons, New York, 1982).

69. M. R. Clover, Private Communication, Los Alamos National Laboratory, 1986.

70. T. A. Mehlhorn, Private Communication, Sandia National Laboratory, 1987.

71. D. Kershaw, UCID-17424, Lawrence Livermore National Laboratory (1977).

72. F. H. Harlow and A. A. Amsden, LA-4700, Los Alamos National Laboratory (1971).

73. B. Badger, et al., KfK-3202, Kernforschung. Karlsruhe and UWFDM-450, Univ. of Wisc. (1981).

74. B. Badger, et al., Univ. of Wisc. Tech Inst. Report UWFDM-625, Kernforschungszentrum Karlsruhe Report KfK-3840, Fusion Power Associates Report FPA-84-4 (1984).

75. B. J. Daly, Phys. Fluids **10**, 297 (1967).

76. W. P. Crowley, UCRL-72650, Lawrence Livermore National Laboratory (1970).

77. D. L. Youngs, AWRE/44/92/23, Atomic Weapons Research Establishment (1981).

78. G. A. Sod, J. Comp. Phys. **27**, 1 (1978).

79. K.-H. A. Winkler, M. L. Norman, and D. Mihalas, in **Multiple Time Scales**, (Academic Press, New York, 1984), J. U. Brackbill and B. I. Cohen, Editors, pp. 145–184.

80. K.-H. A. Winkler, D. Mihalas, and M. L. Norman, Comp. Phys. Commun. **36**, 121 (1985).

ATTENDEES

Dr. Benjamin Arad
Dept. of Plasma Phys.
Soreq Nuclear Research Ctr.
Yavne 70600
Israel

Dr. Andreas Bergmann
Inst. Ang. Phys. THD
Hochschulstr. 4A
D-6100 Darmstadt
Federal Republic of Germany

Dr. Thomas Boehly
Lab. for Laser Energetics
University of Rochester
575 Elmwood Avenue
Rochester, NY 14620

Dr. David K. Bradley
Laboratory of Laser Energetics
University of Rochester
250 East River Rd.
Rochester, NY 14623

Dr. David Cartwright
Los Alamos National Lab.
P. O. Box 1663
Los Alamos, NM 87545

Dr. Chan Choi
Dept. of Nuclear Engineering
Purdue University
West Lafayette, IN 47907

Dr. Carl B. Collins
Center for Quantum Electronics
University of Texas - Dallas
P. O. Box 830688, MS NB 11
Richardson, TX 75083-0688

Dr. Gregory D'Alessio
Inertial Fusion Div.
DP-243 GTN
U. S. Dept. of Energy
Washington, D. C. 20545

Dr. Claude Deutsch
L. P. G. P., Bat. 212
Universite Paris XI
91405 Orsay
France

Dr. R. Paul Drake
Lawrence Livermore National Lab.
P. O. Box 808, L-473
Livermore, CA 94550

Dr. Shalom Eliezer
Plasma Physics Dept.
SOREQ Nuclear Research Ctr.
Yavne 70600, Israel

Dr. Heidi Fearn
Optical Science Center
Univ. of Arizona
Tucson, AZ 85721

Dr. Humberto Figueroa
Univ. of Southern California
Los Angeles, CA 90089-0484

Dr. Jacob Grun
Dept. of the Navy
Naval Research Laboratory
Code 4733
Washington, D. C. 20375

Dr. Martin Gunderson
Dept. of Electrical Engr.-EP
SSC-420
Univ. of Southern California
Los Angeles, CA 90089-0484

Dr. Xian Tu He
Inst. of Applied Physics
   and Computational Mathematics
P. O. Box 8009
Beijing, China

Dr. Heinrich Hora
Univ. of New South Wales
P. O. Box 1
Kensington, New South Wales
Australia

Dr. Stefan Huller
Teor. Quantenelektronik
Inst. fur Angeu. Physik
Hochschulstr. 4A
D-6100 Darmstadt, FRG

Dr. John P. Jackson
Kaman Sciences
1500 Garden of the Gods Rd.
Colorado Springs, CO   80907

Dr. Erk Jensen
Div. PS/RF
CERN
CH-1211 Geneva 21
Switzerland

Dr. Chan Joshi
Univ. of California
Electrical Engineering Dept.
405 Hilgard Ave.
Los Angeles, CA   90024-1594

Dr. Takeshi Kaneda
Dept. of Energy Sciences
Tokyo Inst. of Technology
4259 Nagatsuta, Midori-ku
Yokohama 227, Japan

Dr. D. R. Kania
Lawrence Livermore National Lab.
P. O. Box 808
Livermore, CA   94550

Dr. Koichi Kasuya
Dept. of Energy Sciences
Tokyo Inst. of Technology
4259 Nagatsuta
Yokohama 227, Japan

Dr. Yeong E. Kim
Dept. of Physics
Purdue University
West Lafayette, IN   47907

Dr. Hanoch Kislev
Nuclear Research Center
P. O. Box 9001
Beer Sheva 84, Israel

Dr. J. P. Knauer
Lab. for Laser Energetics
Univ. of Rochester
250 East River Rd.
Rochester, NY   14623-1299

Dr. David J. Mayhall
Lawrence Livermore National Lab.
P. O. Box 808, L-156
Livermore, CA   94550

Dr. George H. Miley
Dept. of Nuclear Engineering
University of Illinois
103 S. Goodwin Ave., 214 NEL
Urbana, IL   61801

Dr. Gu Min
Dept. of Theoretical Physics
University of New South Wales
Kensington 2033
Australia

Dr. E. Minguez
Inst. De Fusion Nuc. (DENIM)
Universidad Politecnica
Jose Gutierrez Abascal, 2
Madrid - 28006, Spain

Dr. Katsu Mizuno
LIC/Davis
Dept. of Applied Science
Univ. of California at Davis
Davis, CA   95616

Dr. Sadao Nakai
Institute of Laser Engr.
Osaka University
Suita, Osaka, 565
Japan

Dr. Keishiro Niu
Tokyo Inst. of Technology
Grad. School at Nagatsuta
4259 Nagatsuta, Midori-ku
Yokohama 227, Japan

Prof. Hansheng S. Peng
Physics & Chemistry Dept.
SW Inst. of Nuclear Physics
P. O. Box 515
Chengdu, China

Dr. Robert R. Peterson
Fusion Technology Inst.
University of Wisconsin
1500 Johnson
Madison, WI   53706

Dr. Mark A. Prelas
Nuclear Engineering Proram
323 Electrical Engineering
University of Missouri
Columbia, MO   65211

Dr. Juan J. Ramirez
Sandia National Laboratories
P. O. Box 5800
Albuquerque, NM   87185

Dr. Joshua Reszick
Science Applications Inc.
1710 Goodrich Dr.
McLean, VA 22102

Dr. Barrett H. Ripin
Head, Space Plasma Branch
Code 4780, Naval Research Lab.
4555 Overlook Ave., SW
Washington, D. C. 20375-5000

Dr. F. Schwirzke
Physics Department
Naval Postgraduate School
Monterey, CA 93943

Dr. Wolf Seka
Lab. for Laser Energetics
University of Rochester
250 East River Road
Rochester, NY 14623-1299

Dr. Hua-Zhong Shen
SW Institute of Nuclear
  Physics & Chemistry
P. O. Box 512-32, Chengdu
China

Dr. Allan Shepp
Avco Research Laboratory
2385 Revere Beach Parkway
Everett, MA 02149

Dr. Erik Storm
Lawrence Livermore National Lab.
P. O. Box 5508, L-481
Livermore, CA 94550

Dr. Enrique Szichman
Plasma Physics Dept.
Soreq Nuclear Research Ctre.
Yavne 70600, Israel

Dr. Frederic Thais
C. E. A.
CEL-V
94195 Villeneuve-St. Georges
Cedex, France

Mr. Wade Williams
Dept. of Nuclear Engr.
University of Illinois
103 S. Goodwin Ave., 214 NEL
Urbana, IL 61801

Dr. Kiyoshi Yatsui
Laboratory of Beam Tech.
Nagaoka University of Technology
Nagaoka, Niigata 940-21
Japan

Prof. H. Yoneda
Inst. for Laser Science
Univ. of Electro-communications
Chofuga-oka, Chofushi
Tokyo 182, Japan

AUTHOR INDEX

Printed in the United States
By Bookmasters